T0186638

Irregularities and Prediction of Major Disasters

Systems Evaluation, Prediction, and Decision-Making Series

Series Editor

Yi Lin, PhD

Professor of Systems Science & Economics
School of Economics and Management
Nanjing University of Aeronautics and Astronautics

Irregularities and Prediction of Major Disasters

Yi Lin
Shoucheng OuYang

CRC Press
Taylor & Francis Group
Boca Raton   London   New York

CRC Press is an imprint of the
Taylor & Francis Group, an **informa** business
AN AUERBACH BOOK

Auerbach Publications
Taylor & Francis Group
6000 Broken Sound Parkway NW, Suite 300
Boca Raton, FL 33487-2742

© 2010 by Taylor and Francis Group, LLC
Auerbach Publications is an imprint of Taylor & Francis Group, an Informa business

No claim to original U.S. Government works

Printed in the United States of America on acid-free paper
10 9 8 7 6 5 4 3 2 1

International Standard Book Number: 978-1-4200-8745-1 (Hardback)

Library of Congress Cataloging-in-Publication Data

Lin, Yi, 1959-
    Irregularities and prediction of major disasters / Yi Lin, Shoucheng Ouyang.
        p. cm. -- (Systems evaluation, prediction, and decision-making series)
    Includes bibliographical references and index.
    ISBN 978-1-4200-8745-1 (hardcover : alk. paper)
    1. System theory. 2. Decision making. 3. Disasters--Forecasting. I. Ouyang, Shoucheng. II. Title.

Q295.L498 2010
003--dc22   2009044549

Visit the Taylor & Francis Web site at
http://www.taylorandfrancis.com

and the Auerbach Web site at
http://www.auerbach-publications.com

Contents

Preface

From different angles and different academic backgrounds, we began to work together on the existing problems of modern science and possible ways to resolve these problems. At that time, Yi Lin was forming the International Institute for General Systems Studies, Inc. In the name of this institute, he invited Shoucheng OuYang to participate in the inaugural international workshop and deliver a speech on his most recent scientific works. However, for various reasons, OuYang did not attend this event but through our initial communications we found our common language and research topics. During the summer of 1997, we met for the first time, in Berlin, Germany, where the 15th World Congress of the International Association of Mathematical and Computer Simulation had organized two special technical sessions, entitled "Concept of Blown-Ups in Nonlinear Systems" and "A Revisit to Lorenz's Chaos," to purposely focus on our joint works. In the years that followed, together we published a series of important research papers, special topic volumes, and monographs.

The first special topic volume, "Mystery of Nonlinearity and Lorenz's Chaos," was published in 1998 by the famous *Kybernetes: The International Journal of Systems and Cybernetics* (Lin, 1998) in the form of a special double issue. This special issue attracted major attention from the world of learning. For that calendar year alone, more than 1,000 scholars participated in various communications and discussions on the topic lines presented in the volume. In July 2000, the World Congress of Scholars in Systems and Cybernetics, held in Toronto, Canada, opened a specific topic section called the "Challenges Facing General Systems in Practice" to present recent development of our works. After that important gathering, we moved to the United States and met D. H. McNeil and others to plan the works *Entering the Era of Irregularity* (OuYang et al., 2002a). (This book was later published by Meteorological Press, Beijing, in 2002. In 2005, this book was out of stock due to a well-received readership.) At the same time, we finished an important paper on the concept of chaos (Lin et al., 2001). In 2002, jointly with Yong Wu, Yi Lin published the book *Beyond Nonstructural Quantitative Analysis: Blown-Ups, Spinning Currents, and Modern Science*, (World Scientific, 2002). Ilya Prigogine, a Nobel Laureate in chemistry, commented highly on this work as the beginning of an important era. Upon OuYang's return to China, we continued our planning on the more important, revolutionary works to follow *Entering the Era of Irregularity* and *Beyond Nonstructural Quantitative Analysis: Blown-Ups, Spinning Currents, and Modern Science.*

Looking back on our respective professional careers and the years we worked together, we individually and separately recognized that in today's world of advanced scientific and technological developments, people are still faced with many unsolvable problems. The very existence of these problems indicates that within the system of modern science there must be problems of unrealisticity or places of imperfection. Spending several decades of our productive years on various experiments and theoretical explorations, we discovered that the key to the many of the problems that modern science has not resolved, is that modern science has not considered changes in materials and events; it is still in the infancy stage of development, focusing on the quantitative manipulations of invariant materials and events.

Because modern science has nothing to do with changes in materials, its works and consequences naturally have nothing to do with evolutions. This end explains why in the present world, where science and technology are in their most advanced stage of recorded history, when we have to predict the future, it always shows how little we know about nature and how ignorant we are about the evolutionary development of our natural surroundings. To address the problem of prediction, contemporary scientists have sufficiently employed the concepts of probability. For example, the commercial weather forecast for tomorrow is often provided as follows: The chance of light rain for tomorrow is about 65%. This sort of forecast of the future often creates various kinds of problems and difficulties for the day-to-day decision makers. For instance, if there were no rain tomorrow, they might arrange an important outdoor activity; if there were rain, the planned activity might have to be rearranged or canceled. In particular, if tomorrow involves a crucial military operation, rain or no rain, then the relevant weather forecast becomes extremely important, because its accuracy and definiteness will involve many lives and the win or loss of a war. However, no matter how the front-line forecaster makes his or her prediction and how big or small a probability is attached to the prediction, the forecast will always be correct and accurate. In particular, in the previous example of 65% chance of light rain, if the rain actually occurs, no doubt, the weather forecast is correct. If the light rain does not occur, the forecast is still considered accurate, because the forecaster only said that the chance of light rain was 65%. As a matter of fact, the specific value of the applied probability is not relevant. What is crucial is that as long as a probability is attached to the predicted future event, the prediction becomes forever correct. It is because even if the probability used is 100%, it still does not theoretically guarantee that the forecasted event will definitely occur. Even if the probability is 0%, in theory it does not exclude the possibility for the 0%-probability event to happen.

This book crystallizes our many years of joint theoretical and practical works. In order to resolve the problem of predicting the unknown future, we analyzed

the scientific and knowledge systems of knowing the world either through analyzing materials or through analyzing quantities, what quantities are and where
they are from, rotational movements of materials, what time is and how it travels
from the past to the future, regularization of information, and the physical significance and the theoretical and practical use of the value of irregular information.
Consequently, we developed the method of digital structuralization of information
that can be and has been employed successfully for the prediction of the occurrence
of forthcoming zero-probability disastrous events.

In our research of the scientific systems of knowing the world either through
analyzing materials or through analyzing quantities, we found that modern science
studies quantitative regularizations and stable time series using either well-posed
equations or statistical methods, pursuing the generality and the abstract uniformity of the form of quantities. Even after realizing the importance of noninertial
systems, modern science has not clearly pointed out the physical significance of variable accelerations and rotations. If materials themselves can change, then objects
can no longer be treated as abstract particles without size and volume, leading to
the problem of internal material structures and the problem of evolution science.
Corresponding to the Western quantitative science, the Eastern world, and especially the Chinese people, began their exploration and knowing the natural world
through nonparticlization of materials — the epistemology of structural transformations of mutual reactions. That explains why the *Book of Changes* — as a classical
theory on evolution — appeared in China, and its teachings have become the moral
standard for the conscientious behaviors of the Chinese people. Through analyzing
materials, Chinese people consider mutual constraints of materials in their expressions of the states and attributes of events. That is, knowing the world through
analyzing materials is about the problem of evolution by emphasizing changes in
material attributes and states, while knowing the world through analyzing quantities investigates movements of invariant materials using the system of formal analysis. Our research sufficiently explains why in the past 300-plus years since modern
science was initially established, it has not been able to produce scientists who can
foretell the future.

In our research on what quantities are and where they come from, we recognized that the philosophical root of quantities can be traced back to the epistemological law that numbers are the origin of all things of the ancient Greek school of
Pythagoras. The latter belief, inherited in most of the contemporary scholarly works,
that numbers are the laws governing all matters of modern science has been greatly
influenced by ancient Greek philosophy. In contrast to this historical development
of the West, Zhan Yin, a famous calculator of the time period of the Warring States
in ancient China, pointed out that there are things that numbers cannot describe.
Plus, China traditionally created its wealth and prosperity through developing agriculture, while the West did so through nomadic trades. So, the concept of quantities has never been as significant in Chinese lives as in the lives of Westerners,
because the Chinese people did not have any pressing need to accurately count

the number of grains of rice or wheat being traded, while the Westerners had the traditional necessity to precisely keep track of the numbers of sheep and cattle that were passed from one hand to another. That might be one of the main reasons at the cultural origin for why modern science did not initially appear in the East. At the same time, quantities are postevent formal measurements, because counts cannot appear before the existence of events or objects. This understanding very well illustrates why when the methods of quantities are employed in the prediction of the unknown future, unexpected events often occur. Because quantities are symbolic labels of the imaginary axes of Euclidean spaces, modern science has to suffer from the problem of quantitative infinity and cannot even find a way around this problem. If the physical world is seen through analyzing materials, each realistic physical space is then curved; without increasing to the quantitative infinity a transitional change will have already occurred. This fact provides an explanation for why after experiencing great difficulties for several thousands of years the ancient Chinese philosophies and scientific methods can still live well in the people and continue their growth and development throughout the ages.

By respectively using the methods of analyzing either quantities or materials, we have shown that the general form of motion of materials is rotation. And, rotations signify an extremely difficult problem for quantities to address. This recognition explains why modern science has tried to avoid the investigation of rotations. Although V. Bjerknes proved his significant circulation theorem in 1898 using nonparticle circulation integrals and pointed out that the fundamental form of fluid movements is rotational motion, with over 100 years of testing and scrutiny, his theorem is still not as well known as it should be in the contemporary world of learning. Considering the fundamentality of rotational movements, we investigated the concept of stirring energies and the physical meanings of the conservation and nonconservation of stirring energy. Our work indicates that this concept can help the first push of modern science transform into the material evolution of the second stirring, and modern science itself into evolution science. To this end, we successfully demonstrate the significance and effects of the rotation of curvature spaces.

As for what time is, a problem modern science still cannot answer, if we escape the control of quantities, we can discover that time in fact is given an expression in events' differences existing along the time order. It does not occupy any material dimension and parasitically dwells in the changes of materials. So, to correctly understand and apply the concept of time, we must describe and provide specific expressions for the differences existing in the changes of materials and events. If materials did not change, there would not be the concept of past and future, so that the concept of time would become meaningless. At the same time, the differences in the changes of materials and events provide a concrete method for how to practically employ the concept of time. Therefore, knowing the world through analyzing quantities is different from that through analyzing materials, and evolution science is different from modern science. This epistemological recognition in essence explains why when modern science, which explores the world through analyzing quantities is employed to investigate the prediction of the unknown future, it has suffered from a great many difficulties and failures. It is because the analysis system of existent quantities of modern science cannot deal with irregular events; plus modern science employs such theories as instabilities, complexities, randomness, and so forth, to eliminate irregular information. This end alone in principle implies that the contemporary understanding of the concept of time still needs to be improved and that modern science has not yet recognized the practical significance and effects of variable materials and variant events.

From the concept of time we realized how to correctly understand the differences that exist in materials and events and the relevant concept of irregular information. Since the time when Shannon initially introduced the concept of information in 1948, this concept has been associated with uncertainties. In particular applications, the school of determinism of modern science eliminates specifics using stabilities, and the school of indeterminism gets rid of small-probability events using stable time series. So, irregular information and peculiar events all disappear in the relevant quantitative manipulations. Speaking more to the point, that is why other than making use of satellites, radar systems, and other highly developed technology to produce live reports, modern science cannot establish true predictions with meaningful lead-times of zero-probability major natural disasters. In fact, both information and events are what have already occurred or are existent. Even if the probability for the information or events to occur is almost zero, they still represent already-occurred or existing events. We cannot simply ignore the existence due to the fact that quantities cannot handle the situation. In order to address this trouble and relevant difficulties experienced by modern science, through analyzing huge amounts of irregular information, we discovered that the appearance of irregular information is closely related to the collision of material rotations. On the basis of this epistemological understanding, we established the method of digitization for dealing with irregular information and showed that this method is effective in predicting transitional changes in events. It is exactly because of the abnormality of irregular information that quantitative methods cannot handle the information. However, the existence of the information determines the forthcoming occurrence of zero-probability disastrous events. That is to say, there is not only the reason for irregular information to appear, but also the mechanism underlying material structures. The significance of this epistemological comprehension has gone far beyond the formality of quantities.

This book is organized as follows: The first chapter, the "Introduction," systematically summarizes the main results, methods, and successful case studies of this book. The second chapter focuses on the discussion of problems existing in the concepts of determinacy, indeterminacy, and complexities and our plans of resolution. The third chapter investigates the fundamental characteristics and problems of quantities, morphological transformations of quantities, and our methods of resolving the problem of quantitative infinity. The fourth chapter studies the historical contributions and existing problems of the system of wave motions of the mechanics, while providing the foundation for introducing the model for rotational movements and noninertial systems. Chapter 5 summarizes the significance and historical contribution of the V. Bjerknes's circulation theorem and the history of the mystery of nonlinearity and the chaos in the relevant contemporary research. At the same time, we propose our method to resolve the mystery of nonlinearity. Chapter 6 provides a relatively detailed account on nonlinear numerical calculations and numerical experiments. By using numerical simulations and practical examples of applications, we show that quantitative manipulations, analytic equations, and statistics are not necessarily the only methods available for solving practical problems. On the basis of this work, in Chapter 7, we formally propose the vision and plan for evolution science that generalizes the scope of investigation of modern science and relevant fundamental concepts. Chapters 8 and 9 investigate the methods and real-life case studies relevant to the proposal of evolution science, where each case study listed signifies a difficult problem for modern science that has taken a long time for the scientific community to consider and is still unsolved in the system of modern science. The case studies in this book explore the research of our theory and methods as applied to the investigations of mid-term and long-term predictions.

Each of us has some connection of varied degrees to areas of the arts, which might explain why with our respective training in the arts, after we entered the realm of natural science, we individually demonstrated our ways of thinking and reasoning different from those of other colleagues. And along with the establishment of our theories and methodologies, we have shown our own individual desires to materialize an organic marriage between natural and social sciences. No doubt, our systemic yoyo model (Lin, 2008b) and the theory and method of digitization (this book) can surely help materialize this multiyear noble dream of ours to remedy the current state of separation of natural and social sciences.

The development of science is motivated by the calls of solving previously unsolvable problems; innovations of technology are stimulated by the pressing demand for meeting practical challenges. In this book, we will not only reveal the calls of the problems existing in the system of modern science, but we will also provide the methodology and technology for resolving the practical challenge of predicting the unknown future. It is our hope that the publication of this book will reach a greater audience of the scientific community, and that the methodology of forecasting zero-probability natural disasters, as introduced in this book, can be further applied and perfected. This book can be used as a reference for scholars and scientific practitioners of many disciplines, such as mathematics, physics, philosophy, meteorology, and other areas of investigation on predictions. If at all possible, we would like to hear from you. The authors can be reached via e-mail: Yi Lin — jeffrey.forrest@sru.edu *or* jeffrey.forrest@iigss.net *or* jeffrey.forrest@yahoo.com *and* Shoucheng OuYang — ouyangsc2000@yahoo.com.cn.

About the Authors

Yi Lin, Ph.D., holds all his educational degrees in pure mathematics and had a 1-year postdoctoral experience in statistics at Carnegie Mellon University. Currently, he is a guest professor of several major universities in China, including the College of Economics and Management at Nanjing University of Aeronautics and Astronautics; a professor of mathematics at Slippery Rock University, Pennsylvania; and the president of the International Institute for General Systems Studies, Inc., Pennsylvania. Dr. Lin serves on the editorial boards of professional journals, including *Kybernetes: The International Journal of Systems and Cybernetics*, the *Journal of Systems Science and Complexity*, the *International Journal of General Systems*, and others. Some of Dr. Lin's research was funded by the United Nations, the State of Pennsylvania, the National Science Foundation of China, and the German National Research Center for Information Architecture and Software Technology. He has published well over 200 research papers and 18 books and special topic volumes. Some of these books and volumes were published by such prestigious publishers as Springer, World Scientific, Kluwer Academic, Academic Press, and others. Over the years, Dr. Lin's scientific achievements have been recognized by various professional organizations and academic publishers. In 2001, he was inducted into the Honorary Fellowship of the World Organization of Systems and Cybernetics. His research interests are wide ranging, covering areas such as mathematical and general systems theory and applications, foundations of mathematics, data analysis, predictions, economics and finance, management science, philosophy of science, and more.

Shoucheng OuYang pursued his higher education in the Department of Meteorology of Nanjing University from 1957 to 1962. Currently, he is a full professor at Chengdu University of Information Technology, a member of the Chinese Association of Geophysics, the International Institute of General Systems, Inc. (United States), and an invited editor of Emerald Journals (United Kingdom), and serves on the editorial or advisory boards of several journals, such as *Applied Geophysics* and *Scientific Inquiry*. Through practice, he established the theory of blown-ups and the method of digitization. Due to its real-life applicability, Professor OuYang's research of the theory and method were funded at various times by many local and national funding agencies and have been employed by some commercial weather forecasting services. He has made over 10 technological innovations, and introduced the theories of second stir and stirring energy. With invitations and relevant funding, Professor OuYang has traveled to Europe, Canada, and the United States to make conference presentations and deliver speeches. As of this writing, he has published more than 10 books, including *Entering the Era of Irregularity* (Meteorological Press, 2002a) and *On Evolution and Informational Digital Diagrams*. The book *Entering the Era of Irregularity* (Meteorological Press, 2002a) has been listed as one of the classics of the fundamental science books by the publisher. He has published over 100 research articles with more than 40 of these papers being collected by such databases as the Science Citation Index (SCI) and Engineering Index (EI) since 1998. Currently, Professor OuYang is working on various general problems using the concepts of uneven heat and information of changes.

Chapter 1

Introduction

Contemporary men would no doubt laugh at the pursuit of the past for the desire to live forever. So, no one would suspect the accuracy of the prediction that every person will die sooner or later. However, why each person has to die or the cause of death has not been addressed scientifically. Although modern science has been in operation for over 300 years, other than the very few scholars of the 20th century (Bergson, 1963; Koyré, 1968; Prigogine, 1980; OuYang et al., 2001a), there does not seem to be anyone else who noticed that modern science is a "science" that is about invariant materials. Even as of this writing, very few people are questioning why modern science cannot explain the physical reason for why people have to die and why modern science limits the concepts of evolution to those of extraordinarily large scales as geological ages or the astronomical times. This end only indicates that the system of modern science does not involve investigations of evolutions of materials; otherwise, there would not be such a need to limit evolutions to extraordinary long-time scales that produce no practical and theoretical significance.

Evidently, the process of human development of birth, growth, sickness, and death; the episodes of weather conditions, such as winds, rains, thunders, and lightning; the flying-over of shooting stars; and the problems of sudden wreckages caused by tsunamis, earthquakes, and so forth are drastically changing events or phenomena that do not belong to the category of extraordinary long-time scales. Surely, any "science" that does not investigate changes in materials or events does not have the ability to foretell the forthcoming changes in materials and events. So, the prediction that people have to die sooner or later in one way or another is not from any theory of modern science. Instead, it is from the empirical experience of people. Evidently, practical problems have to originate in the physical reality. So, some concepts of evolution should not be limited to those of extraordinary long-time scales, that is, whether evolution should or should not be determined by the

1

scales of time. To communicate more clearly, let us define the *evolution of an event* as the difference found in the event along the time order. Due to the common existence of evolution in materials and events, there should be corresponding laws of evolutions and theories and methods for investigating evolutions.

Because modern science does not involve changes in materials, it does not have much to do with evolution (Bergson, 1963; Koyré, 1968; Prigogine, 1980; OuYang et al., 2001). So, some natural questions that arise at this junction include, What kind of science is modern science? and What kinds of problems can it or can it not resolve? To address these questions, it is primarily important for us to know the essence of these questions; we could not confuse events that do not change over time with those that do vary in the time order. Any investigation about the prediction of what has not happened has to involve changes in events and the understanding of when events would not change with time. Otherwise, predicting what had not happened would become meaningless. The so-called prediction must mean that the future will be different from the present or the past. If the future were the same as the present or the past, there would not be any need for the profession of prediction to exist.

In reality, works by Bergson (1963), Koyré (1968), Prigogine (1980), and OuYang et al. (2001 and 2002a) have already shown that modern science denies evolution in materials. If one opens an elementary textbook of modern science, he would realize that since the time when Newton's second law was established, no problems involving changes in materials have been considered. However, what is amazing is that the current widely employed theories of forecasting are developed on the basis of modern science. So, there is a need for us to adequately position modern science to find where the concept of irregular information of modern science is from, what irregular information means, and what significance such information possesses. Such problems as what time is and how we can employ it should be well addressed as problems of physics by providing the corresponding methods instead of simply arguing about them as problems of philosophy at the epistemological level.

Although natural science contains problems of epistemology about nature, it still belongs to the category of productivity needed for human survival. So, the true value of natural science is in its practical use. At the same time, in applications along with the discovery of new problems, incorrect theories are modified. That is why practice is the standard for validating theories. In particular, in order to correct the mistakes existing in an accepted theory, it seems that practical applications are the only powerful means for successfully achieving this end. Otherwise, the momentum of the contemporary stream of beliefs could not be shaken.

Our book entitled *Entering the Era of Irregularity* was written after many years of successful practical applications in order to satisfy the numerous requests from readers of our previously published works (OuYang et al., 2002a; Lin, 2008b). To prove that events are not the same as quantities, it took us more than 40 years of our valuable scientific careers. Because *Entering the Era of Irregularity* focuses on practical applications without going into theoretical arguments much, many readers have sensed our untold theories. They demanded us over the years to publish what has been left behind from that book. That is how the second author of this book decided that we would produce another book following the publication of *Entering the Era of Irregularity* by focusing on the central problem — evolution science and digitization of information. This decision led to the investigation of such epistemological problems as how to position modern science; how to understand irregular information, time, and change; and how to present the changing world using the system of digitization by analyzing materials and events without staying in the realm of eternity. It has been historically shown that the practical significance of science and technology is about constantly challenging human wisdom by solving unsettled problems.

1.1 Analyzing Materials and Events, Analyzing Quantities, and the System of Science

The term *modern science* generally means the contemporary system of science that focuses on quantitative analysis. It was established and perfected by Newton, Einstein, and their followers since the major debate of the 15th century due to the introduction of heliocentricity of Nicolaus Copernicus (1473–1543). Modern science is based on the foundation of separate materials and quantities. When specified to physics, it stands for the system of quantitative analysis developed on the invention of inertial systems, forces, and masses. In terms of the relevant methods, it is about solving well-posed dynamic equations using the quantitative formal logic and statistical analyses of random, stable time series. What is left open are: What is time? What is information and, in particular, irregular information? And how can one make use of irregular information? Although in *Entering the Era of Irregularity* we did not touch on any of these sensitive problems, from our conclusion that quantities are after-event formal measurements, the need to adequately position modern science has become obvious.

Because the modern, scientifically well-posed equations or the statistical methods of stable time series are developed on the ideas of quantitative regularization and stabilization of time series, such a question as what is irregular or small probability information has not been addressed. That is why we have written (OuYang and Peng, 2005; OuYang and Chen, 2006) about the positioning of modern science, why information cannot in general be modified, the end of randomness, and the conclusion of quantitative comparability, and so forth. As a problem of philosophy, the concept of irregular information touches on the central problem of Lao Tzu (time unknown), "Any Tao that can be explained is not the Tao." It is exactly at the very central problems that modern science has walked away from the essence of the multiple varieties of the natural world and that it pursues after the generality and uniformity using the formality of quantities.

To illustrate this end, let us look at Newton's second law of motion as our example:

$$f = ma \tag{1.1}$$

As is well known, this second law in the form of Equation (1.1) has been forced into children's heads since grade schools. However, we have not seen any textbook that clearly points out the fact that Equation (1.1) does not really agree with the reality. It is because in reality, the acceleration a always varies. Now, m is substituted by a quantity due to particlization. So from the assumption that the quantity (mass) of materials is invariant and f is from the Godly eternity, a has to be an invariant constant. For such a constant acceleration, one has to take the average value of the realistically changing acceleration a. That is, to satisfy Newton's second law of the inertial system, the original form of Equation (1.1) should be

$$f = m\bar{a} \tag{1.2}$$

where $\bar{a}$ stands for the mean value of the realistic acceleration. So, at the very start of modern science, in which without assumptions there would not be any science, it has been positioned as the science of average values that have been employed as representatives of generalities. As a matter of fact, there is a need to educate our children that quantities can be used to compute nonexistent matters more accurately than those that do actually exist. That is, average values are not the same as objectively existent events; the probability for any of these average values to appear is smaller than the chance of occurrence of a small probability event.

Besides, in Equation (1.1), if the acceleration a is a realistic variable, even if f stays constant, as a variable m has to vary. That explains why in modern science, one has to assume constant acceleration and needs inertial systems; otherwise, Equation (1.1) would not hold true. As a matter of fact, in Equation (1.1), all of f, m, and a should be variables in order to produce a theory to comply with the physical reality. However, doing so would make Equation (1.1) a nonclosed equation, leading to an unsolvable problem in mathematics. In particular, if f, m, and a are variables, then Equation (1.1) represents exactly a nonlinear equation unsolvable in the system of modern science.

In short, because

$$a = \frac{dv}{dt}$$

where v stands for the speed, the message that can be delivered by Newton's second law, which has been a signature of modern science, is a movement of variable speed with constant acceleration without involving change and origin of f and changes in the material m.

It should be admitted that the system of modern science has already recognized that both variable accelerations and rotations represent noninertial systems (Prigogine, 1980). However, it did not clearly point out the physical meaning of variable accelerations and rotations and did not clearly distinguish inertial and noninertial systems. For instance, the statement "a system that is not inertial is noninertial" is essentially equivalent to not saying anything. When the acceleration a is a variable, then f and m have to vary too. In this case, from Equation (1.1), one has to ask where the force is from and how the material would change. So, let us use this opportunity to introduce a tangible definition for noninertial systems: If the symbols f, m, and a in Newton's second law are all variables, which can also be referred to as variant material or event, then one has a problem of a noninertial system in hand. Because rotation can cause changes in materials and nonlinearity possesses rotationality (OuYang et al., 2002b; OuYang, in press; Lin, 2008b), it follows that rotational movements and nonlinearities all belong to noninertial systems.

No doubt, variable forces will have to affect the epistemology of eternal external forces. If the material is also changeable, then m can no longer be seen as a particle so that the concept of structures for materials has to be introduced. So, the foundation of modern science no longer holds, because at the same time when changing materials destroy the doctrine of particles, they already represent problems of evolution science. Because evolution science is about the investigation of changes in materials, it implies that evolution is a science of noninertial systems. This end provides a clear positioning of modern science.

Specifically, with the invention of physical quantities, modern science does not need to consider structures and attributes of materials so that quantities become its tool of analysis. Placing f outside the object without explaining where the force is from, modern science surely could not answer where the acceleration is located. Due to placing f and a outside the object, modern science loses its foundation needed for physics. So, the essence of separating the object and force is that quantities can exist independently outside the object along with the separation of the object and quantities. With the invention of physical quantities, the analysis foundation of quantitative formal logic is established. However, such a logical foundation should also comply with the epistemological laws of transformation of materials or events. Materials stand for the primary problem of physics. Without materials, there would not be the physics, because forces, accelerations, and other physical quantities are either functions or attributes that dwell on materials. Without materials, such concepts as f and a would not exist. Even if f, a, and other physical quantities could exist outside materials, that would lead to logical paradoxes. At the most fundamental level, modern science tends to experience logical paradoxes and cyclical applications of concepts. For example, in their originality both f and a are really from the object; however, in scientific laws, attributes of materials have been placed outside the underlying materials. Newton treated mass as a common measurement for an object's density and volume, forgetting that density itself also stands for mass. That is, in the foundation of the basic science, a cycle of concepts is employed. Einstein's mass–energy formula $E = mc^2$ recognizes light as a kind of material; however, c is assumed to be the speed of light in a vacuum, forgetting that the definition of *vacuum* implies a void of materials. Since light is a kind of material and can exist in a vacuum, then the vacuum is no longer a true vacuum. That is why at the time when m is 0, Einstein's mass–energy formula also evaporates. Even according to the explanation of modern physics, photon does not have any static mass; the consequence is still an evaporated formula of Einstein's mass–energy for when m is 0. We did not find anywhere that Einstein ever mentioned that Newton's second law is erroneous. However, from his assumption of constant speed of light, we have to conclude

$$C = \text{constant} \Rightarrow \frac{dc}{dt} = a = 0 \tag{1.3}$$

So, we have

$$f = m\,a = 0 \tag{1.4}$$

That is, when the speed of light is a constant, Newton's second law also evaporated, making the entire modern science invalid. Because energy stands for the work done by force, force is equivalent to acceleration; if $f = m\,a = 0$, then the energy as the work of force will no longer exist. So, Einstein's mass–energy formula still evaporates. That is why we treat the mass–energy formula as an energy that does not do any work, leading to the problem of whether or not Einstein's relativity theories actually hold true (OuYang et al., 2002; OuYang, in press).

That is, modern science, which is known for its rigor of formal logic, does not really comply with the formal logic in its fundamental laws. There were such western scientists and Chinese scholars who went to the West to receive the western education who claimed that China did not have modern science or went even as far as to say that China did not have science. In essence these scholars did not understand that it is impossible for a people who based their knowledge system on structural properties of materials (those ancient Chinese from the time of Warring States knew that forces are from the uneven structural gradients of materials [Sun, 2000]) to establish such a concept of force, as a physical quantity, independently outside materials and to employ conceptual cycles. So naturally, the people would not develop such philosophical concepts as the binary object and force or the separation of objects and quantities and the system of modern science. In other words, in order to construct the system of modern science, Newton and his followers had to pay the price of using conceptual cycles and violating some basic concepts of epistemology. Some of these followers never realize that science is about challenging human wisdom through practical applications.

It seems that science is science without the need to add the prefix *modern* to it. However, the phrase *modern science* is not our invention. And even within modern science exists a separation of systems. For example, the quantitative manipulations of Newtonian dynamic equations have been seen as a system of determinacy, while statistics represents a system of quantitative analysis that is indeterminate. So, the so-called science suffers from the problem of inability to be plausible, because modern science advocates the rigor of formal logic; however, systems of determinacy are parts of science, and systems of indeterminacy are also parts of science. Is this phenomenon similar to that of using one's own sword to attract his own shield? The world of learning does not seem to openly discuss this problem. Our understanding is that science is a concept of process; science is not the same as the complete maturity of the science, so that rooms always exist for the system of science to correct and improve itself over time. Second, the value of science is in its practical use, which is different from the value of beliefs. If a part of science is not practically applicable, then sooner or later that piece of knowledge will die out. Third, the reason science itself is referred to as science is its ability to challenge the accepted knowledge and human wisdom. So, the next-step development of science should not be dwindled to the level of arguing about how to phrase a specific concept. Instead, it should focus on the core problem of whether or not the quantitative system can actually resolve the existing problems. The efforts devoted to this endeavor will surely make people look for new theories and methods that possess the desired use value.

Even though the system of modern science suffers from various problems and paradoxes at the fundamental epistemological level and the formal logic, for instance, practical applications of calculus have not complied with the requirement of continuity, it can still be employed to deal with movements of invariant materials, leading to applications in the design and manufacture of durable goods. So, its use value is manifested. In the corresponding Eastern world, especially in China, the way of knowing is based on the nonparticlization — the epistemology of structural transformations of mutual interactions — of materials. The Easterners entered evolution science ahead of the formal logic of the West. That explains why the first push system has been difficult for Chinese people to accept and why these people placed more emphasis on the material morphologies caused by blocked moving materials and relevant changes in the materials' attributes. To this end, the *Book of Changes* (Wilhalm and Baynes, 1967) has been the classic of evolution theory and becomes the standard of knowing and understanding, leading to strange feelings about placing the quantities of the kinds of physical quantities independently outside materials. It is because the first reaction Chinese people have would be how materials mutually constrain each other, leading to the established epistemological theory of mutual existence and constraints of the five elements of yin yang, and the system of quantification — the yin and yang lines (- - and —). With the recent advancement of modern science, the contemporary man might be able to start to understand that the yin yang lines, - - and —, as introduced in the *Book of Changes,* can be used to describe any material or event, including the attributes, states, and transformations of events together with the functions, colors, and quantities of the events. This end has been well shown by modern computer technology. So, the yin yang lines, - - and —, that can directly express the morphology, attributes, and functions of materials and events represent the earliest invention of the currently familiar quantities. As a matter of fact, with these ancient binary numerals, the ancient Chinese had perfected their scientific system of knowing the world through analyzing materials. What happened later is that following generations of the ancient Chinese no longer recognized the structural characteristics and transformationality of events embedded in these yin yang lines, - - and —. Due to their inability to understand the rules of operations of these lines, this method was not further developed over the recent past centuries. Fortunately, digitization led to the invention of high-speed computers, which can directly digitize observational data, and revealed their ability to handle irregular information that cannot be well dealt with in the system of quantitative analysis. This end helped us to uncover a passage to connect modern technology with ancient science and methodology. Consequently, we are able to propose the epistemological foundation of evolution science on which the world will be known through analyzing materials.

Evidently, to know the world through analyzing materials means that one has to focus on the investigation of evolutionary problems of change in materials' attributes and states. Combined with the historical fact of creating prosperity through agriculture, where planting is done in the spring and harvest in the autumn, the changes in seasons led to the early development of Chinese medicine and chemistry (the art of making immortality pills). These inventions were realized in daily life through the introduction of gunpowder, rockets, and tofu, making the Chinese people lead the world in the science and technology of flood control, irrigation, and so forth. In terms of the agricultural achievement, over 7,000 years ago, (according to the archaeological discovery of Cheng San Tou), the Chinese food production had been way ahead of that of the rest of the world. Even at the end of the Qing Dynasty period (about 100 years ago), it was still the case that as long as the area of Hu and Guang, which is the area of the current four provinces of Hunan, Hubei, Guangdong, and Guangxi, had harvest, the entire nation would be satisfied (that is, such a harvest would guarantee the financial need of the entire nation). The establishment and development of Chinese medicine provided the continuous support for the health of the Chinese people for over 5,000 years. The appearance of the *Book of Changes* and its continued presence in the Chinese culture are mainly due to its practical use value. It was only because the rulers of the following generations were content with the already achieved accomplishments that new development in almost any area was not encouraged, so that over time, people were no longer able to note the far-reaching significance and effects of the yin yang lines, - - and —. That is why when modern science initially entered China, combined with the misunderstandings of some scholars produced out of the western education system, the applications and development of the system of evolution science based on the epistemology of knowing the world through analyzing materials was temporarily stopped at the level of intellectual learning. However, the basic system of epistemology of knowing the world through analyzing materials, just as the *Book of Changes* and Lao Tzu, has been continuously practiced throughout the ages by the ordinary Chinese people. Speaking frankly, the reason why we are able to point out in our lectures on modern science that events are not the same as quantities and the existing problems in modern science is not our formal educations received in the organized system of Western knowledge. Instead, it is the edifying influence of the Chinese culture and the Chinese way of thinking.

So, since the very start when we began our formal education of modern science, we had our occasions of casting questions and doubts that our teachers and professors could not address. For example, if f were from God, would that make science a part of religion? If both f and a exist within the inside of objects, how can we write both f and a outside m? (This end indicates that as a fundamental problem of physics, m cannot be particlized.) What is time? And so forth. These and many other problems we ever questioned in our student years surely created great difficulties for our teachers and professors. That is, the quantitative form of the system of separate materials and values of the physical quantities have made the physical quantities escape materials and constitute problems of formality. So, the resultant theories have to suffer from the problems of not agreeing with the realisticity of the states, attributes, and functions of materials and events. Second, the problem of rotation, that is, the numerical invariance of speed, has to become a variable due to the ever-changing directions in the rotational movements. So, each rotation involves variant acceleration and represents a problem of a noninertial system. And, due to the difference in rotational directions because of the duality of rotations (the coexistence of inward and outward spinnings), collisions of different eddy motions have to cause changes in materials so that the varying m can no longer be assumed to be a particle without size and volume. Instead, we have to give expressions to the morphological structures and changes of attributes of materials and events.

Therefore, the essence of modern science is a system of formal analysis based on quantities so that it has to comply with the quantitative calculus of the formal logic. On the other hand, in the system of analysis of evolutions through studying materials, one has to be clear about materials and events instead of merely the relevant quantities. So, most likely the quantitative calculus of formal logic does not hold true here. It can be said that under the epistemology of the *Book of Changes*, identifying quantities with events is a very ridiculous act. Even if we look back by comparing what we have learned in modern science, it is still quite obvious to us that changes in materials have to cause variable accelerations, leading to noninertial systems (according to Newton's second law, when m changes, the acceleration a has to vary too). So, the system of modern science contains theories on invariant materials (Bergson, 1963; Koyré, 1968; Prigogine, 1980; OuYang et al., 2001), and using such theories of invariant materials to investigate changing materials has been faced with difficulties and challenges. That is why modern science is not about evolution science, and why in the past 300-plus years no scientist who could foretell the future was ever produced out of the system of modern science.

The current widely applied principles and techniques of prediction, established on the system of modern science, in essence are monitoring what has already happened. It is not about predicting what is about to occur. This end has been very well illustrated in Laplace's statement (Kline, 1972): "If I know the initial value, I can tell you everything about the future." What Laplace said in fact admits that without knowing the initial value, modern science will have no clue about the future. That is exactly the problem that current weather forecasts, earthquake prediction, and other future-telling face. It has been listed as one of the world-class difficult problems. So, it is a call of our modern time and a must in the development of science for evolution science, by which the world is known through analyzing materials, to appear and develop.

What we have to point out is that science itself is a process, and in its development, conscious behaviors of man have to be part of it. When people did not find a way to deal with events, using quantities to substitute for events could be seen as a human wisdom; after we have found ways to deal with events directly, if people continue to treat events as quantities and try their best to employ quantities without discrimination, then substituting events by quantities would become an intelligence deficiency of mankind. Although the scientific system of knowing the world through analyzing materials has been ridiculed and criticized constantly by contemporary scientists, due to its practical use value, it has been well accepted by the ordinary public as a latent and invisible force that has stood the test of various calamities over the past several thousand years. We have to admit that in the fetters and handcuffs of the system of modern science, it had to be extremely difficult for the French scholars H. Bergson (1963) and A. Koyré (1968) to discover and point out modern science's denial of evolution. As a matter of fact, through their publications they actually prophesied the future direction of development for modern science.

So, based on the convention that science is about challenging human wisdom and its use value, right now seems to be the right time for those scholars who share the same beliefs as ours to analyze and uncover various problems hidden in modern science so that by meeting the need of solving those problems unsolvable in modern science, science will be carried to a higher level.

1.2 What Quantities Are and Where They Are From

Based on what we could find, the introduction of numbers can be traced back to the philosophy that numbers are the origin of all things of Pythagoras of ancient Greece. These people believed that the numbers within the range of 10 possessed mighty powers, representing the primary principles and guidance of the supreme beings, the heavens, and humans. Without knowing numbers and the properties of numbers, one can never comprehend other things. So, this philosophy definitely has greatly influenced the development of the later system of science within which the world is known through analyzing numbers, and quantities are the laws of all matters. However, what has to be pointed out is that the ancient Greeks did not invent the number zero, which was initially invented by Buddhism in ancient India. Later, it was passed on to the Arabs and seen as an invention of the Arabs. There is a need for specialized experts to check into the possible connection between the separation of objects and their attributes in the teachings of Buddhism, to Aristotle's later separation of objects and forces, and then to the separation of materials and quantities of modern science. No matter what is the outcome of such investigations, there seems to be an origin for the epistemology and methodology of separating materials from their attributes and functions in modern science. There is no doubt that knowing the world through analyzing quantities is the religious foundation of modern science. Its use value is mainly found in the areas of design and manufacture of durable goods and the analysis of movements of invariant materials, where quantitative analysis is the tool of operation. However, human wisdom should be much clearer about seeing the existence of problems and comprehending the essence of the problems than just simply following the scientific fads of various times.

Corresponding to the same historical time periods, Zhan Yin of ancient China, who lived in the time of Warring States, pointed out that there are things numbers cannot describe. That explains how later Chinese people treated numbers and why they did not respect numbers nearly as much as the Westerners. Later Gödel established the incompleteness theorems in mathematics, but this was only an event of the 1930s. The reason why the Westerners are more fond of quantities than the Chinese is also due to their use value. Evidently, people whose prosperity depends on nomadic trades have to be serious about the counts of their cows, sheep, and other good-size animals. As for the ancient Chinese people, whose prosperity depended on agriculture, they never had the need to accurately count the number of grains of rice as the Westerners' counting of sheep and cows. What is more important is that due to the need for survival, Chinese people place more emphasis on the attributes of materials, such as functions, tastes, sounds, colors, locations, and so forth, leading to their realization that numbers always suffer from the deficiency of being unable to describe clearly. That is why even in the *Book of Changes*, the ancient Chinese people had started to explore the physical world by analyzing materials. That might be the cultural reason why modern science could not have appeared originally in China. When knowing the world through analyzing materials, one has to face changes in materials. That is why the scientific system of knowing the world through analyzing materials started with noninertial systems with variable materials. So, the difference between the Eastern and Western sciences can be traced to their very beginnings. From these discussions, it becomes easier for us to understand why someone such as Newton could have written *Mathematical Principles of Natural Philosophy* (Newton, 1687). However, what needs to be clear is that Newton's *Mathematical Principles* in essence is about how to use mathematics to describe "natural principles." And because the achievement of calculus that has nothing to do with the specific path taken does not agree with the reality, Newton's

Mathematical Principles cannot really become the realistic physics or the principles of the natural philosophy.

What is quite clear is that quantities are postevent formal measurements. Quantities cannot appear before events. That surely diminishes the hope of using quantities to predict the occurrence of future events. Such statements as "without things, there would not be quantities," "if things do not change, then the quantities would not vary, either," "quantities can be employed to compute nonexistent matters more accurately than those that actually exist," and similar statements are specific realizations of Zhan Yin's claim that there are things numbers cannot describe. In particular, numbers dwell in Euclidean spaces as the formal measurements of the imaginary axes, leading to the problem of unboundedness of the quantitative infinity. However, the objective physical space is curved, where before an event evolves to the quantitative infinity it had already gone through a transitional change. This end reveals the limitation of the quantitative analysis and where the East and the West see things differently.

The composition of human knowledge is directly related to the environments in which people live so that people living in different geographical locations have different ways of knowing things. Different forms of knowing must lead to different cultures. In the culture where not all things can be described by using numbers, Chinese philosophy did not attach natural principles to quantities. Here, not attaching to quantities is not equivalent to not having philosophy and science. In order to know the world through analyzing materials, one has to consider the attributes of materials, such as functions, tastes, sounds, colors, locations, and so forth, and individual materials' specifics and their interactions, leading to the Chinese theory and method of structural analysis of how materials act on each other based on the concepts of five elements and yin and yang. If the contemporary language is applied, interactions of materials belong to noninertial systems. The difference between how the East and the West entered their sciences is that the West made use of the quantities and inertial systems of the first push system, while China of the East structuralized the digits (the yin yang lines) of materials' interactions and formed its scientific system in the cultural form of combined natural and social sciences. In this sense, the *Book of Changes* can be translated into the "theory of evolution" of the contemporary language. That is, Chinese people had long ago entered the era of investigating material evolutions. The movements of variant and invariant materials stand for problems of different characteristics. The epistemologies and methodologies needed to handle these problems have to be different. Due to their formality and generality, quantities provide a method of analysis of the logical calculus for the investigation of movements and oscillations of invariant materials, which constitutes their contributions in the scientific system in which the world is known through analyzing quantities. However, in front of the studies of interaction problems of evolutionary materials, quantities have experienced great difficulties. However, these two classes of problems — movements and oscillations of invariant materials and interactions of evolutionary materials — can be connected by using the concept of rotational stirring energies, leading to the overall system of science.

In short, as the result of our tracing back to the origins of the scientific systems, we find that the science of knowing the world through analyzing materials possesses the system of digital (the yin yang lines) structural transformations, where events are not quantities, and the modern science of knowing the world through analyzing quantities and quantitative manipulations involves only invariant materials, where quantities are used to replace events.

1.3 Rotational Movements

The fundamental form of materials' motion is rotation. The reason why modern science that advocates analyzing quantities in order to know the world avoids rotations is that when quantities are employed to deal with rotations, one experiences difficulty. However, there are those who dare to explore the impossible. For example, V. Bjerknes in 1898 (Hess, 1959) employed nonparticle circulation integrals to establish his remarkable circulation theorem. He was the first person in modern science who pointed out that the fundamental form of fluids' (atmosphere and oceans) motion is rotation (in essence, solids also mainly move in the form of rotation). Later, Saltzman (OuYang et al., 2002a) referred to this result as a major betrayal of the system of modern science. Unfortunately, this important theorem did not catch the attention of the world of learning. Even in the current textbooks of meteorology, it is only briefly mentioned and then forgotten. Our presentation in this book shows that if meteorological science had developed along the thoughts of the circulation theorem, the current situation of weather forecasts would be very different; and not only would meteorological science have truly entered the science of prediction, but also would the entire system of modern science have gone through major reforms. Or at least at the end of the 19th century, modern science would have entered the system of evolution science.

Next, because modern science only limits kinetic energy as the square of speed, including Newton's v^2 and Einstein's c^2, one naturally has to ask what kind of energy the square of angular speed (ω^2) would make. Besides, angular speeds implicitly contain the concept of speeds due to uneven spatial distributions of speeds. Because the square of angular speed is different from that of speed, we introduced the concept of stirring energy (OuYang et al., 2000, 2002; Lin, 2008). Our work shows that modern science missed not only this kinetic energy of rotation, but also material existence in curvature spaces and the physics laws of material transformations. What is important is that whether stirring energy is conserved or not can help the first push of modern science to go across over to variant materials so that the second stir (OuYang et al., 2000) of evolution science is involved. It is exactly because stirring energy contains the kinetic energy of modern science as a special case that modern science can be smoothly expanded into evolution science. That is why we mentioned postmodern science in our book *Entering the Era of Irregularity* without including many details. It can be said that the postmodern science, which deals with mutual interactions, can contain modern science through the investigation of rotations, while directly illustrating the significance and effects of the rotations of curvature spaces (Chen et al., 2005b). Because of this, we will be able to resolve the problem of what time is, which is still an unsettled problem in modern science (OuYang et al., 2001, 2002; Lin, 2008), and that of how to correctly comprehend the meaning of irregular information (OuYang et al., 2001; OuYang and Chen, 2006).

1.4 What Time Is

The modern science of knowing the world through analyzing quantities openly admits that from Isaac Barrow to Newton, to Einstein, and to the current Prigogine, no one has resolved the time problem (Prigogine, 1980). Our discussions in the following chapters and Lin (2008) indicate that such a long-delayed success in addressing the time problem is due to the constraint of quantities. What is worth our attention is that Leibniz, a contemporary of Newton and a cofounder of calculus, once criticized Newton for proposing his numerical time by saying that events are the essence (Kline, 1972). However, such a highly lofty opinion did not catch any attention of the then-scientific community and the following generations. When we emphasize the fact that events are not quantities, we also implicitly appraise Leibniz for his magnificent scientific achievements and opinions that are shown to be far beyond the comprehension of his and the following generations. Under the tight control of the Western quantities, it was very difficult for Leibniz to air his opinion that events are the essence.

In fact, *time* is given an expression by the differences existing along the time order of the essential events. When Shoucheng OuYang, a co-author of this book, was still a college student, during his debate about whether or not Rossby's waves, a fundamental theory of the meteorological science, actually exist, he asked the questions of where physical quantities were from and why they exist, among others, leading to the question of what time is. Although the professor thought OuYang's logic was very strange, he still had to admit that such questions as what time is and where time is located were very thought provoking. The debate was ended with the professor's comment that there were still not any available answers to these questions. However, the desire to answer the question of what time is persisted. After an accidental opportunity of reading the *Book of Changes*, OuYang formed his own philosophical point of view of time and space that happened to be different from those of modern science. What needs to be explained is that we are not used to staying with pure theoretical arguments. Instead, we place more emphasis on practically resolving problems. Only after achieving practical successes do we turn back to establish the relevant theories. So, in our professional careers, it took us a long time to eventually publish such conclusions as, time does not occupy any material dimension and instead it dwells on the changes in materials, and time cannot exist prior to existence, because we were looking for opportunities to practically validate these and other relevant conclusions. The reason we included these conclusions in OuYang et al. (2001, 2002) is our debates with colleagues from different countries, which forced us to employ these conclusions on the structural digitization of irregular information, leading to the publication of the relevant works (OuYang et al., 2005a,b; Zeng and Lin, 2006).

The reason why time does not occupy any material dimension and why time cannot be treated as a parametric dimension as other physical quantities is that when materials and events do not change, time does not exist. That is, time exists in the changes of materials and events. This of course leads to similar questions about certain other physical quantities studied in modern science, such as whether or not such physical quantities such as force, energy, light, and so forth, occupy material dimension and where they come from — because changes in materials and investigations of evolution science will have to pursue why materials occur before anything else. In other words, evolution science has to study the origin and changes in such physical quantities as force; otherwise, it would be impossible to achieve the goal of being able to predict what is forthcoming. Even Aristotle, a forefather of modern science, once pointed out (Apostle, 1966) that physics is about the study of changes in materials and the processes of change, and the capability of mathematics to describe this cannot answer *whys*. The essence of understanding "time" as the differences in materials along the time order is "increase and decrease come in their own times" (hexagram 41, *Book of Changes*) and "all matters resolve with time" (Zhuang Zi [Watson, 1964]). So, the essence of time not occupying any material dimension is that time cannot be treated as a physical quantity or parametric dimension, as what has been done in modern science. When time is treated as a quantitative variable or parameter in an equation, in form it seems that time is considered. However, due to the requirement that mathematical physics equations be well posed or that statistics have a stable time series, the relationship between events and time has already been eliminated. That was why both Bergson (1963) and Koyré (1968) recognized that modern science denies evolution and time becomes a parameter that has nothing to do with change. So, to correctly understand and apply time, one has to provide differences in the changes of materials (if materials do not change, there will not be the past or the future so that time becomes a meaningless concept, which also makes the debate of nearly 100 years within the scientific community about determinacy and indeterminacy unnecessary). These differences may clearly explain how to specifically apply the concept of time. That is, time tells us that we have to admit differences existing along the time order; otherwise, we would not be able to distinguish changes in events. This end involves the system of evolution science of the age, sickness, and death processes of materials. In this sense, we can say that knowing the world through analyzing materials is different from that through analyzing quantities. In other words, evolution science is different from modern science. That is why modern science is only a special case of evolution science or evolution science pushes modern science to undergo revolutionary reforms.

Based on our discussion above, it follows that treating the theoretical system of modern science that only values quantitative analysis as the foundation of prediction science is a major epistemological mistake. In particular, modern science cannot handle irregular events and employs various theories, such as stability, complexity, or stochastics, to eliminate irregular information. Doing so only implies that the scientific community still does not fully understand time and does not recognize the practical significance and function of peculiar events.

So, Leibniz's statement that events are the essence itself has tried to push science toward the investigation of events. In other words, now is the time for modern science to undergo revolutionary changes. The current treatment of irregular information as quantitatively small probability and the elimination of irregular information or events by using averages, filtering, and so forth in essence also get rid of variant events and the information of change. There is no denial that eliminating information of change is the same as getting rid of variant events. The consequence can only make modern science become a system of invariance, pursuing after the heavenly eternity (Prigogine, 1980).

1.5 Irregular Information and Digital Structuralization

To investigate this problem, we have to first make sure we understand what information is. The modern science that is based on knowing the world through analyzing quantities has been in operation for over 300 years so that some concepts have become part of the well-accepted custom. As is well known, it is often difficult to make changes in customs. Evidently, the problem of time reminds us that we need to correctly understand events and differences in events. This end naturally touches on how to correctly understand the problem of information. In particular, in our information age there is a need to adequately give information a name. Otherwise, it would be difficult for us to study irregular information.

The concept of information is closely related to our daily lives. However, modern science did not provide a clear-cut definition for this concept. The reason is that irregular information has brought forward difficulties for modern science. Since the time when Shannon and Weaver (1949) first introduced the concept of information and used the concepts of probability to define the amount of information and the manipulations of information, information has been associated with uncertainty, which helped to intensify the nearly 100-year-old debate between determinacy and indeterminacy within the scientific community. However, in specific applications, the school of determinacy applies stability to eliminate peculiarities, while the school of indeterminacy employs stable time series to get rid of small probability events. Although they walk along different paths, the results are the same: eliminate peculiar events and complexity, forgetting that the generality modern science pursues exactly comes from peculiarities. And the probability for the desired generality to appear is smaller than that of any small-probability event. The so-called generality, just as averages, does not really exist in reality.

In short, we cannot simply ignore irregular information just because quantities cannot handle it. The fact that quantities cannot deal with irregular events reveals the infancy of the system of modern science.

The essence of the debate between determinacy and indeterminacy is that both schools try to sell their own systems of quantitative analysis as the correct system for science. However, in terms of the use value, neither of them can deal with irregular events, which makes the nearly 100-year-old old debate between determinacy and indeterminacy hinder the needed comprehension of events in order to resolve the problem of prediction. So, the theoretical values of these theories in the studies of predictions are lost. Either information or events are facts about what had happened or existed. Even if the probability for appearance is nearly zero, they are still about the already-appeared events, which will not simply disappear just because quantities cannot deal with these events. So, in the development of science, we have met with the problem of how to understand irregular information and how to practically apply it. That is, the statement that events do not contain randomness is not only a challenge to the system of modern science, in which the world is known through analyzing quantities, but also more about establishing a concrete method to deal with irregular events. So, the 21st century will be a millennium during which mankind will show its wisdom by inventing ways to comprehend and deal with irregular or variant events.

To this end, we have empirically analyzed a huge amount of irregular information and discovered that irregular information is closely related to collisions of rotational materials. What is practically meaningful is that by introducing irregular information, in particular, the unusual irregular information of the ultralow temperature existing at the upper layer of the troposphere, we are able to predict the transitionalities existing in the changes of events, including weather phenomena. It is exactly because of its unusualness that quantitative methods cannot deal with it. However, its existence determines the appearance of disastrous weather. After many careful and repeated comparisons, we find that without an ultralow temperature there will not be any convective disastrous weather. So, irregular information not only has its reason but also touches on the mechanism of material structures. Its significance goes beyond the formality of quantities and directly enters the attributes of events. Although people can use quantities to label it, the essential significance is no longer about quantities but simply labels of events. When the symbols of quantities are used to name information, they become names or labels of events. So, these symbols of quantities cannot be identified with quantities and manipulated using the rules of operation of quantities. What has been practiced in modern science is exactly the opposite, leading to the problem of fabricating false events.

So, we believe that since the time of Shannon and Weaver (1949), it has been a major mistake in the study of information to use concepts of probability. Such a mistake is caused by the historically mandatory application of the quantitative analysis. It is like the yin and yang lines in the *Book of Changes*. They can be referred to as digits. In other words, what we see as digits today are exactly the ancient Chinese symbols for the yin and yang lines. They cannot be confused with quantities or numbers. Just as the yin and yang lines do not follow the rules of addition, subtraction, multiplication, and division of quantities or the rules of integration and differentiation, the yin and yang lines or the digits, as known in our modern time, do not satisfy the quantitative calculus of formal logic. Digitization contains quantities and is more about walking out of quantification. It signals that the human knowledge about the world has entered the era of materials and events. So, we repeatedly said at various occasions that digitization cannot be confused with quantification. At this junction, let us repeat one more time that information stands for the signals of already-occurred events so that it has to be deterministic due to the known attributes and characteristics of the events. It cannot be referred to as indeterminate simply because the quantitative analysis cannot handle it. Although information is labeled by using symbols of quantities, these labels are not quantities or numerals. That is why digitized information cannot be treated with operations of quantities.

What needs to be said is that areas of engineering currently treat information the same way as quantities, because what that involves are the design and manufacture of durable goods. However, in specific engineering designs, the risk value, which is empirically obtained through experiments and used as the generality for testing the durability of goods, is not a direct consequence of modern scientific computations. Hence, classifying information as uncertain is a mistake caused by the continued effort of knowing the world through analyzing quantities of the system of quantitative analysis. Irregular information provides us the information of change of the underlying material evolution. It stands for a jump in our epistemological concepts and that science has entered the era of evolution science. And, digitization provides a new system of methods beyond that of quantification along the path of development of science. It constitutes the third methodology along with those of dynamic equations and statistics. And, the method of using figures is the oldest way for people to analyze situations in the scientific history. Only the contemporary digitization makes such ancient method more refined, marking a new development in the scientific system of knowing the world through analyzing materials. It can be seen as a combination of the ancient wisdom and the modern digital technology.

Due to the problem of mathematical nonlinearity, there once appeared a heat wave of nonlinearity in the scientific community during the 1980s. Along with the heat wave, such theories as chaos, solitary waves, fractional dimensions, bifurcations, and so forth were proposed. In fact, if in Newton's second law all f, m, and a are variable, then it becomes a mathematical nonlinear equation. Based on our discussions above, even if modern science has recognized the physical significance of variable accelerations or rotation movements, the current community of mathematicians still considers solving the Navier–Stokes equation as a world-class difficult problem. Its essence is that nonlinear problems touch on and involve the evolution problem of changes in materials. Evidently, solving nonlinear equations using the methods developed for invariant materials of the inertial system in essence indicates that people still do not understand nonlinearity. Many results obtained out of the studies of nonlinear problems by using linearization in substance do not have much practical significance and misguide the future development of science. To this end, we will provide an alternative solution. In fact, we can prove quite straightforwardly that nonlinearity stands for the problem of rotationality of solenoidal fields so that the mystery of nonlinearity is resolved once and for all by using transformations of curvature spaces of physics (OuYang et al., 2002; OuYang and Lin, 2006a; OuYang, in press). Our work indicates that the problem of nonlinearity has been fully resolved; the method of digitization can be employed to deal with irregular information; and there is no further need to invest additional tax dollars to tangle with this problem that cannot be dealt with using quantities.

The physics phenomenon of wave motions maintains materials from any possible damage by transferring energies, which provides a useful playground for the quantitative analysis. That is how modern science has made its historical contributions in the form of inertial systems. However, we could not treat wave motions as the only form of material movements just based on the past glories, because the main and common form of motion of materials is rotation. Because of the duality of rotations and their different spinning directions, at the same time that irregular events are created, damages in materials are also caused. So, in the investigation of the evolution problem of changes in materials, there is an urgent need to reconsider the physical significance of time and irregular information. So, the concept of rotation makes use of the structural digitization of information and helps us to walk out of the realm of quantities and enter into noninertial systems.

The development of science comes from calls of unsettled problems. So, in scientific investigations, the first problem of primary importance is to clearly understand the essence of the question being considered, with theories being secondary, instead of fitting indiscriminately certain methods and theories without considering the attributes of the question. Problems lead to the development of science; the existence of problems challenges the known theories that can no longer resolve new problems and the accepted system of thoughts. Modern science did not produce scientists who can foretell the future. In particular, professors and experts who teach and investigate predictions and forecasting cannot provide meaningful predictions on when torrential rains or earthquakes will occur. No matter how we see it, it has to be a problem that needs to be resolved. The public likes to advocate the value of correct thoughts, while it forgets the huge losses of erroneous reasoning and opinions. Modern science has been in operation for over 300 years. It is time for us to excavate the existing problems within its system by comparing practical problems that people have experienced as difficult to comprehend. Currently, the scientific community is greatly attracted by the fad of ecological and environmental problems. It is surely a good thing. However, just focusing all efforts on the release of carbon dioxide is too biased, because the high dams and highly elevated artificial lakes, constructed to satisfy the constantly increasing need of energy, have obstructed the flow of water in rivers and interrupted the natural exchange of heat. The consequent damaging effects on the ecology are not any less than the release of carbon dioxide. The contemporary people do not seem to understand the effects on the maintenance of ecological balance that green mountains and uninterrupted water flows create. The efforts of modernization that take place in many parts of the world are commonly carried out on the basis of changing the color of the green mountains and interrupting the flow of river currents. Such modernization is no different from cutting off one's own future livelihood.

This book is written to satisfy the numerous requests of the readers of our recent publications regarding the book *Entering the Era of Irregularity* and relevant articles. No doubt, from the first push to the second stir, not only will science itself face the problem of transformation, but more importantly also will our very way of knowing have to go through fundamental changes. In this book, we openly announce the first time that events are not quantities and do not comply with the quantitative calculus of formal logic, that events are not random, that the problem of time challenges modern science, and other fundamental epistemological problems and propositions, and clearly point to the practical applicability of the digitization of information.

No doubt, the development of science has its process. Due to the limitation of the current historical moment, modern science has experienced over 300 years of glory by creating the scientific system of knowing the world through analyzing quantities, where events are replaced by quantities. At the same time, many unsolved problems are left open, constituting obstacles for the future development of science. If we continue to apply the epistemological proposition that quantities are events, then we will have to investigate our conscientious behaviors. To this end, we conclude that the central problem we face currently is a change in how we think and reason.

Of course, changing the accepted system of values and beliefs takes a long time. It cannot be accomplished without going back and forth several times. Also, when facing difficulties and setbacks, it is natural for people to look back and to learn from the mistakes. To this end, we have repeated our practice with the concept of transformations of events. Through digital structuralization of events, we have indeed resolved some practical problems that have been unsolvable by using modern science. Pressured by our colleagues and readers, we published our works ahead of our schedule. When modern science is enjoying extracting eternal (invariant) information out of events, we collect information of change from the events in order to see how the world actually evolves over time.

Science is about challenging human wisdom, and technology is developed to save people's lives. When facing imminent crises, decisions have to be made instantly, and whether or not they are correct decisions will be studied later. The corresponding areas of predicting what is forthcoming are about how to combine both challenging and saving so that these two demands can be met at the same time.

We would like to use this opportunity to express our appreciation to our readers and colleagues for their support and enthusiasm. By holding the "nose" of events to blame the "short tails" of quantities, we will surely cause arguments and debates. However, "short tails" are different from "no tails." Our expectation is all about the future. So, it is our hope that our presentation in this book will play the role of a brick we throw out in the open to attract truly meaningful and beautiful gemstones.

Acknowledgments

The presentation of this chapter is based on Apostle (1966), Bergson (1963), Chen et al. (2005), Kline (1972), Koyré (1968), Lao Tzu (time unknown), Lin (1998, 2008), OuYang (to appear), OuYang and Chen (2006), OuYang et al. (2000, 2001, 2002, 2005), OuYang and Y. Lin (2006), OuYang and Peng (2005), Prigogine (1980), Sun (2000), Watson (1964), and Zeng and Lin (2006).

Chapter 2

Embryonic Deficits of Quantitative Analysis Systems

It is not exaggerating to say that since about 300 years ago when modern science was established, any problem that cannot be dealt with by using science has been classified as indeterministic or stochastic. However, since the concept of indeterministicity was first introduced, nobody seems to have addressed directly the cause of such indeterministicity and studied the meaning and significance of indeterministicity. As a matter of fact, indeterministicity is a philosophicalized concept corresponding to the mathematical concept of stochasticity, which was introduced in relevance to determinacy. Stochasticity and determinacy come individually from the two main methodological systems of modern science — the deterministic logical calculations of the Newtonian dynamic equations and the logical reasoning of the stochastic, statistical mathematics. Their essence still belongs to the mathematical computation of quantitative formal logic, with the former known as determinacy and the later indeterminacy. So, determinant or not is not originated from the underlying materials or events. Instead, it is determined by the mathematical rules of computation employed.

Indeterministicity (or mathematical stochasticity) applies quantities to represent events. However, the events represented by numbers are not identical to the numbers involved. For example, for any so-called random event A, even if it is written using a quantity, it is still a concept of symbols. Instead of A being a number, it is a symbol representing the event or its attributes. That is why the chance for the event A to occur is written as the probability $P(A)$, where $0 \leq P(A) \leq 1$. Since through the probability $P(A)$ event A is completely quantified, following the formal logic of mathematics, determinacy is described by using quantitative models of $P = 1$, which happens to be a special case of the indeterminacy $P \leq 1$. Hence, not only is the concept of stochastic indeterminacy defined quantitatively, but the concept of quantitative indeterminacy becomes a generalization and expansion of that of quantitative determinacy. A practical problem hidden in this process of abstraction is that the event A itself is missing. Speaking more specifically, even though the probabilisticity of the event A might be indeterminate, it does not mean that the event A itself is indeterminable. Also, under the thinking logic of evolution science (see the relevant chapter later), the concept of probability of the event A will have to be redefined. Furthermore, it will be another matter in terms of whether or not event A (or the probability of event A) actually complies with the rules of computation of mathematical probabilities.

Since the time when the stochastic system of indeterminacy was first introduced, the concepts of events and probability of events have been to a large degree mixed together. And, the Newtonian system of dynamic equations has been criticized as the target of determinacy, forgetting that probability of events is not the same as events being probabilistic. That end has led to the over-100-year-old debate between determinism and indeterminism. To understand the essence of this debate, one should see the occurrence of an event from the angle of changes of the event. However, the currently prevalent indeterminacy has undermined the events' variancy by using the events' probabilisticity. So now, it becomes quite clear that no matter whether it is the classical mechanical system or the stochastic statistical system of modern science, the fundamental problem is to avoid or has not touched upon the variancy of events. So, neither the classical mechanical system nor the statistical system has entered the era of evolution science. One prevalent belief seems to be that problems unsolvable by using the classical mechanical system can be resolved by employing statistical methods. Such a belief is indeed formed partially by not fully comprehending the classical mechanics and misunderstandings of the statistics.

All events in the physical world are intertwined somehow, including the connections of space distributions and before-and-after order of time. That is why when we master the rules of change of events, we will be able to predict the determinacy of events after their changes. Even if we call the new events after the changes the occurrence of "the events," they are still deterministic. However, this true determinacy is a concept different from invariancy, and invariancy cannot be seen as determinacy, since in the physical world there does not exist any invariancy.

The so-called indeterminacy should be a nonexisting connection of objective events without any before-and-after order of change. That is why when modern science has reached its state of the art, it has still not touched on the physical essence that eventual changes have on their causes and consequences. That explains why modern science still cannot predict the events after changes. Instead, these new events can only be called probabilistic. Even so, the determinacy of events themselves cannot be altered.

As for the advantage of the prevalent statistical methods, it should be seen as methods of estimating the main tendencies of change using probabilities of events when mankind still has no clue about the rules of change of events. Neither can these methods reveal the rules of change of events, nor are they the same as the indeterminacy of physical events. The probability of events is a human wisdom of mathematics. However, mathematical wisdom is not the same as physical substance. Hence, the probabilisticity of events cannot imply the probabilisticity possessed by the events themselves. In other words, even though the probabilisticity of events contains randomness; it does not mean that the events are random. So, the current problem of whether determinant or not still belongs to the category of human conscious behaviors instead of that of objective changes of events. These objective changes of physical events will not become indeterminate just because of a lack of human understanding of their rules of change.

The prevalent concept of indeterminacy initially comes from the criticism of the deterministic fatalism of the Newtonian dynamic equation systems for its inability of handling irregular events, even though it did not clearly identify what kind of problems the system of dynamic equations is facing. Due to well-posedness requirements, dynamic systems have to be limited by the stability of the initial values. Evidently, the essence of initial value stability is about initial value invariance. The corresponding statement of Laplace that from knowing the initial value he can tell everything about the future has clearly and definitely admitted that the Newtonian system of dynamic equations is a system of invariant events and within this system no more new events will appear. So, there is no need for anyone to talk about the probability for an event to appear in this system. Labeling such a system of invariant events as deterministic is equivalent to drawing a snake and adding feet — not necessary. Also, defining invariancy as determinacy is equal to declaring the nonexistence of deterministicity, since invariancy does not exist in reality. So, the over-100-year-old debate between determinacy and indeterminacy has been an argument without a clear understanding of the essence of the underlying problems.

The reason why the so-called indeterminacy is an advance of the Newtonian system is it gives tacit consent to the variancy of events, on which the concept of occurrence of events can be introduced. Evidently, only because of the existence of eventual changes in reality, the sense of occurrence of events appears. If all events are invariant, then there will not be any new event to occur. So, discussing occurrence of events is the same as studying nonexisting objects. To this end, generalizing events' probability to probabilistic events is only a confusion of epistemological concepts. Or in other words, we propose a different set of concepts for the reader to consider.

The reason we emphasize this problem is that there used to be a prevalent discussion on probabilistic events in the community of learning. For example, some people changed the principle of inaccuracy into that of indeterministicity (Hawking, 1988) without clearly explaining what time is, so that "God does not play dice" is replaced by "God plays dice," leading to the debate between the dead Einstein and the living "Einstein." This example vividly shows that distinguishing events and the probability of events is not an easy task. It has been known since ancient China that *yi wu ce wu, bu fen wu wu*, meaning that when using one object to measure another object, the problem of inaccuracy of measurement always exists. Due to the interaction between the objects employed, the mutual measurements become impossible to determine. That is exactly the inevitability instead of an accidentality of the physical world. The concept of probabilistic waves, which has been influential in a scientific era, of the quantum mechanics stands for a problem of indeterminacy of modern science that deserves our attention. Its essence is the irregular flows (for more details, see later chapters) created by mutual interactions of particles. As for the later probabilistic view of the universe, established on the theory of chaos, it can be seen as a typical point of view of treating events themselves as indeterminate. This discussion reveals the fact that the contemporary scholars who are good at using numbers have ignored the inherent problems existing in quantitative computations. Mathematical equations with large quantities of infinitesimal differences can never be accurately calculated. The computations of indeterminate equations are naturally indeterminate. However, such computational inaccuracy does not lead to the consequence that the underlying events are stochastic.

As for the origin of the concept of indeterminacy, it is from the fact that quantities cannot deal with information of change and the epistemological concept that quantities are the same as events. And, this epistemological concept has been held continuously for at least 300-some years, so that even after the human realm of learning has entered the system of nonparticles, the scientific community is still constrained by quantities in terms of basic methods of solving problems. For example, methodologically, quantitative analysis has been employed to establish the complete dynamic system of wave motions, and the stochastic system is good at the analysis of selecting "vibrating" signals. That is why some scholars hold the belief that quantitative analysis can be employed to deal with any problem, why in quantitative analysis vibrations and rotational disturbances cannot be distinguished, and why in the treatment of signals the irregular eddy currents caused by terrains are still called "mixed waves of terrain and objects," where true irregularities and information of change are basically all filtered out using various methods. The corresponding irregular turbulences are still known as difficult problems that cannot be understood. Even so, the concept of complexity, which used to be seen as a betrayal of modern science, was introduced and widely studied without essential breakthrough in terms of methodology.

The current study of complexity is mainly focused on the variability of spaces with different kinds of parameters as its fundamental methods. Realistic complexity also contains variability of changes, and its corresponding methodology is still very much vacant. As central epistemological problems, we see both the problem of whether or not events are completely equivalent to quantities and the one that quantities cannot successfully deal with changing events. In the world of learning, through practice and application, people seem to have felt the difficulty of employing the system of dynamic equations to handle variant events so that huge amounts of manpower and financial resources have been invested to utilize statistical means of solving the problem. This attempt has also supported the study on indeterminacy. Evidently, no matter whether it is the multiplicity in static spaces or the multiplicity in changes, they are all objective events. So, from the large variability of events, one cannot simply understand or define complexity as a kind of indeterminacy. As for why the objective world shows such a large variability, this is a problem that deserves further and rigorous scientific exploration. To this end, we believe that the large variability the physical world presents has something to do with the rotationality of materials (Lin, 2008). It should not be confused with indeterminacy. Specifically, one should not identify the complexities produced out of quantitative methods as the complexities of the underlying objective problems. Furthermore, one should not make use of these artificial complexities to redirect the study of physics to the direction of indeterminacy or that of probabilistic universe.

To this end, in this chapter, we will focus our attention on specific discussions of indeterminate randomness and complexities induced by its methodology system. Although scientific and technological developments are ultimately about practical benefits, making epistemological concepts clear can help people avoid falling into pointless arguments due to different methods employed. Besides, what needs to be clear is that no matter whether it is the Newtonian system of dynamic equations or the quantitative equations of the stochastic system, in principle, they are all established on the basis of formal quantitative axioms so that they cannot answer *whys*. So, when we realize that the stochastic system cannot explain the causes of matter, in reality, the system of dynamic equations cannot provide the needed *whys*, either. The recognition of determinacy of the classical mechanical system does not imply that this system has provided the causes of events. It is exactly like the situation that Newton himself could not have applied his law of universal gravitation to answer what universal gravitation is or where such gravitation is from.

2.1 Concepts on the Concept of Determinacy

The previous introduction has already briefly talked about the problem of determinacy and indeterminacy. It is because of the historical momentum that the Newtonian dynamic system is seen as determinate. That end involves the problem of how to correctly look at modern science. So, at this junction, we have to systematically discuss the fundamental properties of the classical mechanics.

The currently familiar, classical mechanical system of the modern science was initialized on the invention of the inertial system, the concepts of masses and forces. However, at the very root level, the most elementary concepts are not clearly defined, where the inertial system contains all movements with constant accelerations, and any noninertial system is not an inertial system (or variant accelerations or rotational movements do not belong to any inertial system). In reality, no absolute, static coordinate point in the universe exists. So, we derive the following formal logical explanation about what an inertial system is.

1. The invention of inertial systems is based on the assumption that in the universe, there exists a static coordinate point(s) that is the core of inertia. So, one has to face the problem that the classical mechanics constructed on inertial systems does not really agree with the reality.
2. If we assume that the acting force in an inertial system is real, then realistic acting forces in noninertial systems would have to be imaginary.

The current dynamic system is exactly established on the basis of inertial systems where, when the imaginary is seen as real, the physical reality has to be seen as imaginary. That is why the phrase "imaginary forces" in inertial systems does not mean those forces are imaginary and do not physically exist. Next, let us provide some explanation on physical meanings.

1. Newton's first and second laws of inertial systems are related and should not be stated separately in isolation of each other.
2. In terms of mathematical form of equations, Newton's second law has to impose the condition that the acceleration a is constant; otherwise, $f = ma$ is not a closed equation. However, in reality, accelerations are mostly variable, and a changing acceleration would make m variable so that the second law becomes a nonclosed equation (or indeterminate equation). To maintain the existence of the inertial system of Newton's second law, one has to replace any realistic acceleration by the artificial average $\bar{a}$ that does not mean such an imaginary value actually exists in reality. So, to correctly represent Newton's second law, one should have $f = m\bar{a}$.

This end indicates that since the time when the classical mechanics of modern science was born, it has been a theory of averages so that all phenomena of change have been lost. What is more important is that if m changes, it must cause $f = ma$ to become a nonclosed or indeterminate equation. That means that at the same time when the classical mechanics maintains the inertial system, it limits itself to the study of invariant materials and matters. That also implies that when evolution problems with variable m materials are concerned, one has to deal with issues of noninertial systems. It is because a changing m has to cause the acceleration to change. Conversely, if a is a variable, it would also make m change. So, the physical meaning of inertial systems is constant acceleration a and irrotational movements, which guarantees in motions, the underlying materials do not change. At the same time, the physical meaning of noninertial systems is about variant acceleration or rotational movements, revealing the appearance of evolution science in which variant materials are investigated.

Here, rotational movements can alter the acceleration by changing the spinning direction and differences in spinning directions lead to changes in materials. So, a concrete difference between inertial and noninertial systems is invariant or variable materials. If in movements, the underlying materials do not change or in a certain time frame the materials stay almost constant, then one can introduce an inertial system for his study; otherwise, he has to employ a noninertial system. Evidently, materials in the universe change constantly. That is why inertial systems are special cases of noninertial systems under very specific conditions. This end also directly reveals the fact that nonlinear equations involving rotations cannot be linearized. Consequently, those theories and results obtained by treating nonlinear equations using linearization have to be reconsidered and modified from this new angle.

Based on our discussion above, the physical meanings of inertial systems can be stated as follows: These systems study problems of such movement that involve only average accelerations of invariant materials and allow only changing velocities on the basis of a static coordinate system. This end leads to the recognition that the modern science, established by Newton and his followers, and later by Einstein and others, has not touched on problems of changing materials since its very beginning. Evidently, invariancy of materials is equivalent to "living forever" so that under such invariancy, there is no longer a need to talk about determinacy and indeterminacy. In other words, defining invariancy problems as determinant is equivalent to recognizing the denial of determinacy. After all, invariancy or "living forever" is not an existence investigated in physics. That is why we can conclude: Scientific theories provide explanations to the inexplicable reasons behind baffling problems; theoretical researches, however, explain inexplicably the nonexistent bafflements so that people are more baffled than before!

Therefore, from criticizing the invariancy of the classical mechanics as determinacy, it can be seen that the stochastic system not only does not understand the essence of the mechanics but also chooses a wrong target. And, the reason Einstein has been labeled as an old diehard against indeterminacy is that people did not well clarify the central issue of the debate. If Einstein's statement that "God does not play dice" were intended to mean that the dice themselves were not random, then the determinacy would become from the "God's nonplay," and Hawking's "God plays dice" did not explain clearly whether the randomness is from the "God's play" or the randomness of the dice. There is no doubt that the randomness created by "plays" is different from the randomness embedded in the "dice." That is where the concept of probabilistic universe that was prevalent during the second half of the 20th century came from. If the randomness comes from God's "play," then it is equivalent to acknowledging that Hawking's randomness is man-made, where one only needs to identify "man-made" with "God's play."

In terms of problems of physics, the essence is the investigation of the rules, if any, that govern changes of materials and processes of these changes. And, we need to be very clear that neither of the two main schools of thought of modern science has touched on changing materials and is still wandering around outside the door of evolution science. The reason for this end must have something to do with the conceptual confusion as shown in the excessive debate between determinacy and indeterminacy.

In the latter half of the 20th century, scholars from Belgium, China, France, and the United States individually pointed out the fact that the classical mechanics had not touched on problems involving changing materials. For example, H. Bergson (1963) declared that the system, perfected by Newton and his followers, can address any question, because according to its definition, almost all questions that are not solvable in the system have been labeled as quasi-questions and disregarded. Koyré (1968) commented that the movements described in modern science are those that have nothing to do with time; or speaking more oddly, modern science studies movements that take place without time. From Newton to Einstein, what people have pursued is the eternal heavenly kingdom (Prigogine, 1980). Based on our discovery of the irregular information of ultralow temperatures existing at the top of the atmospheric troposphere, and the consequent establishment and successful applications of the method of directly digitizing events' data so that transitionalities in atmospheric changes can be predicted, we pointed out that for over 300 years of perfecting modern science, what is studied is only about problems of invariancy (OuYang, 1998; OuYang et al., 2002a; Lin, 1998, 2008b). What needs to be pointed out is that before we formally introduced our method to handle irregular information regarding changing materials, we once stated that as long as modern science did not resolve the problem of time, it would not have touched upon the fundamental problems involving changing materials; and, it has in essence shown the fact that modern science is still outside the door of true science.

Other than practice, we have also been greatly influenced by ancient Chinese philosophical thoughts in our process to recognize the fact that modern science is a system of methods dealing with only invariancy. And, there are cultural and historical foundations for us to realize that events are not the same as quantities. On the other hand, it is extremely difficult for such scholars as Bergson, Koyré, Prigogine, and others in the West to discover modern science's denial of evolution simply from their investigations within the quantitative analysis system. What is unfortunate is that even though the opinions of Bergson, Koyré, and Prigogine were published before our works, their results were not well recognized and followed up. One reason for such a lack of attention is that historically, it has been extremely difficult to change the belief system of an era. Also, another reason is that none of these scholars had established a method that could be successfully applied to resolve practical problems that cannot be satisfactorily addressed by using modern science.

What needs to be clear is that invariancy is different from the concept of determinacy. Our purpose of carding the debate between determinacy and indeterminacy is not about who is right and who is wrong. Instead, at the same time when we focus on solving practical problems, we look at furthering the development of science. After all, earthquakes, tsunamis, disastrous weather conditions, and so forth cause huge economic and human life losses. So, there is a very important necessity for mankind to be able to foretell the future. However, the magnificent achievements of modern science do not seem to know much, if anything at all, about foretelling the future and changes in materials.

So, we should be clear that modern science is a system of invariancy and non-evolution. The problem of whether it is determinant or not should be established on the basis of whether or not it addresses problems of change. The cast of the concept of indeterminacy originated from the quantitative incapability of handling irregular information and variant events. And, variant events represent an area modern science has not touched upon. It can also be said that the dynamic system of modern science employs its position of determinacy to extract invariant information out of events in the form of quantities, while the indeterminacy of the statistical system also collects invariant information out of events. Even though these two systems apply different forms to extract invariant information, their consequences do not have much fundamental difference. That is, what is ironic is that the system of randomness has not been collecting the relevant information of the claimed randomness. That is why we declared after publishing our book *Entering the Era of Irregularity* to use our system of digitization to extract information of change without damaging the invariant information.

2.2 Randomness and Quantitative Comparability

2.2.1 Some Background Information

Currently, science and technology have been classified in the category of productivity so that the use value of science and technology has been the focal point of people's attention. When talking about science and technology, people naturally think about the two main systems: the initial-value determinate quantitative dynamic equation and the stochastic statistic methods. However, the debate between these two systems has been going on for over 100 years, and there is the possibility for this debate to continue into the coming years. It is because after all, there are still many unknowns and practical problems waiting for us to fathom and to resolve.

The familiar solution of dynamic equations is constrained by the requirements of well-posedness. The consequent statement that knowing the initial values foretells everything about the future (Laplace) (Kline, 1972) is the same as "the future is equal to the present." That is, the quantitative regularized system, completed by Newton and his followers, is about tracing and monitoring the occurred events. And, from "the future equal to the present" of the dynamic equations, "the future equal to the past" of the statistical methods is evolved and established. That is, the entire modern science does not point to the true future. Evidently, the realistic material world changes constantly. It is neither "the future equal to the present" nor "the future equal to the past."

When studying material movements in curvature spaces, we substitute angular speed for the speed of Euclidean spaces and cast the question that if the square of speed is energy, what is the square of angular speed? Because modern science does not provide us with an answer, we investigate the properties and functions of the kinetic energy of squared angular speed and discover that at the same time when the stirring energy can be transferred and stored through secondary circulations, it can also transfer and store energies so that a true conservation of kinetic energy is achieved. So, we establish the law of conservation of stirring energy. Then, working with our student Chen Gangyi, this work is generalized to macro- and microscopic worlds. More specifically, the form of material existence satisfies the three-ringed transfer of stirring energies. When such a three-ringed transfer of stirring energy is not satisfied, then the underlying materials take the form of evolution, and the transfer of stirring energies is completed through subcirculations. So, subcirculations have to be irregular events, and that is why irregularities of information stand for information of change.

Interestingly, the information of change stands for some of the problems or events that dynamics or statistics, two main methodological systems of modern science, cannot deal with successfully. So, static material existence corresponds to regular information, and irregular information (that is currently so-called random, orderly noise) corresponds to changes of materials, revealing the dynamic changes and their processes of change. In this sense, no matter whether it is the well-posed initial-value invariance and the large probabilitization of historically smooth sequences or completely noisy information, all stand for the determinacy of objective existence. Because of their different forms of appearance, they need to be treated individually with different methods. As for the design of durable products, any irregular information of the "noisy" form stands for signals of change. For the purpose of preventing and reducing the chance of deteriorating as much as possible, one can conditionally modify the irregular information. However, in terms of problems of material evolutions, the irregular information of the noisy kind cannot be modified except for erroneous situations. And, the properties, attributes and principles of change of the irregular information deserve every bit of our attention. To this end, there is a need to reconsider the problem of indeterminate randomness of events. For a detailed discussion on stirring energy and its conservation, please see Chapter 7.

2.2.2 The Problem of Randomness

The concept of randomness comes from the reasoning of formal logic. More specifically, for any given event A, it can either occur or not occur under certain conditions. So, the chance for the event A to occur is denoted by using the concept of probability $P(A)$, satisfying $0 \leq P(A) \leq 1$. In this way, each deterministic model becomes a special case of a stochastic model under the assumption of $P = 1$. That is why the concept of randomness is seen as a generalization of deterministic ones.

As for the physical meaning of determinacy or indeterminacy, or why there exist deterministic and stochastic systems, no one seems to have considered such a problem. The answer to this problem involves how to understand events. The most important contribution of the quantitative analysis system is that the quantitative well-posedness and methods of smooth series provide a whole set of methods for estimation useful for the design and manufacture of products. However, in terms of transfers of information, these methods still cannot deal with random information. That is, the essence of the stochastic system does not really like random. It still tries to extract regular (or regularized), determinant information out of randomness. The essence of this effort is still to collect invariant information.

Considering the history of the world of learning, we spent an excessive amount of effort in the beginning of this chapter discussing the problem of determinacy and indeterminacy. Because of the difficulty of changing the accustomed belief system, many of our colleagues recognize the necessity to emphasize the accuracy of elementary concepts. Especially, most elementary concepts established on the basis of formal logic and reasoning tend to lack practical background. That creates the problem that some of these concepts suffer from their inherent ambiguity, and are unable to distinguish what is right and what is wrong. For instance, our discussion in this chapter has touched on the "event A" not being equal to "the probability $P(A)$ of the event." In other words, the probability $P(A)$ of the event contains indeterminacy, which is not the same as the event A also being indeterminate. By continuously using quantities to substitute for events, a custom of the quantitative system of the physical quantities and parametric dimensions of inertial systems, quantitative computational accuracy will in terms of formal logic be treated as the correctness of the computation; or the quantitative accuracy is identified with the correctness of the event. So, as an epistemological concept, what we emphasize is that the event A and its probability $P(A)$ are two different concepts, a deeper meaning of which has to involve the fundamental problem of whether events, information, their changes and processes of change can be replaced by quantities. Or, stating our opinion directly, we have the following problems: Are events equal to quantities? Do events comply with the quantitative calculus of formal logic? Do events contain randomness? These problems and others surely lead to the problems of how to recognize and handle randomness. In other words, even if events themselves are seen as random, one still cannot use randomness as the ultimate answer to our problems. Currently, prevalent dealings of employing randomness as the ultimate answer to all unsolvable problems are no doubt the same as classifying problems unsolvable in modern science as quasi-problems. There is not denial that the essence of the assumption of stable series in the stochastic system and the requirement of well-posedness of the deterministic solution problem of dynamic systems is to avoid randomness or indeterminacy. That is, the epistemology of indeterminacy does not consider problems of indeterminacy. That constitutes a systemwide embryonic wound of the modern scientific system. That is the very information we would like to deliver in this book.

Also, in the investigation of the stochastic system, if the time-series events are completely random, then the regular (or regularized) information extractable will be extremely poor. So, if randomness is seen as an expansion of determinacy, its purpose is still seeking the hidden determinant regularity instead of studying or pursuing the claimed indeterminacy. The methodology of the indeterminacy system does not have any substantial difference from that of the well-posed system of dynamic equations. Or speaking differently, in the operation of the randomness system, the essence is also to eliminate (or deny) whatever is random, which belongs to the same category of operation of requiring the solutions of mathematical physics equations to satisfy the condition of stability of the well-posedness.

2.2.3 Random Walk and Denial of Randomness

Weng (1996) specifically focused on random walks and provided practical examples to illustrate the concept. Although the concept of randomness was introduced in comparison with determinacy, in applications, it is mostly shown in the analysis of random processes. Evidently, as processes, they have to involve time series, which must show differences in terms of the time order; otherwise, they would not be called processes. At this junction, we would like to emphasize that dynamic equations of inertial systems do not touch on the corresponding nonrandom processes, and the Newtonian system cannot be treated as fatalism, since it is a theory about invariancy that does not involve any change and any process of change.

According to the concept of randomness, the current stochastic system at most contains irregular information in the regularization with the goal of extracting regularized information instead of focusing on the existing irregular information. So, although the two systems have different operational details, the underlying system of thought for what outcomes are sought is the same, with the only difference between "the future equal to the present" and "the future equal to the past." The ultimate goal of studying random processes is still about eliminating randomness while the process is also deleted in the procedure.

As mentioned earlier, complete randomness stands for informational noise. The most studied uniform distribution and equal probability binomial distribution are also information poor. In these cases, what is obtained is not much more than the quantitative averages. And, quantitative averages are typical examples of the problem that quantities can be used to calculate nonexistent matters more accurately than what is actually existent, and the so-called accuracy is not the same as correctness. It is because the probability for an average number to appear is much smaller than that of small-probability events, and such average numbers in general do not realistically exist. What should be the situation is to reject the object of concern based on the extracted information. In the following, let us use the simple example of random walk to illustrate our point in the form of the traditional presentation.

According to the accustomed tradition of quantitative analysis, let us assume that the probability for +1 or –1 to appear is the same. As an introduction to the problem of randomness, many authors use flipping a coin as the example without involving changes in the structure of the coin. If +1 stands for flipping a head, then –1 means obtaining the tail. After flipping the coin a certain number of times, compute the sum of all the numbers recorded or the difference of the absolute values of the recorded values. Use S_n to denote the answer. A relatively elementary illustration is to imagine the walk of a drunken man, assuming that the probability for him to walk forward is the same as that for him to walk backward. Now, the question is, after the drunken man took several steps, where can we locate him?

The first method is to apply the concept of expected values. The conclusion is that we can find the drunken man in the same location where he started his walk. The formula of prediction is given as follows:

$$x_{i+1}^* = x_i \tag{2.1}$$

That is, all the values of $x_{i+1}^*, x_{i+2}^*, \cdots\cdots, x_{i+n}^*$ are the same as x_i so that the drunken man did not leave his starting location. However, the error of prediction increases with n. Although the conclusion is to look for the drunken man at the original location, it will be hard to actually find him at the original location when n value is too large (an ill-posed situation).

The second method is to estimate the square of $S_n(+1, -1)$ as follows:

$$S_n^2(+1, -1) = \left[\sum_{i=1}^{n} (\pm 1) \right]^2 = n(\pm 1)^2 + 2 \sum_{i=1}^{m} \left[(\pm 1)(\pm 1) \right] \tag{2.2}$$

where

$$m = \frac{n}{2}(n-1)$$

and

$$(\pm 1) \cdot (\pm 1) = \begin{cases} +1, & (+1) \cdot (+1), (-1) \cdot (-1) \\ -1, & (+1) \cdot (-1), (-1) \cdot (+1) \end{cases}$$

Since the chance for the drunken man to walk forward or backward is the same, when n is very large, we have

$$\sum_{i=1}^{m}\left[(\pm 1)(\pm 1)\right]\cong 0$$

That is, Equation (2.2) can be rewritten as follows:

$$S_n^2\left(+1,-1\right)=n$$

or

$$S_n^{\pm}\left(+1,-1\right)=\pm\frac{1}{2}\sqrt{n} \tag{2.3}$$

So, the corresponding problem of looking for the drunken man becomes indeterminate as the number of steps n increases indefinitely. Evidently, this indeterminacy is created by the quantitative impossibility of infinity (according to the ancient Chinese, the quantitative infinity stands for a thing numbers cannot address). As for other relevant methods employed to discuss this example, they suffer from the problem of divergence of the approximate solutions, which is exactly an embodiment of our conclusion that computational quantitative accuracy is not the same as quantitative correctness.

As for a realistic drunken man, his motion consists mainly of leftward or rightward sideways walks with his forward or backward shifts being two of many possible components of his rotational motion. With his irregular rotational movements, he falls easily. So, in principle, it is not possible to find the drunken man in his original location. As for the corresponding problem of flipping coins, there are also such details as changes in the flipping direction, changes in flipping strength, and so forth, so the consequent equation becomes an indeterminate quantitative form. Therefore, in essence, the concept of randomness comes from the mathematical form instead of the physical reality. Or in other words, quantitative analysis or computation in principle belongs to the human culture, because materials themselves possess neither zero nor average values. However, quantitative calculations cannot operate without zero or averages; otherwise, the entire mathematical computation would fall apart and become paralyzed. Evidently, zero and averages are human inventions and are not discoveries of some physical reality.

Based on the currently accustomed methods, in stochastic analysis, one often extracts nonrandom information from the purely random (noisy) distribution using, say, $S_n(+1,-1)$. This fact is equivalent to admitting that the stochastic analysis system itself is not about looking for randomness. Instead, it looks after the opportunity to deny randomness through its "stochastic" analysis. So, the system of indeterminacy is not about investigating indeterminacy but about how to eliminate the existing indeterminacy. Hence, the natural problem that is left open is how to understand the purely random noise. Only by realizing the fact that pure randomness (irregular information or events) is the ultimate indication for forthcoming changes, science can truly possess the scientificality of processes. With no change, it is the same as with no process. The so-called random processes are exactly about how to eliminate processes. Therefore, the stochastic system, established on the basis of the concept of randomness, has neither addressed the problem of randomness nor answered what random phenomena are and why there could be random phenomena.

In the following, we will introduce the concept that irregular information is about variant events. At this junction, what needs to be clear is that even if the time-series distribution of noise is considered random, the noise itself does not necessarily stand for a random, indeterminate event. Instead, changes in noise's time series are exactly the first object in the analysis of evolution science. Or speaking differently, when studying problems of evolution, the first task is to extract the essence of purely random (noise) information in terms of changing events. And, next, provide the corresponding methodology to deal with the underlying variant events.

2.2.4 Sample Size and Stable "Roving"

Both uniform and normal distributions are two very important distributions in the stochastic system and represented by using averages and averaged square deviations (variances). And, instable random processes are among the open and less considered problems in the stochastic system. Table 2.1 lists the averages, variances, and coefficients of deflecting state, computed using different-sized samples of real-life observations of west winds data of atmospheric circulations. It also shows the changes of these computed values with sample sizes. Evidently, the computed characteristic values of the statistics are "roving." Changes in the time-orderliness of the sampled series, as shown in Table 2.1, are not specific to this set of data. It can be said that for all sampled realistic long random processes ever studied by us, the involved irregular information almost without exception display changes in their time-order, reflecting the fundamentality of material evolutionary processes. The reason the quantitative analysis system experiences the "roving" of instable "random processes" is the variancy inherently existing in time series. That is also the reason why modern science has not resolved the problem of time. Evidently, without understanding time there would not be any chance for resolving the problem of processes of change. So at this junction, we have to deal with the problem of how to understand events' randomness and the problem of change existing in the time series of irregular noises.

As for the conclusion of the central limit theorem that sample means follow normal distributions, according to Table 2.1, one needs at least sample of sizes between about 2,500 ~ 3,000. That means that the conclusions of some research reports using sample sizes of only 30 need to be reconsidered. Evidently, applying

Table 2.1   Sample Sizes and "Roving" of Stable States

Sample Size	Mean	Variance	Coefficient of Deflecting State
500	205.69	110.75	0.05
1000	201.93	105.28	0.17
1500	209.13	110.21	0.01
2000	210.41	110.18	0.02
2500	210.63	107.50	−0.00
3000	215.53	108.59	−0.00
3500	216.71	106.83	0.02
4000	222.61	108.28	0.04
4500	222.79	107.71	0.065
4749	223.80	108.07	0.08

a certain model to fit real-life observations indiscriminately has to lose some of the innate characteristics of the data. In terms of understanding randomness, the most reliable evidence is still the realisticity of the underlying events. To this end, in the following we will provide some real-life irregular information and the physical phenomena corresponding to the changes in the irregular information.

2.2.5 Digitization of Irregular Information: A Process Physics Principle and End of Quantitative Comparability

In order to fathom realistic irregular information, it seems to be appropriate for us to look at natural disasters as our examples. Evidently, earthquakes and torrential rains are important examples of suddenly appearing natural phenomena. Here, sudden appearance not only represents changes but also transitional changes. So, it can be referred to as nongeneral, specific events. However, such events are not noises without information.

2.2.5.1 Digital Structurization of Irregular Humidity Information and Forecast of Torrential Rains

According to the principle of evolution, each piece of quantitative instable (or irregular) information stands for a problem of transitional changes. So, in terms of the premise of not injuring the irregular information (events do not contain randomness), we established the digitalization method of transformation of time-series data.

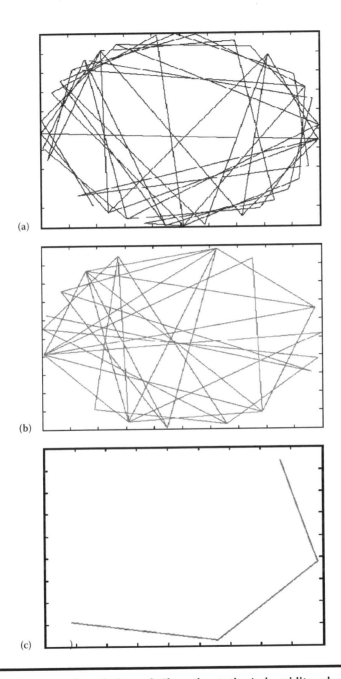

Figure 2.1 Structural evolution of Chengdu station's humidity phase space. (a) Chengdu humidity: July 28. Precipitation: none. (b) Chengdu humidity: July 29. Precipitation: 12.1 mm. (c) Chengdu humidity: July 30. Precipitation: 127.7 mm.

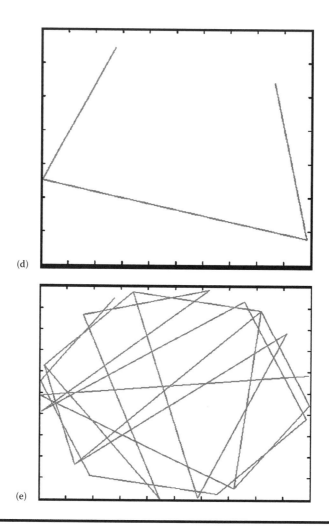

Figure 2.1 **(d) Chengdu humidity: July 31. Precipitation: 63.6 mm. (e) Chengdu humidity: August 1. Precipitation: none.**

During July 28–August 1, 2002, Chengdu station experienced a major torrential rain. According to the traditional methods, it could be said that there was no way to recognize or extract the so-called indeterminate information or the traditional random noise from the available time-series curve (Figure 2.2 is the time-series curve of humidity data on July 28, 2002). However, by employing our digital transformation of the phase space of irregular information (see Figure 2.1a,b,c,d), predicting the said torrential rain is a very easy task (for more details, please consult with later chapters or relevant references in this book). Even for laymen without much rigorous training, they can readily tell from the sequential graphs in Figure 2.1 the difference between before, during, and after the torrential rain.

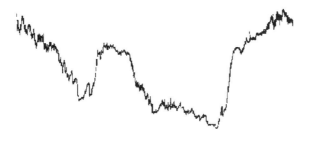

Figure 2.2   The humidity time series of Chengdu station on July 28.

Our purpose of including this example here is to illustrate that any so-called random, irregular information (noise) has not only its reason but also its practical value. Or in other words, even for a completely random process, the noise (without any order or pattern) existing in the instable series may still contain some useful information, which might perfectly reveal transitional changes occurring within the underlying materials so that irregular information (or called purely random noise) represents the information of change of a true process. Evidently, all irregular information that possesses practical values and causes does not stand for any problem of randomness. The philosophical principle that being chaotic implies impending changes perfectly describes the observation that what "chaotic noise" reveals is the approaching or forthcoming event and an event of transitional changes that appears prior to the approaching event. Because of the connection, this philosophical principle becomes a law of process physics and evolution science. It also shows that irregular information appears prior to changes in the underlying event, representing the timely information of the ongoing transformations of subcirculations. So, it can be said that as long as the information does not contain error, then in evolution science there does not exist randomness. The so-called randomness is a by-product of modern science due to the quantitative incapability of handling irregular events.

When we recognize that events are not the same as quantities, we might suddenly realize that the nearly 100-year-old concept of randomness or indeterminacy is nothing but a consequence of an ancient belief that quantities are the origin of all things or numbers are the rules underlying all existence. It is indeed a sequela left behind from misreading the concept of events of the contemporary stream of consciousness of quantities.

2.2.5.2 End of Quantitative Comparability

In modern science, quantitative comparability has been employed as the sole standard of quality or scientificality. This trend has currently spread into areas of social science and humanity. What should be clear is that in quantitative comparability, if seen as numerical comparability, there are only the comparabilities of magnitudes, amounts, or strengths; and only when the difference is within one magnitude (× 10), there might be practical significance. When the difference is of two or more magnitude levels (× 10²), the relevant small (or called infinitesimal) amounts would be eliminated by the practical procedures of quantitative operations, since when quantities of two or more magnitude levels are involved, they meet with the problem of indistinguishability due to measurement inaccuracies and computational errors. That is why quasi-equal quantities are computationally inaccurate and the relevant quantitative comparability loses its significance. This fact has been described by ancient Chinese quantity calculator Zhan Yin (from the time of Warring States about 2,300 years ago) as that not all things can be described by numbers.

The crucial problem we discovered that underlies the chaos theory developed by Lorenz and others during the 1960s to the 1980s is about the computational inaccuracy of quasi-equal quantities. No doubt, treating the phenomenon of quantitative inaccuracy of computation as a new discovery of physics or philosophy (Lin et al., 2001) is against the common knowledge, indicating that the community of learning to a degree did not realize that events are not quantities, or some of the typical examples and cases where quantitative formal logic does not agree with the associations of events. So, quite a few of our colleagues recognize the need to card through the relevant theoretical concepts at the epistemological level. This end explains why revolutions start from changes in the fundamental concepts.

What is important is that quantities can only appear postevent; and as a problem of material evolution, each event is first about structures of materials and second about the attributes of the materials. Quantities are postfact formal measurements and should be characteristics of events listed third or later. A lot of times, quantities cannot distinguish material structures and cannot be employed in making predictions. A simple but intuitive example is the metaphor widely used in weather forecasting that a turning hand implies cloudy weather, while a covering hand indicates rain. Here, the quantity of the turning hand is 1, and that of the covering hand is also 1. Evidently, the quantity 1, representing the magic hand, cannot really tell people whether the hand is turning or covering. So, the quantity 1 cannot warn us whether to expect clouds or rain. That is, quantities cannot address adequately the state and attribute of the event of concern. That is a common knowledge we are all familiar with in daily lives.

What is of historic interest is that in the contemporary world of learning, well over 2,000 years after Pythagoras's time, many scholars are still obsessed with quantities even though quantities can only be used to make after-fact analyses. In order to foretell the future, extraordinary and excessive amounts of manpower and capital have been spent on the tests and experiments of numerical forecasting, producing the consequence of after-weather-event simulations of barometric pressure systems. As of this writing, some scholars are still not clear about the fact that pressure systems are not the cause of the major weather phenomena.

This end must have something to do with the understanding of the fundamental concepts (for more details, please refer to Chapters 8 and 9). More specifically, what is shown is that information is seen as equivalent to quantities, leading to the introduction of such concepts of indeterminacy as chance, randomness, and so forth. We believe that no matter whether the available information is correctly fathomed or not, as long as there does not exist misguided cases in the information, there will be an eventual reason underlying the information. That is why we especially state that within evolution science there is no randomness or the concept that events are not random.

Besides, in predicting evolutions, we are required to know what is behind the scenes based on what is hardly shown, or from almost nothing we need to foretell the coming changes. The current quantitative comparability has already thrown away the quantities that are two magnitude levels smaller, so that the "little shown" or "almost nothing" has truly converted into "nothing." That is one of the reasons why modern science cannot predict major changes from minor differences without mentioning the lack of success in forecasting suddenly appearing events. Evidently, specifics or small-probability events cannot be obtained from generality or large-probability events. So, quantitative comparability as the sole standard of scientificality no longer holds true because of the variancy of materials, and we have pointed out the fact both conceptually and methodologically that modern science has not at all touched upon changes of materials with publication of our theory of blown-ups (Wu and Lin, 2002). Also, it is a mistake to treat problems of prediction as a branch of modern science. Since quantities are different from events, this realization will make modern science go through a major transformation due to more focused attention paid to problems on material evolutions.

Information should be classified into two categories: regular and irregular information. Regular information embodies the stable existence of materials, and irregular information, the evolution process of materials or events. That is, irregular information stands for the signals of change. Modern science employs quantities in the place of events to describe the existence of materials, only applies regular information in the form of average values, and substitutes eventual correctness by quantitative accuracy. However, events are not quantities and do not in general follow the formal logical calculus of quantities. That is why the commonality and the fundamental science of materials and events should be evolution science. In terms of the projected evolution science, the concept of randomness or indeterminacy can only be seen as a by-product that modern science has yet to fully understand. So, randomness rises to the level of a new epistemological problem because of the evolutionality of materials and practical usability of irregular information. Evolution of materials is about processes. That is why irregular events reveal their physical functionality and significance through evolutionary processes. So, evolution science will not consider the concept of randomness. Or in other words, the true significance of randomness is not really about indeterminacy but instead about the underlying changes.

As for the drunken man walking, the physical reason for his seemingly chaotic body movement is quite clear, because the alcohol has interfered with his nervous system in his brain. So, the walking pattern of a drunken man should not be studied in comparison with a sober man. Instead, the walking of the drunken man should have its own specific pattern. That is why when a sober man is observed from the angle of a drunken man, his normal walking pattern also seems random and chaotic.

The phrasing of indeterminacy of the initial-value automorphism of dynamic equations in essence denies the evolutionality of materials. Together with the requirement of stable series in the stochastic system, they are both the quantitative methods describing stable existences. The debate between determinacy and indeterminacy of our modern time is really about the invariancy of "the future = the present" and that of "the future = the past" without much substantial difference. The so-called indeterminacy, as known for quite some time, has been a misconception of the fact that quantities cannot handle quantitatively instable problems under the premise of invariant materials. Its essence is that quantities cannot deal with irregular events. This misconception has prolonged the application of the inappropriate substitution of quantities for events.

The postfact attribute of quantities together with quantitative comparability is not the same as events' comparability so that quantities have to suffer from the problem of inability to distinguish material structures and functionalities. And so, quantities cannot become the analysis method for the investigation of material evolutions. At the same time, what is shown here also ends the scientificality of quantitative comparability. In light of the development of science, modern science should also pay attention to continually modify those concepts and theoretical systems that do not apply in front of practical problems and should always remember where its own embryonic deficits are originated.

Evidently, in our time of information, the first problem we have to address is what information is or what the essence of events is. Only after these problems are well addressed will we have a chance to correctly understand and apply the method of digitization, where digitization cannot be confused with quantification. To this end, it will involve a major change occurring in modern science in terms of scientificality and methodology. Also, when people truly recognize that events are not the same as quantities, they will know what kind of "science" modern science is and what position it will occupy in the history of learning. All the details are omitted here.

In short, when modern science is positioned at the movements without involving changes in the underlying materials, the essence is to embody static figures. Consequently, when variant materials are given concern, the concept of indeterminacy will naturally appear. Due to the need for such indeterminacy, it will be natural to name invariancy as determinacy. When people realize that determinacy and indeterminacy are determined by the variancy or invariancy involved, they will handily understand what the determinacy and indeterminacy are and know how to investigate and employ these concepts.

2.3 Equations of Dynamics and Complexity

The time period when calculus was initially established and Newton posed the dynamic equation of his second law of mechanics — from the end of the 19th century and into the 20th century — can be seen as the most important period in the development of mathematical physics equations. This branch of learning has been so well recognized that all of theoretical physics has been evolving on the basis of these mathematical physics equations. It can be said that without mathematical physics equations, modern science would not exist.

Mathematical physics equations are those mathematical models often seen in physics, including the corresponding individual equations or systems of equations along with their existent boundary-value or initial-value conditions.

The community of learning at that time believed that since laws of physics use differential equations as their mathematical forms, mathematical physics are specifically related to the theory of differential equations. The Noether theorem of transformation groups and its generalizations can provide the mathematical form of the whole series of conservation laws of physics, and these conservation laws of physics are the criteria for judging if a scientific system is complete or not, so mathematical physics equations have almost become the signature and representation of modern science. It can also be said that without the support of mathematical physics equations, modern science would not have much content or could not be seen as a scientific system. So, mathematical physics equations have been listed as one of the courses fundamental to the education of science and engineering majors in colleges and universities around the world.

The development of science goes through periods, and in each historical period a certain "style" is in fashion; just as in the clothing industry, different times feature different styles. During the 19th and 20th centuries, the well-posed system of mathematical physics equations was in fashion in natural sciences. During this time period, many important scientific results and theories were established in the fashion of well-posedness. The method of series expansion used to be popular in theoretical studies. And, in the latter half of the 20th century, due to the development of computers, this method spread into areas of engineering in the form of spectral expansions. And in reality, physics has contributed to the system of modern science with the contents of movements of invariant materials and wave motions of reciprocating vibrations.

In terms of mathematics, mathematical physics equations are a branch of mathematics and can be seen as an area of study crossing the borderlines of mathematics and physics. Its goal is to mathematically describe physical phenomena. Later, it evolved into the dominating area of research in physics — the supporting system of theoretical physics. As history has developed, this area of study has continued along the lines of thinking as follows: Based on experiments and the laws of motions established via experimental observations, ignore all "minor" factors that affect the pattern of movements (the essence is to eliminate factors of change); use the quantitative forms of physical quantities and parametric dimensions; and establish the mathematical models describing the mutual relationship between the physical quantities. Here, establishing the mathematical models that describe the mutual relationship between the physical quantities is the first important step. Later, this step is called mathematical modeling. The second step is to solve the deterministic problem that corresponds to the mathematical model established in step one for its deterministic solution by using mathematical methods and formal logical calculus. The third step is to back-check the solution obtained in step two using the realistic physical phenomenon or event to see if it agrees with the pattern of the physical evolution or verifies the truthfulness of the event.

According to the current belief, the establishment and development of mathematical physics equations can be briefly stated as follows: Relationships between physical quantities in physics are closely related to the theory of differential equations. So, when studying physical systems of finite degree of freedom, use ordinary differential equations of the particle mechanical system; when studying physical systems of infinite degree of freedom, apply the corresponding partial differential equations. The inverse analysis of the historical times in geophysics involves integral equations; the mathematical models employed to study the process of neutron migration correspond to differential-integral equations; quantum mechanics and quantum field theory are filled with operator equations. By treating the invariancy of the classical mechanics of the modern scientific system as determinacy, the classical differential and integral equations are seen as determinant mathematical models with "laws of physics" derived by using the classical quantitative calculus of formal logic of these equations. The indeterminate system, corresponding to the determinacy, is the problem of using probability, statistics, and theory of random processes to deal with indeterminacy. Besides, for those fuzzy systems without clearly definable relationships, fuzzy set theory and fuzzy logic are applied. Evidently, mathematical physics equations are developed on the borderline of mathematics and physics and focus on the study of problems concerned in both of these disciplines. Here, when there is no problem in physics, it does not mean that there is also no problem in mathematics. This is why there is the well-posedness problem of mathematical physics problems. However, the concept of well-posedness did not originate from quantitative formal logic; it substantially constrains any potential changes in quantities. Also, sufficient reason or evidence does not exist to prove that the concept of well-posedness is introduced based on laws of physics.

From this brief historical recall, it can be seen that the development of mathematical physics equations really comes from the progress of mathematics. Together with the three classical classes of mathematical physics equations, none of them has touched on changes in materials other than describing the relationship of movement between physical quantities. The imposed well-posedness condition is only used to make the solution of the deterministic problem and relevant underlying changes controlled within the realm of dislocations of the initial values. The key idea is to use well-posedness to limit the stability of quantitative changes. As a matter of fact, at the first stage of using mathematical physics equations, the essence of "based on the laws of motion established via experimental observations" is not followed without paying any attention to address the question of what the "minor" factors that interfere with the pattern of movement are. Instead, what has been done is that merely based on the logic of mathematics, mathematical deductions are treated as the products of the laws of physics. Therefore, in the establishment of mathematical models, the basic idea is to eliminate the mathematical terms that do not meet the requirements of "standard" mathematical models or to limit the established mathematical models to such a degree that they can be solved by using currently available mathematical methods. In this process of satisfying the need to be able to apply available mathematical techniques, physical concepts, characteristics, and properties have been modified. For example, the disturbance of eddy motions is a concept of physics different from that of the disturbance of vibrations. They represent problems of different kinds. However, they have been uniformly treated as the disturbances of wave motion. And, some publications even openly declared that due to the need to employ certain available mathematical techniques, the well-understood physical waves that transfer energy without transporting masses are modified into such a fuzzy concept as mathematical waves, leading to the mathematical definitions of hyperbolic and dispersion waves.

Evidently, the concepts of indeterminacy and fuzziness themselves do not comply with the original meaning of the well-posedness of mathematical physics equations. The reason for these two concepts to appear and to be considered is due not to physical events themselves but to the incapability of mathematics to deal with the mathematical forms of irregular or variant events. This end is exactly the problem this book is about to address. Also, please notice that the so-called close relationship between the laws of physics and the theory of differential equations is in fact the one-sided wishful thinking of mathematics. Although physical quantities of physics are shown in mathematical differential forms, sufficient evidence does not exist to support the belief that changes in these physical quantities and laws of transformation between these qualities are equal to the formal logic of calculus of differential equations. The result that calculus has nothing to do with specific paths alone has violated the realisticity of physical problems. So, the first step of establishing models does not imply automatically that physical quantities actually follow the formal logical deductions of differential equations. For example, even if we employ quantified physical quantities, in reality, their laws of motions are still closely related to their paths of movement. So, as a problem of modeling, it does not represent a perfect marriage of mathematics and physics. It can also be said that the modeling of mathematical physics equations, even on the basis of formal logical analysis, suffers from some "embryonic deficits." Considering scientific rigor, future development, and the current popularity of mathematical modeling, there is an urgent need to seriously consider what problems are left behind from these "embryonic deficits."

Next, let us look at the problem of back-checking of the third step of mathematical physics equations. The reason why it is a problem is that many consequences of mathematical formal logical deduction do not satisfy the requirements of well-posedness. Even in terms of traditional linear equations, they might contain divergent solutions or nonunique solutions. Historically, no one seems to have considered why such phenomena appear. However, this problem deserves our serious thinking, because a hidden problem behind this one is that of whether or not the mathematical rules of deduction of differential or integral equations actually agree with the physical laws of transformation of the relevant physical quantities. No doubt, if the deduction in the form of mathematical equations agrees with the physical laws of the physical quantities, then the conditions of well-posedness become uncalled for. Conversely, if the conditions of the well-posedness cater to the actual laws of physics, then the ill-posedness derived out of the formal logical analysis in the form of equations implies that the corresponding theory of differential or integral equations does not agree with the laws of physics. In other words, some of the currently popular mathematical physics equations, especially the probabilistic waves of the Schrödinger equation, the gravitational waves of the gravitational field equation, and Rossby waves of meteorological science, do not represent the true faces of the underlying problems of physics. The conservation law of kinetic energy cannot really limit transfers of energy. The fundamental problem, revealed by these problems, has sufficiently shown that the formal logical deduction of mathematical equations does not truly agree with the laws of physical quantities. What is clearly indicated is that as of this writing, sufficient evidence does not exist to prove that the formal logical rules of calculus in the form of equations can be identified with laws of physics. From our discussion thus far, it can be seen that mathematical formal logical deductions can make the originally simple problem of physics become very complicated, a complexity, and mathematics and physics have shown more clearly over time their differences.

What needs to be clear is that the core of physics problems is about changes of materials and the processes of change, while the establishment and development of mathematical physics equations, from the time it started to the most recent times, have fundamentally avoided the relationships of physical quantities involved in changes of materials. Those equations that merely focus on the relationships of movement of physical quantities without involving changes in the underlying materials should not be treated as true mathematical physics equations. The exact location where mathematical physics equations meet with difficulties when involving problems of change is the basic assumption, the desire to meet the requirements of mathematical formal logic, and a lack of knowledge that materials or events themselves and their changes do not necessarily follow mathematical formal logical deductions. Even so, the concepts of indeterminacy and complexity should not have been considered, because the basic idea underlying these concepts is still to pursue after the invariancy of stable series. Hence, even seen as a mathematical physics problem, variancy in materials is indeed the solution for the "congenital deficiency" of mathematical physics equations. That is, problems of changes no longer belong to inertial systems.

Although in form fluid mechanics equations are vector-operator equations of the classical field theory, the components of these equations are still differential equations. The Euler equation of fluid mechanics is one of the main and oldest mathematical physics equations. It is only because of its nonlinearity that the equation has been bothering the community of learning since the year 1757, when it was initially established. However, mathematical physics equations, established for the purpose of describing vibration systems of strings, have laid down the well-posedness system using linearization, which has been a method commonly employed in the study of these equations. Studies on mathematical physics equations have almost filled the entire system of dynamics since Newton introduced his second law so that from the latter part of the 19th century and most of the 20th century, mathematical physics equations became the representative of analysis of natural scientific studies. And the well-posedness theory of linearized string vibrations has been generalized to the study of nonlinear dynamic equations so that mathematics and its applications have reached and permeated all corners and areas of the world of learning, creating the beliefs that quantitative comparability is the sole standard of scientificality, that without mathematical proof nothing can be seen as scientific, and so forth. Even so, many practical problems are still not resolved with the fashionable mathematical modelings and analysis of mathematical physics equations. Worse than that, not the slightest clue about how to attack these problems is obtained. Because of the push for quantitative comparability as the only standard for scientificality, the community of learning gradually starts to see the existing problems in the fundamental concepts and the basic formal logic through studies of events' indeterminacy, fuzziness, complexity, and variancy.

To this end, we once followed the current fashion of modern science and considered a simply nonadiabatic problem by employing such popular methods as linearizing nonlinearities, well-posedness, and the like. Our work based on a simple problem of physics leads to mathematical complexity and fundamental problems about the construction of mathematical physics equations and mathematical modeling.

2.3.1 Well-Posed Transformations of Systems of Nonadiabatic Equations

Continuing the well-posedness system of linearizing nonlinear equations, let us treat the component equations of the Euler operator equation from the classical field theory by using small disturbances and series expansion. What needs to be emphasized is that according to the fashionable formal logic of mathematics, even within the development of mathematics there do not exist such studies that point out whether or not these methods violate the formal logic of mathematics — even though in modern physics these methods have been referred to as magical progress in the study of mathematical physics equations. In fact, the modeling and logic of treating mathematical physics equations can be seen in the problem of gravitational waves of the gravitational field equations of Einstein's general relativity theory. That is also why the entire community of learning during the 20th century went after wave motions. At the end of the 19th century, V. Bjerknes (1898) had shown using nonparticle circulative integrals that mathematical nonlinearity stands for problems of physical rotation. More specifically, the earthly atmosphere and ocean are all spinning eddy currents due to uneven densities. However, the community of learning still employed the methods of small disturbances and series expansion to treat some simplified equations of the Euler operator equation of the classical field theory and obtained the unidirectional retrogressive Rossby waves that exist in a plane. In the forthcoming sections and chapters, we will provide the arguments on why nonlinear equations cannot be linearized. As for now, we apply linearization for the purpose of continuing the accustomed tradition without mentioning the existing problems.

By using the relevant methods of the past, it is not hard for us to have the basic model for being nonadiabatic as follows:

$$\begin{cases} u_t + \left(\bar{u}+u\right)u_x + vu_y + \left(\bar{u}_y - f\right)v = -\left(1/\rho\right)p_x \\ v_t + \left(\bar{u}+u\right)v_x + vv_y + fu = -\left(1/\rho\right)p_y \\ w_t + \left(\bar{u}+u\right)w_x + vw_y = -\left(1/\rho\right)p_z - \left(\rho/\bar{\rho}\right)g \\ u_x + v_y + w_z = 0 \\ \left(\dfrac{\rho}{\bar{\rho}}\right)_t + \left(\bar{u}+u\right)\left(\dfrac{\rho}{\bar{\rho}}\right)_x - \dfrac{N^2}{g}w = \dfrac{dQ}{dt} \end{cases} \tag{2.4}$$

where the bar "—" on top, such as $\bar{u} = \bar{u}(y)$, stands for average value, representing the basic field. (It is a fundamental characteristic of mathematically dealing with practical problems. No doubt, this method of handling problems can be seen as mathematical creativity. However, what is left behind is that mathematical accuracy is not the same as the correctness of the underlying physics problem.) The symbol N^2 represents the Brunt–Väisälä frequency,

$$\frac{dQ}{dt} = K_0 \frac{\rho}{\bar{\rho}}$$

and K_0 a constant. All other notations are commonly used in fluid mechanics and their explanations are omitted here. To convert Equation (2.4) into a system of ordinary differential equations, introduce D'Alembert's transformation of traveling waves:

$$\begin{cases} u = U\left(\xi\right), \quad v = V\left(\xi\right), \quad w = W\left(\xi\right) \\ p = P\left(\xi\right), \quad \left(\dfrac{\rho}{\bar{\rho}}\right) = \pi\left(\xi\right) \\ \xi = kx + my + nz - \nu t \end{cases} \tag{2.5}$$

Substituting Equation (2.5) into Equation (2.4) leads to

$$
\begin{cases}
\dfrac{dU}{d\xi} = \dfrac{1}{AB}\left(a_1 U + b_1 V + c_1 \pi\right) \\[2mm]
\dfrac{dV}{d\xi} = \dfrac{1}{AB}\left(a_2 U + b_2 V + c_2 \pi\right) \\[2mm]
\dfrac{d\pi}{d\xi} = \dfrac{1}{ngB}\left(a_3 U + b_3 V + c_3 \pi\right)
\end{cases}
\tag{2.6}
$$

where

$$
\begin{cases}
A = k^2 + m^2 + n^2,\ B = -v + \left(\bar{u} + U\right)k + mV \\[1mm]
a_1 = kmf,\ b_1 = \left(m^2 + n^2\right)\left(f - \bar{u}_y\right),\ c_1 = kng \\[1mm]
a_2 = -f\left(k^2 + n^2\right),\ b_2 = -km\left(f - u_y\right)\left(f - \bar{u}_y\right),\ c_2 = mng \\[1mm]
a_3 = -kN^2,\ b_3 = -mN^2,\ c_3 = -ngK_0
\end{cases}
$$

Now, Equation (2.6) is a first-order autonomous system of ordinary differential equations. Here, the method linearization is applied without considering the mathematical meanings and differences between linear and nonlinear equations. Now, a natural question is: Can nonlinear equations be linearized? If nonlinear equations cannot be linearized, then the results of Equation (2.6) and the following formal logical deductions will become meaningless and useless. That is a sign of immaturity of mathematics itself. However, when this linearization was initially done, it was the fashion of the community of learning. And, since the time of Einstein, one of the purposes of applying series expansion is to eliminate higher order (nonlinear) terms. There is no doubt that applying D'Alembert's transformation of traveling waves is for series expansion. By taking the Taylor series expansion of the right-hand side at near the point $(U, V, \pi) = (0, 0, 0)$ and keeping the first-order terms, we have:

$$
\begin{cases}
\dfrac{dU}{d\xi} = \dfrac{1}{A\left(-v + k\bar{u}\right)}\left(a_1 U + b_1 V + c_1 \pi\right) \\[2mm]
\dfrac{dV}{d\xi} = \dfrac{1}{A\left(-v + k\bar{u}\right)}\left(a_2 U + b_2 V + c_2 \pi\right) \\[2mm]
\dfrac{d\pi}{d\xi} = \dfrac{1}{ng\left(-v + k\bar{u}\right)}\left(a_3 U + b_3 V + c_3 \pi\right)
\end{cases}
\tag{2.7}
$$

So, Equation (2.7) is the system of linear ordinary differential equations corresponding to Equation (2.4). That is, a system of infinite degree of freedom is transformed into a system with only a finite degree of freedom that belongs to the system of classical particle mechanics. Evidently, the magical progress made in the study of mathematical physics equations is nothing but reformulating the originally nonlinear models into linear ordinary differential equations and performing a second round modeling in the name of simplifying equations. After that, analyze the resultant mathematical problems using deductions of the mathematical form.

2.3.2 Characteristic Equations and Eigenvalues

Assume that $|v| \gg |k\bar{u}|$ (for at least two magnitude levels). Now, substituting the specific expressions of $a_i, b_i, c_i (i = 1, 2, 3)$ into Equation (2.7) produces

$$
\begin{cases}
\dfrac{dU}{d\xi} = -\dfrac{1}{vA}\left[kmfU + \left(m^2 + n^2 \right)\left(f - \bar{u}_y \right)V + kng\pi \right] \\[2mm]
\dfrac{dV}{d\xi} = -\dfrac{1}{vA}\left[\left(k^2 + n^2 \right) fU + km\left(f - \bar{u}_y \right)V + mng\pi \right] \\[2mm]
\dfrac{d\pi}{d\xi} = -\dfrac{1}{ngv}\left(kN^2 U + mN^2 V + ngK_0\pi \right)
\end{cases}
\tag{2.8}
$$

This system of equations can be reduced into a third-order linear differential equation by individual differentiation and combination. Then, organize its characteristic equation as follows:

$$
\lambda^3 + \frac{-AK_0 + km\bar{u}_y}{Av}\lambda^2 + \frac{n^2 f\left(f - \bar{u}_y \right) - mkK_0\bar{u}_y + \left(k^2 + m^2 \right)N^2}{Av^2}\lambda
$$
$$
+ \frac{-n^2 fK_0\left(f - \bar{u}_y \right) + kmN^2\bar{u}_y}{Av^3} = 0
\tag{2.9}
$$

For convenience of expression, let us take

$$
a = \frac{-AK_0 + km\bar{u}_y}{Av}
$$

$$
b = \frac{n^2 f\left(f - \bar{u}_y \right) - mkK_0\bar{u}_y + \left(k^2 + m^2 \right)N^2}{Av^2}
$$

$$
c = \frac{-n^2 fK_0\left(f - \bar{u}_y \right) + kmN^2\bar{u}_y}{Av^3}
$$

So, Equation (2.9) can be rewritten as:

$$\lambda^3 + a\lambda^2 + b\lambda + c = 0 \tag{2.10}$$

Equation (2.10) is a third-degree algebraic equation of the general form. So, let us introduce the conventional mathematical transformation

$$\lambda = \lambda' - \frac{a}{3}$$

which helps to change Equation (2.10) into the following:

$$\lambda'^3 + p\lambda' + q = 0 \tag{2.11}$$

where

$$p = -\frac{a^2}{3} + b, \quad q = \frac{2a^3}{27} - \frac{ab}{3} + c$$

So, by using Cardano's formula (see Mathematics Manual Group, 1979) we obtain the eigenvalue solutions:

$$\begin{vmatrix} \lambda_1 = \lambda_1' - \dfrac{a}{3} = \sqrt[3]{-\dfrac{q}{2} + \sqrt{D}} + \sqrt[3]{-\dfrac{q}{2} - \sqrt{D}} - \dfrac{a}{3} \\[4mm] \lambda_2 = \lambda_2' - \dfrac{a}{3} = \omega_1 \sqrt[3]{-\dfrac{q}{2} + \sqrt{D}} + \omega_1^2 \sqrt[3]{-\dfrac{q}{2} - \sqrt{D}} - \dfrac{a}{3} \\[4mm] \lambda_3 = \lambda_3' - \dfrac{a}{3} = \omega_1^2 \sqrt[3]{-\dfrac{q}{2} + \sqrt{D}} + \omega_1 \sqrt[3]{-\dfrac{q}{2} - \sqrt{D}} - \dfrac{a}{3} \end{vmatrix} \tag{2.12}$$

where

$$\begin{vmatrix} \omega_1 = \dfrac{-1 + \sqrt{3}\,i}{2} \\[4mm] D = \left(\dfrac{p}{2}\right)^3 + \left(\dfrac{q}{2}\right)^3 = \dfrac{1}{4}\left(\dfrac{2a^3}{27} - \dfrac{ab}{3} + c\right)^2 + \dfrac{1}{27}\left(-\dfrac{a^2}{3} + b\right) \end{vmatrix} \tag{2.13}$$

Taking

$$s = \frac{a}{3}$$

leads to

$$D = s^3 c - \frac{1}{12} s^2 b^2 - \frac{1}{2} sbc + \frac{1}{4} c^2 + \frac{1}{27} b^3 \tag{2.14}$$

By substituting

$$s = \frac{a}{3}, b, c$$

into the corresponding expressions, letting

$$C_0 = n^2 f \left(f - \bar{u}_y \right) + \left(k^2 + m^2 \right) N^2$$

and then organizing the resulting expression according to the polynomial in K_0, then Equation (2.14) can be rewritten as

$$D = E K_0^4 + F K_0^3 + G K_0^2 + H K + I = f \left(K_0 \right) + I \tag{2.15}$$

where the corresponding coefficients are given below:

$$E = \frac{n^2 f \left(f - \bar{u}_y \right) \bar{u}_y}{27 A v^6} - \frac{k^2 m^2 \left(\bar{u}_y \right)^2}{12 \times 9 A v^6}$$

$$F = \frac{kmN^2 f \left(f - \bar{u}_y \right) \bar{u}_y}{18 A^2 v^6} - \frac{k^3 m^3 \left(\bar{u}_y \right)^3}{27 A^3 v^6} + \frac{C_0 - 2AN^2}{27 A^2 v^6} km\bar{u}_y$$

$$G = -\frac{kmN^2 \left(f - \bar{u}_y \right)^2}{18 A^2 v^6} - \frac{k^2 m^2 n^2 f \left(f - \bar{u}_y \right) \left(\bar{u}_y \right)^2}{18 A^3 v^6} + \frac{2 C_0 k^2 m^2 \left(\bar{u}_y \right)^2}{27 A^3 v^6}$$

$$+ \frac{n^2 f \left(f - \bar{u}_y \right)}{2 A^2 v^6} \left[\frac{1}{2} n^2 f \left(f - \bar{u}_y \right) - \frac{C_0}{3} \right] - \frac{C_0^2}{4 \times 27 A^2 v^6} - \frac{k^4 m^4 \left(\bar{u}_y \right)^4}{4 \times 27 A^4 v^6}$$

$$H = \frac{k^3 m^3 (\bar{u}_y)^3}{27 A^4 v^6} \left[\frac{1}{2} C_0 - n^2 f \left(f - \bar{u}_y \right) - 3AN^2 \right] + \frac{C_0 k m \bar{u}_y}{27 A^3 v^6} \left(\frac{1}{2} C_0 + \frac{9}{2} AN^2 \right)$$

$$+ \frac{k m n^2 f \left(f - \bar{u}_y \right) \bar{u}_y}{A^3 v^6} \left(\frac{C_0}{6} - \frac{1}{2} AN^2 \right) - \frac{C_0}{9 A^3 v^6} k m \bar{u}_y$$

$$I = \frac{k^2 m^2 (\bar{u}_y)^2}{27 A^4 v^6} \left[k^2 m^2 N^2 (\bar{u}_y)^2 - \frac{C_0^2}{4} - \frac{9}{2} A C_0 N^2 + \frac{27}{4} A^2 N^4 \right] + \frac{C_0^3}{27 A^3 v^6}$$

Evidently, from the complexity of the coefficients of Equations (2.12) and (2.14), the solutions of the characteristics could be also very complicated. The solutions of the corresponding system of ordinary equations and the system (Equation 2.4) using the method of small disturbances should also be extremely complicated. Although the entire process of deduction above is not about the difficulty of mathematics, it reveals the fact that from mathematical equations, the complexity of quantitative form can be produced. Evidently, if it is not due to the special addiction for mathematical deductions or the desire to understand the essence of the problem, it will be very difficult for anyone to perform the afore-described operations. It is because even though we obtained the results and even if the results were extremely accurate or correct, it would still be impossible for us to apply the results. Or speaking frankly, after spending huge amounts of time and energy to solve mathematical physics equations, instead of simply pursuing complexity and finding out roughly what factors affect the complexity, people would naturally want to produce the consequent quantitative outcomes in order to show the meaning and effect of quantitative comparability as the only standard for scientificality.

However, the afore-described outcomes can at best tell us roughly that the factors that had caused the disturbances in wave motions are the nonheat insulation (where K_0 is only a simple situation involved linearly in the density ρ), the fluid's share ($\bar{u}_y$), the fluid's vertical stability parameter (the Brunt–Väisälä frequency N^2), and so forth. And, due to the requirements of well-posedness, the effects of these factors can only be a propagation of oscillation of the initial values. As a matter of fact, without solving Equation (2.8), we can also pinpoint the main influential factors from the physical problem itself or the structure of the equations.

Since we have produced the previous results, there is no harm for us to analyze the specific cases of both $K_0 = 0$ (heat insulation) and $K_0 \neq 0$ (nonheat insulation).

2.3.2.1 The Case with Heat Insulation

Assume that $K_0 = 0$, that is the nonadiabatic case. We have

$$D = I = \frac{k^2 m^2 \left(\bar{u}_y\right)^2}{27 A^4 \mathbf{v}^6} \left[k^2 m^2 N^2 \left(\bar{u}_y\right)^2 - \frac{C_0^2}{4} - \frac{9}{2} A C_0 N^2 + \frac{27}{4} A^2 N^4 \right] + \frac{C_0^3}{27 A^3 \mathbf{v}^6} \quad (2.16)$$

2.3.2.1.1 When No Shear Exists

If $\bar{u}_y = 0$, that is when the fluid does not have any shear, then

$$D = \frac{C_0^3}{27 A^3 \mathbf{v}^6} = \left[\frac{n^2 f^2 + N^2 \left(k^2 + m^2\right)}{3 A \mathbf{v}^2} \right]^3 \quad (2.17)$$

Substituting this expression into Equation (2.12) produces

$$\lambda = \begin{cases} 0 \\ \sqrt{3} \, i \sqrt{n^2 f^2 + N^2 \left(k^2 + m^2\right)} \\ -\sqrt{3} \, i \sqrt{n^2 f^2 + N^2 \left(k^2 + m^2\right)} \end{cases}$$

Evidently, when $N^2 \geq 0$, λ has one zero root and two pure imaginary roots. When $N^2 < 0$ and $n^2 f^2 > \left| N^2 \left(k^2 + m^2\right) \right|$, then λ has also a zero root and two imaginary roots, which represent an oscillation of the center point.

If $N^2 < 0$ and $n^2 f^2 < \left| N^2 \left(k^2 + m^2\right) \right|$, then λ has three real roots, showing a node point or a saddle point, containing both convergent stable and divergent instable trajectories. This end indicates that even after linearization, the resultant wave motion systems still cannot completely satisfy the theory of well-posedness. It can also be said that well-posedness is only such a very special and strong condition that it can make the outcome equivalent to nonexistent. Evidently, as a well-posed problem of mathematical physics equations, each instable trajectory, produced out of mathematical deductions, has to be disregarded because it does not satisfy the requirements of well-posedness. So, following the principle of well-posedness, what is discussed must have revealed the unreliability of the mathematical formal logical reasoning. Conversely, after selecting the convergent stable trajectory as the solution of this mathematical physics equation based on the well-posedness conditions, the physical significance of this solution is a dislocation of the initial value, which is only a special case under special or very strict conditions. This solution does not possess the desired generality and forms a denial of the mathematical formal logic reasoning. It implies that the mathematical deductions of the second step of mathematical physics equation become meaningless.

Surely, this mathematical model only expresses simple physical information. The application of mathematics makes the simple situation extremely complicated. It seems necessary to understand that multiplicity is different from the current popular complexity. The former is a product of natural evolution, while the latter often explicitly contains artificiality created out of applying mathematical treatments. This end also presents another reason for the epistemological points of view of randomness and indeterminacy to appear, where the essence is not about the factors of the physical events but instead the existence of ill-posed nonconvergent trajectories in the standard equations obtained by using mathematical formal deductions. So, if mathematical formal logic is very closely related to laws of physics, then the conditions of well-posedness have violated the laws of physics. Conversely, if the conditions of well-posedness of mathematical physics equations agree with experimental observations, then mathematical formal logical deductions are not the same as laws of physics.

Surely, from the previous process of deductions, the reader can readily see the mathematical complexity. Pushing problems of concern to complexity is equivalent to not being able to provide an answer. Science is called by the name of *science* exactly because it can help simplify complicated situations and problems so that plausible answers or explanations can be potentially provided. Based on this spirit of science, the concept of complexity does not belong to the realm of science.

2.3.2.1.2 When Shear Does Exist

If $\bar{u}_y < 0$, then

$$C_0 = n^2 f\left(f - \bar{u}_y\right) + \left(k^2 + m^2\right) N^2 \qquad (2.18)$$

So, the condition of either $D \geq 0$ or $D \leq 0$ is equivalent to

$$k^2 m^2\left(\bar{u}_y\right)\left[k^2 m^2 N^2\left(\bar{u}_y\right)^2 + \frac{27}{4} A^2 N^4\right] + AC_0^3 \geq C_0 k^2 m^2\left(\bar{u}_y\right)^2\left[\frac{C_0}{4} + \frac{9}{2} AN^2\right] \quad (2.19a)$$

or

$$k^2 m^2\left(\bar{u}_y\right)\left[k^2 m^2 N^2\left(\bar{u}_y\right)^2 + \frac{27}{4} A^2 N^4\right] + AC_0^3 \leq C_0 k^2 m^2\left(\bar{u}_y\right)^2\left[\frac{C_0}{4} + \frac{9}{2} AN^2\right] \quad (2.19b)$$

So, we have the following: When $D > 0$, λ contains conjugate complex roots, becoming the focal point, and divergent trajectory; when $D = 0$, λ contains three real roots, (for detailed discussion, see Section 2.3.2.1.1); when $D < 0$, λ contains root 0 of multiplicity three.

The discussion for the case of $\bar{u}_y > 0$ is divided into two situations. If $f - \bar{u}_y > 0$, then we have

$$N^2 \geq 0, \quad C_0$$

$$N^2 < 0, \quad n^2 f \left(f - \bar{u}_y \right) > \left| N^2 \left(k^2 + m^2 \right) \right|, \quad C_0 > 0$$

$$N^2 < 0, \quad n^2 f \left(f - \bar{u}_y \right) < \left| N^2 \left(k^2 + m^2 \right) \right|, \quad C_0 < 0$$

So, based on the conditions in Equation (2.19), our discussion follows similar to that in Section 2.3.2.1.2. All details are omitted here. If $f - \bar{u}_y < 0$, we have

$$N^2 \geq 0, \quad n^2 f \left(f - \bar{u}_y \right) < \left| N^2 \left(k^2 + m^2 \right) \right|, \quad C_0 > 0$$

$$N^2 \geq 0, \quad n^2 f \left(f - \bar{u}_y \right) > \left| N^2 \left(k^2 + m^2 \right) \right|, \quad C_0 < 0$$

$$N^2 < 0, \quad C_0 < 0$$

So, based on the conditions in Equation (2.19), our discussion follows analogous to that in Section 2.3.2.1.2. All details are omitted here.

2.3.2.2 The Case without Heat Insulation

If $K_0 \neq 0$, that is the case of nonheat insulation or the system is open and exchanges with its external environment, then we have

$$D = EK_0^4 + FK_0^3 + GK_0^2 + HK + I$$

If $\bar{u}_y = 0$, we have

$$C_0 = n^2 f^2 + \left(k^2 + m^2 \right) N^2, \quad E = \frac{n^2 f^2}{27 A v^6} > 0, \quad F = 0$$

$$G = \frac{1}{A^2 v^6} \left(\frac{1}{4} n^4 f^4 + \frac{1}{6} C_0 n^2 f^2 - \frac{C_0^2}{4 \times 27} \right), \quad H = 0, \quad I = \frac{C_0^3}{27 A^3 v^6}$$

So,

$$D = EK_0^4 + GK_0^2 + I \qquad (2.20)$$

When $C_0 > 0$, we have: If $G \geq 0$ or

$$\frac{1}{4} n^4 f^4 + \frac{1}{6} C_0 n^2 f^2 \geq \frac{C_0^2}{4 \times 27}$$

then $D > 0$. That is, λ has a real root and a pair of conjugate complex roots. When $G < 0$ or

$$\frac{1}{4} n^4 f^4 + \frac{1}{6} C_0 n^2 f^2 < \frac{C_0^2}{4 \times 27}$$

if $EK_0^4 + I > |GK_0^2|$, then $D > 0$ and λ has a real root and a pair of conjugate complex roots. If $EK_0^4 + I = |GK_0^2|$, then $D = 0$ and λ has a zero root of multiplicity three. If $EK_0^4 + I < |GK_0^2|$, then $D < 0$ and λ has three real roots.

For the case of $C_0 < 0$, the discussion is similar and is omitted. If $\bar{u}_y < 0$, then we have

$$C_0 = n^2 f \left(f - \bar{u}_y \right) + \left(k^2 + m^2 \right) N^2$$

and

$$D = EK_0^4 + FK_0^3 + GK_0^2 + HK + I$$

So as not to make our discussion too tedious, we will only look at the explanation for the case of $C_0 > 0$. The condition corresponding to $E \to I$ is given in the following. The conditions for either $E \geq 0$ or $E \leq 0$ are:

$$Af \left(f - \bar{u}_y \right) n^2 \geq \frac{1}{4} mk \left(\bar{u}_y \right)^2, \quad Af \left(f - \bar{u}_y \right) n^2 \leq \frac{1}{4} mk \left(\bar{u}_y \right)^2$$

The conditions for either $F \geq 0$ or $F \leq 0$ are:

$$k^2 m^2 \left(\bar{u}_y \right)^2 \geq A \left[\frac{3}{2} n^2 f \left(f - \bar{u}_y \right) + C_0^2 - 2AN^2 \right]$$

or

$$k^2 m^2 \left(\bar{u}_y\right)^2 \leq A\left[\frac{3}{2} n^2 f\left(f - \bar{u}_y\right) + C_0^2 - 2AN^2\right]$$

The conditions for either $G \geq 0$ or $G \leq 0$ are:

$$Ak^2 m^2 \left(\bar{u}_y\right)^2 \left[\frac{1}{2} n^2 f\left(f - \bar{u}_y\right) + \frac{1}{2}\left(k^2 + m^2 - 3n^2\right)N^2\right] + \frac{27}{4} A^2 n^4 f^2 \left(f - \bar{u}_y\right)^2$$

$$\geq \frac{1}{4} A^2 C_0 + \frac{1}{4} k^4 m^4 \left(\bar{u}_y\right)^4$$

or

$$Ak^2 m^2 \left(\bar{u}_y\right)^2 \left[\frac{1}{2} n^2 f\left(f - \bar{u}_y\right) + \frac{1}{2}\left(k^2 + m^2 - 3n^2\right)N^2\right] + \frac{27}{4} A^2 n^4 f^2 \left(f - \bar{u}_y\right)^2$$

$$\leq \frac{1}{4} A^2 C_0 + \frac{1}{4} k^4 m^4 \left(\bar{u}_y\right)^4$$

The conditions for either $H \geq 0$ or $H \leq 0$ are:

$$Ak^2 m^2 \left(\bar{u}_y\right)^2 \left(\frac{C_0}{2 \times 27} + \frac{AN^2}{9}\right) + AC_0\left[\frac{C_0}{2 \times 27} + \frac{1}{6} AN^2 + \frac{1}{6} n^2 f\left(f - \bar{u}_y\right)\right]$$

$$\geq \left(\frac{k^2 m^2}{27} + \frac{1}{2} A^2 N^2\right) n^2 f\left(f - \bar{u}_y\right)\bar{u}_y - \frac{1}{3} AC_0^2$$

or

$$Ak^2 m^2 \left(\bar{u}_y\right)^2 \left(\frac{C_0}{2 \times 27} + \frac{AN^2}{9}\right) + AC_0\left[\frac{C_0}{2 \times 27} + \frac{1}{6} AN^2 + \frac{1}{6} n^2 f\left(f - \bar{u}_y\right)\right]$$

$$\leq \left(\frac{k^2 m^2}{27} + \frac{1}{2} A^2 N^2\right) n^2 f\left(f - \bar{u}_y\right)\bar{u}_y - \frac{1}{3} AC_0^2$$

The conditions for either $I \geq 0$ or $I \leq 0$ are:

$$k^2m^2\left(\overline{u}_y\right)^2 N^2\left[k^2m^2\left(\overline{u}_y\right)^2 + \frac{27}{4}A^2N^2\right] + AC_0^3 \geq k^2m^2\left(\overline{u}_y\right)^2 C_0\left(\frac{C_0}{4}C_0 + \frac{9}{2}AN^2\right)$$

or

$$k^2m^2\left(\overline{u}_y\right)^2 N^2\left[k^2m^2\left(\overline{u}_y\right)^2 + \frac{27}{4}A^2N^2\right] + AC_0^3 \leq k^2m^2\left(\overline{u}_y\right)^2 C_0\left(\frac{C_0}{4}C_0 + \frac{9}{2}AN^2\right)$$

So, we have: If E, F, G, H, and I are all greater than 0, then $D > 0$ and λ has a real root and a pair of conjugate complex roots; if E, F, G, H, and I are all smaller than 0, then $D < 0$ and λ has three different real roots; if E, F, G, H, and I are all equal to 0, then $D = 0$ and λ has root 0 of multiplicity three; if the values of E, F, G, H, and I are both positive and negative, then the situation with λ becomes very complicated. To limit space, all the details are omitted.

2.3.3 Transformation between a Phase Trajectory Equation and a Phase Trajectory of a Center Point

When purely imaginary roots exist, let $U = U_0 \cos\lambda\xi$ and $V = U_0 \sin\lambda\xi$. That is equivalent to assume in Equation (2.9) that

$$\frac{-AK_0 + km\overline{u}_y}{Av} = 0 \quad \text{and} \quad \frac{-n^2 fK_0\left(f - \overline{u}_y\right) + km^2N^2\overline{u}_y}{Av^3} = 0$$

Now, the third equation in Equation (2.8) can be rewritten as

$$\frac{d\pi}{d\xi} - \frac{K_0}{v}\pi = \frac{kN^2}{ngv}U_0 \cos\lambda\xi + \frac{mN^2}{ngv}U_0 \sin\lambda\xi \tag{2.21}$$

Equation (2.21) is a first order ordinary differential equation with solution given below:

$$\pi = \frac{v^2N^2}{ngv\left(K_0^2 + v^4\right)}\left[\left(k\lambda + \frac{mK_0}{v}\right)U_0 \sin\lambda\xi - \left(m\lambda + \frac{kK_0}{v}\right)U_0 \cos\lambda\xi\right] + Ce^{\frac{K_0}{v}\xi} \tag{2.22}$$

Substituting this expression into the first equation in Equation (2.8) and noticing $V = U_0 \sin\lambda\xi$ leads to

$$\frac{dU}{d\xi} = -\frac{1}{Av}\left\{-kmdU + \left(m^2 + n^2\right)\left(f - \bar{u}_y\right)U_0 \sin \lambda\xi + \frac{kN^2}{A\left(\frac{K_0^2}{v^2} + \lambda^2\right)}\right.$$

$$(2.23)$$

$$\left. \times\left[\left(k\lambda + \frac{mK_0}{v}\right)U_0 \sin \lambda\xi - \left(m\lambda + \frac{kK_0}{v}\right)U_0 \cos \lambda\xi\right] + Ckmge^{\frac{K_0}{v}\xi}\right\}$$

where C is an integration constant. For convenience of expression, let us take

$$\alpha_1 = -\frac{kmf}{Av}, \quad \gamma_1 = k\lambda + \frac{kK_0}{v}, \quad \gamma_2 = m\lambda + \frac{kK_0}{v}, \quad \varepsilon_2 = -\frac{Ckng}{Av},$$

$$\beta_1 = -\frac{\left(m^2 + n^2\right)\left(f - \bar{u}_y\right)}{Av}, \quad \varepsilon_1 = -\frac{kN^2}{Av^2\left(\frac{K_0^2}{v^2} + \lambda^2\right)}$$

So, Equation (2.23) can be rewritten as

$$U_\xi = -\alpha_1 U - \beta_1 U_0 \sin \lambda\xi + \varepsilon_1 U_0 \left(\gamma_1 \sin \lambda\xi - \gamma_2 \cos \lambda\xi\right) + \varepsilon_2 e^{\frac{K_0}{v}\xi}$$

If its coefficients are constant, then its solution is

$$U = \frac{1}{\alpha_1^2 + \lambda^2}\left[\varepsilon_1 \gamma_2 \alpha_1 - \left(\beta_1 + \varepsilon_1 \gamma_1\right)\lambda\right]U_0 \cos \lambda\xi$$

$$-\frac{1}{\alpha_1^2 + \lambda^2}\left[\left(\beta_1 + \varepsilon_1 \gamma_1\right)\alpha_1 + \gamma_2 \varepsilon_1 \lambda\right]U_0 \sin \lambda\xi + \frac{\varepsilon_2}{\frac{K_0}{v} - \alpha_1}e^{\frac{K_0}{v}\xi}$$

$$(2.24)$$

In order to help with our intuition, let us take

$$\begin{cases} L = \left[\varepsilon_1 \gamma_2 \alpha_1 - \left(\beta_1 + \varepsilon_1 \gamma_1\right)\lambda\right]/\left(\alpha_1^2 + \lambda^2\right) \\ M = \left[\left(\beta_1 + \varepsilon_1 \gamma_1\right)\alpha_1 + \varepsilon_1 \gamma_2 \lambda\right]/\left(\alpha_1^2 + \lambda^2\right) \\ R = \varepsilon_2 \Big/ \left(\frac{K_0}{v} - \alpha_1\right) \end{cases}$$

then Equation (2.24) becomes

$$\left(1 - L\right)U + MV = \mathrm{Re}^{\frac{K_0}{v}\xi} \tag{2.25}$$

By introducing

$$\tilde{U} = U - \frac{R}{1-L} e^{\frac{K_0}{v}\xi}$$

Equation (2.25) becomes

$$\tilde{U} + \frac{M}{1-L}V = 0$$

Squaring this equation produces

$$\tilde{U}^2 + 2\frac{M}{1-L}\tilde{U}V + \frac{M^2}{\left(1-L\right)^2}V^2 = 0 \tag{2.26}$$

Similarly, we have

$$\tilde{V}^2 + 2\frac{M'}{1-L'}U\tilde{V} + \frac{M'^2}{\left(1-L'\right)^2}U^2 = 0 \tag{2.27}$$

where

$$L' = \frac{\left(\alpha_2 - \varepsilon'\gamma_2\right)\beta_2 + \varepsilon_1'\gamma_1\lambda}{\beta_2^2 + \lambda^2} \quad \text{and} \quad M' = \frac{\left(\alpha_2 - \varepsilon'\gamma_2\right)\lambda + \varepsilon_1'\gamma_1\beta_2}{\beta_2^2 + \lambda^2}$$

the inside coefficients, are defined as follows:

$$\alpha_2 = \frac{\left(k^2 + n^2\right)f}{Av}, \quad \beta_2 = \frac{km\left(f - \bar{u}_y\right)}{Av}, \quad \varepsilon_1' = -\frac{mN^2}{Av^2\left(\frac{K_0^2}{v^2} + \lambda^2\right)}, \quad \varepsilon_2' = -\frac{mngC}{Av}$$

and the corresponding transformations are

$$\tilde{V} = V - \frac{R'}{1-L'} e^{\frac{K_0}{v}\xi}, \quad R' = \frac{\varepsilon_2'}{\dfrac{K_0}{v} - \beta_2}$$

So, Equations (2.26) and (2.27) are the phase trajectory equations with dislocations under the condition of nonheat insulation. Their discriminants below

$$\delta\tilde{U} = \begin{vmatrix} 1 & \dfrac{M}{1-L} \\ \dfrac{M}{1-L} & \dfrac{M^2}{\left(1-L\right)^2} \end{vmatrix} \quad \text{and} \quad \delta\tilde{V} = \begin{vmatrix} 1 & \dfrac{M'}{1-L'} \\ \dfrac{M'}{1-L'} & \dfrac{M'^2}{\left(1-L'\right)^2} \end{vmatrix} \qquad (2.28)$$

are hyperbolic. This implies that nonheat insulation can make the originally elliptic trajectory under the condition of heat insulation dislocated into hyperbolic trajectory. Or, in other words, this result that is only obtained for the special case of purely imaginary roots can destroy the conditions of the traditional well-posedness of mathematical physics equations due to the simple situation of nonheat insulation. This end reveals how fragile mathematical conditions, such as those of well-posedness, can be.

Even if we take $\xi = 0$ in Equation (2.25), we also have

$$\left(1-L\right)U + MV = R \qquad (2.29)$$

Squaring this expression produces

$$U^2 + 2\frac{M}{1-L}UV + \frac{M^2}{\left(1-L\right)^2}V^2 = \frac{R^2}{\left(1-L\right)^2} = C_1^2 \qquad (2.30)$$

From the constants made of the coefficients, we have

$$DU = \begin{vmatrix} 1 & \dfrac{M}{1-L} & 0 \\ \dfrac{M}{1-L} & \dfrac{M^2}{\left(1-L\right)^2} & 0 \\ 0 & 0 & C_1^2 \end{vmatrix} = \frac{M^2}{\left(1-L\right)^2}C_1^2 - \frac{M^2}{\left(1-L\right)^2}C_1^2 = 0$$

That is, we have

$$\delta U = \begin{vmatrix} 1 & \dfrac{M}{1-L} \\[3mm] \dfrac{M}{1-L} & \dfrac{M^2}{(1-L)^2} \end{vmatrix} = 0 \qquad (2.31)$$

Similarly, both DV and δV also satisfy

$$DV = 0 \quad \text{and} \quad \delta V = 0 \qquad (2.32)$$

That is, the condition $\xi = 0$ can lead to centerless trajectories of straight-line type (for details, see Mathematics Manual Group, 1979), which is equivalent to the case that the trajectories experience transformations based on the situation of the node points. Even for the phase trajectories of straight-line type, there are also the cases of either convergent stable or divergent instable node points. This fact indicates that the center points of the closed curves under the condition of heat insulation can be transformed into open node points under the condition of nonheat insulation. This phenomenon is referred to as that adiabatic periodicality can be destroyed under nonadiabatic situation.

Looking at Equation (2.16), if we assume $K_0 = 0$ (heat insulation), in form Equation (2.25) can be changed into Equation (2.29). The detailed discussion is similar to that of Equations (2.28) and (2.31) with different meanings. That is, similar mathematical forms are not the same as the same physical essences. And, from the references we know that such examples are plentiful. No doubt, from $K_0 = 0$ to $K_0 \neq 0$, a change in terms of energy has occurred. That corresponds to changes from periodicality to nonperiodicality. Here, energy transformation is indeed a fundamental problem of physics.

So, the well-posedness of linear systems of mathematical physics equations, as a topic of mathematics, is also a special case among special cases. Even for linear systems, it still does not have generality. Even when applied to the design and manufacture of durable products, numerical computations of mathematical physics equations still cannot directly provide applicable results. Only after repeated testing and experiments, considering the specific circumstances involved, relatively realistic parameters can be found and used to adjust the results so that they could potentially be useful. Our example shows that such a simple effect as nonheat insulation can easily mix up the well-posed instability of mathematics. Evidently, nonheat insulation is an objective existence instead of a quasi-problem. Besides, science is about how to reach simplicity. Making simple scenarios more complicated does not go along well with the spirit of science.

The achievement of our tedious computations is that modern mathematics can derive complexities from simple physics problems through mathematical deductions so that even if the results obtained out of computers are accurate or correct, these results still lose their potential benefits of practical applications due to the complexities involved. So, quantitative comparability, the only standard for scientificality, leads to complexity due to formal logic calculus. What needs to be pointed out is as a matter of fact that beyond our example above, most equations do not satisfy the well-posedness of mathematical physics equations. So, a natural question arises: Have mathematical physics equations, as a borderline study between mathematics and physics, truly married these two disciplines because the three classical types of equations "hold true" so that their applicability can be generally employed in all areas of learning? Also, the so-called holding true or success of the three typical equations is only limited to the study of propagation of wave motions. Besides, it still awaits further consideration on whether or not diffusing heat conduction equations are truly about propagations of wave motions.

In short, mathematical physics equations, as a borderline study of mathematics and physics, have been active for over a century. Not only is it the support of theoretical physics but it is also seen as the symbol of the system of modern science. However, neither sufficient theoretical arguments nor practical evidence exist to show that the theory of mathematical differential equations can represent the laws of physics.

In terms of problems of changes, especially those about constant changes, one can no longer continue to apply the risk values of product designs as the measurement of success. Instead, success has to be measured using what actually happens. So, to this end, the achievements of mathematics as of this writing are still helpless. Also, the original equations of fluid motions are nonlinear. We have shown (Wu and Lin, 2002; Lin, 2008) that the fundamental attribute of nonlinearity is rotationality (see later sections and chapters), while modern mathematics and any "perfect systems" that chase after quantitative forms cannot resolve problems involving rotations. That is why we established our digitalization (a nonquantification) method in order to face problems of material evolutions.

Other than transforming through dislocation adiabatic elliptic trajectories into hyperbolic trajectories, nonheat insulation can also create dislocation transformations of straight-line trajectories. This result was not expected before we did our tedious computations. Another problem one should be clear about is that the result that a simple physics proposition can lead to complexity because mathematical formal logical deductions reveal the fact that laws of physics and mathematical models may not be in concert in terms of mathematical physics equations and the inconsistency between mathematical formal logic calculus and the requirements of well-posedness. Therefore, a perfect mathematical modeling cannot be stopped at simply describing physical quantities in differential equations. Instead, its focus should be at showing whether or not laws of physics truly follow the deduction rules of differential and integral equations.

The rotationality of nonlinearity not only destroys the well-posedness condition, but also does not comply with the requirement of continuity of the calculus system, on which mathematical physics equations are developed. In later sections and chapters, we will go into more detail on this. Our purpose of mentioning this fact here is that nonlinear problems cannot be seen as indeterminate or complex, either, and that our previous discussion has touched on the linearization of nonlinear problems, where our example has shown that even after linearizing a nonlinear problem (for the purpose of simplifying the problem), a simple problem actually becomes complicated, leading to complexity.

It should be recognized from our discussion that in the study of mathematical physics equations, the need to introduce the condition of well-posedness is not due to the need to solve physics problems. Instead, mathematical equations, even after being linearized, also suffer from the ill-posedness problem. One should first check out the reason why mathematical equations suffer from this problem and then consider using equations to model physics problems; otherwise, mathematical physics equations themselves carry "embryonic deficits." What we like to express directly is that mathematical physics equations, as a borderline study between mathematics and physics, did not magically mature; instead, this study did not fundamentally resolve the problem of connection that grafts the disciplines along the borderline. Especially, it requires further and deepened discussion about whether or not laws and patterns of physics are the same as the basic theory of differential equations.

What is more important is that mathematical physics equations that continue the classical mechanical system are only limited to the relationship of movements between physical quantities and cannot be deepened to the level of study of material rotationality. Because material rotationality has to cause injuries and damages in materials, that leads to changes in materials, leading to establishment of evolution science. The entire modern science, including mathematical physics equations, provides a playground for mathematics in its attempt to deny evolutionality and to maintain invariancy of materials, which is embodied in the "success" of the system of wave motions of physics. And, the functionality of the physics of wave motions is to maintain the invariancy of materials through transfers of energy in the form of energy propagation by borrowing the form of mathematics to describe wave motions. Now, the realisticity of material evolutions and problems of rotationality existing in evolutions make physics face noninertial systems and mathematics experience nonlinearities. This end clearly points out that modern science is only about the existence physics (without changes), as described by I. Prigogine, waiting to enter the process physics of variant materials.

What needs to be pointed out is that process physics is not a concept we first introduced. Instead, it was from Aristotle's statement that physics is such a discipline that studies changes in materials and processes of change. That is, physics in terms of its original purpose according to Aristotle should be the foundation of evolution science. It was only due to the difficulty of quantitative analysis in handling changes in materials that scholars invented inertial systems and physical quantities and achieved successes in computing locations of celestial bodies where excessively large distance and time scales are involved. However, these correct computations of celestial body locations and movements are not the same as those computations foretelling changes in materials.

Since multiplicity and complexity, which are different from using quantitative means to make simple matters complicated, are the original faces of nature, a natural question is why the natural multiplicity and complexity exist. Exploring and addressing the causes and processes of natural multiplicity and complexity will make a true physics. Modern science not only avoids this problem, other than making simple matters complicated, it also pushes the responsibility of providing explanations for complex phenomena to complexity. The result is that no answer to anything is provided. No doubt, when faced with the natural multiplicity and complexity, the first and most important matter is about how to extract the information from multiplicity and complexity, how to analyze the events involving multiplicity and complexity, and how to uncover the laws of physics governing these events.

In other words, pointing out the fact that modern science has not touched upon evolution is not trying to play down modern science; instead, by doing so, we can correctly position modern science by exactly knowing its strengths and weaknesses so that more efforts and investments can be directed to solve currently unsolvable problems. For example, it is well known that weather changes constantly without any clear pattern. However, contemporary meteorologists just like to apply indiscriminately the theory of invariancy in their study of changes. So, some prediction problems that can be relatively easily resolved have become internationally difficult issues. In essence, it is because of the misunderstanding of "monitoring" as "forecasting" that the research on the prediction of natural disasters has not been making much progress. That also shows a need for us to clearly position modern science and its existing problems. The development of science comes from calls of solving more currently unsolvable problems instead of modifying the problems in order to fit the existing theories. This later phenomenon is exactly why and how people classify problems unsolvable in modern science as quasi-problems.

Acknowledgments

The presentation in this chapter is based on Chen and OuYang (2005), Chen et al. (2005a,b), Hawking (1988), Li and Guo (1985), Lin (1998, 2008), Lin et al. (2001), Mathematics Manual Group (1979), OuYang (1984, 1994, 1998, in press), OuYang and Chen (2007), (OuYang et al. (2002a, 2005b), OuYang and Peng (2005), Prigogine (1980), Wang (1982), Weng (1996), Wu and Lin (2002), and Wu and OuYang (1989).

Chapter 3

Attributes and Problems of Numbers and Their Morphological Transformations

When talking about numbers, people naturally think about mathematics. This phenomenon shows how much number computations since their elementary school years have left impressions in people's memories. However, mathematics is only about quantities. And, cultures in both the West and the East almost all started to explore the environmental world from shapes. For example, the *Book of Changes* (Wilhalm and Baynes, 1967) of the East starts off on the basis of morphological structures of things by introducing the concept of *yao* written in "- -" and "—". And the famous "river map" that has been in existence for over 6,000 to 7,000 years also applies morphological structures to represent the numbers 1, 2, ... 10 and sum of 40; and the fraction 7/11 indicated in the figure exactly and implicitly shows the rate of the circumferences of circles (or the irrational number π) of the transformation between curvature spaces and Euclidean spaces. At the time when Euclid's elementary geometry was published, or a little ahead of the time of Socrates of ancient Greece, there was a school of learning named Pythagoras. It believed that all things were originated in (rational) numbers, that all the numbers within 10, meaning less than 10, possessed the mighty influential power, and that numbers represented the foremost principle and guidance that governed everything in the universe. So, to them, without understanding (rational) numbers and their properties it would be impossible to know anything else, such as mutual relationships between things.

Objectively speaking, any thought or opinion is formed under the influence of its background and existent environment. Such a belief that the school of Pythagoras had held strongly is closely related to the then-historical environment and cultural circumstances. Based on the literature we can find, it can be shown that at that time period, Greece was located at the center of routes of commercial trade between Europe and Asia, which naturally brought about the corresponding economic prosperity and the formation and development of cultures. No doubt, commercial trades had to push the development of numbers and their computations. At the same time, those nations whose economic prosperity depended mainly on nomadic trades would pay much closer attention to the numerical accuracy than those nations whose economic prosperity depended largely on agriculture. So, compared to Chinese people, whose economic prosperities have largely been based on agriculture, Western people are relatively more fond of quantities due to their need for survival, which is within the range of comprehension. As for promoting quantities to the level that could dominate everything, it could have been a combined result of the cultural and religious backgrounds of that time. The specific verification of this end we leave to historians to finish. Even in today's world, we can still sense that Western scholars are indeed more sensitive either consciously or unconsciously to quantities than Chinese scholars. So, it is quite natural without much, if any, resistance for them to introduce and apply physical quantities and parametric dimensions in their explorations of nature.

No doubt, as for the ancestors of Chinese people, whose economic prosperities had always been established on agriculture, they paid more attention to weather and climate changes. This is because the food production during good years and disastrous years could reach such a degree that it would become almost impossible to numerically calculate. That was why Zhuang Zi of the time of Warring States pointed out *wu, liang wu qiong* (there are so many things in the world that it is impossible to account for them using numbers), and his contemporary quantity calculator Zhan Yin also believed that *shu you suo bu dai* (quantities always have the deficit of being unable to correctly describe events or things). The *Book of Changes* shows the opinion and method on how to employ the morphology of events or matters to understand the world, which was also originated from studying changes in objective matters and events, producing long-lasting influence in history. Under the current atmosphere of knowledge structure, if someone uses quantities to substitute for events, he will have to experience great obstacles in China. This fact might be the reason and cause for the formation of two different Western and Eastern knowledge systems.

As for *shu you suo bu dai*, meaning that quantities always have the deficit of being unable to correctly describe events or things, it can be explained using the concept of implicit transformation between figures and numbers. For example, let us look at the irrational number $\sqrt{2}$ of mathematics. If we use the concept of numbers to illustrate figures, we have to face the typical *shu you suo bu dai* or *wu, liang wu qiong*. On the other hand, if we use figures to illustrate numbers, we can clearly and definitely answer what $\sqrt{2}$ is by using the hypotenuse of the right triangle with both legs being 1. Here in mathematics, a definite situation of figures becomes an unexplainable problem of indeterminacy using numbers. That is why in Chapter 2 we especially emphasized that indeterminacy is not originated from events, but instead quantities themselves suffer from the inability to clearly explain things. So, the Pythagorean belief that numbers are the origin of all things stands for an indeterminate problem of difficulty of explanation. Conversely, using figures to describe things represents a very simple deterministic problem. Or in other words, the essence of the doctrine that numbers are the origin of all things is using *figures*, such as the symbol $\sqrt{2}$, to show that quantities do not really occupy any material dimension. That is why we have Lao Tzu's *dao zhi wei wu, qi zhong you xiang*, meaning that Tao is about things, in which there are figures, and our own nonparticlization of things. Evidently, the nonparticle attribute of things has to led to the conclusion that events are not equal to quantities. When it is combined with the premise of materials being the first (without things, there will not be any number), we conclude that quantities are not the origin of all things and they are not the reasons of laws of all things, either. So, in Chinese history, in which *wu, liang wu qiong* and *shu you suo bu dai* have been well rooted, it was impossible for schools like those of Pythagoras, for the quantitative axioms of Galileo's dialogue of two kinds of sciences, and later for the belief that quantities are the laws of all things, as described in Newton's *Mathematical Principles of Natural Philosophy*, to appear.

As the origin of one of the two main civilizations of the world today, the *Book of Changes*, the most important among the six classics of the ancient Chinese literature, does not recognize numbers; instead it only shows how to treat quantities. The concept of yin and yang implicitly contains the numerals in the binary system and shows that both yin and yang possess much mightier and incomparable covering power than the (natural) numbers within 10 of the Pythagorean school. That is, yin and yang can be combined in different ways to include all things in nature and to represent the ups and downs of man in human society. That is why they seem to be more incomparable than Pythagorean (natural) numbers when they are placed side by side. However, both yin and yang are introduced on the basis of the material figurative structures of *yuan qu zhu wu, jin qu zhu sheng* (at a distance take the objects, and nearby take the bodies themselves of the objects). So, quantities can only appear after materials or events first come about. And, when pursuing the causes for changes in materials or events, it must be about studying the structures, attributes, functionalities, and so forth of the materials or events instead of going after their consequent quantities.

From the "river map" of ancient China (about 6,000 to 7,000 years ago), the directions in Figure 3.1 have been changed to accommodate the current orientation with the top as north and bottom as south (it used to be south on the top and north on the bottom), where the sum of the six black dots and one white dot on

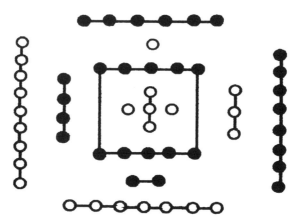

Figure 3.1 The river map.

top is seven, the sum of the eight black dots and three white dots on the right hand side is 11. That is the mysterious structural ratio 7/11 that appeared earliest in history. What needs to be pointed out is that for the directional orientation of south on top and north on bottom of the original river map, if the earthly mutations and dislocations of the poles are the causes of earthly disasters, then over 6,000 to 7,000 years ago there might have appeared a major mutation and pole shift of a 180° turnaround. From the pyramids in Egypt, we can all see the relationship of transformation between figures and quantities. What is interesting is that they all involve the structural ratio 7/11 of figures and quantities without specifically giving the nonterminating repeating decimal representation (*shu you suo bu dai*) 7/11 = 0.6363.... Evidently, ancient peoples had already understood the determinacy of the structural ratio 7/11 and how to apply it conveniently, while the quantitative 0.6363.... not only shows the indeterminacy of quantities but also the impossibility for practitioners to apply such quantities as 0.6363... As illustrated by the pictographic characters of ancient Egypt, the time length of construction of each pyramid is measured by using the linear measurement unit of wrist foot, which was defined to be the length from the tip of the middle finger to the elbow joint of the leader. The height of the pyramids is 280 wrist-feet and base length is 440 wrist feet. The ratio of the height and base length is exactly 280.440 = 7/11. What is extremely interesting is that $(7 \times 2\pi) \cong 11 \times 4$ or $\pi \cong 22/7$. That fact at least shows that as early as 6,000 to 7,000 years ago in China or as late as 4,500 years ago in Egypt, people had already known about the ratio of a circle's circumference to its diameter. The meaning of a deeper level is about the concept of the transformation relationship between curvature spaces and zero-curvature (Euclidean) spaces, revealing the indeterminacy of quantities and the fact that since quantities are dependent on Euclidean spaces, it becomes impossible for them to produce the numerical accuracy of the corresponding curvature spaces. That explains that when using quantities to describe problems of physics in curvature spaces, quantities always suffer from the problem of computational inaccuracy and indeterminacy. That is where Zhuang Zi's *wu, liang wu qiong* and Zhan Yin's *shu you suo bu dai* come from.

The phenomenon of Egyptian pyramids is not an isolated situation but also includes Shao-Hao Tomb, located in Shandong, China, and Central American Mayan artifacts. The mysterious structural ratio of 7/11 can be traced back to the river map of ancient China over 6,000 to 7,000 years ago and is also the foundation for the establishment of the *Book of Changes*. What attracts people's attention is the practical functionality of this structural ratio 7/11, which is embodied in the fact that after over 50 centuries since the time when the Egyptian pyramids were constructed, these man-made structures are still standing in front of us; and other than some erosion and damage caused by water, wind, and other factors, what is most important is their ability to withstand earthquakes. That end also reflects the natural collapsing functionality of the spinning earth itself. As for the relevant, detailed discussion of the river map, please consult our book *Entering the Era of Irregularity* (OuYang et al., 2002a). The details are omitted here.

According to the research of modern medicine, the structural ratio 7/11 is also found on human body structures. If a man's width ratio of his waist and shoulders and a woman's width ratio of her waist and buttocks are about 7/11, then the person will not only be healthy but also have a handsome or beautiful body with a stronger immune system and reproduction ability than the those without the 7/11 bodily structural ratio. This example shows that the structural ratio is better suited to represent the attributes and functionalities of materials, the fact that structures possess the straight-to-the-point intuition and determinacy, the superficiality of quantitative forms, and where quantitative indeterminacy is from. So, our understanding of science should not only be limited to quantitative computations, but instead it should be *shan suan zhe bu yong chou ce* (Lao Tzu), meaning that the one who is truly good at calculating does not need any tools for his computation. That is, the meaning of *shan suan zhe bu yong chou ce* is deep. At the least it can be understood as computational accuracy not being the same as realistic correctness.

The fundamental characteristics or the central problems of modern science employ physical quantities, parametric dimensions, and quantities to substitute for materials and events and translate the relevant problems of concern into mathematical equations of inertial systems, emphasizing the belief that equations are eternal. Speaking more straightforwardly, the central problem of modern science, since the time of Newton, is to inherit the belief of the Pythagorean school, that (rational) numbers are the origin of all things, while constructing nonelementary modern science and quantitative comparability as its sole standard for scientificality on the basis that numbers are the laws of all things. Although mathematics itself involves problems of figures and quantities, problems of figures have been lowered to the level of elementary, nontheoretical problems more closely related to experiences when compared with the high-level, nontrivial scientific theories of the eternal equations and quantitative comparabilities. Problems that can be simply explained by using the figurative structures of materials nowadays have to be transformed into problems of quantities so that they become inexplicable complexities (see Chapter 2), or to a degree, only being unable to illustrate clearly by using quantities is considered scientific.

Expanding on this point of view, based on the formulation of modern science, the concept that science needs assumptions has been formed. The essence of this concept is about replacing events by using quantities and transforming shapes or figures into quantities, even if to achieve this end one has to change the original problem of study or make the theoretical study totally irrelevant to the physical reality. It has become a fashion that without explicitly listing equations of inertial systems, no work can be seen as scientific. Especially when faced with nonlinear equations, where no quantitative methods can help to solve the equations, instead of analyzing potential problems existing in the quantitative analysis system, huge amounts of capital and manpower are still invested to push for numerical models and quantitative computations. To this end, let us analyze the following quantitative computation of a simple mathematical model to see what problem the so-called quantitative comparability faces when dealing with nonlinear equations. For our purpose, let us look at a second order mathematical model as follows:

$$\frac{du}{dt} = ku^2$$

Its corresponding difference equation is

$$u_{n+1} = u_n + ku_n^2 \Delta t$$

Evidently, the numerical computation of this difference equation has to experience the following problem of calculations.

1. The problem of computing nonsmall effective values:
If we take $\Delta t = 10s$ (second), $k = 10 \ m^{-1}$ (m is meter, which is taken to be the realistic precision of effective values with one decimal digit) and take the initial value u_0 = 20.1 m/s, then it is not hard to obtain the following by numerical computations:

$$u_1 = 20.1 + 10 \cdot (20.1)^2 \cdot 10 = 40401$$

$$u_2 = 40401 + (40401)^2 \cdot 100 = 1785680100$$

$$\cdots$$

$$u_{n+1} \rightarrow \infty$$

This end implies that the quantitative computation of the nonlinear equation experiences the instability of fast numerical growth of nonsmall quantities so that the resultant quantities approach infinity, or as stated by Zhuang Zi, *wu, liang wu qiong*, indicating that one experiences the problem of inability to do much when using quantities to compute materials. Modern science refers to this instability as ill-posedness; its essence is really the indeterminacy of quantitative incapability, or called quantitative indeterminacy of mathematical equations. That is why mathematical physics equations need the constraint of well-posedness so that the condition of stability is required. So, our analysis has clearly shown that the so-called stability in the study of mathematical physics equations is nothing but about disallowing quantities to grow indefinitely. Consequently, changes in quantities are also limited. The corresponding law of philosophy that quantitative changes lead to qualitative changes is not materialized in the study of solving mathematical physics equations. Instead, the study becomes a denial of changes because of the need to maintain the initial-value stability. That is to say, the signature achievement of modern science (or the magic construction of mathematical physics equations) merely stands for an external expansion of the given initial values without fundamentally touching upon problems of changes.

2. The problem of numerically computing decimal effective values:

If we take $\Delta t = 10s$ (second), $k = -0.1 \ m^{-1}$, which is taken to be the realistic precision of effective values with one decimal digit place, and take the initial value $u_0 = 0.1 \ m/s$, then we have

$$u_1 = 0.1 - 0.1 \cdot (0.1)^2 \cdot 0.1 = 0.0999$$

$$u_2 = 0.0999 - (0.0999)^2 \cdot 0.01 \approx 0.0999$$

$$u_3 \approx 0.0999 \approx 0.1$$

$$\cdots$$

Evidently, small effective values reveal the fact that nonlinear quantitative computations are only about minor changes around the given initial values. Or in other words, the quantitative computations of nonlinear equations that are made up of small effective values can lead to far smaller values than the effective value range so that the nonlinear terms become useless and contribute nothing to the computations. However, these noneffective values have been graphically shown using geometric figures leading to the concept of attractors of the chaos theory. Or in other words, the greatest achievement of the chaos theory is that it has shown vividly that indeterminacy is really from the quantitative computational inaccuracy of mathematical equations. As a matter of fact, our work has shown that even out of linear equations, similar results can be obtained.

What is more important is that careful readers can see that when dealing with the quantitative computations of small effective values, in form one can handle nonlinear equations by using the method of linearization. However, in this case, one has to require that the initial values be small too and the consequent outcomes are almost entire error values (beyond the range of effective values). The real significance of such computations is not about the so-called attractors; instead, it touches on the problem of feasibility of mathematical linear modeling using simplified equations. That is, at the time when nonlinear equations are linearized, one has to require that the initial values be so small that they are close to the level of error values; otherwise, the nonlinear equations cannot be linearized. So, at the time when nonlinear equations are linearized, in essence one is no longer in need of any of the equations in his study. Therefore, we have been constantly pointing out that nonlinear equations cannot be linearized on the basis of both theoretical explanations and realistic computations. Some individuals do not like and/or are unwilling to accept this conclusion because it implies that the many "great" achievements of the 20th century produced out of linearizing nonlinear equations would potentially become imaginary, beautiful illusions. There is no denial of the fact that all equations, including those linear ones found in modern science, are almost all from the linearization of nonlinear equations. Based on our discussions, we believe that if chaos theory is ever recorded in the book of scientific history, it will be seen from the angle of helping us to understand quantitative determinacy and indeterminacy.

The explanation for why materials or events are seen as objective reality should not be given only on the basis of the theory of particles that are needed for applying quantities and the passive pushes of inertial systems. Also, we should notice that behind such commonly accepted requirements in modern science as stability, smoothness, evenness, continuity, and so forth the essence is that at the same time of employing quantitative analysis methods, quantitative changes are not allowed. Even so, the quantitative analysis system has to admit the existence of such phenomena as irregularity, singularity, folding, sharp turns, bifurcation, saddle points, and so forth. All these phenomena have become the weak links in modern science.

Especially, quantities suffer from their own problem of infinity, which quantities can neither address satisfactorily nor know any method to deal with it successfully. To this end, our studies indicate that when studying and employing quantities, one has to face this problem of infinity instead of trying to avoid it. In the following, we will systematically illustrate the incompleteness of quantities and several urgent problems that have to be resolved by using fundamental characteristics of quantities.

3.1 Incompleteness of Quantities

At the eve of the birth of modern science, human conceptual thinking was at a fork along scientific history. Even after grouping the existent thoughts and beliefs of the time, there still existed at least two different ideas of development and epistemological understandings. The schools of the Middle Ages, following Aristotle's teaching, believed that the task of science was to explain why things happen by uncovering the root reason underlying changes in events and causes for changes in materials. So, the study of physics was positioned at the investigations of changes in materials and processes of change. They pointed out that no matter what values mathematical formulas had, they could not explain material *whys* other than describing the underlying matter. They also deepened their studies to "qualitative" levels. On the other hand, schools with Galileo as their representative looked for laws (in essence, formal "quantitative axioms") governing changes in materials, leading to the introduction of physical quantities without asking necessary whys. What was opposite of the Aristotle schools is that Galileo schools did not pursue whys. That is why in his "dialogues between two new sciences," Galileo openly claimed that the cause of the acceleration experienced by falling objects is not a necessary part of investigation. He selected a group of measurable concepts so that their measures (quantities) could be connected using formulas. These concepts included distance, time, speed, acceleration, force, mass, and the like. The essence of his work is to lead physics to the direction of study of physical quantities and parametric dimensions. So, it can be said that after clearing the path from pursuing whys, Galileo schools turned on the green light for Newton's *Mathematical Principles of Natural Philosophy* and determined the direction of development for Western science. That is, schools of Galileo's "quantitative axioms" guided Newton to establish the system of modern science; and so, Galileo was the giant beneath Newton's feet.

Of course, as an elementary stage of scientific development, modern science, established on the basis of quantitative analysis, has indeed found its applications in the design and manufacture of durable goods. However, because of these successes, the functions of quantities are also exaggerated to such a degree that the Pythagorean school's belief that (rational) numbers are the origin of all things becomes the current belief that quantities are the laws of all things, leading the popular belief that quantitative comparability is the sole standard of scientificality. This end shows the tendency of moving away from the realistic objectivity. There is no doubt that because of the pursuit of quantitative analysis after "quantitative axioms," the desire of seeking whys in the study of physics has been lost so that physics has entered the inertial systems on the basis of continuity. On continuity calculus was established, while the assumption of continuity becomes an operational obstacle of calculus. It is because if the material world is continuous, between any two points, there must be infinitely many continuous points making calculus never being able to reach from one point to the other. Historically, that was the second crisis in the foundations of mathematics. However, to this end the community of mathematicians did not really provide an essential explanation (Lin, 2008). In practical applications, calculus has never followed the required continuity. Even so, calculus still cannot deal with discontinuous changes in materials and the irregularity appearing in changes. The root cause for this incapability of calculus is that the differentiation and integration operations in calculus are governed by and played out according to the rules of invariancy of the original functions. That is why in practical applications, calculus does not follow the requirement of continuity. Instead, it "jumps" over the infinitely many continuous points by using invariancy, which constitutes the theoretical foundation for why difference equations can be used in place of differential equations. So, the practical value of calculus is its disconnection with paths so that material invariancy is maintained. Because of this end, the methods of calculus can serve inertial systems and become the first push of the slaving effects.

What should be said is that mathematics is not limited to the part of quantities only. It also contains the part of figures. For example, Euclidean geometry was established before the part of quantitative mathematics. Due to the later belief that quantities were the essence in the development of mathematics, figures become a concept of shapes and determinacy. That is why in analytic geometry, "figures" are still written in the form of quantities so that shapes are transformed into numbers while forgetting that Galileo's "quantitative axioms" were about forms without asking whys. That is why we see the current popular belief that quantitative comparability is the sole standard for scientificality without the corresponding introduction of figure comparability as a measurement of scientificality, leaving behind a serious problem in the structure of knowledge. Especially, although the writing format of Newton's *Mathematical Principles of Natural Philosophy* looked after that of Euclid's *Elements of Geometry* (even though Newton openly claimed that Euclid's geometry did not really help him in his formulation of his own work), Newton wanted to emphasize the fact that the quantitative analysis system is the theoretical foundation of classical mechanics. However, Newton and his followers did not notice such a fact that the classical mechanics of inertial systems and the consequent theories and methods, such as mathematical physics equations, were established exactly inside Euclidean spaces, leaving behind the problem of ill-posedness of unbounded quantities. So, if one calmly walks through classical mechanics, started from Galileo and completed by Newton, it will not be hard for him to see the following:

1. From the fact that quantities are postevent formal measurements, it follows that the post-event attribute of quantities implies that quantities are not a concept at the most basic level and cannot point to answers to *whys*. Evidently, without things, there would not be any number; without changes in things, quantities would not be changing, either. That is, quantities cannot foretell the future, or quantities cannot be employed to predict. And, it also clearly indicates that the formality of quantities is about avoiding the substantiality of events, which is embodied in the fact that quantities cannot address the states and attributes of events. For instance, both the turning hand and the covering hand — in the statement that a turning hand stands for clouds, while a covering hand represents rain — stand for quantity "1"

without clearly spelling out the state of the hand; and of course, this quantity "1" cannot tell people the attributes and functionalities of the event. Also, we are led to the problem of epistemology: The philosophical law that quantitative changes lead to qualitative changes is not materialistic.

As a matter of fact, Galileo's "quantitative axioms" have already clearly expounded the formality of quantities. It was only after Newton that the formality of quantities has been promoted to laws. When pressed for what causes universal gravitation, Newton used the formality of axioms to avoid a direct answer. As for Einstein, he seemed to have recognized the formality of quantities. However, he realized this too late in his life. That was why at an old age, he believed that it was very possible that physics was not established on the concept of fields, that is, it was not on entities of continuity; if it were so, all his work … including the theory of gravitation … and all other modern physics would be all gone.

2. Modern science used to limit the concept of evolution to the time scales of astronomy and the changes of geological eras, which indirectly declares that modern science does not study evolutions. However, such evolutionary changes as that of birth-age-sickness-death, the flashing appearance and disappearance of shooting stars, constantly changing weathers, and so forth are commonly seen phenomena of change in our lives. When compared with changes of the astronomical scales or geological measures, these commonly seen evolutions are of tiny scales. So, the only existing invariancy of the world is change. This fact has to force us to ask why people did not recognize that modern science only investigates formal invariancy of materials or events. As a matter of fact, H. Bergson (1963) already pointed out that all has been clearly spelled out in classical mechanics: Change is nothing but a denial of evolution, with time being only a parameter that is not affected by the change it describes; the picture of this stable world (a world free of evolutionary processes) is still the product of theoretical physics; the dynamic system, established by Newton and his followers, such as Laplace, Lagrange, Hamilton, and others, can answer any question, since all unanswerable questions have been eliminated as quasi-problems. Prigogine (1980), the founder of the 20th century Brussels school of learning, also believed that as for the most founders of classical science, including Einstein, science is nothing but a test; it wants to go over the realistic world and reach an extremely feasible, Spinoza world where no time exists; its essence is to pursue an eternal heaven.

From our discussions, one can see that it is very difficult for Westerners to see this end. At the very least, they recognize that formality is not the same as realistic existence, even though they did not successfully provide an operational method to handle the situation. No doubt, in a world without time, no one can enter the science of evolution. So, we see the need to systematically describe the incompleteness of quantities and provide the following general discussion.

3.1.1 Indeterminacy of Quantities Themselves

As a problem of pure quantities, if the state of shape or the structural ratio of events is not applied, irrational numbers and nonterminating repeating decimals are indeterminate numbers. For example, $\sqrt{2}$ and 7/11 are quantities to which accurate quantitative comparability cannot be given. The realistic comparability comes exactly from the structure or state of shapes. No doubt, we are familiar with the fact that the decimal representation of number π cannot be produced exactly. If we apply a sequence analysis using the binary number system, the random characteristics of this number can be seen readily. However, even if we were a little child, we still could not simply treat number π as a random indeterminacy, since the ratio of the circumference and the diameter of a circle is quite deterministic. And, similarly, $\sqrt{2}$ can be determined by using the hypotenuse of a right triangle. As for the repeating decimal number 7/11 = 0.6363..., although it is named as a rational number, it is also a number that is impossible to finish writing in the form of a decimal number. Besides, since quantities depend on the axes in Euclidean spaces, the concept of ill-posedness of mathematics has to be introduced due to the unboundedness of quantities. That is why since the ancient time of China, *wu, liang wu qiong* has clearly stated that there are problems quantities just cannot help to resolve. That also explains why later Chinese were not easily falling into the trap of quantities. Over 2,000 years ago, famous quantity calculator Zhan Yin warned us that *shu you suo bu dai* (not all things can quantities address). Evidently, in terms of quantitative comparability, specific, operationally applicable structural ratios not only possess the determinate rigor but also reveal the secrets of irrational numbers and nonterminating and repeating numbers. What is important is that the structurality of shapes can help to transform quantitative indeterminacy into structural determinacy. That is, quantities suffer from the weakness of converting problems of determinacy into problems of indeterminacy. Especially, the indeterminacy of irrational numbers is really from the determinacy of figurative comparability.

When people substitute events with quantities and talk about the probabilistic universe, they should not forget that quantities themselves inherently contain indeterminacy, and operations of quantities can lead to indeterminacy and complexity (see Chapter 2 for more details).

3.1.2 Artificiality of Quantities

What needs to be clear first is that physics does not investigate nothing, as what Lao Tzu said, *dao zhi wei wu* (Tao is about objects) instead of *dao zhi wei shu* (Tao is about quantities). The quantity 0 in mathematics is an artifact. It is an invention instead of a discovery of man. According to the scientific literature, the credit of inventing 0 has been given to Babylonians. However, colleagues from India we met in 2000 claimed that it was Indians who invented 0. Since we could not locate any relevant references, we finally consulted with some well-learned scholars. What we were informed is that the Buddhist sutra proves the Indian invention of 0, including that the Arabic numerals, as so called today, were also Indian inventions, which later became Arabic invention due to plunders of wars. In the seventh century, they were passed on to Europe and became the foundation of their quantitative culture.

There is no doubt that even though physical zero does not exist, it becomes the "king" of all numbers, because it helps to form the quantitative carrying system. Without the quantity 0, the entire mathematics system would become paralyzed.

In the operations of quantities, it is assumed that when quantity 0 is multiplied to any chosen number, the result is 0. Or, it can be said that quantity 0 can make all numbers disappear, which leads to a formal logical contradiction in mathematics. That is, any two different quantities become the quantity 0 when they meet with (are multiplied by) 0 so that the magnitudes of quantities cannot really be distinguished, or called the formal logical contradiction of identity. For instance, $1 \neq 2$. However, when multiplied by quantity 0, both numbers 1 and 2 become 0. Now, following the formal logic of mathematical operations, one can produce $1 = 2$. Similarly, the quantity infinity, invented in mathematics, comes from dividing a nonzero number by quantity 0. The consequence is that any chosen number ($\neq 0$) becomes the quantity infinity. Once again, all quantities can no longer be distinguished from each other. So, the rigor of the quantitative analysis system violates the basic rules of calculus of formal logic. What is more inexplicable is that theoretical physics, which is made up of physical quantities and parametric dimensions, shows its scientificality by using quantities to replace events. So, modern science has to suffer from the problem of making all events disappear by following the formal logical calculus of quantities, when the physical quantities, used to substitute for the events of concern, are multiplied by quantity 0. That is why we concluded that events do not follow the formal logical calculus of quantities. In other words, formality cannot be used as a substitute for objective existence. What deserves more of our thinking is that since quantity 0 is not an objective existence, quantitative infinity must stand for something nonexisting too. Then, how can such a nonexistence be described as ill-posed? In modern science, we did not see any formal explanation about the definite meaning of quantitative infinity or any denial about the existence of quantitative infinity. In recent scientific literature and many books, we have often seen such phrases as "the infinity of universal time and space"; Einstein's universe is originated from an infinitely large amount of energy and infinitely small volumes. However, quantitative infinity is introduced using quantitative formal logic on the basis of the physically nonexistent quantity 0. No doubt, the purpose for modern science to employ quantitative analysis is not for the investigation of nonexistences.

In the mathematical operations involving the quantitative infinity, quantities are required to be bounded in the name of well-posedness of mathematical physics equations. The purpose for imposing this requirement is that by requiring quantities to be bounded, one does not have to answer what the quantitative infinity is without thinking about violating the rules of formal logic, consequently producing invariant translations or clones of the stability of the initial values. This end makes modern science reveal its bottom line. That is, after over 300 years of development, all modern science can tell us is still the eternal heaven that expands the original initial values. So, to study the corresponding problems of nonlinear mathematics, which should have been successfully resolved over 200 years ago, one still has to first look at how to understand what the quantitative infinity is over 200 years later.

As an extended explanation, it is because of its infinite magnitude that the quantitative infinity steps out of Euclidean spaces; and at the same time when it walks out of the quantitative system, it enters the structure of curvature spaces. Or in other words, for the quantitative analysis system, because of the introduction of the quantitative infinity, the completeness of this system is destroyed, which is unexpected by all the founders of modern mathematics and physics since the time of Galileo.

Quantitative infinities of Euclidean spaces correspond to transitions in curvature spaces, making quantities face problems of nonparticle structures. This end explains why quantitative analysis meets with difficulties in resolving nonlinearities, since nonlinearity corresponds to transitions in curvature spaces. That is why nonlinear problems cannot be linearized. If a nonlinear problem is linearized, it is equivalent to modify transitions in higher dimensional spaces to local phenomena in lower dimensional spaces, or artificially force transitions into local plane or line problems. Or speaking differently, nonlinearity stands for problems of the Riemann spaces; linearization is treating problems of Riemann spaces as problems of Euclidean spaces. So, the substance of nonlinear problems touches at least upon the spinning direction and structural characteristics of curvature spaces. And, nonlinearity cannot be understood as an approximation problem of using linear quantities. Quantities' incapability of resolving problems of transitions means that quantities cannot address problems of structures, because mere changes in directions are already about structures. Also, in rotational movements, because of the existence of duality in spinning directions, conflicting rotation directions surely cause subeddies to appear. That is why in the physical world, irregular events must exist. That is the realistic universe in which mankind exists and where the principles of nature are from.

3.1.3 The Problem of Computational Inaccuracy of Quasi-Equal Quantities

It might be due to differences in living environments and cultural origins that people learn and explore the world in different ways. In China, there have always been such beliefs that when looking at an object from my angle, the object is constrained by me; when an object is measured by using another object, one cannot really tell the difference between the objects. If we illustrate this statement using modern language, then it means that when applying a human sensing organ to measure an object, then the object has to be limited by the ability of the human organ; if using another object to make the measurement, then these two objects will have to interfere with each other so that the measurement cannot be accurate.

This inaccuracy of mutual interference is an inevitable or objective reality, which is the Chinese expression of Heisenberg's principle of measurement inaccuracy, and substantially answers at the height of epistemology the reason for the measurement inaccuracy. In terms of quantities, there is also the corresponding problem of computational inaccuracy. For example, there are two quantities x and y with quasi-equal magnitudes. Then, we must have $x - y \cong 0$. However, approximately equal to 0 is not the same as being 0. So, these quantities are called quasi-equal. Evidently, these quasi-equal quantities could come from either the quantitative measurement inaccuracy of these quantities themselves or the ineffective difference between the effective values of these quantities. This difference could be the result of the realistic values or the involved error values. Evidently, when the realistic values and the error values are the same, computational inaccuracy is produced. To this end, the situation is called large quantities with infinitesimal differences. In practice, it is impossible to find two absolutely identical things so that so-called realistic quantification does not exist. However, such a common knowledge, in the eyes of great contemporary physicists, becomes the principle of indeterminacy (S. Hawking generalized Heisenberg's principle of measurement inaccuracy to the principle of indeterminacy). This end no doubt reveals the thinking logic of Plato's static figures in contemporary science. Science has entered our modern time, while the thinking behind science is still stopped at an unrealistic way of reasoning of the far distant past. That of course explains why conceptual chaos has to be produced in modern science. Besides, Lao Tzu of ancient China is already a familiar figure in the world today. His *xi, yi, wei bu ke zhi jie* is not the same as that of indeterminacy as studied by S. Hawking and others but an inevitability of objective events. The publication quantity of Lao Tzu's work is only next to the Holy Bible in the West. So, his teaching must have gained a large degree of acceptance.

No doubt, *xi, yi, wei bu ke zhi jie* already implicitly contains "large quantities with infinitesimal differences." For example, *xi* means those sounds human sensing organs cannot hear clearly; *yi*, those things human visual organs cannot see clearly; and *wei*, those objects human sense of touch cannot feel clearly. Although relative, they all stand for problems of objective reality. So, *xi, yi, wei bu ke zhi jie* is not only limited to quantities. Its meaning covers not only physics but also implications of philosophy.

Using today's language, almost all mathematical equations, including algebraic, differential/integral, and difference equations used in numerical computations, involve this kind of problem, leaving behind the deceivability of quantitative computations. As for some people's belief that equations are eternal, it does seem that they are not serious or careful about what they say. For example, the once-hot chaos theory of the 20th century is a farce of the scientific community caused by either ignoring the quantitative computations of large quantities with infinitesimal differences or the ignorance of the problem of quantitative computational inaccuracy. Applying previous knowledge of the past to guide the development of science is about seeking guidance in the overall direction instead of fitting indiscriminately what the past knowledge talks about in the format of fitting axioms. The fact that such a piece of common knowledge or elementary concepts caused major problems in the hands of scientists has to mean that it has something to do with people's desire to pursue the formality of quantitative axioms. It is just like the situation when Newton introduced the quantitative axiom named the law of universal gravitation without being able to answer what universal gravitation is or where universal gravitation is from. That is the same situation as in modern science, where no matter how many equations are employed, people still cannot answer *whys* or what laws are. For example, where are probabilistic waves from? Or why are there probabilistic waves? What is a gravitational wave? How can there be any unidirectional retrogressive waves on the plane? Who really had the luck to physically observe a true "solitary" wave?

As a writer, Shakespeare could clearly tell that merchants from Venice would never be able to cut off 1 pound of meat, while scientists have been playing with the computational meaningless difference of quasi-equal quantities. That is the reason we pointed out that no equation with equal quantities exists.

If we borrow Aristotle's claims that physics is the discipline that investigates changes in materials and processes of change, that formulas, no matter what values they have, cannot provide explanations other than descriptions, and that the fundamental task of science is to explain why things happen, while digging up the cause for changes in materials, then other than schemed numerical computations, there are still a lot of things scientists could do. And the first works scientists should consider doing are addressing what problems numerical schemes can really resolve. Or, the statements of Aristotle are a very meaningful topic of research. If mathematical formulas only had the functionality of description without any chance to provide answers to whys, then it would be a good time for us to seriously consider the purpose of schemed numerical computations.

3.1.4 Regularization and Large Probabilitization of Quantitative Variables

Calculus has been listed in history books as one of the three major scientific achievements of the 17th century. One of these achievements was the opinion that each fluid can be seen as a continuum of solids with low resistance. That was one of the reasons that continuity is a basic assumption of modern science. The second achievement is the establishment of conservation laws of energy. Here, the conservation of kinetic energy needs to satisfy not only the Noether theorem of fixed quantities of variations under the group of infinitesimal transformations and the differential equations of mathematical models, but also the assumption of continuity. The third achievement is the establishment of calculus on the assumption of continuity. Evidently, the core of these three major achievements of the 17th century is about continuity. As a matter of fact, applications of calculus have not strictly followed the requirement of continuity, leading to the second crisis in the foundations of mathematics.

The root reason for the crisis to appear is that if the requirement of continuity has to be satisfied, calculus cannot be employed. It is because there must be infinitely many points between any two distinct points so that one could never reach the second point if one started off from the first point no matter how fast he traveled, and that related computations could not be carried out. That is, in applications, calculus is employed with jumps instead of continuity. The reason calculus can be applied in practice with jumps is that an invariancy of the original function on the corresponding intervals is assumed. The essence of the corresponding difference equations is to apply jumps to accommodate the necessary computations; otherwise the entire calculation could not be carried out. For more detailed discussions, please consult Lin (2008a). So, the truth or purpose and the functionality of continuity are about invariancy, which forms the core of modern science. Now, the reader can understand why calculus needs to be unrelated to specific paths. This end explains what Einstein (1976) also worried about:

> I believe that it is very possible that physics is not established on the concept of fields; that is, it is not established on continuous entities. If it is so, then all my theories … including the theory of gravitation … even together with all other modern physics will be gone.

Now, it becomes clear why since the end of the last century we have been time and time again informed that the modern scientific system is about methods of estimation useful for the design and manufacture of durable goods instead of an epistemology, while pointing out that prediction science is not a branch of modern science, that the fundamental theory of Rossby long waves does not hold true, and that there is a need to walk out of the nonevolution scientific system.

Next, as a problem about practical applications, because of the smoothness condition of calculus or the requirement of stable series of the statistics of the stochastic system, the so-called quantitative analysis system of variables cannot truthfully deal with irregular information regarding variables. And when faced with discontinuous irregularities, just as what Einstein worried about, the modern scientific system runs into the risk of routing all along the line. Even if we do not discuss the debate between determinacy and indeterminacy now (see Chapter 2 for more details), the statistical methods of the concept of randomness can only be applied to situations of large probabilitization. The consequent effect is roughly the same as that of applying the Newtonian system of methods based on regularities.

What is important is that since the time when modern science was born, irregular or small-probability events have not been really touched upon. All problems unsolvable in the Newtonian system are labeled as quasi-problems and ignored from further consideration. The stochastic system of statistics should be good at handling small-probability events. However, within that system, the problem of what small-probability events are cannot be answered; and problems involving instable small-probability events are excluded from serious consideration. In short, the fate of irregular or small-probability events in the system of modern science is being disregarded. To us, because of the major torrential rains that occurred in the North China Plain in 1963, we were assigned the task of finding out whether or not the same scale disaster would possibly occur in the second largest basin along the Songhua River in northeastern China. (If it were so, the then Jilin City would be entirely under floodwater). Since then, our fate has been intertwined with the "miserable misfortune" of irregular or small-probability events. In the following decades, because of the discovery of ultralow temperatures existing at the top of the troposphere right before disastrous weather conditions, our scientific points of view have been forever changed. We consequently recognize that events are not equal to quantities, events do not comply with the formal logic of quantities, and randomness in events does not exist. That is the reasoning with which we published our book *Entering the Era of Irregularity* in 2002.

Corresponding to our works, Bergson (1963), Koyré (1968), and others also applied various satirical languages to point out the nonevolutionality of modern science. However, the scientific community did not become consequently awakened, except for S. Hawking and several others who severely criticized Newton's determinacy while walking along with the popular indeterminacy. And Einstein became an old diehard against randomness because of his statement that God does not play dice. So, from S. Hawking's "God plays dice" and other related statements, the concept of a probabilistic universe was established.

It might be possible that people's conceptual recognition have something to do with their way of analysis or their knowledge structures. However, what is important is that no matter what opinion, way of thinking, or theory it is, it should focus on the practical value of resolving problems and bringing forward new understandings about nature. So, the development of science is from answering calls of problems. Theories and empty talk that does not have any practical value of application often produces immeasurable economic and human intelligence losses. For example, even if the chance for an event to occur is seen as random, it is still not the same as saying that the event itself is random. What is important is that the epistemological point of view of random events did not really resolve the problem of random events. On the other hand, physical events are not random; they provide information of change before the change actually occurs. That is why irregular or small-probability information does not represent a quasi-problem; instead it possesses the practical value of applications in analyzing evolutionary transitional changes. Evidently, since irregular or small-probability information can help resolve problems, why do we still hold onto some of the traditional concepts that we know do not really work in practice? That is, we have traveled along a path that betrays some of the concepts of modern science. The core of our theory is that events are not random, "God does not play dice," materials and events do evolve, and multiple faces of the world are not the same as a probabilistic universe.

The reason the regularization and large probabilitization of modern science ignore irregular or small-probability events is that quantities and their calculus rules of formal logic cannot handle these events. Or in other words, regularization and large probabilitization come from quantification; and leaving behind irregular, small-probability events is also a helpless choice of quantification. That end leads to such an interesting epistemological problem that generality is extracted out of specifics, or special events are the mother of general events; without specifics, no generality can be formulated. However, modern science has pushed for disregarding the "mother" in order to "sell" the scientificality of regularization and large probabilitization.

If one goes deeper to investigate why modern science is embodied in regularizing or large-probabilitizing events, he will see that its core idea or purpose is to maintain the invariancy of the generality. That is why from Newton to Einstein, and their followers, they have pursued the heavenly eternity (Prigogine, 1980).

Evidently, if events are not random, then one has to research their causes and processes of change. For this purpose, irregular or small-probability events involve the evolutionary process of material rotations. Since there is duality in rotations, different rotational directions will have to produce the formation of subeddies. The occurrence of subeddies at the same time indicates the weakening of the "mother" eddies, which explain how the subeddies are created. No doubt, when compared with their "mother" eddies, the subeddies stand for irregular or small-probability events. And from the angle of the original field of rotation, these events also seem to be orderless and nonperiodic. However, if one stands at the level of the sub-eddies, the events have patterns and follow certain rules of operation, which not only shows the causalities of the events but also the process of evolution. So, not all determinant problems can be simply determined in the form of quantities. To understand these problems that are not determinable by using quantities, one has to look deeper into the structures and properties of the underlying materials. That is like the situation of telling sugar and salt apart, where one cannot determine which is which simply by quantitative forms. It can also be said that the science of quantitative comparability simplifies the world into one tone, which is the central epistemological problem of modern science. As a matter of fact, if the world were indeed one-toned, there would be true orderlessness, because in such a one-toned world, everyone would look one-toned and identical so that no one could have his own identity and tell one person from another.

3.1.5 The Problem of Nonisomorphic Equal Quantities

In this section, we will look at the problem of equal–quantitative nonisomorphism, because in the physical world there are materials of different attributes that involve the same quantities. For example, the ordering directions of fingers and toes of human left and right hands and feet in their mirror images are opposite of each other, possessing differences in terms of order structures. L-lactic acid and D-lactic acid are different only in terms of their optical rotations; and the difference in their optical rotations causes their difference in attributes. In chemistry, methyl ether and ethyl alcohol have completely different properties due to their order differences in their structures. So, equal quantities can mean totally different attributes.

Einstein once pointed out that modern physics originated in the invention of inertial systems and the concept of forces. Even if we do not talk about the problems existing along with inertial systems and the concept of forces, the concept of mass, defined as a physical quantity of materials, has separated the study of physics from the principles of materials. So, our concept of nonisomorphic equal quantities in essence questions some of the fundamental concepts of physics. Or in other words, the "quality" of material attributes cannot be identified with the numerical magnitudes of physical quantities.

Evidently, understanding materials only from the angle of quantities cannot naturally resolve the problem of equal–quantitative nonisomorphism, while the order of causality is reversed. That is why since modern science was born, it has not been able to address *whys*. From Galileo's mathematical formulas of the form of axioms to Einstein's eternity of equations, all avoided the exploration of causes and essences behind physical problems. So, the sole standard of scientificality of quantitative comparability in essence is a mathematical scientific standard without involving attributes of materials and cannot be a standard of physics. Although Aristotle's fundamental understanding of physics is different of that of Galileo, his existence of formal forms does not have much fundamental difference from Galileo's mathematical formulas of axiomatic forms. It can also be said that they are produced out of almost the same origin in terms of the quantitative culture, which led to the later substitution of quantities for events.

As another original flow of human knowledge, Chinese people on the other hand, emphasized nonisomorphic attributes of materials instead of the relevant quantities. Although the *Book of Changes* does not contain signs of calculus of formal logic, it does have a way of reasoning that reflects the existence of quantities in transformations of events without replacing events by quantities. The principles of design of all of the digitization methods of state transformation and irregular information phase-space transformation, and so forth, that we invented and successfully applied, initially for meeting the pressing needs of solving urgent real-life problems were later found to agree with the ideas of structural transformations presented in the *Book of Changes*. And, thousands of tests have shown the practical validity of our methods, while pointing to the facts that using quantities to make predictions is a major mistake made in modern science and that direct structural transformations of events, instead of mere human experiences, follow a certain set of rules and can be passed on to new generations so that they can be applied to save people and to reduce economic losses from natural disasters.

Therefore, it is found that structural comparability possesses more scientific value of application than quantitative comparability. Besides, in quantitative comparability, one has to face the problem of quasi-equal quantities so that even if models in the form of mathematical equations can be established, he may still find himself in the difficult situation of being unable to find the quantitative solution, just as in Shakespeare's play when merchants in Venice can never cut off a pound of meat exactly.

What deserves our attention is that the concept of equal–quantitative nonisomorphism challenges the quantitative equations of our modern time. It is because equations involving equal quantities have to produce results of equal quantities. However, quantitative results cannot tell people what things are concerned with. This end reveals that the well-adopted form of quantitative equations cannot answer calls of specific events. In other words, at the time when people are celebrating their miracle invention of mathematical physics equations, if these equations are seen as a "machine," then people suddenly realize that this machine can only recognize quantities without being able to tell what is behind the quantities. Since the time when classical mechanics was born, the methods of nondimensionalization or unification of dimensions have been applied in order to establish the quantity-transfer machine; otherwise, the theory and methods of the theory of mechanics cannot find the ground to show their power. In the next section, we will look at problems the method of nondimensionalization suffers from.

3.1.6 *The Problem of Nondimensionalization*

Nondimensionalization can also be called unification of dimensions. Evidently, without nondimensionalization, the mechanics, developed since the time of Newton, will have trouble operating. That is why some scholars treat nondimensionalization as one major invention of modern science and a core reason why the development of modern science has been pushed forward. In particular, in any chosen system of dynamic equations, even if each of the equations stands for a relationship of some physical quantities, these equations are still faced with the problem of not having the same dimension. So, all the dimensions have to be unified or nondimensionalized; otherwise, the entire system of equations could only be a decoration without any chance to be operated upon. Consequently, the solution of the system of physical quantity equations cannot tell people specifically what things are. For example, since they have different dimensions, apples and pears cannot be added directly. When they are unified at the dimension of fruits, apples and pears can indeed be added together. However, the result of addition with the dimension of fruit cannot really tell what is produced — apples or pears.

When still in college, we consulted with various professors about how to write scientific research papers. Here is what we learned. Step 1: Use the method of nondimensionalization to establish and simplify equations. Step 2: If the equations obtained contain terms of different magnitudes, ignore the terms that are two or more magnitude levels smaller than other terms. Step 3: Analyze the resulting equations obtained from ignoring infinitesimal terms to see if they are one of the typical types of equations well studied in mathematical physics equations. Step 3.1: If a match is found, directly apply the procedure of solutions of the typical equation, and discuss the solution in terms of the original problem. Then the planned research paper is finished and can potentially be judged as a pretty good paper. Step 3.2: If the established equations cannot be solved using one of the methods well established for solving the typical equations, then discussions of the physical meanings of each term of the equations can be used as the final research paper. Step 3.3: If the established mathematical model is ill-posed, then assume that the model has a solution of certain form. After substituting the assumed solution into the model, the analysis of the physical meaning of the results can also be counted as a research paper. So, whether or not the physical meanings truly agree with the reality, no one seems to bother and check. For example, the well-known Rossby long-wave theory in meteorological science was produced in this fashion.

When we questioned whether or not such sequence of activities could be considered scientific research, the answer we obtained was, "Of course!" However, when we pressed for "what unidirectional, retrogressive waves look like" or "how unidirectional waves can exist on a plane," the only reply we ever received was, "This internationally known theory you also dare to question? How reckless are you?" Evidently, scientific achievements are about how to explain epistemological whys. Years later, we finally published our work, proving the invalidity of the Rossby wave theory. This little story also helps to explain why we have been more interested in solving practical problems. This effort, along with many other scientific endeavors, led us to the observation that scientific research is about how to explain the bafflement of the existing problems; however, scientific research inexplicably explains some nonexisting bafflements, making people further baffled.

Evidently, formal analysis is not the same as specifics of events; and quantitative accuracy is not the same as the correctness of events. However, as of this writing, people are still using quantitative interpolations for refinements of events. Quantities appear postfact so that they become after-event, 20/20 hindsight. That is the true essence of quantities, just like what Galileo said in his form of quantitative axioms that quantities cannot answer whys, and that Newton's law of universal gravitation does not tell why there is universal gravitation or what universal gravitation is.

However, the reason why science is so named is that it pursues whys, leading to deepened epistemological understandings. Evidently, the achievement of modern science is to make quantification applied everywhere by using nondimensionalization. However, doing so has also made science forever stay outside the doors of scientific whys.

The development of science is originated in the calls of problems. In our renewed study of figures, we find that the concept of figures does not stand for the problem of the so-called determinacy; instead, it involves the exploration of whys behind problems in its deterministicity, and that figurative comparability shows processes of events so that evolutions of materials are brought in. It can also be said that figurative comparability embodies the problems of epistemology.

No doubt, scientific technology is about its value of use, while in the values of applications, there are also theories of epistemology. If we walk through the development of the history of modern science, it can be seen that the fundamental concepts and elementary beliefs were not very good at the start. In particular, during the 300-plus years after Newton, people have been fond of the development of quantitative mathematics while forgetting about Galileo's advice — quantitative axioms cannot address whys. Many quantitative laws in the form of quantitative axioms were initially established to address whys. However, instead of answering the whys, they in other areas of science have led to inexplicable theories.

Maybe some of our readers feel that we have spent too much in the way of efforts and time on picking out weaknesses and deficiencies of the quantitative analysis system, since this system indeed has its advantages and usefulness. Similar to currencies, quantities can be applied commonly and show the ability of deduction of formal logic. No doubt, they are convenient tools for people to use. The reason for us to emphasize the problems of quantities is mainly for the purpose of utilizing these tools appropriately. That is why we have the following:

1. One must pay attention to the fact that, like currencies, the value of quantities is at the supporting materials (goods). Without the supporting goods (materials), the currencies would be like useless papers, while the corresponding quantities would be even more useless than the useless papers.
2. Currencies possess deceivability, and so do quantities. The direct deceivability is reflected in that they can compute nonexistent entities more accurately than the actually existent objects, creating false information and leading to erroneous theories and methods.

Everyone knows that round things can roll easily, while square objects tend to be more static. If we expand this metaphor to physics, it means that "round" is from the rotational changes of the "roundness" of curvature spaces, and "square" is the static invariancy of the "squareness" of Euclidean spaces. Quantities are the staticness resting on the "squareness" of Euclidean spaces. That is why as of today, modern science still stays at the stage of the static figures of the Plato kind. However, although currencies have the same kind of commonality as that of quantities, they are still dependent on goods so that they can walk out of Euclidean spaces. That is why currencies move due to the "roundness" of circulations, leading to the creation of the modern industry of finance. On the other hand, the motion of the "roundness" of excessive circulations of money will have to lead to "imaginary trades" like the imaginary castle in the air of quantities. That is why there are economic crises or bubble economies. Correspondingly, some scientific theories are created out of "imaginary trades" of quantities that do or will walk toward "bubble science," such as chaos theory, prediction profession, and so forth. These areas of learning have already been trapped in "bubble science." The reason these areas have not shown any potential "burst" is that area monopolies, for various reasons, have helped to have them well packaged.

After operating for over 300 years, why could not more progress be made in certain areas of learning? That is why we see the need to provide an appropriate positioning for the system of modern science so that we can be clear about those problems modern science can resolve and those it cannot help much. This work will surely help future generations of scientific workers to direct their effort and time in more useful paths that will be beneficial to mankind. Or in other words, in the development of science, one should also be cautious and recognizant about potential scientific crises and the possible appearance of bubble sciences. Just analogous to the situation that people do not like financial oligarchs to manipulate the economy, scientists will not like "scientific oligarchs" to fool around with "bubble sciences."

Although the substance of scientific technology is about practical uses, epistemological problems often determine the underlying concepts, which in turn involve beliefs. So, changing concepts is a very difficult task to accomplish. In particular, the quantitative analysis system of modern science has been in operation for over 300 years; and due to the human nature that preconceived ideas keep a strong hold, quantities have basically become the only method and way of thinking. The consequent scientific standard of quantitative comparability is no longer limited to natural sciences and has been permeated into all corners of social sciences and philosophy. That is why the philosophical law that quantitative changes bring forward qualitative changes has become a well-accepted axiom. From Galileo's proposition of axioms to Newton's *Mathematical Principles of Natural Philosophy*, although no one openly promoted that quantities are the origin of all things, the essence has been to push that quantities are the laws of all things. No doubt, even among the great minds of scientific history, those masters of science have also been back and forth with epistemological fundamentals. Although Galileo and Aristotle held opposite positions in terms of elementary concepts and beliefs, they at least recognized that mathematical formulas could not provide answers to whys. However, Newton and others of later times even placed mathematics above philosophy, including Einstein, who even believed the eternity of equations. That belief helped to establish and widely apply mathematical physics equations in the natural sciences. Even more than that, the latter situation has gone to the extreme that when a problem is unsolvable theoretically no matter how urgent it is and how it exists objectively, the problem is labeled as a quasi-problem and ignored.

Because we were pressed for solving practical, desperate problems, we had to ponder the problems facing the theories and methods of modern science in order to resolve our practical and pressing problems. In our exploration of whys, our work touched upon such problems as what quantities are, where quantities are from, within what ranges quantities can be employed correctly, what difficulties quantities face, and so forth. So, people have to be aware of the existence of bubble sciences and consequent damages.

We have summarized problems of quantities from six different angles with emphasis placed on clarifying some misunderstandings and misguiding concepts, providing some elementary explanations and discussions on the most urgent problems. For example, our discovery that quantities are postfact formal measurements was initially published due to our helplessness at the historical time. However, its essence was to deliver our opinions on the position of modern science. The key message delivered in this discovery is that quantities, as postfact formal measurements, can neither provide answers to whys nor provide predictions of the future. However, presently, people are still throwing in huge amounts of capital and manpower in quantitative analysis hoping to capture the future from the postevent quantities. Also, through transformations between figures and quantities, in this section, we will show that the unboundedness of quantities that has been around for several hundred years is not simply a meaningless ill-posed problem; instead, it represents an important problem of evolutionary transitionalities that has been missed by modern science.

3.2 Folding and Sharp Turning in Mathematical Models

Since we propose to reconsider the problem of shapes and figures, then there is a need for us to look at the problem of what quantitative infinity is. This quantitative entity has been avoided by modern science since the time when it was initially born. In the present system of quantitative analysis, quantitative infinity has been listed as a singularity without providing any clear and concrete meaning to it. That is, we would like to know what the ill-posed, mathematically meaningless problem, as so called among the quantitatively comparable concepts, really is.

3.2.1 Indeterminacy in Mathematical Descriptions of Linear Reciprocating Points and Practical Significance of Corresponding Figurative Comparability

By introducing the concept of mappings, all collinear reciprocating points can be seen as implicit transformations of a moving point on a curve in curvature spaces. That is, if the moving point on the curve has coordinates (x_1, x_2) and the corresponding point on the plane is (y_1, y_2), then the transformation can be given as follows:

$$\begin{cases} y_1 = f_1(x_1, x_2) \\ y_2 = f_2(x_1, x_2) \end{cases} \tag{3.1}$$

That is how we have the ideal relationship between the sphere and plane in Figure 3.2, where the point p' on the plane corresponds to the point p on the surface of the sphere, which is mapped to singular reciprocating points of the plane. This corresponding relationship can be written as

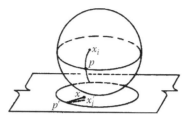

Figure 3.2   Implicit transformation between the plane and a surface.

$$\begin{cases} y_1 = x_i^2 \\ y_2' = x_i' \end{cases} \tag{3.2}$$

No doubt, if a point x_i on the curve travels from the side north of the equator to the south side, then the corresponding point x_i' on the plane moves toward the equator, showing a reciprocating singular point on the equator, while the point x_i on the curve travels continuously from the northern hemisphere to the southern hemisphere. Evidently, from the angle of the special plane concepts, the linear observer standing at the location of the second equation in Equation (3.2) will see a quantitative infinity on the shadow plane of the higher-dimensional sphere. And, in the traditional sense of quantities, this observed quantitative infinity is meaninglessly ill-posed. However, in essence, this meaningless ill-posed quantitative infinity stands for a meaningful transition in the curvature space and is determinant. So, the indeterminate form, as obtained in the specialized observations and control of the Euclidean space, is not the end of our effort to understand nature. Instead, it represents a blind territory or erroneous area of the system of quantitative analysis, which conceals the essence of the nonlinearity of the original mathematical model (the first equation in Equation 3.2) and makes human epistemology trapped in perplexity so that as of today, people are still using quantities to name deer as "horse" and expanding the concept of quantitative infinity to relevant concepts of physics and epistemology. For example, in handbooks of philosophy, we can easily see such a statement as "the time and space of the universe are infinitely large," and calculus also uses the concept of infinitesimals as its foundation. So, the greatness of Lao Tzu is clearly shown in his teachings from over 2,000 years ago: *zhi ze wang, qu ze quan* (being straight will lead to distortion, and being curved will cover all) and *da yue shi, shi yue yuan, yuan yue fan* (large means gradually shrinking, gradually shrinking implies far away, and far away represents return).

The concept of infinity in essence is a product of the specialized observations and control of Euclidean spaces. The corresponding systems of quantities and calculus need the concept of "infinity" because they are also products of Euclidean spaces. If we walk out of quantities and look back at them from a higher-dimensional curvature space, we find that quantities are merely shapes of higher-dimensional spaces transformed into entities (called numbers or quantities) of lower-dimensional spaces. That is why the quantitative analysis of the Newtonian first push system is only a game of play of finite domains in Euclidean spaces. So, shifts in directions in the "rotations of being round" of higher-dimensional spaces cause the appearance of "infinities" in lower-dimensional Euclidean spaces, leading to a problem unsolvable by using the quantitative methods of Euclidean spaces. These "infinities" are not realistic nonexistence. That is why the problem of indeterminacy is also a product of space transformations, hinting that the problem of nonlinearity cannot be seen as a quantitative problem of lower-dimensional spaces. That is why we pointed out that mathematical nonlinearity stands for problems of structures, which brings us into curvature spaces.

3.2.2 Transformation of the Shapes and Quantities of Sharp Turning

On the plane, the semicubic parabolic mathematical model is a typical example for describing sharp turnings and can lead to the problem of singularity in mathematical derivatives. By introducing an implicit spatial transformation, we have the following mapping relationship:

$$\begin{cases} y_1 = x_1^2 + x_1\, x_3 \\ y_2' = x_2' \end{cases} \tag{3.3}$$

Evidently, the first equation in Equation (3.3) is nonlinear, while the second equation in Equation (3.3) is linear. The top part of Figure 3.3 shows a sharp turning, while the bottom graph depicts the projected line drawings on the plane. No doubt, point p_0 corresponds to point p_0' and is a sharp turning point, while $p_0'q_0'$ corresponding to p_0q_0 stands for the folding. In the top graph of Figure 3.3, when the point x_j travels through point p_0, the plane point x_j' travels to point p_0', becoming the quantitative singularity of the sharp turning. In other words, the ill-posed point in the Euclidean space stands exactly for the transition point in the curvature space. That is, transitions of curvature spaces meet a deficiency of quantitative ability to describe. Up to this point, it becomes very clear that the so-called problem of

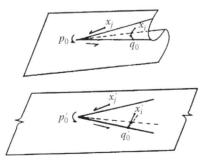

Figure 3.3   Sharp turning (top) and folding (bottom).

quantitative infinity is a reflection of the problem of quantitative incompleteness. And, quantitative infinity is neither a physical existence nor a problem of epistemology. So, the conclusion that the time and space of the universe are infinitely large is produced out of misguidance of quantities and is an indiscreet statement. As for the current popular effort to quantitatively solve nonlinear mathematical models, people are still trying to constrain quantitative instability to stability so that the requirements of well-posedness of linear equations can be extended. This, no doubt, represents a major mistake in terms of both epistemology and methodology.

Evidently, as a problem of mathematical "shapes," it should be a major part of mathematics. Even if it is considered as a branch of mathematics, it should still be considered prior to quantitative mathematics. The problem is that after Galileo and Newton, quantifications in the form of mathematical formulas and equations have gradually become the mainstream of thoughts, later extended into the studies of physics, and become the system of science supporting discussions of philosophy. Such quantifications have led to the law of philosophy that quantitative changes lead to qualitative change that has mistakenly reversed the underlying causal relationship, and such slogans as "quantitative comparability is the only standard for scientificality," causing confusion in many fundamental concepts. In particular, theoretical physics is not about principles of materials and instead is completely dominated by laws of quantities. That is why such a statement as "the time and space of the universe are infinitely large" can enter handbooks of philosophy with dignity.

To this end, pressed by meeting our need to resolve transitional changes of the kind that at extremes, matters will move in the opposite direction, we introduced the concept of transforming shapes of curvature spaces and solved successfully some of the problems unsolvable before by using quantities. Our work and successes indicate that the so-called quantitative indeterminacy can be converted back to physical determinacy by using shapes and figures. That not only resolves the epistemological problem of ill-posedness but also provides a method for practical operations.

What is interesting is that the problem of nonlinearity that scholars had chased during the later part of the 20th century is exactly the problem of transitional changes, as described by the sharp turning and folding of our nonlinear mathematical model. The then-popular effort of studying sharp turning and folding was to find the characteristics of nonlinearity of branch solutions or broken solutions, while our discussion emphasizes the virgin territories of quantitative analysis, involving the epistemological problems of what physical meaning quantitative unboundedness has, its potential applications, and what the essence of mathematical nonlinearity is.

3.3 Blown-Ups of Quadratic Nonlinear Models and Dynamic Spatial Transformations

The key to the implicit transformation between shapes and quantities, as discussed earlier, is that the quantitative infinity of Euclidean spaces is not an objective physical reality. On the basis of this epistemological understanding, let us look at the dynamics of quadratic mathematical nonlinearity so that a deeper understanding of quantitative infinity can be reached. And, from the following quadratic nonlinear mathematical model, the problem of what infinitesimals are physically comes into being.

The standard mathematical model of the quadratic form is given as follows:

$$\dot{u} = u^2 + pu + q \tag{3.4}$$

When $p^2 - 4q = 0$, we have

$$u = -\left(\frac{1}{t+c} + \frac{p}{2} \right) \tag{3.5}$$

Corresponding to $c > 0$ and $c < 0$, we have the hyperbolas I and III in Figure 3.4 that can be transformed to each other in a similar fashion as the spatial transformations of the hyperboloids of the tire in Figure 3.5. No doubt, this has very clearly illustrated that the "infinity" of the quantitative deductions of nonlinear models means exactly a problem of transitional changes in curvature spaces. It is because between the hyperbolas I and III, there are not only quantitative differences in their initial values but also discontinuities. These discontinuities we call blown-ups. So, even in terms of mathematical properties, nonlinear mathematical models no longer satisfy the linear existence theorem of initial values. This end also suggests that when combined with physical realities, many difficult problems of mathematics can be easily resolved. For example, the reason why the Navier–Stokes equation has been listed as one of the seven most difficult problems in mathematics is because

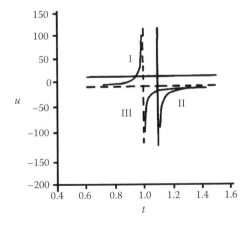

Figure 3.4 Quadratic form and hyperbola.

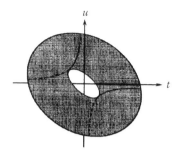

Figure 3.5 A hyperboloid and its image.

its nonlinearity makes it difficult to produce its analytic solutions by sticking to the requirements of well-posedness of mathematical physics equations. What needs to be noted is that nonlinear equations do not satisfy the existence theorem of initial values of the linear system. So, the proposition of satisfying the well-posedness of linear equations itself is already a situation violating the quantitative formal logic. As a matter of fact, by pointing out the facts that there exist quantitative instabilities in the deterministic solution problem of nonlinear equations, and that by using the nonlinear form of nonsmall quantities, quantitative unboundedness can quickly result, the mystery of quantitative nonlinear equations has been resolved. Since nonlinear equations implicitly contain transitional changes of curvature spaces, the essence of the problem of nonlinearity becomes very clear.

Figure 3.5 was originally constructed by D. H. McNeil and consists of hyperbolas on a "tire." In other words, by projecting the tire's curved surface space onto the plane, one obtains exactly the quantitative infinity approached by the hyperbolas. And, if the inner hole of the tire is seen as an ideal point, that point will stand for an infinitesimal.

In order to explain the structural morphology of mathematical quadratic forms, let us for now cite the results that hold true for general nonlinear equations (for more detailed discussions, please consult Section 3.6):

When $p^2 - 4q > 0$, then

1. If $\left| \dfrac{p}{2} + u \right| < \dfrac{1}{2}\sqrt{p^2 - 4q}$, then

$$u = -\frac{1}{2}\sqrt{p^2 - 4q}\ th\left(-\frac{1}{2}\sqrt{p^2 - 4q}\ t - \frac{1}{2}A_0 \right) - \frac{p}{2} \qquad (3.6)$$

2. If $\left| \dfrac{p}{2} + u \right| > \dfrac{1}{2}\sqrt{p^2 - 4q}$, then

$$u = \frac{1}{2}\sqrt{p^2 - 4q}\, cth\left(-\frac{1}{2}\sqrt{p^2 - 4q}\, t - \frac{1}{2}A_0\right) - \frac{p}{2} \tag{3.7}$$

When $p^2 - 4q < 0$, then

$$u = \frac{1}{2}\sqrt{4q - p^2}\, \tan\left(\frac{1}{2}\sqrt{4q - p^2}\, t + \frac{1}{2}A_0\right) - \frac{p}{2} \tag{3.8}$$

Therefore, the structural morphology of the mathematical models of the quadratic form in Equations (3.6) through (3.8) can be illustrated using two-dimensional dynamic transformations between a circle and a straight line of the plane.

3.3.1 The Implicit Transformation between a Circle and a Tangent Line

In Figure 3.6, point p_i travels along the circle, and p'_i is the point corresponding to p_i on the straight line. Since the set of points p_i is one-to-one corresponding to the set of points p'_i, the top polar point N of the circle should correspond to the point N' that is infinitely far away on the straight line (this point cannot be labeled on the graph), the infinity on the line, which corresponds to the transitional point N on the curve. Or in other words, in the process of point p'_i traveling from quantitative ∞ to $-\infty$, the corresponding point p_i moves from and to the transitional point N on the circle from two different directions. The point N' on the line corresponding to point N is the singularity of quantities that depend on the straight-line axis. Conversely, the tangent point u_0 between the circle and the line represents an infinitesimal point on the line.

No doubt, if one uses the condition of well-posedness to limit the unboundedness of quantities, he will surely eliminate the underlying transitionality. Evidently, the quadratic mathematical model used above is only a simply nonlinear model. However, it has already revealed the fact that the mathematical concept of unboundedness is incomplete. And, from the angle of the formation of mathematics, it can be seen that the implicit transformation between the circle and straight line in Figure 3.6 has clearly illustrated that the mathematical condition of well-posedness does not apply in the study of nonlinear mathematical models.

3.3.2 The Implicit Transformation between a Circle and a Secant Line

This situation can be considered in two possibilities.

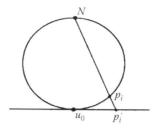

Figure 3.6   The implicit transformation between a line and a circle.

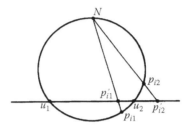

Figure 3.7   The singular transformation of curves.

3.3.2.1 Movement on the Minor Arch

According to Figure 3.7, if the point p_{i1} on the curve is limited on the arch between u_1, u_2, then the corresponding point p'_{i1} on the straight line can only move within the line segment $u_1 u_2$. So, the movement of point p'_{i1} stands for the smooth and bounded solution of Equation (3.6); the corresponding point p_{i1} cannot pass through point N, and so the movement of the corresponding point p'_{i1} does not contain any singularity. In other words, within a local finite range, one can apply well-posedness as the constraining condition for quantitative stability.

What we would like to mention at this junction is that the solution of the problem of solitary waves, which was motivated by the solitary water peak observed in a river in 1834 by J. S. Russell, a British engineer, can be obtained easily by differentiating Equation (3.6) once without the need to introduce tedious assumptions and to go through complicated mathematical deductions. As for the phenomenon of the 1960s, which was treating solitary waves as a characteristic of nonlinearity as an individual, isolated situation in the study of nonlinear problems. The true problem of nonlinearity is the transitionality accompanying the mathematical singularity of nonlinearity instead of the solitary-wave kind of smoothness.

3.3.2.2 Movement on the Major Arch

If a dynamic point p_{i2} moves along the major arch outside the line segment $\overline{u_1 u_2}$, then the corresponding point p'_{i2}, accompanying that fact that the point p_{i2} can travel through the point N, has to experience a transition from quantitative infinity to $-\infty$. So, once again, singularities on the straight number line are not realistic existence. And, from Equation (3.7), one can realize how to correctly understand the mathematical problem of the so-called nonlinearity.

3.3.3.3 Implicit Transformation between a Circle and a Disjoint Line

From the mathematical quadratic form in Equation (3.8), it is not hard to see that

$$\sqrt{4q - p^2} + \frac{A_0}{2} = \left(\frac{1}{2} + 2k\right)\pi, \quad \left(k = 0, \pm 1, \pm 2, \cdots\cdots\right) \tag{3.9}$$

Evidently, in this case, since the circle and the straight line do not intersect, point p_i can pass through point N as many times as needed without any constraint. In terms of the straight number line, there will naturally appear repeated singular indeterminacy. Our simple transformation between shapes and quantities discussed above should have clearly resolved the concept of quantitative singular indeterminacy and explained the instability of nonlinearity and the essence of quantitative unboundedness. In particular, the pursuit after quantitative comparability as the only standard for scientificality since the middle of the 20th century indicates that people have forgotten the true meaning and functionality of mathematical "shapes." And, by forcing the use of quantities, originally clear and simple problems become complicated and erroneous concepts are introduced, which is well shown by the incorrect recognition that the time and space of the universe are infinite. In particular, by devaluing the "shape or figure" analysis to determinacy, deterministic accuracies become the correctness that is against the reality without really understanding what determinacy means. Just as the ratio π of a circle's circumference and its diameter comes from the ratio of a curvature space over a Euclidean space, quantities cannot correctly describe the problem of transition in curvature spaces. As soon as quantitative instability is eliminated, nonlinearity is gone. Speaking more specifically, quantitative instability is a forewarning for a realistic curvature space to move closer to a transitional change. Its appropriate method of analysis should be that of inversely transforming zero-curvature spatial quantities back into the structures of a certain curvature space and then analyzing the resultant structures in order to determine the nature of the forthcoming transitional changes. We should notice that realistic curvature spaces do not have to be perfectly circular. This is still an open problem requiring further and careful investigation.

In this section, our focus is to look at the concepts developed in our analysis of mathematically quadratic nonlinear models on the basis of the discussion in Section 3.2 so that we once again confirm that the so-called ill-posedness in quantitative mathematics stands for points of transition of transitional changes occurring in curvature spaces. Quadratic models are those mathematical models that are commonly employed in applications. In this section we use the opportunity to illustrate the fundamental difference between the catastrophe theory and chaos theory, while showing that the research of the catastrophe theory has not truly touched upon the essential characteristics of nonlinear mathematical models. In other words, the current studies on the characteristics of nonlinearity have been stopped at the consideration of branch solutions and/or the jumps of broken solutions without entering the study of the characteristics of quantitative instability and unboundedness of nonlinearity.

3.4 The Dynamic Implicit Transformation of the Riemann Ball

There is no doubt that the Riemann ball is often used in the complex analysis of mathematics. Its purpose and applications are often about the typical transformation between curvature spaces and Euclidean spaces (Figure 3.8). This example vividly illustrates the correspondence between the point N of the North Pole on the sphere and the infinitely far away point on the plane, where the projection of the South Pole is mapped right onto the origin O on the plane, which is the infinitesimal point of the plane. So, the concept of infinite (includes both infinity and infinitesimal) of Euclidean spaces should not be treated as one of physics or philosophy. What is interesting is that Lao Tzu more than 2,000 years ago had taught about *yuan yue fan* (far way indicates return), while in the recent 300-some years, "far way" even has become "infinity."

So, being contemporary or fashionable does not necessarily mean being more advanced. Now, we can see a need to clear up some of the epistemological concepts. In so-called quantitative science, there is indeed a need to clarify the incompleteness of quantities.

If the Euclidean plane is transformed back to the sphere (Figure 3.8), then it can be seen clearly that the point N that helps quantities to change from quantitative $+\infty$ to $-\infty$ (or the 0 point where a negative infinitesimal changes to a positive

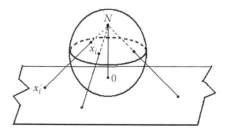

Figure 3.8 **The implicit transformation of the Riemann ball.**

infinitesimal) is exactly the transition point of the transitional changes. Although the concept of blown-ups of our blown-up theory was initially introduced based on studies of nonlinear equations, the relevant epistemological concepts of physics and philosophical thinking came from the idea of implicit transformations between curvature spaces and Euclidean spaces. In this example, the so-called quantitative indeterminacy of the narrow oberv-control space delivers the message of structural determinacy of a more general space through the dynamic, implicit transformation of a blown-up in directions. Or in other words, the reasons we use *blown-up* as another phrase for *transitional change* are, first, even in forms, any realistic event does not actually either continue its initial states (values) or consist of a sequence of clones of the initial states; instead, it will mostly be the situation that from minor details, it is difficult for one to know major changes. Second, we like to transform the accustomed narrow oberv-control of quantities to the general concept of structures and move onto the scenario that without seeing the large picture, it will be different to see the details so that we can materialize the theory of blown-ups and its practical applications. Third, continuing the ancient, several thousand-year-old epistemology of China that at extremes, matters will evolve in the opposite direction, we cannot simply blab about nonsense; instead we have to do careful analysis and thinking.

Especially, as applications of directions of material structures, it is exactly an important tool that the quantitative analysis system of modern science does not make use of. That is why in our practical applications, we have constructed our inversed informational structure. As a concept of physics, it should be noticed that order is a problem about structures.

It should be admitted that the implicit transformation of the Riemann ball is an age-old problem of mathematics, indicating that contemporary scholars have already clearly understood the true meaning of quantitative unboundedness. And, the implicit transformation of the Riemann ball also shows that the essence of mathematical nonlinearity is about transitionality. So, according to the epistemological concept of implicit transformations between shapes and quantities, it can be said that in at least the 19th century, the mathematical problem of nonlinearity had been resolved. Unfortunately, later scholars just forced themselves to fit problems of nonlinearity into the shoes of quantitative descriptions, leading to the trap — the mystery of nonlinear — they planned for themselves. To this end, looking at the current situation of how quantitative operations are applied, we also emphasize the fact that the unboundedness of quantities is not a realistic physical existence. Since the corresponding mathematical quantitative analysis has been employing the concept of ill-posedness to avoid or constrain quantitative instability, the fundamental intuition of mathematics has been violated. And, as of this writing, the logic and method of mathematical well-posedness are still applied to deal with quantitative instability, or using problems such as computer spills caused by quantitative instabilities to artificially limit quantitative instability. What was and is still applied in the community of learning is to no longer use problems to modify relevant theories or computational schemes; instead, it is to modify the problems of concern based on the classical theories or computational schemes. Consequently, problems of transitional changes have been difficult or impossible to resolve.

That is why we employ the concept of implicit transformations between shapes and quantities and map quantities back into shapes (or figures); by using inversed ordering structure, we can resolve — and have been easily resolving — transitional changes. As a system of methodology, our method belongs to a third category beyond modern science, called digitization of events.

3.5 Whole Evolution of Bifurcation Mathematical Models and Nonlinear Elasticity Models

It should be seen that either the bifurcation model or the nonlinear elasticity model is a special case of the general nonlinear model (see the following discussions in this chapter). In this section, we first look at these two models before the general case as the reason that they are some of the mathematical models well studied in the contemporary research of nonlinear problems and are once again considered as the origin for the catastrophe theory to appear. As a matter of fact, these studies focused on the multiplicity of values or the sudden changes of the type of jumps by looking at the characteristics of change of the "equilibrium states" with changes in parameters or by imposing locally continuous quantitative stability. Although neither multiplicity of value nor piecewise continuity satisfies the requirements of well-posedness, these studies did not touch on the characteristics of singularity of nonlinearity. To this end, we will focus on the whole evolutionary characteristics of mathematical singularity based on what we had done in the previous four sections. Or, we will reconsider the fundamental characteristics of these mathematical models from the angle of structural transitionality, while showing that our blown-up theory is different from the existing catastrophe theory. In the following, we try to avoid as many figures of specific situations as possible without affecting the smoothness of our discussion.

3.5.1 Standard Bifurcation

The standard bifurcation model of the bifurcation type can be written as follows:

$$\dot{X} = \frac{dx}{dt} = x^3 + \mu x \tag{3.10}$$

where μ is a parameter. Based on the discussion in Section 5.3, Equation (3.10) is a special case of third-order nonlinear models. And, from Theorems 1 and 2, it follows that when $\mu > 0$ or $\mu < 0$, the model all contains evolutionary transitional changes — blown-ups. All the details are omitted here.

3.5.2 *The Standard Bifurcation Model of Saddle, Node Points*

The standard bifurcation model of saddle, node points is

$$\begin{cases} \dot{x} = x^2 + \mu \\ \dot{y} = \pm y \end{cases} \tag{3.11}$$

where μ is a parameter. Evidently, Equation (3.11) is also a special case of the quadratic mathematical models studied in Chapter 6. When $\mu \geq 0$ or $\mu < 0$ and $|x| > \sqrt{-\mu}$, this model contains transitional changes of the whole evolution — blown-ups.

3.5.3 *The Standard Hofe Bifurcation Model*

The standard Hofe bifurcation model can be written as follows:

$$\begin{cases} \dot{x} = -y + x\left[\mu + \left(x^2 + y^2\right)\right] \\ \dot{y} = x + y\left[\mu + \left(x^2 + y^2\right)\right] \end{cases} \tag{3.12}$$

where μ is also a parameter. Rewriting Equation (3.12) using the polar coordinate system leads to

$$\begin{cases} \dot{r} = r\left(\mu + r^2\right), & \dot{r} = \dfrac{dr}{d\theta} \\ \dot{\theta} = 1, & \dot{\theta} = \dfrac{d\theta}{dt} \end{cases} \tag{3.13}$$

where $\dot{\theta} = 1$ stands for the movement along the trajectory at a constant angular speed. Notice that Equation (3.13) is a third-order nonlinear model similar to the bifurcation model of the bifurcation type in Equation (3.10).

What needs to be pointed out is that, originating in the successes of the catastrophe theory of the late 20th century, the contemporary research of nonlinear mathematical models has focused on bifurcation models. Although the catastrophe theory mainly investigates nonlinear models, the problems considered by the theory are merely those that experience sudden quantitative changes or stepwise jumps due to changes in parameters within limited quantitative ranges without involving transitional changes in whole evolutions. So, there are fundamental changes between our blown-up theory and the catastrophe theory. In principle, the catastrophe theory still did not walk out of the continuity system of initial values by maintaining the extensionality of the original functions using stepwise and piecewise jump functions, while our blown-up theory considers transitional changes totally different from the initial value systems of the original functions, where the concept of curvature spaces is employed to reveal the underlying structural changes. In terms of differences in specific operations, the catastrophe theory has not considered the unboundedness of quantitative instability, while the blown-up theory uses quantitative unboundedness to reflect transitional changes in the noninitial value systems. So, at the height of epistemology, the blown-up theory has completely altered the belief that quantitative infinity is not an objective existence, showing that it is instead a sign for transitional changes in physical problems. In the blown-up theory, quantitative instability cannot be eliminated and is seen as a necessary path leading to transitional changes. Therefore, it shows that the current, various methods developed to eliminate quantitative instability in essence have removed the vitally important transitional changes, which is the reason the profession of predictions has not made substantial progress in the past 100-plus years.

3.5.4 Nonlinear Elasticity Models

Traditionally, elasticity models have always been related to reciprocating vibrations. And since the famous Taylor wave motion equation was established for string vibrations, the community of physicists of the 20th century can be seen as a "kingdom" of wave motions. From microscopic to macroscopic points of view, from solids to liquids, almost every corner of physics produced a whole series of wave motion equations, reaching such a state that without wave motion equations, a study would not be treated as scientific. It can also be said that the reason modern science is known as science is the support of the system of wave motions. Without wave motions, the core theories and methods of modern science would be all gone. For example, at the microscopic level of fundamental particles, one can see the concept of probabilistic waves in quantum mechanics; at the macroscopic level of gravitational fields, one has the concept of gravitational waves of the general relativity theory; or in fluid mechanics, as a study of mesoscopic atmosphere and ocean, revolutionary flows of eddy currents are also treated as some unidirectional retrogressive Rossby waves. Other areas of learning have had a wide variety of representative works. For details, please consult Chapter 4.

Even from the beginning of our careers we could not accept the theory of plane unidirectional retrogressive waves, when we had our first chance to experience the asymmetric break-offs in metal materials of the laboratories. Since then we have started to rethink the problems about the system of wave motions. What we found is that each wave motion is a movement of relatively evenly structured materials under such a stress that the materials will not be destroyed. As soon as the material under stress is structurally uneven or damaged so that it can no longer recover to its original state, there will be no wave motion. Or in other words, in any motion involving materials with uneven structure or uneven acting force, the commonly seen phenomenon is an asymmetric break-off. In short, unevenness will alter or destroy the system of wave motions and is a stopper for the phenomenon of wave motions to appear. And, this stopper of wave motions can perfectly appear in mathematically nonlinear elasticity models.

The general nonlinear elasticity equation can be written as

$$\ddot{X} = x^3 - \sigma x \tag{3.14}$$

where

$$\ddot{X} = \frac{d^2 x}{dt^2}$$

σ is a constant. So, Equation (3.14) is also a special case of second-order nonlinear mathematical models. Integrating it once produces

$$\dot{X} = \frac{1}{4} x^4 - \frac{1}{2} \sigma x^2 + h_0 = F \tag{3.15}$$

where h_0 is the integration constant. So, for those readers who have a certain amount of mathematical knowledge, it can be seen that if F has a root of multiplicity 4, roots of multiplicity 2, or distinct roots, then Equation (3.14) contains transitional changes of quantitative singularity — blown-ups. (For more details, please consult Section 5.3.) It should be noted that this kind of change is not the same as local, sudden changes of piecewise functions; instead it is about transitional changes in the whole structure, constituting the truly common phenomenon seen in the evolutions of almost all materials, since materials with uneven structures are common. So, the evolutionality of damages due to "singularities" is the consequently common phenomenon with differences only in time scales involved. So, unevenness in materials or in acting forces causes break-offs in materials due to irreversible bending, destroying the so-called wave motionality of elastic materials. The reason for this end to hold is that unevenness in materials' structures has to change the so-called first-pushing forces into the effects of moments of the forces so that the existing reciprocating vibrations are destroyed by the resultant spinning movements. Therefore, this discussion completely uncovers some of the problems existing along with the classical mechanical system. That is why we pointed out that wave motions are those movements that keep the materials undamaged, which betrays the beliefs of modern science and walks into the realm of evolution science.

3.6 Eight Theorems on Mathematical Properties of Nonlinearity

Since problems of physics have entered curvature spaces, the corresponding mathematical equations have to inevitably involve nonlinearity. So, there is a need to understand the mathematical characteristics or properties of nonlinear equations. As we have mentioned earlier, it is Newton's second law of mechanics that first involves nonlinear equations in modern science:

$$f = m a \tag{3.16}$$

where if f, m, a are all variables, then this equation is nonlinear. In scientific history, Newton treated f as being heavenly fixed and m particlized as the quantity of the object so that it becomes heavenly invariant. Therefore, the acceleration a has to be a constant, leading to the assumption of inertial systems. That is why there are inertial systems in modern science.

Considering the heat wave of research on nonlinearity of the latter part of the 20th century, in the following, let us look at the fundamental characteristics and properties of nonlinearity from the angle of pure mathematics.

3.6.1 Fundamental Characteristics of General Nonlinear Equations

Although the dynamic system of physics comes from Equation (3.16), due to the custom in mathematics, most of the system is summarized by using mathematical models, derived out of Equation (3.16) as follows:

$$a = \dot{u} = \frac{du}{dt} = F \tag{3.17}$$

The essence is to treat the mass as a constant quantity of the material by taking 1 as its value. In this formula, u is generally named as a variable of movement and is specifically known as the speed. So, Equation (3.17) is Newton's second law, which is also known as the general form of mathematical models in the Lagrange language. Evidently, for generality, F can take the different form function of u. For example, if F takes the form of a polynomial of certain order, then one can discuss the mathematical characteristics of nonlinearity of second- or third-order polynomials.

3.6.1.1 Second-Degree Polynomials

If F is seen as a second degree polynomial, then we have

$$F(u) = u^2 + pu + q \tag{3.18}$$

where *p, q* are constants. From Equations (3.17) and (3.18), it follows that

$$\dot{u} = u^2 + pu + q \tag{3.19}$$

Evidently, when *p, q* are different values, Equation (3.19) has solutions of different forms. Let us consider this end in three cases.

When $p^2 - 4q = 0$, Equation (3.19) can be written as

$$\dot{u} = \left(u + \frac{p}{2} \right)^2 \tag{3.20}$$

Integrating Equation (3.20) provides

$$u = -\left(\frac{1}{t + A_0} + \frac{p}{2} \right) \tag{3.21}$$

where A_0 is a constant. From Equation (3.21), it follows that when $A_0 > 0$, *u* changes with quantity *t* and is continuous within finite time interval (see curve II in Figure 3.4); when $A_0 < 0$, then the model contains blown-ups of transitional changes (see curves I and III in Figure 3.4). Notice that blown-ups do not stand for sudden changes of the jump type of problems in stable state. Instead, they represent transitional changes of the noninitial value system while bringing forward new structures of different properties. *Sudden changes* means stepwise changes of the initial value system without altering the underlying attributes. It only represents the difference between stable changes and fast changes.

When $p^2 - 4q > 0$, Equation (3.19) can be written as follows:

$$\dot{u} = \left(u + \frac{p}{2} \right)^2 - \frac{1}{4}\left(p^2 - 4q \right) \tag{3.22}$$

Integrating Equation (3.22) produces

$$\ln \left| \frac{u + \dfrac{p}{2} - \dfrac{\sqrt{p^2 - 4q}}{2}}{u + \dfrac{p}{2} + \dfrac{\sqrt{p^2 - 4q}}{2}} \right| = \sqrt{p^2 - 4q}\, t + A_0 \tag{3.23}$$

From Equation (3.23), it follows that if the movement is only limited to the bounded local changes within

$$\left| u + \frac{p}{2} \right| < \frac{\sqrt{p^2 - 4q}}{2}$$

$$u = \frac{1}{2}\sqrt{p^2 - 4q} \ th\left(-\frac{1}{2}\sqrt{p^2 - 4q} \ t + \frac{1}{2}A_0 \right) - \frac{p}{2} \tag{3.24}$$

Equation (3.24) is the smooth solution within the limited range. What is interesting is that differentiating Equation (3.24) once can lead to the solitary wave solution, which has been bothering the scientific community for over a century. So, the so-called solitary wave is in fact a special case of nonlinear equations and cannot represent a general characteristic of nonlinearity.

If the movement is of the overall unboundedness of the kind of

$$\left| u + \frac{p}{2} \right| > \frac{\sqrt{p^2 - 4q}}{2}$$

then we have

$$u = \frac{1}{2}\sqrt{p^2 - 4q} \ cth\left(-\frac{1}{2}\sqrt{p^2 - 4q} \ t - \frac{1}{2}A_0 \right) - \frac{p}{2} \tag{3.25}$$

Equation (3.25) shows that the movement can evolve into transitional blown-ups, and when $A_0 > 0$, a blown-up occurs at

$$t = t_b \left(= -A_0 \Big/ \sqrt{p^2 - 4q} \right)$$

Both Equations (3.24) and (3.25) show that a same nonlinear model can have solutions of different forms under different conditions. The general characteristic of nonlinear equations is the discontinuity of the quantitative form and the continuous smooth solutions are only special cases. And, as special cases of locally continuous and smooth solutions, they are still not harmonic waves but the well-known solutions. Since the general characteristic of nonlinearity is not satisfying the requirement of continuity, it forms the fundamental cause for the difficulty of solving nonlinear equations. That is, nonlinear equations cannot satisfy the traditional initial value existence theorem of mathematical physics equations.

When $p^2 - 4q < 0$, Equation (3.19) can be written as

$$\dot{u} = \left(u + \frac{p}{2} \right)^2 + \frac{1}{4} \left(4q - p^2 \right) \tag{3.26}$$

Integrating Equation (3.26) produces

$$u = \frac{1}{2} \sqrt{4q - p^2} \ \tan \left(\frac{1}{2} \sqrt{4q - p^2} \ t + A_0 \right) - \frac{p}{2} \tag{3.27}$$

By taking

$$\frac{1}{2} \sqrt{4q - p^2} \ t + A_0 = \frac{\pi}{2} + n\pi$$

n = a whole number, then Equation (3.27) describes multiple periodic blown-ups.

3.6.1.1.1 Examples of Applications

For the convenience of application and understanding, let us use the once-famous population model of the chaos theory as our example,

$$\dot{x} = \frac{dx}{dt} = -\lambda x^2 - x + 1$$

Taking $\lambda > 0$ and $-4\lambda < 1$ produces

$$\frac{1}{1 + 4\lambda} \ln \frac{2\lambda x + 1 + \sqrt{1 + 4\lambda}}{2\lambda x + 1 - \sqrt{1 + 4\lambda}} = t + c \tag{3.28}$$

where c is a constant. Through some detailed manipulations, we have

$$x = \frac{\left[\left(1 - \sqrt{1 + 4\lambda} \right) e^{(t+c)\sqrt{1+4\lambda}} - \left(1 + \sqrt{1 + 4\lambda} \right) \right]}{2\lambda \left(1 - e^{(t+c)\sqrt{1+4\lambda}} \right)} \tag{3.29}$$

Taking $\lambda < 0$ and $-4\lambda > 1$ leads to

$$\frac{2}{\sqrt{4\lambda - 1}} \tan^{-1} \left(\frac{-2\lambda x - 1}{\sqrt{4\lambda - 1}} \right) = t + c \tag{3.30}$$

By taking $1 + 4\lambda = 0$, we have

$$x = \frac{1}{\lambda(t+c)} - \frac{1}{2\lambda} \tag{3.31}$$

The three situations considered in Equations (3.29) through (3.31) all indicate that with time, the evolution contains blown-ups of transitional changes. Evidently, the population evolution or production described by the population model is neither the chaos as claimed in the chaos theory nor an unconstrained explosion of a biological population. Instead, it presents a problem for the population of concern to decrease or extinction. This example indicates that nonlinear physical problems show that at extremes, the matter will evolve in the opposite direction so that transitionalities existing in material evolutions are clearly revealed. That is one of the reasons for us to establish the theory of blown-ups.

3.6.2 Eight Theorems about Third- and Second-Degree Nonlinear Models

3.6.2.1 The Third-Order Nonlinear Model

The general form of the third degree polynomial looks like

$$F = u^3 + pu + q$$

Inserting this expression into Equation (3.17) produces

$$\dot{u} = u^3 + pu + q \tag{3.32}$$

The mathematical properties of Equation (3.32) involve the following theorem:

Therom 1. If u_1 is a root of F of multiplicity three, let $F = (u - u_1)^3$. Then the following holds true:

$$u = u_1 \pm \left(A_0 - 2t\right)^{-2} \tag{3.33}$$

where A_0 is a constant. Equation (3.33) stands for a problem of two-pole blown-ups. When it is generalized to the case of a root of multiplicity n, then Equation (3.32) contains $(n-1)$ pole blown-ups.

Theorem 2. If $F = 0$ has a real and a pair of conjugate complex roots, then Equation (3.32) contains blown-ups.

Proof: Assume that u_1 the unique real root of $F = 0$, then F can be written as follows:

$$F = \left(u - u_1\right)\left(u^2 + p_1 u + q_1\right) \tag{3.34}$$

where p_1, q_1 are constants, satisfying $p_1^2 - 4q_1 < 0$. So, Equation (3.32) becomes

$$\dot{u} = \left(u - u_1\right)\left(u^2 + p_1 u + q_1\right) \tag{3.35}$$

Expanding Equation (3.35) using partial fractions leads to

$$\frac{1}{\left(u - u_1\right)\left(u^2 + p_1 u + q_1\right)} = \frac{A}{u - u_1} + \frac{Mu + N}{u^2 + p_1 u + q_1} \tag{3.36}$$

where

$$A = \left(\beta^2 + \gamma^2\right)^{-1}, \quad M = -\left(\beta^2 + \gamma^2\right)^{-1}, \quad N = -\alpha\left(\beta^2 + \gamma^2\right)^{-1},$$

$$\alpha = p_1 + u_1, \quad \beta = \frac{p_1}{2} + u_1, \quad \gamma = \sqrt{q_1 + \frac{p_1^2}{4}}$$

Substituting these expressions into Equation (3.34) and integrating the resultant equation provide us with

$$\frac{1}{u^2 + p_1 u_1 + q_1} \ln\left|\frac{u - u_1}{\sqrt{u^2 + p_1 u + q_1}}\right| + \frac{N - \dfrac{p_1 M}{2}}{\sqrt{q_1 + \dfrac{p_1^2}{4}}} \tan^{-1} \frac{u + \dfrac{p_1}{2}}{\sqrt{q_1 - \dfrac{p_1^2}{4}}} = t + A_0 \tag{3.37}$$

Evidently, Equation (3.37) indicates that within finite ranges of time t, multiple blown-ups will appear.

Theorem 3. If $F = 0$ has two equal real roots u_1 and a single root u_2, then assume $u_1 > u_2$ (for the case of $u_2 > u_1$, similar results also follow). When $u > u_1$ or $u < u_2$, for the overall unbounded movement, Equation (3.35) contains blown-ups; however, for the limited movement within the local range $u_1 > u > u_2$, there will not be any blown-up.

Proof: Based on the given conditions, F can be written as follows:

$$F = \left(u - u_1\right)^2 \left(u - u_2\right)$$

Accordingly, Equation (3.32) can be written as

$$\dot{u} = \left(u - u_1\right)^2 \left(u - u_2\right) \tag{3.38}$$

By employing the method of partial fractions, we have

$$\frac{1}{\left(u - u_1\right)^2 \left(u - u_2\right)} = \frac{A}{\left(u - u_1\right)^2} + \frac{B}{\left(u - u_1\right)} + \frac{C}{\left(u - u_2\right)}$$

where

$$A = \frac{1}{u_1 - u_2}, \quad B = -\frac{1}{\left(u_1 - u_2\right)^2}, \quad C = \frac{1}{\left(u_1 - u_2\right)^2}$$

Substituting these expressions into Equation (3.38) and integrating the result provide that

$$-\frac{A}{u - u_1} + C \ln \left| \frac{u - u_2}{u - u_1} \right| = t + A_0 \tag{3.39}$$

where A_0 is a constant. If $u > u_1$ or $u < u_2$, then Equation (3.39) becomes

$$-\frac{A}{u - u_1} + C \ln \frac{u - u_2}{u - u_1} = t + A_0 \tag{3.40}$$

Equation (3.40) means that within finite time range of t, there exist blown-ups.

If $u_1 > u > u_2$, that is, if u moves within a finite range, then Equation (3.39) can be written as

$$-\frac{A}{u - u_1} + C \ln \frac{u - u_2}{u_1 - u} = t + A_0 \tag{3.41}$$

Notice that the difference between Equation (3.41) and Equation (3.40) is that when $u_1 > u$, the denominators in the second terms are different. Substituting the corresponding expressions of A, C into Equation (3.41) provides

$$u = \frac{u_2 + u_1 e^{\frac{(u_2-u_1)}{(u_1-u)}} e^{(u_1-u_2)^2(t+A_0)}}{1 + e^{\frac{(u_2-u_1)}{(u_1-u)}} e^{(u_1-u_2)^2(t+A_0)}} \tag{3.42}$$

When $t \in [0, \infty]$, Equation (3.40) does not contain any blown-up. That is, Theorem 3 holds true.

Theorem 4. If $F = 0$ has three distinct real roots u_1, u_2, u_3 (without loss of generality, assume $u_1 > u_2 > u_3$), then:

if $u > u_1$ or $u < u_3$, then Equation (3.32) contains blown-ups;

If $u_2 > u > u_3$ or $u_1 > u > u_2$, then Equation (3.32) does not contain any blown-up.

Proof: Assume that the three distinct real roots of $F = 0$ are u_1, u_2, u_3, then Equation (3.32) can be rewritten as

$$\dot{u} = \left(u - u_1\right)\left(u - u_2\right)\left(u - u_3\right) \tag{3.43}$$

Integrating this expression using partial fractions leads to

$$A \ln|u - u_1| + \ln|u - u_3| - B \ln|u - u_3| = \frac{t + A_0}{C} \tag{3.44}$$

where

$$A = \frac{u_2 - u_3}{u_1 - u_2}, \quad B = -\frac{u_1 - u_3}{u_1 - u_2} C$$

and A_0 is a constant.

If $u > u_1$ or $u < u_3$, then Equation (3.44) can be written as follows:

$$\ln \frac{\left(u - u_1\right)^{\frac{u_2 - u_3}{u_1 - u_2}} \left(u - u_3\right)}{\left(u - u_2\right)^{\frac{u_1 - u_3}{u_1 - u_2}}} = \frac{1}{C}\left(t + A_0\right) \tag{3.45}$$

So, when the overall movement is unbounded, Equation (3.45) contains blown-ups and multiple values; if $u_2 > u > u_3$ or $u_1 > u > u_2$, that is when the movement is local, then no blown-up will occur. So, Theorem 4 holds true.

3.6.2.2 Second-Order Nonlinear Models

Considering the generality, let us consider a second-order, one-variable nonlinear model as our example of discussion. That is, we have

$$\ddot{u} = \frac{d^2 u}{dt^2} = u^3 + bu^2 + cu + d \tag{3.46}$$

where b, c, d are constants. Using the method of reducing the order, let us rewrite Equation (3.46) as the following system of first order equations in one variable:

$$\begin{cases} \dot{u} = v \\ \dot{v} = u^3 + bu^2 + cu + d \end{cases} \tag{3.47}$$

Taking the integral of the second equation in Equation (3.47), we have

$$v = \frac{1}{4} u^4 + \frac{b}{3} u^3 + \frac{c}{2} u^2 + du + h_0$$

where h_0 is a constant. Taking $\dot{u} = v = E$ provides us with

$$\dot{u} = E = v = \frac{1}{4} \left(u^2 + b_1 u + c_1 \right)\left(u^2 + b_2 u + c_2 \right) \tag{3.48}$$

where $b_1, c_1; b_2, c_2$ are also constants. So, we have the following theorems:

Theorem 5. If the equation $E = 0$ has a root of multiplicity four, then Equation (3.48) contains blown-ups.

The proof is similar to that of Theorem 1 and is omitted here.

Theorem 6. If the equation $E = 0$ has two distinct root of multiplicity two, the Equation (3.48) can be written as

$$E = \frac{1}{4} \left(u^2 + b_1 u + c_1 \right)^2 \tag{3.49}$$

and

when $b_1^2 - 4c_1 < 0$, the model contains blown-ups, see Equation (3.26) for more details;

when $b_1^2 - 4c_1 > 0$, then the overall unbounded movement becomes a problem of blown-ups, and the local bounded movement is not a problem of blown-ups, see Equation (3.22) for more details.

Theorem 7. If $E = 0$ has only one root of multiplicity, the Equation (3.48) contains blown-ups.

Proof: For convenience, assume the root of multiplicity is 0. So, Equation (3.48) can be rewritten as

$$\dot{u} = u\sqrt{u^2 + b_2 u + c_2} \tag{3.50}$$

Taking the transformation

$$w = \frac{1}{u}$$

makes the previous equation become:

$$-\frac{dw}{\sqrt{c_2\left(w + \dfrac{b_2}{2c_2}\right)^2 + \dfrac{1}{4c_2}\left(4c_2^2 - b_2^2\right)}} = dt \tag{3.51}$$

1. If $c_2 < 0$ and $b_2^2 - 4c_2 > 0$, integrating Equations (3.51) and (3.18) provides us

$$-\sin^{-1}\frac{w + \dfrac{b_2}{2c_2}}{\sqrt{\dfrac{1}{4c_2}\left(b_2^2 - 4c_2\right)}} = \sqrt{-c_2}\left(t + A_0\right) \tag{3.52}$$

   where A_0 is a constant. Since $w \to 0$, $u \to \infty$, Equations (3.51) and (3.48) contain blown-ups, which are periodic.

2. If $c_2 > 0$ and $b_2^2 - 4c_2 < 0$, integrating Equation (3.18) leads to

$$-\ln\left[\sqrt{c_2}\left(w + \frac{b_2}{2c_2}\right) + \sqrt{c_2\left(w + \frac{b_2}{2c_2}\right) + \frac{4c_2^2 - b_2^2}{4c_2^2}}\right] = \sqrt{c_2}\left(t + A_0\right) \tag{3.53}$$

Similar to the discussions before, Equation (3.53) contains blown-ups. That is, Theorem 7 holds true.

3.6.2.3 The Nonlinear Problem of nth-Degree Polynomial and Theorem 8

For the nonlinear problem of nth-degree polynomial, even though we cannot provide the exact solution, we can still analyze its mathematical properties and its physical significance. On the field of real numbers, from the theorem that each polynomial can be uniquely factored, the autonomous system of Equation 3.16 can be written as follows:

$$\dot{u} = a_0 + a_1 u + \cdots + a_{n-1} u^{n-1} + u^n \tag{3.54}$$

which can be factored into

$$\dot{u} = F = \left(u - u_1\right)^{p_1} \cdots \left(u - u_r\right)^{p_r} \left(u^2 + b_1 u + c_1\right)^{q_1} \cdots \left(u^2 + b_m u + c_m\right)^{q_m} \tag{3.55}$$

where $p_i (i = 1, 2, \ldots, r)$, $q_j (j = 1, 2, \ldots, m)$ are positive whole numbers, satisfying

$$n = \sum_{i=1}^{r} p_i + 2 \sum_{j=1}^{m} q_j, \quad n = \sum_{i=1}^{r} p_i + 2 \sum_{j=1}^{m} q_j, \quad \Delta = b_j^2 - 4c_j < 0 \quad \left(j = 1, 2, \ldots, m\right)$$

Without loss of generality, if $u_1 \geq u_2 \geq \ldots \geq u_r$, then the mathematical properties of Equation (3.54) are listed in the following theorem:

Theorem 8.

1. When $F = 0$ of u_i ($i = 1, 2, \ldots, r$) does not have any real root, then blown-ups exist;
2. If $F = 0$ does not real root(s), there are the following two scenarios:

   When n is even, if $u > u_1$, then blown-ups exist; if $u < u_r$, no solution exists. When n is odd, no matter whether $u > u_1$ or $u < u_r$, there are blown-ups.

Proof:

1. If the equation $F = 0$ does not have any real root, then Equation (3.55) can be written as

$$\dot{u} = \left(u^2 + b_1 u + c_1\right)^{q_1} \cdots \left(u^2 + b_m u + c_m\right)^{q_m} \tag{3.56}$$

From $\Delta = b_j^2 - 4c_j < 0$, it follows that $u^2 + b_j u + c_j$ has a minimum value

$$\alpha_j = \frac{1}{4}\left(4c_j - b_j^2\right) > 0$$

The condition of estimation for Equation (3.56) is given by

$$\dot{u} \geq \beta_0\left(u^2 + b_1 u + c_1\right) \geq \beta_0\left(u + \frac{1}{2}b_1\right)^2 > 0 \tag{3.57}$$

where $\beta_0 = \alpha_1^{q_1-1} \cdot \alpha_2^{q_2} \cdots \alpha_m^{q_m}$. If u is a monotonic increasing function, from Equation (3.57), it follows that

$$u \geq -\frac{1}{2}b_1 + \frac{u_0}{1 - u_0\beta_0 t} \tag{3.58}$$

where u_0 is the initial value. So, changes in u contain evolutionary, transitional blown-ups. That is, statement (1) in Theorem 8 holds true.

2. If the equation $F = 0$ does have real roots u_1 ($i = 1, 2, \ldots, r$), then we have:
   a. When n is even, since when

$$q = 2\sum_{j=1}^{m} q_j$$

   is even,

$$p = \sum_{i=1}^{r} p_i$$

   is also even, so, for either $u > u_1$ or $u < u_r$, we always have $\dot{u} > 0$. That is, u is a monotonic increasing function, which contradicts with $u < u_r$. That means that in this case, there is no solution.

   Now, we prove that when $u > u_1$, there exist blown-ups. Take $u_1 \geq u_2 \geq \ldots \geq u_r$, then Equation (3.55) can be rewritten as follows:

$$\dot{u} \geq \beta_1\left(u - u_1\right)^p \tag{3.59}$$

where $\beta_1 = \alpha_1^{q_1} \cdot \alpha_2^{q_2} \cdots \alpha_m^{q_m}$. Solving the inequality in Equation (3.59) gives

$$u \geq u_1 + \frac{1}{\sqrt[p-1]{-\left[A_0 + (p-1)\beta_1 t\right]}} \tag{3.60}$$

where A_0 is a constant, which can be determined by using the initial value. Evidently, Equation (3.60) contains blown-ups of transitional changes.

b. When n is odd, since q is odd, p is also odd. So, when $u > u_1$, we have

$$\dot{u} > 0 \tag{3.61}$$

So, u is a monotonic increasing function, satisfying

$$\dot{u} \geq \beta_1 \left(u - u_1\right)^p > 0$$

Solving this equation and taking the (+) branch provide the following:

$$u \geq u_1 + \frac{1}{\sqrt[p-1]{-\left[(p-1)\beta_1 t + A_0\right]}} \tag{3.62}$$

so, u contains blown-ups.

When $u < u_r$, we have $\dot{u} < 0$. So, u is a monotonic decreasing function, satisfying

$$\dot{u} \leq \beta_1 \left(u - u_1\right)^p < 0$$

Solving this inequality and taking the (–) branch provide

$$u \leq u_r - \frac{1}{\sqrt[p-1]{-\left[(p-1)\beta_1 t + A_0\right]}} \tag{3.63}$$

So, no matter whether it is an increasing or decreasing function, there are always blown-ups of evolutionary transitional changes. That is, the statement 2 in Theorem 8 holds true. Together with the argument earlier, we have shown that Theorem 8 holds true.

3.6.3 Some Explanations

Our initial purpose of studying these eight theorems can be summarized as follows.

The proofs and results of these theorems indicate that the traditional linear system treats invariancy as the generality, leading to the conclusion that from knowing the initial values, one can know everything about the future (Laplace) and that modern science cannot provide the difference between the past, the present, and the future (Einstein). On the other hand, nonlinearity reveals changes with blown-ups of transitional changes as the generality, making the "generality" of invariancy as a special case. Consequently, the entire related epistemological system of concepts is flipped upside down. That is, the traditional generality becomes a special case of evolutions, leading to the conclusion that even if we know the initial values, we still do not know much or anything about the future. Since the traditional special case becomes the generality, it can help provide information on the difference between the past, the present, and the future. So, Einstein himself and the past all become the nonreturnable past.

As influenced by V. Bjerknes's circulation theorem, we generalized variable acceleration to the (n–1)th order. Changes in accelerations represent a problem of physics of variant accelerations so that nonlinearity belongs to noninertial systems. That fact tells us that nonlinearity cannot be linearized and the essence of linearization is to eliminate evolution. Therefore, many conclusions obtained by using linearization have to be reconsidered.

The well-posedness theory of basic mathematical physics equations limits instabilities naturally existing in quantities as well as invalidates the law of philosophy that quantitative changes bring about qualitative changes that is produced out of modern science. Quantitative instabilities come from transitions in the events of curvature spaces; and as a matter of fact, before an instability actually happens, the corresponding events have already gone through transitional changes. Combined with the posteventness of quantities, it can be seen that the law that quantitative changes bring about qualitative changes of modern science is not a materialistic law.

As an extended explanation, we should notice that from formal quantitative analysis, physical quantities could be introduced. And, the symbol f, as the force in Newton's second law, does not really occupy any material dimension, since from the fact that forces come from unevenness in gradients of materials' structures (Lin, 2007), only materials occupy material dimensions. So, we should take $m = 1/\rho$ and $f = 1$, which helps to write Equation (3.1) as $\dot{u} = \rho(u)$. On this basis, we can generate our discussion and produce results similar to what we have obtained earlier so that evolutionality can be directly revealed.

As for the mathematical properties of nonlinear equations, the first is the betrayal from continuity. The next is that nonlinearity reflects physical changes of transitionality and that mathematical well-posedness is not a product of the formal logic of mathematics.

3.7 Conclusions

Our discussion in this chapter is mainly about the problem of quantitative unboundedness — the quantitative infinity and infinitesimal, the concept of implicit transformation between quantities and shapes, while touching upon the problem of mathematical nonlinearity and the relevant fundamental concepts of physics. Our results can be summarized as follows.

Quantitative infinity or infinitesimal are products of quantitative formal measurements of low-dimensional zero-curvature spaces, because the concepts of lines and planes in low-dimensional zero-curvature spaces belong to the limited narrow systems of imaginary spaces. That is why quantitative unboundedness is not a physical reality, and the process of change from quantitative instability to quantitative unboundedness can be revealed by using implicit transformations between quantities and shapes. It has been shown that this process of change is exactly the process for materials to get closer to transitional changes and that it enters the territory of evolution science at the same time it betrays modern science.

Evidently, the concept of implicit transformations between quantities and shapes reveals the unboundedness of quantities, which is exactly the problem that at extremes, matter will evolve in the opposite direction. The so-called indeterminacy of quantities should be correctly renamed as the determinacy of transitionality in curvature spaces. It also illustrates that mathematical nonlinear models implicitly contain materials' structurality so that nonlinearity has walked out of the particlization or the belief that quantitative comparability is the only standard for scientificality. So, classical mechanics and its consequent theoretical system of wave motions, as the foundation of modern science, can only be a theoretical system under the concept of narrow zero-curvature spaces. Its core or characteristic is to maintain invariancy. That is why the theoretical system of wave motions cannot resolve problems of material or event evolutions. And that is why prediction is a branch of evolution science. Or in other words, the problem of shapes and quantities, after Galileo published his *Two New Sciences*, once again pushes science to a fork along its path of development, where a decision needs to be made.

The so-called quantitative instability of nonlinearity and the ill-posedness of quantitative infinity cannot be seen as meaningless concepts; instead they represent signals for the underlying materials or events to approach transitional changes. Therefore, the constraining conditions, such as that of well-posedness of mathematical physics equations, stability of computational mathematics, and so forth, are only needed to limit changes in quantities so that the classical mechanical system can be maintained. So, the essence of nonlinearity is a betrayal of the well-posedness system of the traditional mathematical physics equations in the form of quantitative instability. What one needs to be cautious about is that the so-called quantitative instability reflects perfectly the path through which materials and events are approaching transitional changes.

As to whether or not the development of science has come to another crossroad, or what the developmental direction of science is, the interested reader can surely think on his own. However, our learned experience from practice and theory has taught us that real-life problems, which urgently wait for resolutions, seem to be the attraction pointing to the future developmental direction of human exploration of nature. Also, we should notice that the second crisis in the foundations of mathematics has evolved into a crisis of the entire system of modern science (Lin, 2008). That is, without continuity, the modern palace of science cannot have been constructed. With continuity, modern science cannot move forward without difficulty, making the modern scientific system a castle in the air that waves with the wind.

Acknowledgments

The presentation of this chapter is based on Bergson (1963), Chen and OuYang (2005), Chen, et al. (2005a,b), Einstein (1976, 1997), English and Feng (1972), Eves (1986), Feyerabend (1992), Hawking (1988), Koyré (1968), Lao Tzu (time unknown), Li and Guo (1985), Lin and OuYang (1996), Lin (1998, 2007, 2008a,b), Lin et al. (2001), Mason (1989), OuYang (1984, 1998a, 1998b, in press), OuYang and Chen (2007), OuYang et al. (2005a, 2002a, 1998), OuYang and Peng (2005), Prigogine (1980), Wilhalm and Baynes (1967), Wolf (1985), and Wu and Lin (2002).

Chapter 4

Achievements and Problems of the Dynamic System of Wave Motions

The system of wave motions of the first push has been seen as the representative of one of the most magnificent achievements of classical mechanics or the system of dynamics. In the entire 20th century after the concept of Taylor string vibration was initially introduced, the kingdom of wave motions has included almost all topics of modern science so that in modern science, wave motions have been listed as a topic that covers the widest range of problems. The generality of wave motions includes the movements of solids, fluids, and gases, leading to the studies in the fundamental particles of quantum mechanics, such as the concepts of probabilistic waves, light waves, and gravitational waves in the theory of gravitations, and the famous unidirectional retrogressive Rossby waves in meteorological science, and so forth.

The modern science developed on the quantitative analysis is most known for its investigation of the phenomena of wave motions in physics. And, solving deterministic problems of mathematical physics equations, which appeared in history at almost the same time as that of calculus, have become famous due to their wave-motion solutions. In order to make studies of all physics problems suitable for employing mathematical physics equations, due to the "universality" of the capability of quantitative analysis, not only have the diffusion equations established for

125

disturbances been identified with heat-conduction equations, but also have their wave-motion solutions been found. Although each area of natural science and every corner of engineering touch on problems of wave motions, as a simple problem of intuition of physics, wave motions embody energy transfers in the form of reciprocating movements of the material media without involving any disposition of the material itself. That is why we once pointed out that wave motions prevent the underlying materials from being damaged by allowing energy transfers so that quantitative analysis can be applied.

However, in order to use the quantitative analysis system in the investigation of physics, wave motions are specifically defined as a general concept. In particular, to keep conditions and limitations to the minimum and to meet the need of employing mathematical methods, wave motions are defined as disturbances of any kind, including sudden changes in quantitative values and distortions of extreme values. That is why in modern science, physical phenomena from macroscopic to microscopic levels all belong to the category of wave motions. And, because of the employment of mathematical techniques, the physical classification of wave motions is also done mathematically, using such names as hyperbolic waves and dispersive waves.

In particular, by treating any disturbance as a wave motion, rotational disturbances are also included in the system of wave motions. As a matter of fact, even within inertial systems of the first push, it has been admitted that rotations are not problems of these systems. Although the quantities of rotational disturbances can be the same as those of wave disturbances, the physical meanings of these two classes of disturbances are different. In reality, translations, vibrations, and rotations are three forms of movements with translations and vibrations appearing in rotations so that they are special cases of rotations. The movements with truly fewer or no constraints and limitations are those of rotations. That is why modern science suffers from some quite unrealistic problems that appear in the investigation of physics, as the consequence of treating all movements as wave motions.

In other words, because of its generality, the concept of wave motions seems to have forgotten about its physical, intuitive, and simple contents: Each wave motion stands for such a reciprocating movement of particles that do not leave their points of equilibrium. Evidently, if all movements were waves, then we would never be able to move away from our original homes; instead, the most that we could do would be to move back and forth around the homes.

Even if we borrow the mathematical definition of dispersive waves or dissipative waves, there is no doubt that we will still avoid the problem by following the quantitative axioms of physics that require us to provide explanations for the mechanism for the cause and processes of dispersions or dissipations. However, by following the quantitative axioms, we have reached the goal of limiting quantitative instable growth in the name of physics. But, as a problem of physics, it is exact quantitative instabilities by which evolutionary transitions are signaled ahead of time. The reason the concepts of dispersions and dissipation were used by the system of wave motions was to maintain the propagation of the original waves without changes.

From the angle of mathematics, the abstract generality derived from the recent concept of wave motions reflects the fact that the physically intuitive concept of wave motions cannot describe whole movements well and cannot include the investigation of spinning movements in this physically intuitive concept. That motivated the mathematical abstraction of the concept of wave motions, leading to such a concept of wave motions that is first macroscopic and abstract, and then about describing a problem of physics without in essence touching on the substance of physics. That is why we refer to it as that of mathematical waves.

No doubt, the developmental history of the scientific system of wave motions has sufficiently shown the process of removing physical essences in the investigations of physical problems. So, wave motions, an originally simple and intuitive concept of physics, have become an unclear and fuzzy concept in terms of the physical meanings involved due to the need to apply mathematical techniques. Evidently, if people pursued the physical meanings of the concept, the relevant mathematical techniques would have lost their capability of description. Although no one has even seen what a plane, unidirectional retrogressive wave is, what a gravitational wave is, and what probabilistic wave is, these have still become well-known theories in the world of learning by simply applying indiscriminately the system of wave motions, leading to such a consequence that no one seems to question whether or not these concepts could really answer and resolve any problem. When looking at the history of science in this light, it also shows the way that many of the well-known scientific theories were established by making sure mathematical formal logical calculus could be applied without even looking at the properties and the underlying principles of transformation of physics, because the main capability of mathematics can only be applied to wave motions without involving other forms of motions. That is how the world of learning has had the custom of chasing after fads, where the fashion of wave motions has indeed conducted the fashionable movement of the 20th century.

Due to the past fashion and the current ongoing trend of applying the system of wave motions, in this book we have to review the mathematical techniques employed in the study of wave motions by looking at the classical vibrations in strings and the relevant mathematical descriptions. Our presentation explains how from the three major linear systems, established on the classical vibrations of strings, such problems as vibrations, heat conductions and diffusions, equilibriums between physical quantities, and so forth can all be addressed by using the solutions in the form of wave motions out of the mathematical physics equations involving second-order derivatives. That no doubt has made its important contribution to applications of mathematics and the appearance of wave-motion physics in the 20th century. At the same time, that also exaggerated the importance of vibrations in the system of wave motions and helped push for wide employment of linearization of nonlinear equations. All these consequences inevitably lead to problems in applications and in the theoretical studies of physics. In particular, changes in curvatures occurring in realistic movements in fact touch upon the twist-ups of mutual interactions so that natural disasters and events can be investigated from a brand-new angle.

4.1 The Classical Vibration and Wave Motion System

Vibrations of strings are a familiar class of mathematical examples of wave motions. Our purpose of reviewing this class of models is of multiple folds.

Look at how these models are established through modeling. In other words, we like to see how mathematical techniques are used to model wave motions.

Look at the basic ways of establishing mathematical techniques and its consequent effects on the system of wave motions, problems existing in the process of modeling, and how to deal with these existing problems.

In terms of the structure of the system of wave motions, the Taylor equation of string vibration not only signals the establishment of wave-motion equations but also is considered a major historical event in the rise of the investigation of mathematical physics equations in the 20th century. It can be said that without the introduction of the Taylor equation of string vibration, the appearance of mathematical physics equations, as a branch of mathematics, and the development of theoretical physics would be slowed or even developed in a different direction so that the structure of modern science would be different.

Although the Taylor equation of string vibration is quite simple in terms of its structure, it has presented the basic construction of the system of wave motions and the basic process of applying mathematical techniques, which were later furthered, polished, and generalized to the studies of other areas of science, leading to, for example, wave-motion equation of fluid, Bessel equation of variable coefficients, the Schrödinger equation (established by directly introducing wave functions), electric field equation, electricity transmission, and the later weak nonlinear equations involving waves, such as shock waves, solitons, gravitational waves, and so forth. And, the characteristic functions of linear systems are all given in the form of wave motions. That is why out of linear mathematical physics equations, one has to produce the science of wave motions. No doubt, linearization of nonlinear and weak nonlinear equations inevitably leads to solutions of wave motion problems. So, the elegance of the wave motions of the 20th century has to become the elegance of the mathematical physics equation. This can perhaps tell how and why wave motions have become so widely studied and explains why we need to pay a visit to the equation of string vibrations.

The establishment of the familiar Taylor equation of string vibration is done based on the following assumptions:

- ▪ The string is elastic and made of even-mass material. So, cases of uneven and/or inelastic materials are not part of this study. Here, even materials can only be ideal and nonexistent in reality.

- When the string is in its state of equilibrium, it can only be at the stressed straight-line position. In reality, this assumption could hold true only for imaginary special cases involved within short distances.
- The string experiences a pulling force and its own weight only.
- Other than small horizontal vibrations, the string does not contain any other forms of motion.

No doubt, even for our discussion of a very simple string vibration, the imposed conditions are very strict. And, these strict conditions have already touched upon the conceptual problem of understanding movements of materials. In other words, from the very start, the so-called commonality of wave motions is given by listing very strict constraints. Other than the previous assumptions, the following critical problems also exist.

First, the assumption of evenness of the string did not seem to be any problem at the historical moment when it was introduced. However, when looking at it today, this assumption does not agree with reality. Not only is its form not rigorous, but also does it touch on the problem of physical essence of whether or not materials' evenness or unevenness satisfies the first push. Evidently, if the materials are even, then one is looking at a problem of physics of the first push system; if the materials are uneven, then he is dealing with the problem of whether or not the physics of the first push system actually holds true. It is because if the material the string is made up of is uneven (which should be the common situation), then the unit mass (density) of the string will no longer be constant, which will inevitably cause interactions between the variable density and the external acting force. No doubt, the commonality of mutual reactions is spinning movements instead of string vibrations. So, the assumption of the string's evenness itself is not about pursuing the common form of physical motions; instead, it is about limiting the investigation within the range of mathematical linearity due to the need of mathematical modeling. So, the so-called system of string vibrations is established on the assumption that the underlying materials cannot be damaged. In other words, the commonality of wave motions itself is an expansion of the thinking logic of vibrations in solids within the system of invariant materials. To this end, please do not conclude that we are trying to deny the problem of wave motions. Instead, we only wish to cast our doubts on the commonality of wave motions.

Second, the study is only limited to small, horizontal vibrations of the string, where at any point on the string, the amplitude of vibration needs to be so small that the angle formed by the tangent line and the axis is nearly zero. Evidently, it is hard for people to comprehend this requirement, since the tangent line's near-zero angle of inclination is equivalent to no obvious difference or the vibration of the string is so small that it is nearly not moving at all. However, from our discussion below, it can be seen that this assumption is introduced in order to apply some necessary mathematical techniques.

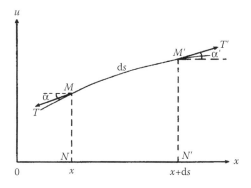

Figure 4.1   The basic model for string vibration.

Third, the string vibration is limited to horizontal direction only. In other words, when the string vibrates, it cannot sway from left to right; instead it can only move up and down vertically to the *x*-axis in Figure 4.1. If it allows swaying motions, the resultant movement will become an irregular vibration where mathematical techniques will have to meet with problems they cannot resolve.

The mathematical process of modeling of the string vibration: Based on the basic assumptions, let the amplitude of the string vibration be *u* (*MN* in Figure 4.1), the horizontal coordinate of a point is *x*, and the location of *t* is *M*. Then, $u = u(x,t)$. If the corresponding length of the arched string is $ds = \overline{MM'}$, ρ stands for the line density of the arch, and the stresses at both ends at *M*, *M'* are respectively *T*, *T'*. These stresses balance the recovering forces for the string to return its original equilibrium state and the string's weight.

The mathematical expression for the stressed string: The problem of a stressed string at time moment *t* is considered essentially by introducing Newton's second law of the first-push system:

$$f = m\,a$$

where *m* stands for the mass of the unit arch *ds* and *a* the acceleration of the unit arch *ds*. The corresponding component of the tension along the tangent direction is:

$$-T\cos\alpha + T'\cos\alpha'$$

The corresponding combined force of the component along the *u* direction and the gravitation (where it has already assumed that the acceleration of gravity *g* is perpendicular to the *x*-axis) is

$$f = -T\sin\alpha + T\sin\alpha - \rho\,g\,ds$$

Since it has been assumed that the vibration is small and the magnitude is equal to the angle of inclination of the tangent line:

$$\alpha = \alpha' \cong 0$$

we can use the series expansion of trigonometric functions:

$$\cos \alpha = 1 - \frac{\alpha^2}{2} + \frac{\alpha^4}{4} - \cdots\cdots$$

Now, a commonly used mathematical treatment is to ignore the quadratic or higher power terms. And, the component of the acting force along the x direction is

$$\begin{cases} \cos \alpha = \cos \alpha' \cong 1 \\ T = T' \end{cases} \tag{4.1}$$

Besides, from the elementary facts of trigonometric functions, it follows that when $\alpha = \alpha' \cong 0$, we have

$$\begin{cases} \sin \alpha = \dfrac{tg\alpha}{\sqrt{1 + tg^2\alpha}} \cong tg\alpha = \dfrac{\Delta u(x,t)}{\Delta x} \cong \dfrac{\partial u(x,t)}{\partial x} \\ \sin \alpha' \cong tg\alpha' = \dfrac{\partial u(x+dx,t)}{\partial x} \end{cases} \tag{4.2}$$

So, from the Pythagorean theorem and the approximation relationship of differences, we have

$$ds^2 \cong \Delta x^2 + \Delta u^2 \cong dx^2 + \left(\frac{\partial u}{\partial x}\right)^2 dx^2$$

According to the assumption that the vibration is small, by ignoring the second-order term of

$$\frac{\partial u}{\partial x}$$

we have

$$ds^2 \cong \Delta x^2 \tag{4.3}$$

By introducing the acceleration

$$a = \frac{\partial^2 u}{\partial t^2}$$

of the unit arch ds of mass ρds, we can obtain the mathematical expression of the acting force in the vertical direction from Equations (4.1) through (4.3), which is exactly the combined force of vertical component of the string's shear stress force and the gravitation, as follows:

$$f = -T \sin \alpha + T \sin \alpha - \rho g \, ds \tag{4.4}$$

The corresponding Newton's second law for this special case of string vibration is

$$f = \rho \, ds \frac{\partial^2 u}{\partial t^2}$$

So, $T = T'$ in Equation (4.1) leads to

$$f = \rho \, ds \frac{\partial^2 u}{\partial t^2} = T \left[\frac{\partial u(x + dx, t)}{\partial x} - \frac{\partial u(x, t)}{\partial x} \right] - \rho g \, ds \tag{4.5}$$

Since the second-order derivative can be written as

$$\left[\frac{\partial u(x + dx, t)}{\partial x} - \frac{\partial u(x, t)}{\partial x} \right] \cong \frac{\partial}{\partial x} \left(\frac{\partial u}{\partial x} \right) dx = \frac{\partial^2 u}{\partial t^2}$$

substituting this expression into Equation (4.5) produces

$$\left(T \frac{\partial^2 u}{\partial x^2} - \rho g \right) dx = \rho \frac{\partial^2 u}{\partial t^2} + g$$

From the assumption that the stress acceleration of elasticity

$$a = \frac{\partial^2 u}{\partial t^2} \gg \text{the acceleration of gravitation } g$$

(in modern science, the symbol ">>" stands for *greater than two magnitude levels* and is a commonly applied symbol with the quantitative comparability in inertial systems), and the requirement that the function u is twice continuously differentiable with respect to x,t, then we have:

$$\frac{\partial^2 u}{\partial t} = \alpha^2 \frac{\partial^2 u}{\partial x^2} \tag{4.6}$$

where

$$\alpha^2 = \frac{T}{\rho}$$

Equation (4.6) is the famous Taylor equation of string vibrations, which can be expanded to the form of three-dimensional space. What should be noted is that the deterministic problem corresponding to the wave motion Equation (4.6) has not only the characteristic value but also the characteristic function:

$$\lambda^2 = \frac{n^2 \pi^2}{L^2} \quad (n = 1, 2, 3, \ldots) \tag{4.7}$$

$$X_n(x) = B_n \sin \frac{n\pi}{L} x \quad (n = 1, 2, 3, \ldots) \tag{4.8}$$

Since in these formulas L stands for the length of the string, π the ratio of circumference and the diameter of a circle, B_n coefficients, from Equations (4.7) and (4.8), it follows that λ is periodic and $X_n(x)$ takes the form of vibration of harmonic waves.

So, at this junction, the reader can see that the classical Taylor equation of string vibration has quite clearly shown the fact that inertial systems, masses, and forces of the Newtonian first push system are human inventions, where the mass m is given quantitatively by the product ρds of the line density ρ and the unit arch ds, and the stress of the string T and the weight $\rho g ds$ are also given in the form of quantities. And he can also vividly see the specific process of modeling of the problem of the classical string vibrations and how mathematical techniques are applied. The essence of the evenness assumption is to avoid any twisting effect and its consequent rotations due to interactions caused by unevenness. The assumption of small vibrations is required by the need to apply the method of linearization (ignoring the quadratic term). And the requirement of vertical vibrations eliminates any potential irregular movements that cannot be addressed mathematically. Here, the constraint of mathematical linearity embodies the mathematical techniques of form of wave motions, which can also be referred to as the mathematical character-istics of the system of wave motions. These characteristics not only laid down the foundation for mathematical physics equations, but also led to the establishment and construction of the theoretical physics of wave motions.

So, the Taylor equation of string vibration not only is a representative of model-ing using mathematical physics equations, but also places wave motions, as physical phenomena in the form of mathematics, in the palace of science, leading to the three major classical equations of mathematical physics — the hyperbolic equations (such as the Taylor equation of string vibration), the parabolic equations of heat conduc-tion, and the Poisson or Laplace equations that represent equilibrium relationships.

However, what needs to be noted is that it is indeed a forced analogy for the equations of heat conduction and density diffusion to take the same form, leading to wave propagations out of their deterministic problems. Evidently, the diffusion of density is a problem about transformations of eddy currents, while the corre-sponding mathematical model did not reveal the existence of eddy currents, leaving behind an extremely difficult problem of disturbances for modern science. Or in other words, as an intuitive phenomenon, diffusion of density is completed through transformations of disturbances. Especially, the wave motions of energy transfers cannot complete transfers of pf materials. In particular, in the later development of mathematical physics equations, many models established in the system of wave motions by using linearizing of nonlinear equations also suffer from the problem of treating eddy currents as wave motions, which inevitably leads to problematic consequences. That is why the wide-range generality of mathematical waves is not the same as the physical generality of wave motions.

Therefore, the outstanding achievement of the wave-motion physics of the 20th century simply missed the more general eddy movements under the cover of the system of wave motions. In reality, it can be shown using the quite simple method that nonlinear dynamic equations stand for the problem of rotationality of solenoi-dal fields.

No doubt, from the process of modeling and establishing the equation of string vibration, people can at least see that the mathematical description of wave motions and the corresponding mathematical physics equations are some of the products of specific constraints. Slightly loosening up some of the conditions will make the mathematical description of wave motions no longer valid. So, we would seriously think about the mathematical ability of description, as mentioned by Aristotle; and in mathematical descriptions, there are also many forced analogies of mathematics itself. In particular, the essence of physics is to address causes of events, while math-ematical descriptions cannot provide answers to *whys*. So, in terms of the essence of physics, its current research has not reached its planned goals. In terms of the concrete problem we have been looking at, as long as the string is not even (which should be the common situation), even if the vibration is small, its characteristic values will no longer be periodic and the property of harmonic waves of its characteristic function will inevitably be destroyed. So, one can see at least that in applications of mathematical techniques, he has to face the following:

Modern science has to experience that without assumptions, there will not be any science.

The essence of mathematical techniques is to make a rough estimate of the underlying physical problem. In other words, at the same time when people pursue the quantified accuracy, they create fabricated problems of physics so that in many situations, the events of concern are modified and false information is created.

For example, because of the inertia of the traditional thinking logic and the establishment of the classical linear equations, people gradually applied the relevant ideas and methods in the studies of nonlinear movements of fluids, fundamental particles, and the theory of gravitation (Einstein). That is why in the community of learning of the 20th century, the belief that wave motions commonly exist appeared.

Evidently, after the three classical equations of wave motions are established, equations of fluid mechanics are surely among those considered next. Since the viscosities of water and air are relatively small, shear stress forces cannot be seen as internal to the structures of fluids. So for fluids, one should not follow the process of establishing the Taylor equation of string vibration to create equations of wave motion for fluids. After Euler in 1755 introduced his equation of fluid movement, Navier and Stokes added a dissipation term so that their new equation has been referred to as N-S equation. That is how nonlinear mathematical models were initially produced, leading to the mystery of nonlinearity in mathematics and physics. In particular, after the Coriolis force of rotation was added into the N-S equation, the resultant equation has been studied by using linearization and other methods, gradually revealing the deficiencies of mathematical techniques and general concepts of the current system of wave motions and showing confusion in physical concepts, and further and further departing from the original realistic problems.

In the following, to explain things more clearly, let us use the process of simplifying equations of fluid mechanics into equations of wave motions. At this junction, one should be careful that simplifying an equation would not alter the properties of the equation. Even for mathematical properties, none of them can be modified in the name of simplification. And, the currently well-applied method of linearizing nonlinear equations has substantially changed the mathematical properties of the equations involved. Even as pointed out here, in the following, we will still employ the current method and see what is missing from the mathematical process of simplification.

First, according to the custom, let us assume that no friction exists and the pressure and temperature are not related; then the Euler equation and its corresponding system of equations are

$$
\begin{cases}
\dfrac{\partial \vec{v}}{\partial t} + \vec{v} \cdot \nabla \vec{v} = -\dfrac{1}{\rho}\nabla P \\[2mm]
\dfrac{\partial \rho}{\partial t} + \nabla(\rho \vec{v}) = 0 \\[2mm]
P = f(\rho)
\end{cases}
\tag{4.9}
$$

where $\vec{v}$ stands for the current speed vector, ρ the fluid density, P the intensity of pressure of the fluid, and ∇ the Hamilton operator. In terms of the physical significance, Equation (4.9) describes a physical problem with constant temperature (or insulation) and constant entropy. Even so, the mathematical model of Equation (4.9) still involves nonlinear equations. Due to historical reasons, people did not recognize until recently that nonlinear dynamic equations stand for problems of rotationality of solenoidal fields and cannot be linearized. Due to the historical inertia of thinking logic, it is a natural consequence for people to linearize nonlinear problems. Based on the format of thinking of modern science that without assumptions, there will not be any science, people have tried if at all possible to find reasons for employing linearization. That is why it is required that Equation (4.9) describes such a fluid that has little viscosity under small disturbance and weak unevenness (unevenness or strong unevenness will have to produce irregular rotations). No doubt, the purpose of imposing these assumptions is for the convenience of linearization. Or in other words, the simplification of Equation (4.9) can be carried out in a similar fashion as what is done for the mathematical modeling of the classical string vibrations.

That is, using the method of introducing small disturbances into Equation (4.9) and ignoring the higher order terms produce

$$
\begin{cases}
\rho = \bar{\rho} + \rho' \\
P = \bar{P} + f'(\rho)\rho' \\
\vec{v} = \vec{v}'
\end{cases}
\tag{4.10}
$$

where $f'(\rho)$ is a linear function only in ρ. Inserting Equation (4.10) into Equation (4.9) and ignoring small terms of second and higher orders, we obtain the corresponding equation of disturbance. For the convenience of presentation, in the following, we omit the symbol "'" from each term of the following equations. So, we have

$$\begin{cases} \dfrac{\partial \rho}{\partial t} + \overline{\rho} \cdot \nabla \vec{v} = 0 \\[2mm] \dfrac{\partial \vec{v}}{\partial t} + \dfrac{f'(\overline{\rho})}{\overline{\rho}} \nabla \rho = 0 \end{cases} \qquad (4.11\text{a})$$

Taking

$$\frac{\partial}{\partial t} - \overline{\rho}\,\nabla \qquad (4.11\text{b})$$

produces

$$\frac{\partial^2 \rho}{\partial t^2} = f'(\overline{\rho})\nabla^2 \rho = \alpha^2 \nabla^2 \rho \qquad (4.12)$$

where $\alpha^2 = f'(\overline{\rho})$. Evidently, Equation (4.12) is a wave motion equation of the fluid that looks the same as the Taylor Equation (4.6) of string vibrations except that it is written in the three-dimensional form. So, this equation can also be rewritten in the form of other variables or physical quantities. For example, if we assume that in Equation (4.10), P and ρ are linearly related, we have

$$\frac{\partial^2 P}{\partial t^2} = \alpha_1^2 \nabla^2 \rho \qquad (4.13)$$

where α_1^2 and α^2 are also linearly related. If we take

$$\frac{\partial}{\partial t}(4.11\text{b}) - \frac{f'(\overline{\rho})}{\overline{\rho}}\nabla(4.11)\text{b}$$

we can also obtain:

$$\frac{\partial^2 \vec{v}}{\partial t^2} = \alpha_2^2 \nabla^2 \vec{v} \qquad (4.14)$$

where $\alpha_2^2 = f'(\overline{\rho})$. If we introduce the speed potential $U, \vec{v} = \nabla U$, then we have

$$\frac{\partial^2 U}{\partial t^2} = \alpha_3^2 \nabla^2 U \tag{4.15}$$

where $\alpha_3^2 = \alpha_2^2$.

So, Equations (4.12) through (4.15) are respectively equations of fluid wave motion in the variables or physical quantities ρ, P, $\vec{v}$ and U. At the same time, it also indicates that after nonlinear equations are linearized, they become wave-motion equations similar to those describing vibrations in solids. So, that explains why the world of learning later on mimicked this process by following the historical belief that each fluid is a continuum of low-resistance solid (which has been considered as one of the three major scientific achievements of the 17th century), which was supported by the phenomenon that wave motions appear in fluids under small disturbances.

It should be pointed out that the small disturbances in wave motions of fluids, as described above by using mathematics, are equivalent to ripples in fluids. In other words, wave motions in fluids are only one form and not the only form of fluid movements. This form of movement is not the major form of fluid movement, either. This end indicates that the wave-motion equations using mathematical linearity can expand the recovering elasticity of solids into nondamaged continuous movements in fluids.

What is unfortunate is that all the analytical analysis of the equations of fluid motion since the time of Euler has not revealed the main or common fluids' mathematical property — the mathematical properties of nonlinearity. Due to the mathematical successes on linear problems, a fixed thinking logic of mathematical linearity was established, on which it has been perceived that mathematical nonlinearity is simply expansions of linear problems, or nonlinear problems can be approximated by using linear estimations. That is not only an epistemological mistake but also a theoretical nonsense, which has been continued for over a century and still influences how some people think, leading to such concepts as gravitational waves in general relativity theory, probabilistic waves in quantum mechanics, and others that have affected various fields of learning, such as shock waves, solitons, trapped waves, chaos, wavelets, fractional dimensions, and so forth. Even as of this writing, some people still believe that fundamental particles are all waves, leading to the eternal and commonly existent wave motions in the universe.

As a matter of fact, even in terms of mathematical forms, nonlinear equations do not satisfy the linear initial-value existence theorem. Or, speaking in ordinary language, the solution of a nonlinear equation in general cannot be obtained by stably expanding or the extrapolating of initial values, which is also stated as follows: Nonlinear equations do not satisfy the well-posedness theory of linear equations. Evidently, in terms of physics, mathematical nonlinearity corresponds to the concept of mutual reactions outside of the first push. Speaking more straightforwardly, fighting against each other is different from the concept of being beaten up on the first push. If nonlinear problems are linearized, it is like modifying an active physical fight against each other into being beaten. In other words, the basic epistemology as reflected by the physical phenomenon of mutual reactions should be about how to walk out of the first-push system, which stands for the fundamental problem of how to understand the natural philosophy after it was initially considered over 300 years ago. To this end, it has to involve the problem of whether or not the commonality of wave motions of the first-push system is truly the commonality of materials' movements in the universe. Or speaking to the point, have the nonlinear forms of mathematics described the commonality of wave motions?

Next, nonlinear problems of mathematics touch upon the familiar concept of inertial systems, which stands for a fundamental problem that has appeared ever since the time when the modern scientific system initially appeared. Although the mechanics, or in particular the fluid mechanics, has treated the Coriolis force, caused by the rotation of the earth, as imaginary in order to maintain inertial systems of the Newtonian particle mechanics (or stated as expanding inertial systems to include the mechanics of spinning fluids), and the noninertial system of the spinning earth has been considered in form, the problem of variable accelerations has been ignored or forgotten. In fact, variant accelerations stand for problems of noninertial systems. If we introduce the Coriolis force, also called the Coriolis acceleration, then the equation of movement of the spinning fluid in the Lagrange language is

$$\frac{d\vec{v}}{dt} + 2\bar{\Omega} \times \vec{v} = -\frac{1}{\rho}\nabla P - \vec{g} + \vec{F} \tag{4.16}$$

where $\vec{v}$ stands for the current speed vector, $\bar{\Omega}$ the angular speed vector of the earth rotation, ρ the density of the fluid, ∇P the pressure (or the gradient force of the cubic pressure) of the fluid, $\vec{g}$ the acceleration of gravitation in the vector form, and $\vec{F}$ the friction vector. If we move the second term on the left-hand side of Equation (4.16) to the other side, then we obtain the Coriolis force. So, we should notice that Equation (4.16) is an equation of fluid motion with Coriolis acceleration added. If we only analyze the left-hand side of Equation (4.16), and rewrite it as

$$\frac{d\vec{v}}{dt} = -2\bar{\Omega} \times \vec{v} \tag{4.17}$$

Even if we assume that the right-hand side $\bar{\Omega}$ is a constant, since $\vec{v}$ of the right-hand side is not a constant, Equation (4.17) has to be a problem of variant acceleration of noninertial systems. Evidently, in this case, we can no longer assume $\vec{v}$ is a constant. If we assume $\vec{v}$ on the right-hand side of Equation (4.17) is a constant, then the $\vec{v}$ of the left-hand side is also a constant and

$$\frac{d\vec{v}}{dt} = 0 \tag{4.18}$$

So, the corresponding Equation (4.16) is no longer true. Establishing the equation of motion with the Coriolis force introduced means that we have already entered a noninertial system.

For convenience, let us introduce the one-dimensional equation of fluid motion in Euler language; then Equation (4.18) can be written as follows:

$$\frac{\partial u}{\partial t} + u\frac{\partial u}{\partial x} = 0 \tag{4.19}$$

Similarly, even if we assume that

$$\frac{\partial u}{\partial x}$$

in Equation (4.19) is a constant, then since u can no longer be assumed to be a constant, Equation (4.19) has to become a problem of variant acceleration of noninertial systems. If the u in the second term is taken to be a constant, then the u in the first term has to be a constant too. So, we must have

$$\frac{\partial u}{\partial t} = 0 \tag{4.20}$$

Now, the corresponding Equation (4.19) is no longer true. This analysis has sufficiently shown that nonlinear dynamic equations are no longer problems in inertial systems. However, the current community of learning has not recognized this problem.

This discussion indicates that if nonlinear dynamic equations are linearized, then one has artificially changed problems of noninertial systems to ones of inertial systems. In other words, nonlinearity cannot be linearized. However, the popular, simplified mathematical models, created by linearizing nonlinearities in the 20th century, are consequences of a conceptual mistake. Those who have made this mistake include such big names as Einstein. Therefore, many theories established in the movement of expanding the system of wave motions by using linearization of nonlinearities have to be reconsidered.

Third, the equations of movement of spinning fluids, similar to that in Equation (4.16), should be named as N-S-C equations, because with or without the Coriolis term, the relevant physical meanings are completely different. Only in terms of quantitative analysis will one have to face the following problems. For instance, from only considering a horizontal movement, one has the following terms that are nearly equal in terms of their quantitative magnitudes:

$$\left| 2\bar{\Omega} \times \bar{v} \right| \approx \left| \frac{1}{\rho} \nabla P \right| \tag{4.21}$$

that is what is called quasi-equal quantitative magnitudes. Or in other words, the magnitude of the first term in Equation (4.16) is smaller than the second term or the first term on the right-hand side by at least two magnitude levels. That is

$$\left| \frac{dv}{dt} \right| \ll \left| 2\bar{\Omega} \times \bar{v} \right| \quad \text{or} \quad \left| \frac{1}{\rho} \nabla P \right| \tag{4.22}$$

Even those with only some elementary knowledge of mathematics also know that Equation (4.16) stands for a problem of quantitatively computing large quantities with infinitesimal differences. Or in other words, even if Equation (4.16) indeed holds true, it is still a formula that is not operational in quantitative computations. It is because if a term on the right-hand side of Equation (4.22) contains an error of one magnitude level, then the left-hand side will experience an error of two magnitude levels.

Therefore, even if one only considers problems of quantities, it does not mean that all equations are meaningful. And, quantitative analysis also needs to deal with specific circumstances. It seems that after the quantitative analysis system has been in operation for over 300 years, it is a good time for us to think about problems existing in this system.

Evidently, the computational inaccuracy of mathematical equations in essence challenges the eternality of equations. Besides, inertial systems of modern science also meet with the challenges of nonlinearity and noninertial systems, providing hints about how to recognize and treat problems of nonlinearity.

Also, as for problems involving celestial body movements, what deserves our thinking is whether or not the angular speed of the celestial body's rotation can be seen as constant just because its quantitative changes are quite small or minor. One needs to be clear that rotational movements are not only caused by acting forces of the first push; instead, they involve the correspondence between nonfirst pushes and problems of nonlinearity — the problem of moments of forces.

For example, changes in the angular speed of the earth's rotation are relatively small in the time scale used currently so that the nutation of the earth is also relatively small. However, even though the nutation is relative small, the nutation moments should not be seen as small. After all, the diameter of the earth or any celestial body is very large; and so, in the study of changes in rotations the effects of moments of forces should always be considered. What needs to be noted is that due to differences in spinning directions, rotations have to cause changes in materials' structures. Therefore, taking $\bar{\Omega}$ as a constant cannot help to represent the phenomena of natural disasters. Considering the development of science and the need to resolve problems, forces cannot be the ultimate reasons for changes occurring in materials. That is why studies and theories of moments of forces and of materials' structures should be constructed within the range of active knowledge explorations and should be treated as the foundation of human knowledge. Also, each inertial system is always constructed on a static reference frame, while in nature there does not exist any static object that can be used as a reference, and being static or equilibrium are only relative in comparison with high-speed spinning celestial bodies. So, moments of forces of rotations and noninertial systems are the most effective ways of work chosen by nature instead of inertial systems of the first push. Inertial systems are introduced due to the human need to replace events by physical quantities and to employ quantitative analysis where the universe is seen as static and objects and forces are separate.

So, noninertial systems should be a concept physics especially emphasizes with the realization that in nature, objects and forces are not separate from each other. And, moments of forces and rotationality of materials should be considered in the forthcoming new era of process physics.

4.2 Mathematical Waves and Related Problems

The previous discussion has mentioned that modern science has treated wave motions as a general topic such that almost all areas of science and engineering study wave motions and that the theory of wave motions has been seen as the highest level of achievement of science. That was the reason why in the middle of the 20th century, the general definition of wave motions applied as few constraints as possible with as much generality as possible: A wave motion is a recognizable signal that possesses a speed of propagation from one part of the media to another; it can be a disturbance of any attribute such that even if the quantitative value or extreme value experiences sudden changes, its disturbance can still be clearly recognized with its position determinable at any time moment; even if the signal is distorted with different quantitative values, it is still recognizable.

Evidently, the purpose for introducing this concept in this way is for the convenience of employing mathematical methods and techniques without thinking much about the relevant physical properties. What is more important is that physically intuitive, vibrant, reciprocating straight-line disturbances are generalized to unidirectional disturbances of eddy currents, including all forms of movements. This definition of wave motions exactly makes different disturbances indistinguishable and confuses properties of various disturbances. No doubt, its essence is to implicitly lay down the foundation for the introduction at the start of the 20th century of such concepts of modern physics as probabilistic waves, gravitational waves, shock waves, trapped waves, unified field theory and gauge fields, and so forth.

It is exactly because of the need for employing mathematical techniques that wave motions are defined generally so that the linearity of reciprocating vibrations is confused with the nonlinearity of eddy motions at the price of mixing up the difference between inertial and noninertial systems. So, the logic thinking of Euclidean spaces is smoothly carried over into curvature spaces. That is why some scholars believe that the concept of probabilistic waves is the foundation for understanding microscopic physics and can be generalized to all fundamental particles, leading to the epistemology that all fundamental particles are waves, all movements are wave motions, and so forth.

4.2.1 Transformation between Mathematical Hyperbolic Wave Motions and Nonlinear Flows

Mathematical hyperbolic waves use hyperbolic differential equations to describe the underlying wave motion. In mathematics, hyperbolic equations have their clear definition and have nothing to do with whether or not these equations have explicit solutions.

No doubt, although mathematical hyperbolic wave motions might have solutions of different forms, the given equations determine the forms of the solutions, including the traveling waves of the equations of one-dimensional wave motions, the standing waves of the two-dimensional wave equations, the three-dimensional spherical waves, and so forth. And, their mathematical forms are carried over to the studies of nonlinear equations. In particular, after the method of traveling waves developed for solving linear equations is applied in the investigation of nonlinear equations, nonlinear hyperbolic equations are also seen as wave equations. For instance, although the solutions of one-dimensional nonlinear (which used to be called quasi-linear) equations can be obtained by using the method of traveling waves, the properties of the obtained solutions have been substantially altered.

4.2.1.1 Nonlinear One-Dimensional Flows and Their Transitional Changes

In the system

$$\begin{cases} u_t + u u_x = 0 \\ u_0(x,0) = F(x) \end{cases} \tag{4.23}$$

take $u = F(\xi) = F(x - ut)$; then we can let

$$u - F(\xi) = 0 \tag{4.24}$$

Differentiating Equation (4.24) with respect to t produces

$$u_t - F_\xi \cdot (-u - u_t t) = 0$$

or

$$u_t = -\frac{F_\xi u}{1 + F_\xi t} \tag{4.25}$$

This equation indicates that even if the function u is continuous, it is the flow speed instead of the wave speed. And, the changes in the flow speed u_t can experience discontinuity. We have to note that flow speed is a concept about the movement of the fluid, while wave speed is a concept about the propagation of energy without any movement of the fluid.

By differentiating Equation (4.24) with respect to x, we have

$$u_x - F_\xi \cdot (1 - u_x t) = 0$$

or

$$u_x = \frac{F_\xi}{1 + F_\xi t} \tag{4.26}$$

Equation (4.26) indicates that even if the function u is continuous, the spatial distribution u_x of u can still contain discontinuity. And, multiplying Equation (4.26) by u produces

$$uu_x = \frac{F_\xi u}{1 + F_\xi t} \tag{4.27}$$

It means that even if we use the method of traveling waves to analyze the properties of the solution of a nonlinear equation, there are still fundamental differences between nonlinear and linear equations.

As for solving the deterministic problem of differential equations, we have to follow the rules of operation that govern these equations. To this end, we can directly differentiate the first equation in Equation (4.23) with respect to x and produce

$$\begin{cases} (u_x)_t + u(u_x)_x + (u_x)^2 = 0 \\ t = 0, u_x(x, 0) = u_{0x}(x) \end{cases} \tag{4.28}$$

Using characteristic equation or the transformation of the Euler and the Lagrange languages,

$$\frac{du}{dt} = 0, \quad \frac{du}{dt} = \frac{\partial u}{\partial t} + u \frac{\partial u}{\partial x} = 0$$

The first equation in Equation (4.28) can be rewritten as follows:

$$\frac{\partial}{\partial t}(u_x) + u \frac{\partial}{\partial x}(u_x) + (u_x)^2 = 0$$

If writing it together with its initial values, then Equation (4.28) can be written as

$$\begin{cases} \dfrac{d}{dt}(u_x) + (u_x)^2 = 0 \\ t = 0, \, u_x(x,0) = u_{0x}(x) \end{cases} \tag{4.29}$$

By using the method of separate variables, it is not hard for us to get the solution as follows:

$$u_x = \frac{u_{0x}}{1 + u_{0x}t} \tag{4.30}$$

where

$$u_{0x} = \frac{\partial u_0}{\partial x}$$

Or its nonlinear term is

$$uu_x = \frac{uu_{0x}}{1 + u_{0x}t} \tag{4.31}$$

That is, Equations (4.30) and (4.31) are similar to Equations (4.26) and (4.27).

If we apply the method of separate variables specifically, for Equation (4.23), we can take

$$u(x,t) = A(t)U(x)$$

$$U_0(x) = A(0)U(x) \tag{4.32}$$

Substituting Equation (4.32) into Equation (4.23) and letting

$$\dot{A} = \frac{dA}{dt}$$

produces

$$U\dot{A} = -A^2 U U_x$$

Now, separating the variables leads to

$$\frac{\dot{A}}{A^2} = -U_x = -\lambda \tag{4.33}$$

By noticing the initial value condition (4.32) and $\lambda = U_x$, and considering the function U_x as the spatial distribution so that it can be assumed to be a constant, we have

$$\dot{A} + \lambda A^2 = 0$$

or

$$\dot{A} + U_x A^2 = 0. \tag{4.34}$$

Integrating this expression with respect to the separate variables and taking $A(0) = A_0$ lead to

$$A = \frac{A_0}{1 + U_x A_0 t}$$

Point-multiplying the expression by U_x and taking $A_0 U_x = u_{0x}$ produce

$$u_x = \frac{u_{0x}}{1 + u_{0x} t} \tag{4.35}$$

Similarly, we can obtain

$$uu_x = \frac{uu_{0x}}{1 + u_{0x} t} \tag{4.36}$$

No doubt, Equations (4.35) and (4.36) are the same as Equations (4.30) and (4.31). And because of this, we point out that the nonlinear problem meets transitional changes that are named as blown-ups. Since

$$u_{0x} = \frac{\partial u_0(x)}{\partial x}$$

it stands for such a term that indicates whether the initial value is convergent or divergent. If $u_{0x} < 0$, we have a convergent problem of physics; if $u_{0x} > 0$, we have a physical divergence and the dimension of

$$u_{0x} = \frac{\partial u_0(x)}{\partial x}$$

is $[s^{-1}]$, one over second. So, we have the following:

First, if the initial value is convergent, that is, $u_{0x} < 0$, then Equation (4.35) becomes

$$u_x = \frac{u_{0x}}{1 - |u_{0x}|t} \tag{4.37}$$

So, $|u_{0x}| = t^{-1}$ and if $\left\| u_{0x} \right| t \right| = 1$, then Equation (4.37) stands for a problem of mathematical singularity. As a problem of physics, the spin of the earth will not stop, so when $\left\| u_{0x} \right| t \right| > 1$, the initial convergence will have to be changed to divergence. That is a blowing-up change — blown-up — of the initial value.

Second, if the initial value is divergent, then we have

$$u_x = \frac{u_{0x}}{1 + |u_{0x}|t} \tag{4.38}$$

That is, $u_{0x} > 0$. So, as time $t \to \infty$, the divergence of the initial value will surely approach zero.

What should be clear is that as a problem of mathematics, due to the requirement of well-posedness, if $u_{0x} < 0$, the situation is only limited to $\left\| u_{0x} \right| t \right| \to 1$, and is called a mathematical problem of singularity or ill-posedness. As a problem of physics, after all, mankind cannot constrain the spin of the earth. So, the concept of blown-ups is originated in physical problems instead of the requirement of mathematics. The relevant discussion also illustrates that mathematical nonlinear problems do not satisfy the linear existence theorem of initial values and represent a brand-new class of problems for mathematics. That is, the mathematical theory of linear well-posedness can no longer be inherited in the study of nonlinear problems.

Also, one-dimensional nonlinear initial value problems can be analyzed completely by using currently available methods of mathematics producing the corresponding mathematical properties. What is important is that what nonlinear initial value problems reveal is the non-initial-value automorphic transitional changes, which not only explains the philosophical law that at extremes, matters will evolve in the opposite direction, but also agrees with the underlying physical changes of transitionality. However, what nonlinear mathematical models represent is flows instead of the waves of linear mathematical models.

If nonlinear problems are linearized, although the consequences satisfy the principle of well-posedness, they violate physical reality. So, it should be clear that theories that agree with reality should not be limited to only the one-dimensional mathematical forms. However, it is a pity that all two- or higher-dimensional nonlinear mathematical models suffer from mathematical difficulties. However, the properties of one-dimensional nonlinear mathematical models should present some of the fundamental characteristics of nonlinear mathematical models.

In short, mathematical linear models show mathematical waves, which are not suitable for nonlinear models. Or speaking with concrete details, since one-dimensional nonlinear mathematical equations can produce the properties of their solutions by using the method of traveling waves, it can be concluded that nonlinear hyperbolic equations are not mathematical waves.

4.2.1.2 Nonlinear Two-Dimensional Flows and Their Transitional Changes

By introducing a flow function, the equations for 2-D fluid dynamics can be written as follows:

$$\begin{cases} (\Delta\psi)_t = J(\Delta\psi, \psi) \\ \Delta\psi(x,t)\big|_{t=0} = \Delta\psi_0 \end{cases} \tag{4.39}$$

If in Equation (4.39)$_1$, we add $J(f, \psi)$, $f = 2\Omega\sin\phi$, where ϕ stands for the earthly latitude, then we obtain the general equation of eddy motion for a spinning atmosphere, where

$$J(A,B) = \frac{\partial A}{\partial x}\frac{\partial B}{\partial y} - \frac{\partial A}{\partial y}\frac{\partial B}{\partial x}$$

is the Jacobi operator. No doubt, $\psi(x, y, t)$ is a two-dimensional flow function,

$$u = -\frac{\partial\psi}{\partial y}, v = \frac{\partial\psi}{\partial x}$$

$\Delta\psi = \nabla^2\psi = \varsigma$, the vorticity in the vertical direction. Equation (4.39) represents the Cauchy initial value problem of a two-dimensional spinning fluid. In form, this problem is equivalent to the two-dimensional equation of fluid mechanics in Euler language. Considering that this equation comes from the fluid mechanics of continuous media, we have to introduce the assumption that $\psi_0(x,y)$ is continuously differentiable on $-\infty < x, y < +\infty$ and

$$\left|\psi_0(x,y)\right| \leq k, \, k = \text{const.} \tag{4.40}$$

Now, assume that $\psi(x,y,t)$, for $t \in [0,T](T\langle\infty); \, x, y \in (0,L)$ is uniformly continuous. And for the convenience of discussion, let us take

$$\begin{cases} x, y = 0, \, \Delta\psi(x,0,t) = \Delta\psi(x,0,t) = 1 \\ x, y = L, \, \Delta\psi(L,y,t) = \Delta\psi(x,L,t) = \Delta\psi_L \end{cases} \tag{4.41}$$

and, without loss of generality, let us introduce the separate variables as follows:

$$\psi(x,y,t) = A(t)\Psi(x,y),$$
$$\psi_0(x,y,0) = A(0)\Psi(x,y) \tag{4.42}$$

where we assume that $A(t)$ is positive for $t \in [0,T](T\rangle 0)$. Substituting Equation (4.42) into Equation (4.39) and some simplification produces

$$\Delta\Psi\frac{dA}{dt} = A^2\left[\Psi_y(\Delta\Psi)_x - \Psi_x(\Delta\Psi)_y\right]. \tag{4.43}$$

Evidently, the two terms within [ ] in Equation (4.43) are essentially the same. For convenience, let us take the first term for our discussion. For the convenience of presentation, take

$$\frac{dA}{dt} = \dot{A}$$

So, we have

$$\Delta\Psi\dot{A} = A^2\Psi_y(\Delta\Psi)_x \tag{4.44}$$

By using condition Equation (4.40) and factoring the not expanded variables, let us take

$$\frac{\dot{A}}{A^2} = -\lambda,$$

$$(\Delta\psi)_x + \frac{\lambda}{\Psi_y} \Delta\Psi = 0 \tag{4.45}$$

From Equations (4.45) and (4.42), it follows that

$$A = \frac{A_0}{1 + \lambda A_0 t} \tag{4.46}$$

When $1 + \lambda A_0 t = 0$ or $t = t_b$, we have

$$\lambda = -\frac{1}{A_0 t_b} \quad \text{or} \quad t_b = -\frac{1}{\lambda A_0} \tag{4.47}$$

Substituting this expression into Equation (4.46) and then point-multiplying the result by $\Delta\Psi$ produce

$$\Delta\psi = A(t)\Delta\Psi(x, y) = \frac{A_0 \Delta\Psi}{1 + \lambda A_0 t} = \frac{\Delta\psi_0}{1 - t/t_b} = \frac{\zeta_0}{1 - t/t_b} \tag{4.48}$$

where ζ_0 stands for the initial vorticity. No doubt, Equation (4.48) uncovers a transformation of the eddy current; and when $t \geq t_b$, a blown-up of transitional changes occurs. What deserves our attention is that this change is reversible.

If the initial vorticity satisfies $\zeta_0 > 0$, then when $t < t_b$, the movement will be a continuation of the positive vorticity. When $t = t_b$, the movement experiences blown-ups of transitional changes. When $t > t_b$, since $(1 - t/t_b)$, the movement changes from the initial $\zeta_0 > 0$ to the vorticity of $\zeta_0 < 0$. If the initial value is a negative vorticity $\zeta_0 < 0$, then the changes will be opposite of what has been described above. So, the nonlinear problem of two-dimensional eddy currents is different from that of one-dimensional convergent flows. That is, one-dimensional convergent flows can be transformed to divergent flows, and divergent flows cannot be inversely transformed back to convergent flows, while for two-dimensional eddy currents, positive and negative rations can be transformed into each other.

This is a very important problem, since the beginning of the 20th century, people have chased the system of wave motions, reaching such a frenzied level without exaggeration that no one seemed to care about having the original physical problems changed to mathematical problems so that the concept of mathematical waves could be introduced. If we investigate the reason behind these frenzied pursuits, as discussed above, we find that it is because mathematics has met with quantitatively ill-posed problems so that quantitative changes have to be controlled by artificially imposing the requirement of well-posedness. Consequently, our understanding of nonlinearity becomes limited. The conclusions obtained above can also be shown by using vector operations that nonlinearity stands for the rotationality of solenoidal fields.

4.2.2 Mathematical Dispersive Wave Motions

Mathematical dispersive waves are different from hyperbolic waves, where dispersive wave motions are introduced to describe the type of solutions of equations. For linear systems, the solution of the following form exists:

$$\varphi(x,t) = A e^{i(kx-\omega t)} \tag{4.49}$$

where A stands for the amplitude of vibration, k the wave number, and $\omega = W(k)$ the frequency. Here, x, k can be either scalars or vectors. However, in general, $W(k)$ is a real-valued nonlinear function in k. What needs to be noted is that hyperbolic and dispersive waves are not exclusive of each other. For example, the Klein–Gordon equation

$$\varphi_{tt} - \varphi_{xx} + \varphi = 0$$

is both hyperbolic and dispersive. However, such examples of nonexclusiveness are only isolated cases and we should not confuse the differences of these two classes of wave motions.

For a linear equation of constant coefficients

$$P\left(\frac{\partial}{\partial t}, \frac{\partial}{\partial x}\right)\varphi = 0 \tag{4.50}$$

if Equation (4.49) is a solution of Equation (4.50), then we can substitute Equation (4.49) into Equation (4.50). Since the coefficients in the equation are constant, by eliminating the amplitude of vibration and exponents, we have

$$P(-i\omega, ik) = 0 \quad \text{or} \quad G(W, k) = 0 \tag{4.51}$$

Therefore, the relationship between ω and k is dispersive. No doubt, the general solution of a linear problem can be constructed by using special solutions.

Since Equation (4.49) takes the complex form, when necessary, we can take the real part:

$$\operatorname{Re}\varphi = |A|\cos(kx - \omega t + \eta), \quad \eta = \arg A$$

Take

$$\xi = kx - \omega t \tag{4.52}$$

where ξ is referred to as the phase space. If ξ = a constant, then we have parallel planes, representing plane waves. The gradient of ξ with respect to the space variables is the wave number. For example, for the three-dimensional case, we have

$$grad\xi = \left(\frac{\partial \xi}{\partial x_1}, \frac{\partial \xi}{\partial x_2}, \frac{\partial \xi}{\partial x_3} \right) = \left(k_1, k_2, k_3 \right)$$

For the one-dimensional case, we have

$$grad\xi = \frac{\partial \xi}{\partial x} = k$$

The positive direction of the gradient is the same as the normal direction of the plane. If $\xi = kx - \omega t$ = a constant, differentiating with respect to time t produces

$$k\frac{dx}{dt} - \omega = 0 \quad \text{or} \quad \frac{dx}{dt} = \frac{\omega}{k}$$

So, we can define the phase speed as

$$c \equiv \frac{\omega}{k} \vec{k}_0 \tag{4.53}$$

where $\vec{k}_0$ stands for the unit vector along the direction of k. Take $\omega = W(k)$ and substitute it into Equation (4.53). We have

$$c = \frac{W(k)}{k} \vec{k}_0 \tag{4.54}$$

which represents that the phase speed is a function of the wave number k.

As for the equation of wave motion

$$\varphi_{tt} - c_0 \varphi_{xx} = 0$$

its dispersion relationship is $\omega = W(k) = \pm c_0 k$. So, we have

$$|c| = \left| \frac{\omega}{k} \right| = \left| \frac{\pm c_0 k}{k} \right| = c_0$$

That is, the phase speed is the same as the wave speed and has nothing to do with the wave number k. Generally, the phase speed c is related to the wave number k such that a different wave number causes a different phase speed. No doubt, since the phase speeds are different, the form of motion of the wave motion becomes dispersed and is called *dispersion*. In terms of the equation of wave motions, if the phase speed is constant, then there will not appear any dispersion. Evidently, for dispersive waves, one has to eliminate the scenario of constant phase speed and require:

1. $\omega = W(k)$ is a real-valued function;

2. $\det \left| \dfrac{\partial^2 W}{\partial k_i \partial k_j} \right| \neq 0$, (for the one-dimensional case, $W_{k_i k_j} \neq 0$).

That is, we require that $W(k)$ is a real-valued nonlinear function in k. So, the wave motion that satisfies the above conditions is referred to as a dispersive wave, and the corresponding equation as a dispersive equation. There are many examples of dispersive equations, for example, the Klein-Gordon equation

$$\varphi_{tt} - \alpha \varphi_{xx} + \beta^2 \varphi = 0, \quad \omega = \pm \sqrt{\alpha^2 k^2 + \beta}$$

the equation of the vertical wave of an elastic pole

$$\varphi_{tt} - \alpha \varphi_{xx} + \beta^2 \varphi_{xxtt} = 0, \quad \omega = \pm \frac{\alpha k}{\sqrt{1 + \beta^2 k^2}}$$

and the linearized KdV equation

$$\varphi_t + \delta \varphi_x + v \varphi_{xxx} = 0, \quad \omega = \delta k - v k^2$$

are all dispersive equations. However, the equation of heat-conduction (heat-diffusion)

$$u_t - \alpha u_{xx} = 0\,(\alpha > 0), \quad \omega = i\alpha k^2$$

and the linear equation

$$u_t + \beta u_x = 0, \quad \omega = \beta k$$

are not dispersive equations, because $\omega = W(k)$ is respectively a complex relationship and linear real-valued function. What should be noticed is that complex numbers are good at describing vibrations; and mathematical linearity focuses on quantities without transitionality. So, dispersion is shown by $W(k)$ as a real-valued nonlinear function in k. Its essence is to show the transitionality of nonlinear functions. In other words, the requirement of dispersion of linear equations that $W(k)$ is a real-valued nonlinear function in k has already indicated that the physical mechanism of dispersive waves is not about the "dispersion" in the form of the wave motions, instead, it is about transformations. This fact can be seen more clearly in the following nonlinear equations. So, from the previous example, we can see that the following:

In real-coefficient linear equations that contain only even-order derivatives or only odd-order derivatives, there must be real dispersive relationships.
In real-coefficient linear equations with both even- and odd-order derivatives, no real-dispersive relationship exists.

Besides, for the Schrödinger equation

$$i\hbar \frac{\partial \varphi}{\partial t} + \frac{\hbar^2}{2m}\varphi_{xx} = 0$$

although it contains both even-order and odd-order derivatives, this equation also contains real-dispersive relationship

$$\hbar\omega = \frac{\hbar^2 k^2}{2m}$$

It is because this equation contains a complex coefficient. What should be said is that the complex coefficient is not the ultimate reason. The fundamental reason for causing the dispersion is the transformations of spinning disturbances (see Section 4.4).

Dispersive equations can only lead to dispersive relationships of the polynomial kind.

4.2.3 The General Dispersive Relationship

The fact obtained above, that dispersive equations can only lead to dispersive relationships of the polynomial kind, naturally gives rise to the following problem: What kinds of operators can produce the general dispersive relationship? To this end, let us introduce the following integral-differential equation:

$$\varphi(x,t)_{tx} + \int_{-\infty}^{\infty} K(x-\xi)\varphi(\xi,t)_\xi \, d\xi \tag{4.55}$$

where K stands for a known function. If this equation has a solution of the form of wave motions as in Equation (4.49), substituting Equation (4.49) into Equation (4.55) leads to

$$\omega = k \int_{-\infty}^{\infty} K(x-\xi) e^{-ik(x-\xi)} d\xi = k \int_{-\infty}^{\infty} K(\varsigma) e^{-ik\varsigma} d\varsigma \tag{4.56}$$

where $\varsigma = x - \xi$. Equation (4.56) is a dispersive relation, which is determined by the function $K(\varsigma)$. If we rewrite Equation (4.56) as follows:

$$C(k) = \frac{\omega}{k} = \int_{-\infty}^{\infty} K(x) e^{-ikx} dx \tag{4.57}$$

then the right-hand side of Equation (4.57) is the Fourier transformation of $K(x)$. By using the inverse transformation formula, we have

$$K(x) = \frac{1}{2\pi} \int_{-\infty}^{\infty} C(k) e^{ikx} dk \tag{4.58}$$

So, for the general dispersive relationship, for an arbitrarily chosen $C(k)$, $K(x)$ should be the inverse Fourier transformation of $C(k)$. This fact indicates that $K(x)$ must be a piecewise smooth function and absolutely integrable on $(-\infty,\infty)$. In particular, if we take

$$C(k) = C_0 + C_2 k^2 + \cdots\cdots + C_{2m} k^{2m}$$

the Equation (4.58) implies that

$$K(x) = C_0 \delta^{(0)}(x) - C_2 \delta^{(1)}(x) + \cdots\cdots + (-1)^m C_{2m} \delta^{(m)}(x)$$

So, Equation (4.55) becomes

$$\varphi_t + C_0 \varphi_x - C_2 \varphi_{xxx} + \cdots + (-1)^m C_{2m} \varphi_{x^{2m+1}} = 0$$

When $C(k)$ is taken as a Taylor series in k, the corresponding equation is a differential equation containing all orders of derivatives.

Evidently, the previous dispersive relation only holds true for constant-coefficient linear differential equations. As for linear variable-coefficient differential equations, we can generally consider separable variables of the form:

$$\varphi(x,t) = e^{-i\omega t} X(kx), \quad \omega = W(k)$$

where $X(kx)$ should be the vibration function or called a harmonic wave function, taking, say, the Bessel function, and so forth. That is why this class of solutions can be referred to as dispersive wave motions and the corresponding equations the dispersive equations. However, what we covered here still cannot contain the complete definition of dispersion. But, there is one point that is clear: As for the Bessel function, $X(kx)$ should also be smooth.

As for nonlinear dispersive waves, they represent unsettled problems in the study of wave motions. In the current study of mathematical physics problems, most dispersive relationships are linearized.

As for problems involving uneven masses, they are treated as wave motions and introduce the following solutions of the form of wave motions:

$$\varphi(x,t) = \alpha \cos \xi \tag{4.59}$$

where α, ξ are all functions in x,t. Now, the local wave number and local frequency can be defined by

$$k(x,t) = \frac{\partial \xi}{\partial x}, \ \omega(x,t) = -\frac{\partial \xi}{\partial t} \tag{4.60}$$

Continuing the convention:

$$\omega = W(k) \tag{4.61}$$

If we eliminate ξ in Equation (4.60), we have

$$k_t + \omega_x = 0 \tag{4.62}$$

Substituting Equation (4.61) into Equation (4.62) leads to

$$k_t + \frac{dW}{dk}k_x = 0 \tag{4.63}$$

Define the group speed

$$C(k) \equiv \frac{dW(k)}{dk} \tag{4.64}$$

Then the local wave number $k(x,t)$ satisfies the following nonlinear hyperbolic equation

$$k_t + C(k)k_x = 0 \tag{4.65}$$

No doubt, Equation (4.65) represents a typical problem of nonlinear transitional changes (blown-ups). Even if it is treated as a wave motion solution, mathematical singularities also exist. That is when $t \to t_b$, the wave number $k \to \infty$ or the wavelength $L = 2\pi/k \to 0$, the wavelengths approach breaking-ups. That is, the wave nonlinearity reveals, disappear by walking to their breaking-ups. The essence is flows cause wave motions to disappear. If we consider the common existence of uneven-mass materials, then the physical essence of the so-called dispersion of wave motions is that flow motions cause wave motions to disappear.

So, speaking in the language of the theory of wave motions, nonlinear and linear dispersive waves are fundamentally different. No doubt, as problems of fluids, linear dispersion is shown in the Coriolis parameter, while without introducing the Coriolis term, one obtains no dispersion. Other than special, continuous even-mass objects, the essence of the nonlinear dispersion is not a weakened equilibrium of the linear form; instead, it goes repeatedly through transformations of the breaking-up form. For example, the following is a linearized equation for the absolute vorticity

$$v_{txx} + \bar{u}v_{xxx} + \beta v_x = 0, \quad \omega = \bar{u}k - \frac{\beta}{k} \tag{4.66}$$

which contains all odd-order derivatives. So, it should have real-valued dispersive motion. In particular, if $\bar{u} = 0$, then

$$\omega = -\frac{\beta}{k} \quad \text{or} \quad \omega k = -\beta$$

What needs to be noted is that

$$\beta = \frac{\partial f}{\partial y} \frac{\partial \psi}{\partial x}$$

where $f = 2\Omega \sin \phi$. So, the effect of the Coriolis term is rotation. So, even for the dispersive waves of linear equations, the essence is still that due to rotations of eddy currents, the dispersion is completed.

What should be noted is that due to the need for employing mathematical treatments, in current investigations of mechanics the angular speed Ω of the earth's rotation is still assumed to be a constant. In particular, in the currently applied theory of meteorology the assumption of constant Ω is still continued. The reasoning behind this assumption is that changes in this angular speed are relatively small. As a matter of fact, in the study of the atmospheric movements the initial reason this angular speed Ω is introduced is it is no longer measured by using our daily scales or microscopic scales of measurements. In terms of problems of rotations, it has become problems of moments of forces. In the theory of mechanics that is still continued in theoretical research and practical applications, not only are changes and origins of forces not yet considered, but also is the large radius of the earth not noticed so that even under the earthly condition of relatively small mutations, the relevant moments of forces are still extremely large. Therefore, changes in the earthly nutations often trigger natural disasters including weather disasters and earthquakes.

As for the dispersion problem of mathematical waves, it also involves the following problems:

1. If we take $\bar{u} = 0$ and

$$\omega = -\frac{\beta}{k}$$

then ω and k are hyperbolas with equal axes. And, due to $k > 0$, $\omega(k)$ or $C(k) < 0$. So, one has inversely propagating dispersion effects. Or, in other words, the dispersion effects of the dispersive wave are felt from the lower streams to the upper reaches instead of being felt down the stream. That is why the Rossby waves, which are unidirectional retrogressive plane waves, of meteorological science become very strange objects in mathematics and physics. Even if we apply the generalized concept of mathematical waves, one-dimensional waves are also two-directional. Besides, being unidirectional is the concept of flows; and even for the flow direction of a spinning eddy current, no matter how frequently the direction changes, the direction of flow at any given moment of time is still unidirectional. What is important is that since 1940 when Rossby proposed his theory of Rossby waves, as of this writing, no one has observed any unidirectional retrogressive wave.

In its essence, the purpose of scientific research is to provide explanations for what is unknown about some existing phenomena. However, from this discussion, we have seen that in the theoretical research, people have inexplicably settled some bafflement about nonexisting phenomena so that people become more confused. Here, the Rossby waves, conducted by pursuing mathematical waves, are a typical example of such an inexplicable explanation about some nonexisting physical phenomena. It is, as of this writing, still considered as a basic theory of meteorological science, which is indeed inexplicable.

2. If $\bar{u}$ is taken to be a constant, as done as in current meteorological science, and assume that $\beta = 0$, then we have

$$C = \frac{\omega}{k} = \bar{u} = \text{const.}$$

The phase speed has nothing to do with the wave number k, and the wave motion becomes nondispersive. Or in other words, the dispersion effects of linearized equations of fluid movement come entirely from the effects of earth's rotation of $\beta \neq 0$, which in essence is equivalent to denying the dispersive waves of Euler's original equation of fluid mechanics. Also, vorticity equations themselves contain the characteristics of rotations instead of the dispersion of nonwave motions.

3. Mathematical dispersive waves not only are a stretched concept developed on the basis of the concept of wave motions, but also explain only in form that the amplitudes of disturbances decrease in the fashion of divergence without providing any reason for the divergence. Whether or not the ultimate answer to the weakening of disturbances can be addressed by describing it using phase speeds and proving it has something to do with the wave numbers is still debatable. The core of the concept of dispersion is nothing but transforming aggregated carries to the overall, whole carry. However, in the transformations of mathematical waves, one faces the problem that faster waves cannot overtake the lower waves in the front. It is just like in the one-dimensional situation that a faster wave cannot overtake a slower wave in the front so that one-dimensional dispersions can never materialize the desired property of dispersion. In terms of fluids, the so-called aggregate dispersion comes from the dispersions of the flow movement; and the aggregation and dispersion of the energy of wave motions are always the post-effects of the aggregation and dispersion of the flow movements. Therefore, even if the theory of wave motions is correct, it still cannot be employed to foretell what has not happened. Instead, our focus needs to be on providing the causes for the aggregations and dispersions of flow movements.

4.3 Linearization or Weak-Linearization of Nonlinear Equations

There are many examples of linearization or weak linearization of nonlinear dynamic equations. In order not to make our presentation too long, we will constrain ourselves to some of the examples used in college textbooks to provide our discussion directly.

4.3.1 Rossby Equation and Theory of Rossby Waves of Fluid Mechanics

4.3.1.1 The Fundamental Characteristics of Spinning Fluids

No doubt, the fluids on the earth include at least the atmosphere and oceans. Their movements cannot escape the effects of the earth's rotation. And, the unevenness in the materials of the fluids has to produce twisting forces causing the fluids to be stirred up. Even for the so-called noncompressible fluids, uneven temperatures can still produce uneven densities, causing the appearance of twisting and stirring forces so that these fluids will no longer follow the Newtonian system of the first push. Even based on the traditional beliefs, one still has to admit that over 99% of all materials in the universe are liquid materials. So, speaking relatively, the amount of solids that mankind observes on the earth can be considered next to nothing; and under certain conditions these solids can be transformed into fluids (Lin and Qiu, 1987). That might be the reason why in his book *Tao Te Ching* over 2,500 years ago, Lao Tzu obtained his clue from fluids and proposed *da dao wu xing*, the mighty Tao does not have any shape (the river located near where Lao Tzu was born is still named today *wo he*, rotational river). Or in other words, without understanding fluids, the amount of human knowledge is nearly zero, on which we cannot *wu de da dao*, (we cannot truly comprehend the mighty Tao). From this end, one can see the position and importance of the properties and theories of fluids in our understanding of the physical world.

So, a natural question arises: Can properties and theories of solids be generalized to those of fluids without much modification? Evidently, at both the levels of practice and theoretical investigations, fluids and solids are fundamentally different. So, when the properties and theories of solids are expanded to the studies of fluids, we have met with difficulties and challenges.

In terms of the fundamental properties of solids, even solids are not continua and are made up of loose particles. Liquids and gases, as fluids, show more clearly their looseness in their structures than solids. Even based on macroscopic intuitions, discontinuity should be common and absolute, and solids can experience break-offs and separated states under certain conditions. So, continuity is only a concept or method of handling situations under relative comparisons instead of a physical reality. That is why Einstein's worry that if physics is not constructed on continuity, then all of it will be gone is not just some frightening words to raise alarm.

There is no doubt that at the early stages of modern science, the entire theoretical foundation was constructed on the assumption of continuity. The three representatives of the 17th century scientific achievements are (1) treating fluids as continua of low-resistance solids; (2) introduction of conservation laws of energy in the form of speed; and (3) the establishment of calculus. From that time on, the world of learning has identified fluids as solids. Even today when the development of science has been fast moving, the progress in the study of fluid science has been so extremely slow that many age-old problems have become more and more difficult to attack instead of being closer to their eventual resolutions. Even in terms of formal logic, the concept of treating fluids as continua of low-resistance solids suffers from contradictions of logic. Since solids can be easily placed in fluids, it means perfectly that in the macroscopic sense, all fluids, which have no shape and accept other objects, should not possess the imaginary continuity. This end is indeed our key foe to correctly understanding fluids.

The purpose of treating fluids as continua of low-resistance solids is for the introduction of the assumption that under certain conditions, fluids possess the pressure effects of elasticity just like solids so that the effects of wave motions can be discussed by using pressures of elasticity and the mathematical techniques of differential equations. Even so, it does not mean that there are differences between the properties of fluids and those of solids and fluid movements are limited to those of wave motions. It can also be said that even if movements of fluids do have characteristics of wave motions, it still does not mean that the dispersion or dissipation of mathematical waves has resolved the physical mechanism of the dispersion or dissipation of fluids' wave motions. Therefore, as a problem existing in our scientific exploration, if the wave motions and pressure effects of elasticity of solids cannot cover all forms of movements of materials, in particular, if the form of movement of wave motions cannot represent the common or main characteristics of movements, then we have to consider the problem of how to understand the essence of fluids' movements. As a matter of fact, rotational movements in fluids should have been the common phenomenon seen in our daily lives; otherwise, there would not be a need for Lao Tzu of more than 2,500 years ago, who established the specific epistemology of curvature spaces by pointing out fluids' *wu zhuang zhi zhuang, huo da dao wu xing* (the state of no shape or the mighty Tao without form) and *qu ze quan, zhi ze wang, yuan ze fan* (when it curves, it will be complete; when it is straight, it will be crooked; when it is far away, it will return). Even at the end of the 19th century under the influence of modern science, by applying simple nonparticle circulation integrations, V. Bjerknes in 1898 (Hess, 1959) established the mathematical circulation theorem on the basis of quantitative analysis, while showing that due to density unevenness, eddy currents in the atmosphere and oceans are formed. As of today, by applying the mathematics of the system of wave motions, physicists still cannot prove the invalidity of Bjerknes's circulation theorem. What is interesting is that in today's world of learning, when the problem of nonlinearity, which is also called the mystery of nonlinearity, is still the fashion or center of discussion, people do not realize that this so-called mystery of nonlinearity had been resolved at the end of the 19th century by Bjerknes. What is a pity is that Bjerknes's work did not catch much attention in the scientific community and is also being forgotten in the area of meteorological science, and even Bjerknes himself did not recognize the significance of his circulation theorem.

As a matter of fact, the Bjerknes's circulation theorem has already revealed that unevenness in materials can cause mutual interactions instead of the first push of the Newtonian system. This result not only shows that the first push system of modern science does not possess the perceived commonality, but also uncovers what mathematical nonlinearity is. Or in other words, the first pushes and mutual interactions are problems from two totally different scientific systems, where mutual interactions can be vividly described by using the word *confrontation*.

The physical significance of confrontation should not be overlooked, because the chance of hitting the center of mass of any chosen object is very small so that the general consequence of confrontation should be rotational movements. Therefore, mathematical principles established on the system of first push are not the same as principles of nature. Besides, rotations are originated from structural unevenness in materials; quantities cannot be used to replace these materials' structures and cannot represent the attributes of materials. So, there is a need to reconsider such a concept as mass as a quantity of material.

It was indeed a very important event for the circulation theorem to appear at the end of the 19th century. It might be because its establishment and conse-quences were too much ahead of their time, or because people were more fond of applications of mathematics in physics — although the circulation theorem had already resolved the problem of mathematical nonlinearity — that people have still been infatuated with wave motions so that problems of rotations have been ignored. In particular, in the community of meteorologists, the circulation theorem has been employed to explain the weather phenomena of small-scale land-and-sea breezes. However, more than 40 years later, scholars in meteorol-ogy spared no effort nor anything else but pushed for the theory of Rossby waves of the scale of more than several thousand kilometers on the assumption of even atmosphere so that the prediction of disastrous, zero-probability weather condi-tions has become impossible.

No doubt, modern science, as a system of wave motions, should have made it clear that mathematical techniques could not violate the physical, realistic mean-ings. However, instead, it has emphasized wave motions whose characteristics are to maintain the imaginary invariancy of materials in the form of transferring energies. So, continuing the discussion in the previous section, let us look at some additional problems met by the linearization or weak linearization of nonlinearity so that a clearer epistemological concept can be produced. Due to the specifics involved, we will use real-life examples to illustrate our points.

4.3.2 Linearization of the Euler Equation and Problems with Rossby Waves

4.3.2.1 Fundamental Characteristics of Spinning Fluids

Evidently, the rotationality of celestial bodies has to make materials on these bodies unable to escape from the fate of rotation. Correspondingly, the earthly oceans and atmosphere have to be affected by the earth's rotation. Also, as fluids themselves, they additionally have to experience forces of stirring movements due to density unevenness. As for noncompressible fluids, uneven temperatures can trigger twisting forces. By these twisting forces, together with the effects of speeds and directions of movements, friction, and so forth, spinning movements are created. So, rotational movements are the common form of motion in materials. Now, a natural question arises: Since fluids and solids have fundamentally different properties and satisfy different theories, studying fluids' attributes and essences should be the central task of the investigation of fluids. In terms of the basic properties of materials, even solids with definite shapes are also continua that are made up of loose particles. This character of looseness is shown more clearly in fluids, and gases, in particular. Even under the macroscopic intuition, the character of discontinuity is the commonality and absoluteness. So, the continuity of materials is only a methodology under relative concepts instead of physical reality.

It can be said that the earliest observed rotationality of macroscopic fluids is from the weather maps of meteorological science. Figure 4.2 is the flow chart of the average January ground-level wind field. No doubt, even for laymen without any knowledge of meteorology, the rotationality of eddy currents in the atmosphere can also clearly be seen. However, based on the concept of wave motions formally introduced in the 20th century, the contemporary community of meteorologists places the rotationality by isograms of the pressures and/or geopotentials (Figure 4.3). By doing so, other than making the rotationalities of the current field disappear, disturbances, shown in maps similar to that in Figure 4.3, are referred to as atmospheric wave motions.

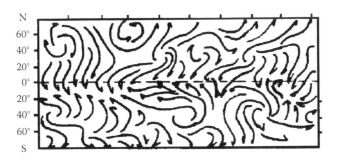

Figure 4.2 **Average January surface circulations.**

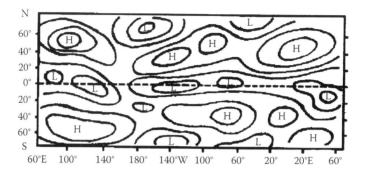

Figure 4.3 Average January barometric pressure field.

According to Rossby's own statement, he established the presently well-known theory of Rossby long waves by simplifying (linearizing) the equation of fluids mechanics based on the hint that during the 1930s, disturbances of wave motions, shown in maps similar to the altitude fields of the isobaric surfaces as in Figure 4.3, were called atmospheric long waves. Later, his followers listed this theory as one of the basic theories of meteorology. As of this writing, some scholars in the profession of weather forecasting refer the nonclosed curves in maps like Figure 4.3 to as waves, and the closed curves as eddies. To this end, we have to question: How can waves and eddies appear at the same time on the same chart? Or, how can the eddy currents in Figure 4.2 become a mixture of waves and eddies after the flow chart Figure 4.2 is transformed to isograms of pressure fields or altitude fields? To address these questions, one has to deal with the problem of how to understand the fundamental properties and characteristics of spinning currents. That is the reason we propose to reconsider the theory of Rossby waves.

Based on the basic theories of meteorological science, spinning fluids possess all the characteristics of classical mechanics and form a branch of classical fluid mechanics; and the investigation of weather is a branch of modern physics. So, the basic theories of classical mechanics have been carried over into meteorological science without much added consideration. Because the 20th century was the realm of the physics of wave motions, meteorological science naturally was established on the foundation of wave motions. Although Bjerknes at the end of the 19th century pointed out in the area of meteorology the solenoidal effects of the atmosphere and his circulation theorem showed that the atmosphere and oceans are rotational eddy currents, the community of meteorologists stored these works away and 40 years later pushed for the theory of Rossby's long waves. Since fluids cannot be without the characteristics of rotation, the corresponding fluid mechanics treats the effect of rotation as an influencing parameter on wave motions. Or in other words, the reason why the present mechanics of spinning fluids is seen as a branch of classical mechanics is that, at least in form, the mechanics of spinning fluids is a continuation of the mechanics of wave motions without altering the properties of wave

Figure 4.4   Plane flow chart of spinning fluids.

motions of the movements studied. However, it is exactly this parameter of rotation that has already changed the fundamental characteristics of classical mechanics, leading to the problem of how to correctly understand the characteristics of fluids' movements, as shown in the transformation from Figure 4.2 to Figure 4.3. This end touches on the laboratory results of spinning dishes of fluids widely observed since 1970s. One must note that the community of meteorologists has believed that the experiment of spinning dishes of fluids is the best evidence for the existence of Rossby long waves.

Figure 4.4 is a plane flow chart of spinning fluids (water) constructed based on the video records of the experiment of spinning dishes of fluids by the Institute of Atmospheric Physics of Chinese Academy of Sciences (1993) and the Meteorology Department of Japan (1972). For documented details, see Hide (1953). What is very interesting is that from the video of fluids' movement, different people can observe different patterns. For example, almost all front-line meteorologists who make day-to-day weather forecasts, laymen, or scientists from the Institute of Ocean Studies unanimously saw eddy currents instead of waves, while scholars in meteorological science, in particular, those who are in theoretical meteorology, declared that they were waves after brief moments of consideration. These scholars called the eddy leaves that surround the central polar circle, which is the ice pocket at the center of the rotating dish and the center circle in Figure 4.4, as a Rossby wave and treated these eddy leaves as the experimental evidence for the validity of the theory of Rossby long waves.

No doubt, different patterns observed have to cause debates on the underlying beliefs. If the arrows of the flow lines are removed, the disturbances of the remaining curves in form seem to be like waves. However, *like* is not the same as *is*. The obvious difference is that the water currents in the spinning dish are unidirectional flows along curve paths instead of reciprocating vibrations of the fluids. Even based on the concept of wave motions of classical mechanics, each plane wave propagates in all directions. However, the water currents in the spinning dishes are not flowing in all directions. In particular, as morphology changes, what is shown in Figure 4.4 is not only a changing morphology but also eddy currents with a variable number of leaves. Along with changes in temperatures of the polar circle in the center and the periphery (the equator circle), one can observe eddy currents of two leaves, four leaves, five leaves, six leaves, and so forth. As for the eddy currents with three leaves, they possess different strengths, positions, and formations. What deserves our attention is that in spinning dish experiments, no one has ever seen unidirectional retrogressive waves other than eddy currents in either the clockwise or counterclockwise directions traveling along with the overall eastward-moving fluid in the dish.

As for the entire experiment of spinning dishes, it can be summarized as follows. As the temperature difference between the polar circle and the periphery decreases, the number of eddy leaves increases. If the temperature difference disappears, the water in the spinning dish flows around the polar circle as a large whirlpool. That is, the concept of polar eddies of synoptic meteorology, a branch of meteorological science, agrees with the experiment of spinning dishes. When the temperature difference increases, one can observe water currents flow almost along the meridian lines. In essence, throughout the entire experiment, the water in the dish is an eddy current flowing around the polar circle. Along with changes in temperature differences between the polar circle and the periphery and the speed of rotation of the dish, different snake-shaped rotational currents around the polar circle appear with upstream effects without conserved downstream effects. And, no overall vibrant waves are ever observed.

Although experiments cannot replace physical reality, they are still closer to reality than theories. What the spinning dish experiment proves is that it cannot be employed as the evidence for the existence of Rossby waves. Instead, what is shown is that side temperature differences can cause eddy currents to appear and that along with changes in the side temperature difference and in the rotational effects of the dish, the shape of the overall polar eddy current changes, revealing a variant eddy current with a different number of eddy leaves. What deserves our special attention is that this experiment cannot verify the theory of Rossby waves established on the β-effects. It is because the β-effect stands for that of change of the parameter $f = 2\Omega \sin \varphi$ of the earthly rotation along the geographical latitude, where Ω is the angular speed of the earth's rotation; and φ, the geographical latitude; and for small-scale (about the scale of 10 kilometers) movements, the rotational effect of the earth can be ignored. As for the spinning dish experiment, the size of the dish is about 40 to 50 cm in diameter; so the rotational effect of the earth can be completely eliminated. If we believe that the spinning dish experiment verifies the existence of Rossby waves, then it will be erroneous to ignore the rotational effect of the earth from the small-scale equations established in meteorological science. On the other hand, if the spinning dish experiment did not confirm the β-effect but instead presented Rossby waves, then the Rossby equation established for large scales in atmospheric science would not hold true. So, if the spinning dish experiment is used as the evidence for the existence of Rossby waves, it in fact becomes a denial of the theory of Rossby waves. It can also be said that the spinning dish experiment has nothing to do with Rossby waves and that it actually denies validity of the theory of Rossby waves in the form of experiments.

Also, in the scientific history of the classical mechanics, Rossby waves are known for their uniqueness to fluid mechanics for being plane, unidirectional, and retrogressive, and once cited in the theory of oceanography. However, they did not get their confirmation from the spinning dish experiment. Even if we treat the disturbances of the eddy currents of varying shapes as waves, they are still not unidirectionally retrogressive; instead, they are eastward-moving eddy currents circling around the center pole.

4.3.2.2 The Rossby Equation and the Theory of Rossby Waves

On the basis of calling the serial 4000~8000 km disturbances in the atmospheric pressure field or the altitude field of the midlatitudes or the midaltitudes atmospheric long waves, C. G. Rossby in 1940 (Hess, 1959) established the long-wave formula, which was later named as Rossby waves, by linearizing the vertical vorticity equation of the mechanics of spinning fluids. Later, this theory was considered unique by meteorologists, which holds true only for the atmosphere, and is extended into the study of oceanography.

By introducing the assumption of horizontal nondivergent even-mass positive pressures, Rossby rewrote the equation of large atmospheric vorticity as follows:

$$\varsigma_t + u\varsigma_x + v\varsigma_y + \beta v = 0 \tag{4.67}$$

where $\varsigma = v_x - u_y$ is referred to as the vertical vorticity,

$$\beta = \frac{\partial f}{\partial y}, \, f = 2\Omega\sin\varphi$$

known as a parameter for earth's rotation with Ω representing the angular speed of the earth, φ the latitude. Based on the conventional methods of the wave-motion system of time in the community of physicists, Rossby applied linearization and introduced the following assumptions by considering that the main airstreams of the atmosphere are west winds along the latitudinal direction and that the wind speeds in the longitudinal direction are smaller than those in the latitudinal direction,

$$u = \bar{u} + u', \ v = v' \tag{4.68}$$

where $\bar{u}$ stands for the average speed of airstreams along the latitudinal direction and the terms that represent the magnitudes of disturbance. Substituting Equation (4.68) into Equation (4.67) and ignoring the second-order disturbances lead to

$$\varsigma_t + \bar{u}\varsigma_x + \beta v = 0 \tag{4.69}$$

where for convenience, " ' " is omitted. Assume that Equation (4.69) has a solution of harmonic waves. Substituting this assumption into the previous expression produces

$$c = \bar{u} - \frac{\beta}{k^2 + n^2} \tag{4.70}$$

where $\bar{u}$ stands for the average speed of flow in the latitudinal direction, k,n are respectively the wave numbers in the plane along x, y directions, c the speed of propagation of plane waves. Equation (4.70) is the well-known theory of Rossby long waves in the community of meteorology. And, when $\bar{u} = 0$, Equation (4.70) stands for the propagation speed of a unidirectional retrogression. That is why in the community of meteorology, it only holds true in fluid mechanics that unidirectional retrogressive Rossby long waves are considered unique.

In the following, let us point out some of the fundamental problems existing with the theory of Rossby long waves.

Even for the mathematical theoretical form, the theory of Rossby long waves initially comes from Rossby's modification of the classical fluid mechanics equation by adding the term of deviation force (called the Coriolis term) caused by the earth's rotation so that the original equation becomes one about a large movement of rotational fluids. Then, this resultant equation is transformed into an equation of vorticity. Evidently, as an elementary concept, vorticity is a measure of rotation, while wave motions are a concept about linear reciprocating movements. Although the method of nondimensionalization or unification of dimension is applied by introducing physical quantities, even if we assume that the quantitative values of rotational disturbances are the same as those of amplitudes of wave motions, mathematically speaking, disturbances of eddy motions are those of flows along curves, while disturbances of wave motions are those along straight-line flows. That is why eddy motions are different from wave motions. Although at the end of the 20th century, in order to apply mathematical techniques, people have expanded and exaggerated the concept of wave motions by using general wave motions; mathematical concepts are after all not the same as concepts of physics. The belief that a disturbance means a wave motion (see Section 4.2) is exactly the consequence of generalizing the concept of wave motions so that eddy and wave motions are confused with each other and their physical essences are hidden. The applications of mathematics in this aspect could be seen as misusing mathematical techniques. As for the essence of physics, rotation stands for a problem of noninertial systems, while the exaggerated wave motions have confused noninertial system problems with those of inertial systems. Wave motions are the elastic pressure effects of classical mechanics of the first push, but eddy motions are the effects of stirring forces of rotations under mutual interactions and belong to noninertial systems. Therefore, before Rossby transformed eddy motions into wave motions, he had already confused problems of noninertial systems with those of inertial systems. So, even looking from the angle of the elementary knowledge of modern science, Rossby had made a mistake at the elementary level. However, it is a pity that there are still scholars in the area of meteorology who have not recognized this reality.

At the end of the 19th century, V. Bjerknes established his circulation theorem, which has been known to almost everyone in meteorology. This result indicates that due to uneven densities, each atmospheric system of several ten kilometers long is a rotational eddy current. As of this writing, the community of meteorologists still explains the phenomena of land–ocean breezes by using this circulation theorem. So, as a scholar in the area of meteorology, Rossby must have known this famous theorem. However, he still treated earth's atmosphere as having positive pressures. No doubt, the requirement of the atmosphere of positive pressures in essence is to require an atmospheric uniformity in the vertical direction so that the density of the atmosphere becomes even, which is against the elementary knowledge of the atmosphere. If the earth's atmosphere were even, then any investigation on weather changes and weather forecasts would become meaningless. As for the noncompressibility of seawater, it does not mean that seawater is even, because differences in temperature can make seawater uneven. Rossby long waves can cover a range of 4,000~8,000 kilometers. By crossing such a great range, no matter whether the atmosphere or an ocean, it cannot possibly be even. Even land–ocean breezes of several tens of kilometers can be explained based on the uneven densities by using the circulation theorem, so how can anyone imagine the existence of even atmosphere or ocean of several thousand kilometers in length?

No doubt, contradictory conclusions should not appear in any discipline of science. Both Bjerknes's circulation theorem and Rossby's long waves are considered parts of the foundations of meteorological science. However, these two results are contradictory to each other. The former states that either the atmosphere or the ocean is made up of rotational eddy motions, and the latter claims that the atmosphere is made up of vibrant wave motions. What a confusion these results have created!

Also, in terms of mathematical logic, there are problems with regard to Equation (4.67) that was given by Rossby. The correct equation of vorticity derived from fluid mechanics should be

$$\varsigma_t + u\varsigma_x + v\varsigma_y + \beta v = -\left[\left(\frac{\partial}{\partial x}\left(\frac{1}{\rho}\right)\frac{\partial p}{\partial y}\right) - \left(\frac{\partial}{\partial y}\left(\frac{1}{\rho}\right)\frac{\partial p}{\partial x}\right)\right] \quad (4.71)$$

where $\rho = \rho(x, y)$. Evidently, Equation (4.67) is derived from Equation (4.71) by assuming that ρ is a constant or p is a constant. That is assuming either that the atmosphere has even density or that the ocean has even density. However, when ρ is a constant, the equations of horizontal movements corresponding to the original fluid are respectively given as follows:

$$\begin{cases} u_t + uu_x + vu_y - fv = -\dfrac{1}{\rho_c} p_x \\[2mm] v_t + vu_x + vv_y + fu = -\dfrac{1}{\rho_c} p_y \end{cases} \tag{4.72}$$

and

$$\begin{cases} u_t + uu_x + vu_y - fv = 0 \\[1mm] v_t + vu_x + vv_y + fu = 0 \end{cases} \tag{4.73}$$

Evidently, by cross-differentiating Equations (4.72) and (4.73) and then considering the difference of the results, we can derive the Rossby equation in Equation (4.67) after introducing the assumption of no divergence. However, Equations (4.72) and (4.73) have different physical meanings;

$$\begin{cases} -fv \cong -\dfrac{1}{\rho_c} p_x \\[2mm] +fu \cong -\dfrac{1}{\rho_c} p_y \end{cases} \tag{4.74}$$

or, the gradient force of barometric pressures are roughly the same as the deviation force (called Coriolis force) in terms of quantitative magnitudes, but $u_t + uu_x + vu_y$ or $v_t + vu_x + vv_y$ is smaller than either the gradient force of barometric pressures or the deviation force by two magnitude levels. In other words, Equation (4.72) is a mathematical equation involving large quantities with infinitesimal differences. Now, Equation (4.73) delivers the message that the magnitude of either $u_t + uu_x + vu_y$ or $v_t + vu_x + vv_y$ is quasi-equal to that of the deviation force, or the magnitude of the deviation force becomes roughly the same as that of the acceleration of the underlying movement, where the fundamental properties of rotational fluids of the earth are artificially altered.

No doubt, if Equation (4.72) holds true, one must have $\beta v \cong 0$, because the term $2\bar{\Omega} \times \bar{v}$ of deviation force is greater than the term of acceleration by two magnitude levels. Next, Equation (4.67) is equivalent to either Equation (4.72) or Equation (4.73) in a different form. Now, even if we apply pure mathematical deduction without considering the relevant physical meanings, Equation (4.67) cannot be derived out of either (4.72) or (4.73), constituting a contradiction of mathematical formal logical calculus. So, in essence, there is no longer any value for any continued discussion on the Rossby equation.

Also, what needs to be noted is that the right-hand side of Equation (4.71) stands for the solenoidal term of the stirring motion, and βv comes from the vorticity $2\bar{\Omega} \times \bar{v}$ of the earth's rotation. So, based on the traditional point of view of mechanics, Equation (4.72) is feasible in terms of both mathematical and physical logic and shows a quasi-balance between the atmospheric deviation force and the stirring force that is made up of the density and the gradients of the barometric pressures and that Equation (4.71) represents a stirring source of eddy motions and a vorticity equation of rotational movements. However, we have to point out that Equation (4.71) is not the Rossby equation.

As a consequent effect of the theory of Rossby waves, in meteorology, closed curves on isobaric (or contour) surfaces are called *eddies* instead of closed isobaric lines *waves*. No doubt, these phrases themselves have been confused with some elementary concepts: On the plane, how can there be an eddy current that circulates around a center with its surrounding filled with reciprocating waves so that the fluid is not like a static pond of dead water, but it is not flowing, either? Evidently, when talking about moving fluids on the rotational earth, due to changes in moving directions, these fluids have to be rotational eddy currents. Although when looked at from the angle of whole evolution, the flowing movement could be blocked and show a local vibration, this vibration stands for waves of blocked flows. That is, the effect of wave motions is dependent on the movement of flows. And, the waves of blocked flows can also change their directions with the movement of flows or become eddy currents, or disappear in eddy currents. For example, it has been observed that ocean breakers can evolve into cubical whirlpools. So, even for the purpose of applying mathematical techniques, one still cannot identify fluids with solids. When compared to solids, fluids seem to contain more principles of essence of natural sciences.

What needs to be noted is that Bjerknes's circulation theorem appeared before Rossby's long waves. What is more important is that in terms of the future development of science, the contribution of this circulation theorem is not limited to the area of meteorology, because at the very least, at the end of the 19th century, this theorem had resolved the mystery of nonlinearity. Also, it should be seen that the circulation theorem has already fundamentally pointed out that nonlinearity cannot be linearized. If the development of science had gone along the direction of the circulation theorem, current modern science might not have been trapped in its pursuit of linearization and produced mathematical waves with the system of wave motions greatly exaggerated.

If slightly more respect is paid to the elementary concepts of physics, from Equation (4.67), it can be readily seen that the βv term is from the $2\bar{\Omega} \times \bar{v}$ term of the original equation. No doubt, $2\bar{\Omega}\times\bar{v}$ represents the movement of the eddy current caused by the rotation of the earth. So, no matter what mathematical technique is applied, the essence of rotation of the earth should not be altered. Consequently, the βv term can only stand for the effect of rotation. As a matter of fact,

$$\beta v = \frac{\partial f}{\partial y}\frac{\partial \psi}{\partial x}$$

represents exactly a vorticity, where ψ is the flow function. And in particular,

$$\beta = \frac{\partial f}{\partial y} = \frac{1}{r}2\Omega\cos\varphi$$

shows that the so-called β-plane is only a special case and holds true only when $r \to \infty$ or

$$\varphi = \frac{\pi}{2}$$

However, the actual radius of the earth is finite (and in the actual curvature space the quantitative infinity does not exist; this problem has been well discussed in Chapter 3), while

$$\varphi = \frac{\pi}{2}$$

stands for the special problem of the polar areas on earth. It has to be pointed out that the purpose of introducing approximate β-plane is the consideration of the areas of low latitudes in order to prevent $\sin\varphi \approx 0$, which will cause the quantitative magnitudes of the meteorological winds of the earth's rotation to become infinity. In other words, at midlatitudinal areas on earth, any good approximation for the β-plane no longer exists. However, the theory of Rossby long waves has been limited to applications of the midlatitudinal areas by Rossby himself. So, it can be expected that it will be difficult for the Rossby theory to produce valid consequences in practical applications.

In meteorology, people have been interested in developing physical explanations out of the theory of Rossby long waves, because the physical mechanism of the theory is the β effect. Or, the reason Rossby long waves exist in the atmosphere is that the parameter $f = 2\Omega\sin\varphi$ of the earth's rotation changes with the latitude without illustrating why air particles travel northward or southward. As a matter of fact, heat exchange in fluids is accomplished in the form of turbulences. The uneven heat distribution between the south and north has to cause the air to move southward and northward in the form of turbulences. And, accompanied with the effect of the deviation force (the Coriolis force), rotations have to occur. The subsequently created subeddies in comparison with the polar eddy make the polar eddy change its shape and experience disturbances in the form of traveling-snake flows. What needs to be noted is that the disturbances in the form of traveling-snake flows in the upper altitude has nothing to do with the β effect and are created by the earth's rotation and uneven atmospheric temperature differences between the south and north. Although their form is similar to that of disturbances of wave motions, their essence is not the same as wave motions. They are the result of interactions between rotational fluids and temperature difference between the interior and exterior. And, temperature difference corresponds to traveling-snake flows. As for a realistic fluid on earth, the formation and development of the distorted eddy currents are completely determined by the earth's rotation and temperature differences between the polar areas and the equator, which also influence the interaction between the polar eddy and revolutionary current of the subeddies. Combined with earth's nutation, movement of the poles, angular speed, and differences between seas and lands, terrains, the stirring effects of temperature and pressure, and so forth, sequences of rotational flows are formed and eddies of different scales are created. So, all these conduct the birth–death evolution of eddy currents and corresponding weather conditions, such as cloudy, clear, rainy, dewy, or disastrous weathers. It can be said that without transformations of eddy currents, there will not be the constant weathers. And, constant change is exactly a problem the invariant system of modern science cannot describe. We must note that the design and manufacture of durable goods represent a problem that is fundamentally different from the problem of constant changes. Since all problems of prediction have been seen as a branch of modern science with the monitoring of extrapolating initial values used as the corresponding method of application, that explains why as of this writing, the prediction science still cannot resolve transitional changes. What is extremely important is that almost all disasters occur during transitional changes. So, the prediction science has not made much substantial progress. In 1963, Shoucheng OuYang discovered the abnormal information of ultralow temperatures that appear ahead of disastrous weather conditions near the top of the troposphere. By looking at the problem of transformation of eddy currents in fluids, after over 30 years of practical tests, he has proved the relationship between the ultralow temperature information and disastrous weathers. At the same time when he proposed the prediction method of digitization, he established the theory of blown-ups. After practical application, it has been found that this method can substantially improve the accuracy of forecasting disastrous weathers by using realistic transitional changes and the current international rate of roughly 10% of accuracy in forecasting disastrous weathers has been improved to over 70%, including misguided cases.

In terms of the macroscopic characteristics of rotational fluids, no matter how the eddy currents of the traveling-snake form at midlatitude are distorted, the polar eddy always appears in the sky above the polar areas. This end indicates that the rotational effects of the earth cannot be replaced and cannot be ignored. And, the polar eddies always appear in the form of distorted eddies, while the corresponding solenoidal stirring effects caused by temperature differences between the south and north cannot be eliminated either. Since the solar radiation on earth is not evenly distributed, the temperature differences between the south and north or the uneven heat distribution caused by terrains or other reasons always exist objectively. So, the parameter of earth's rotation, as the physical mechanism for Rossby long waves to exist as so reasoned by Rossby himself, suffers not only from confused concepts of physics but also from unreasonable deductions of mathematical formal logic.

In particular, the spinning dish experiment did not verify the validity of Rossby long waves; instead it shows that rotational fluids under the effect of temperature differences do not need the β effect to form distorted eddy currents circulating around the polar area. And, the disturbances in the form of traveling snakes of the distorted eddy currents are not waves, but instead, they are distorted revolutionary currents circulating around the polar eddy.

Besides, in the Rossby equation, other than the β effect, no other effect of the temperature difference between the south and north exists. However, in the spinning dish experiment, if the temperature difference does not exist, only a large, circular whirlpool will appear, without any disturbing, circulating current of the traveling snake form. So, it can also be clear that wave motions are the effects of elastic pressures of the first push of classical mechanics, while eddy motions stand for the stirring effects of interactions of the second stir of the mechanics of rotations. Even as generalizations of the concept of wave motions, applications of mathematics should not change eddy motions into wave motions. In other words, no matter what mathematical technique is employed, if an eddy motion is transformed into a wave motion, the result will violate the objective reality and betray the basic spirit of science.

No doubt, wave motions in physics represent a concept of morphological changes in the media, where each particle of the media does not travel away from its equilibrium point. That is the classical concept of wave motion in physics. And, each wave motion propagates at least bidirectional (for one-dimensional case) or multidirectional (for two- or higher-dimensional cases). The corresponding so-called unidirectionality itself is a characteristic of flows, where each particle of the fluids travels along the direction of the flows without being limited to regional reciprocating movement. That is the reason we questioned Rossby waves. Although in *rotational slow waves*, the current name for Rossby waves, contains the meaning of spinning, the essence is just another new term for spinning waves. Evidently, whirlpools are whirlpools, and waves are waves. The reason for joining whirlpools with waves is that some scholars just cannot give up the concept of waves.

Our strictly separating the concepts of eddies and waves is about understanding the basics of fluids. For example, why has the problem of fluids been so difficult that it could not be resolved after a long time? What are specific problems about nonlinearity or noninertial systems? How can material evolutions be understood? What is time? How can these related problems and concepts be applied to resolve practical problems? Aiming at addressing these and other problems, and by emphasizing the difference between eddies and waves, we have been able to forecast disastrous, zero-probability weather conditions using the system of eddy motions and achieved a rate of success of over 70%.

4.4 Nondimensionalization of the Two-Dimensional Navier–Stokes–Coriolis Equation and the Problem of Solving the Rossby Equation

4.4.1 The Nondimensionalization of the Two-Dimensional Navier–Stokes–Coriolis Equation

As a matter of fact, the theoretical problems of modern meteorological science are developed and applied on the basis of equations of fluid mechanics of the mechanics of modern science. What is different from the equations of classical mechanics is that the scales of measurements used in the studies of geo-atmosphere and oceans are much greater than those used in classical mechanics with the effects of earth's rotations included. It was Coriolis who first included the deviation force (later named the Coriolis force) caused by the earth's rotation in the Navier–Stokes equation of fluid mechanics. That is why we name this equation the Navier–Stokes–Coriolis equation. Without considering the term of viscosity of the original equation, the two-dimensional form of the equation looks as follows:

$$\vec{v}_t + \vec{v} \cdot \nabla_b \vec{v} + 2\vec{\Omega} \times \vec{v} = -\frac{1}{\rho} \nabla_b p \qquad (4.75)$$

where the arrow "→" on top of a variable stands for vectors,

$$\nabla_b = \left(\vec{i} \frac{\partial}{\partial x} + \vec{j} \frac{\partial}{\partial y} \right)$$

a two-dimensional operator. Now, introduce the nondimensionalization variables:

$$(x, y) = L(x', y'), \quad \vec{v} = U \vec{v}' \qquad (4.76)$$

$$P = \rho \Omega U L \, p', \quad t = \frac{t'}{\Omega} \tag{4.77}$$

Substituting Equations (4.76) and (4.77) into Equation (4.75) and omitting the symbol " ′ " produce

$$\vec{v}_t + \varepsilon(\vec{v} \cdot \nabla \vec{v}) + 2\vec{k} f \times \vec{v} = -\nabla p \tag{4.78}$$

where $\vec{k}$ stands for the unit vector along the vertical direction,

$$\varepsilon = \frac{U}{\Omega L} \approx 10^{-1} - 10^{-2}$$

So, by using quantitative magnitude levels, the nonlinear terms in ε, the component equation can be rewritten as

$$\begin{cases} u_t - f\,v = -p_x \\ v_t + f\,u = -p_y \end{cases} \tag{4.79}$$

By directly cross-differentiating the two equations in Equation (4.79) and then considering the difference of the resultant equations, or ignoring the nonlinear terms of small magnitudes, taking the computation of vorticity, and introducing the assumption of no divergence, we have

$$\varsigma_t + \beta v = 0 \tag{4.80}$$

No doubt, Equation (4.80) is the linearized, nondimensionalized Rossby equation without the basic flow $\bar{u}$. If the basic flow term and the component of the vorticity along the northeast direction are added, we obtain Equation (4.69). As Rossby did, by assuming that the solution of Equation (4.80) exists in the form of harmonic waves, substituting such a solution into Equation (4.80) leads to

$$c = -\frac{\beta}{k^2 + n^2} \tag{4.81}$$

If considering the fact that the atmosphere itself would not take averages, then Equation (4.81) explains where the typical Rossby plane, unidirectional retrogressive waves are from. In particular, considering the quantitative magnitudes used in the scale analysis in meteorology, the first term on the left-hand side of Equation (4.79) is smaller than either of the other two terms by two magnitude levels. After ignoring this smaller term, we produce the relationship formula for the winds of earth's rotation,

$$\begin{cases} f u = -p_y \\ f v = p_x \end{cases} \tag{4.82}$$

Substituting Equation (4.82) into Equation (4.80) leads to

$$\nabla_b^2 p_t + \beta p_x = 0 \tag{4.83}$$

This is the Rossby equation of the barometric pressure field according to the original logic of Rossby, produced by using comparisons of quantitative magnitudes and the method of nondimensionalization. No doubt, this way of producing the desired equation is much cleaner and more straightforward than the original method Rossby used through linearization. Also, other than introducing the same assumption of no divergence as Rossby did, we do not need the assumptions of positive pressure, even atmosphere, and so forth, together with the employment of linearization. It is because from the angle of physical quantities, quantitative magnitude comparisons can be used directly to eliminate the nonlinear term without introducing the assumption that the quadratic nonlinear term is relatively small, as Rossby did. Accordingly, when nondimensionalized, the angular speed $\bar{\Omega}$ of the earth's rotation is treated as a quantity only without involving the rotationality of the original vector. This end no doubt has shown that the essence of the Rossby equation contains no rotationality. Or in other words, the so-called Rossby waves are established on the basis of an irrotational atmosphere. That explains why he could "logically" transform the Rossby waves of the flow field into those of the barometric and/or altitude fields. Although our mathematical treatment seems to be more "beautiful" than Rossby's original deduction, in terms of eliminating rotationality, what we did matches perfectly well with the establishment of Einstein's general relativity theory on the basis of irrotational Riemann geometry. So, in essence what has been shown is that the key of the system of wave motions and linearization is about irrotationality. In recent years, some scholars had renamed Rossby waves as eddy slow waves. Although they had shown some changes in their way of thinking, it still shows their fondness of and their unwillingness to give up the concept of waves and their confusion of eddy and wave motions under the influence of mathematical waves.

4.4.2 Problems with Nondimensionalization

If we carefully analyze the quantitative deduction process of nondimensionalization, we can readily find the following.

First, if we introduce the flow function, $\varsigma = \nabla^2 \psi$ and let

$$v = \frac{\partial \psi}{\partial x}$$

then Equation (4.80) can be rewritten as follows:

$$\nabla_b^2 \psi_t + \beta \psi_x = 0 \qquad (4.84)$$

What must be clear is that the Rossby wave solution of the Rossby equation is not obtained by using the conventional method of solving mathematical physics equations in the form of deterministic problems of differential equations. Instead, it is assumed that the unknown function ψ of Equation (4.84) is harmonic. Then by substituting this harmonic function into Equation (4.84), an analysis of frequency leads to the now-well-known solution. No doubt, for Equation (4.84) to hold true mathematically, we must guarantee that the flow function ψ has higher than second-order continuous derivatives. However, each stable field of eddy currents must be made up of eddies of opposite directions — clockwise and counterclockwise — leading to discontinuities appearing between these eddies spinning in opposite directions. Or in other words, not only can the desired condition that the flow function ψ has continuous second-order derivatives not be guaranteed, but also may the continuity of the original flow function not be warranted. Besides, there is no way for the geo-atmosphere, as it is situated in a curvature space, to possess no rotationality. So, its rotation has to produce discontinuities.

Second, if we take $p = \psi$, then we have Equation (4.83). Similarly, if we take $p = \phi$, we can obtain the Rossby equation of the altitude field. However, even on the basis of mathematical formal logic, one can discover that Equation (4.83) is obtained by substituting the relationship formula Equation (4.82) of the earth's rotation of the meteorology, where the derivative with respect to time has been already ignored, into Equation (4.80) and by making use of Equation (4.82). Evidently, mathematically speaking, the method of making use of the derivative term with respect to time after ignoring such a term earlier is not allowed, constituting a problem of misusing formal logic. In other words, if Equation (4.82) holds true, then Equation (4.80) will no longer hold true.

Besides, from the mathematical analysis of second-order differential equations (see the relevant references), the relationship between ψ and ϕ has been known:

$$f\nabla^2 \psi + \nabla f \cdot \nabla \psi + 2J(\psi_x, \psi_y) = \nabla^2 \psi \qquad (4.85)$$

where $J(\ ,\ )$ stands for the Jacobi operator, and $J(\psi_x, \psi_y)$ is a nonlinear term. By using a static approximation relationship, ϕ can be transformed into p. In order to transform ψ into ϕ (or p), one has to eliminate the nonlinear term and assume that

f = const. However, when f is a constant, the β effect does not exist; therefore, no Rossby wave exists.

Third, since 1940 when Rossby published his research on long waves, no paper in the area of meteorology on the mathematical properties of Equations (4.84), (4.80), or (4.69) has been seen. All works in this area have been walking along the same lines as outlined above — first assuming the existence of a harmonic solution, then substituting this solution into the equation to produce the desired wave speed formula. That is why we feel obligated to check Equation (4.84) (and its different forms in Equations 4.80 and 4.69) to see whether or not it actually has a harmonic solution as our discussion of its mathematical properties. Evidently, as a linear mathematical physics equation, no matter which method is employed, the solution should be the same. Or in the discussion of the mathematical properties of the equation, one should follow the conventional methods of mathematical physics equations. So, let us apply the method of separate variables and introduce

$$\psi = A(t)\Psi(x, y) \tag{4.86}$$

Substituting Equation (4.86) into Equation (4.84) produces

$$\dot{A}\nabla^2\Psi + \beta A\Psi_x = 0 \tag{4.87}$$

where

$$\dot{A} = \frac{dA}{dt}$$

Now, Equation (4.87) can be written as follows with the variables separated:

$$\frac{\dot{A}}{\beta A} = -\frac{\Psi_x}{\nabla^2\Psi} = -\lambda$$

For the convenience of expression, denote

$$\lambda = \frac{1}{k^2}$$

Then we have

$$\begin{cases} \dot{A} + k^{-2}\beta = 0 \\ \nabla^2\Psi - k^2\Psi_x = 0 \end{cases} \tag{4.88}$$

Now, we introduce the polar coordinate system and consider the circular symmetry so that Ψ is only related to r, Equation (4.88) can be written as

$$r^2\Psi'' + (1 - k^2r)\, r\Psi' = 0 \tag{4.89}$$

where

$$\Psi'' = \frac{d^2\Psi}{dx^2}, \; \Psi' = \frac{d\Psi}{dx}$$

Evidently, Equation (4.89) is now a variable coefficient differential equation. By introducing the substitution,

$$\Psi = ru \tag{4.90}$$

we then have

$$\Psi' = u + ru; \; \Psi'' = 2u' + ru'' \tag{4.91}$$

Substituting Equation (4.91) into Equation (4.89) leads to

$$r^2u'' + (3 - k^2r)ru' + (1 - k^2r)u = 0 \tag{4.92}$$

No doubt, Equation (4.92) is not a typical Euler differential equation with variable coefficients. From the properties of variable-coefficient differential equations, it follows that even the solutions of the equations of the Euler type are not simple harmonic waves. Considering that in the establishment of the Rossby waves of the original Rossby equation, the amplitude of the disturbances of the wave has been assumed to be very small, in order to make Equation (4.92) transform into a typical Euler equation, let us assume that both $3 - k^2r$ and $1 - k^2r$ are constants in order to analyze the properties of the solution. That is,

$$\alpha_1 = 3 - k^2r; \quad \alpha_2 = 1 - k^2r \tag{4.93}$$

Evidently, Equation (4.93) is an artificially imposed condition under the assumption that the amplitude of disturbance is small. Even so, it can help reveal the fundamental mathematical properties of Equation (4.92). Now, Equation (4.92) can be rewritten as follows:

$$r^2u'' + \alpha_1 ru' + \alpha_2 u = 0 \tag{4.94}$$

Introduce the transformation

$$r = e^R \tag{4.95}$$

or $r^{-1} = e^{-R}$, then we have

$$\frac{du}{dR} = \frac{du}{dr}\frac{dr}{dR} = \frac{du}{dr}e^R$$

or

$$\frac{du}{dr} = e^{-R}\frac{du}{dR} = r^{-1}\frac{du}{dR}$$

$$\frac{d^2u}{dr^2} = r^{-2}\left(\frac{d^2u}{dR^2} - \frac{du}{dR}\right)$$

Substituting all these expressions into Equation (4.94) leads to the following constant coefficient differential equation:

$$\frac{d^2u}{dR^2} + (\alpha_1 - 1)\frac{du}{dR} + \alpha_2 u = 0 \tag{4.96}$$

Its characteristic equation is

$$m^2 + (\alpha_1 - 1)m + \alpha_2 = 0$$

So, the condition for Equation (4.96) to have a harmonic solution is that m is a complex root. That is,

$$m_{1,2} = \frac{(\alpha_1 - 1) \pm \sqrt{(\alpha_1 - 1)^2 - 4\alpha_2}}{2}$$

or

$$(\alpha_1 - 1)^2 - 4\alpha_2 < 0$$

Substituting Equation (4.93) into this expression produces

$$\left[(3 - k^2 r) - 1\right]^2 - 4(1 - k^2 r) < 0$$

After some manipulations, we obtain

$$k^4 r^2 < 0 \quad \text{or} \quad k^2 r < 0 \tag{4.97}$$

Since $r > 0$, we must have $k^2(\lambda) < 0$. Evidently, $k^2(\lambda) < 0$ contradicts the condition

$$\lambda\left(=\frac{1}{k^2}\right) > 0$$

under which the second-order differential equation has a nonzero solution. And, from $k^2(\lambda) < 0$ and Equation (4.88), it follows that A (or Ψ) diverges exponentially. If $k^2(\lambda) > 0$, then from Equation (4.97), it follows that m is not a complex root. So, Equation (4.88) does not have any solution. Or in other words, the Rossby equation, obtained by using linearization, does not have any solution if one uses the conventional methods of solving mathematical physics equations. Or, it can be called an ill-posed equation. So, it is not hard to see that the assumption, as Rossby himself introduced, that the Rossby equation has a solution of the harmonic form does not have its needed mathematical basis. Although in the earlier discussion, we assumed that α_1, α_2 are constants, if we did not make such an assumption, Equation (4.92) would be a much more complicated differential equation with variable coefficients so that even if it had a "harmonic" solution, it would not be a simple harmonic wave. Besides, as a mathematical problem itself, the basic solutions of boundary value problems of Laplace equations in general are not problems of simple harmonic waves. So, it can be said that the method of pre-assuming the existence of harmonic waves does not satisfy the fundamental principles of mathematics. That is why in mathematics, for ill-posed equations, from assumed harmonic solutions, it does not naturally follow that these harmonic solutions actually exist. For mathematical physics equations, one cannot violate the requirement of well-posedness; otherwise, it is considered misusing mathematics.

No doubt, since it is about a physical phenomenon, it should of course have to agree with the relevant physical reality. On a plane, there is no way for a unidirectional retrogressive wave to appear. Unidirectionality is a fundamental characteristic of flows. By ignoring this common sense, the community of meteorologists in the past half century has pursued the ever unseen and inexplicable Rossby long waves and did not recognize that barometric pressure systems are phenomena posterior to weather events. As a matter of fact, people have fully understood neither the essence of the problems of predicting weathers and earthquakes nor what kind of science modern science is. That is why as of this writing, people still did not recognize that modern science could not serve as the theory for the study of changes in materials, including weather phenomena and occurrences of earthquakes. In other words, when modern science is seen as the theoretical foundation of prediction science, it is exactly those experts who cannot make practical predictions. For example, studies on barometric pressure systems have been considered the pride of meteorology; however, the theoretical experts of meteorology just somehow cannot explain why strong convection weathers, such as torrential rains, appear ahead of the relevant pressure systems.

In particular, *unidirectionality* means flows and must naturally cause the directionality for the flow field. Combining with the duality in spinning directions of rotational flow fields, discontinuities have to appear due to collisions of different directionalities involved. So, in the range of 4,000 to approximately 8,000 kilometers in the upper air above the earth's curved surface, it is impossible for a differential continuous flow field that satisfies the Cauchy–Riemann condition to exist. Besides, it is exactly because of the discontinuities in the flow field that conduct the transitional changes in atmospheric evolutions, leading to disastrous weather conditions.

Fourth, for the purpose of pursuing the nondimensionalization of the quantitative system of particle mechanics, people can easily and formally derive the Rossby equation of the system of linearity, which shows the intriguing power of nondimensionalization. In other words, the key technique that has made modern science successful and that helps to quantify the system of modern science is the nondimensionalization after first introducing physical quantities in the theory of particles. Without nondimensionalization or unification of dimensions, even with the introduction of physical quantities, classical mechanics would not be established and developed. So, nondimensionalization or unification of dimensions is the key element for modern science to even become science. However, along with the deepening and specialization of the development of science, the excessive expansion of quantification, as conducted by nondimensionalization or unification of dimensions, has to be revealed. It is because events after all are not quantities, formal unification is not the same as the specifics of events. That is why quantitative principles are not the same as those of physical materials and events. From that, what is uncovered with specific situations is how modern science has departed from reality; that is also the essential reason for the incompleteness of quantities or mathematics.

The reason why we have used the method of nondimensionalization to derive the Rossby equation is to illustrate that mathematicality is not the same as physicality. For example, by applying the nondimensionalization of Equations (4.76) and (4.77), we changed the original acting term (with dimension)

$$-\frac{1}{\rho}\nabla p$$

of the density and the gradient force (this term should have been called the stirring force term of the pressure gradient forces caused by changes in density), into a nondimensional $\nabla p'$. Through the process, the original physical solenoidal rotation effects are changed to quantitative gradient pushing forces. And, in the nondimensionalized f', β', the angular speed $\bar{\Omega}$ of the earth's rotation no longer appears so that the physical characteristics of rotation have been eliminated.

Also, we must note that the uniqueness of the theory of Rossby long waves is its unidirectional retrogression. That characteristic of Rossby long waves has been seen neither in the objective world nor in the laboratory, including the experiment of spinning dishes. In particular, the spinning eddy currents observed in the spinning dish experiment do not have any relation to the β effects of the earth's rotation. So, the opinion that the spinning dish experiment can evidence the theory of Rossby long waves means that the attributes of the rotational fluids in the spinning dishes have not been fully understood. Along with the heat tide of pushing for the system of wave motions in the 20th century, it is very understandable for scholars, including Rossby, to mimic classical mechanics in order to keep up with the scientific fad of wave motions. However, the spirit of science is about finding mistakes and problems and providing new approaches to correct the mistakes and to resolve the problems. That is, by introducing even geo-atmosphere, by using linearization and employing the method of harmonic waves, all the then-accustomed methods of logic thinking from mimicking the development of classical mechanics, Rossby obtained the unrealistic unidirectional retrogressive waves. This end in fact hints to us to even question the system of classical mechanics. Unfortunately, not only did the communities of fluid mechanics and meteorology not raise any question, but instead glorified the unidirectional retrogressive waves as being unique to the movements of fluids without recognizing their violation against physical reality. On all the daily weather maps of flow fields or the pictures of the flowing fluids in spinning dishes of the laboratory, what is observed is circulating currents of distorted eddies. And, even the eddy currents at mid to high altitudes are not retrogressive. That is, from real-life observations or from laboratory experiments, Rossby long waves have not been practically observed.

No doubt, the Rossby equation as in Equation (4.83), describes an even and nondimensionalized barometric pressure fields, and no longer possesses the mechanism for rotations. It is because the only cause for changes in the flow field is the βp_x with β being a constant. Notice that for the nondimensionalized β, it no longer contains the earth's rotationality Ω with p_x being only the gradient push along the x-direction. Also, being even has to naturally eliminate the solenoidal stirring effects and the elasticity of the fluids. Or in other words, even fluids are noncompressible so that the corresponding fluids cannot be pushed. That is to say, each two-dimensional fluid with its z-directional rise and fall disallowed cannot be moved horizontally by pushing forces. So, what is evidenced by the experiment of spinning dishes is exactly that the rotation of the spinning dish, which mimics the rotation of the earth, creates the whirlpool in the fluid in the form of the polar eddy; and combined with the interaction of the solenoidal terms of the temperature and pressure differences, a circulating flow of distorted polar eddy current is obtained. That is, the experiment of spinning dishes of fluids just cannot be used as the evidence for the existence of Rossby long waves.

Also, we need to note that in his process of deriving the Rossby waves, Rossby assumed that the disturbances are small. However, in the experiment of spinning dishes, the involved disturbances are not small, where the amplitude of the disturbances can approach the polar circle from the equator (see Figure 4.4). No doubt, the purpose of introducing the assumption of small disturbances will not be materialized for circulating currents of eddy motions; an even push cannot trigger any uneven disturbance; and even fluids cannot be moved by even pushes. That is why only stirs can cause uneven disturbances; however, that is already a rotational eddy motion.

As a problem of physics or a property of physics, eddy motions and wave motions represent movements of completely different attributes. Eddy motions are originated from rotations of materials or mutual reactions of uneven stirs, and can transfer materials and energies at the same time, while wave motions stand for the morphological changes caused by reciprocating, straight-line movements of materials. They can only propagate energy without transporting materials. Evidently, without being able to transport materials, how can there be any input of water vapors, and how can traveling clouds plant rains? In particular, eddy motions involve moments of forces of materials' directional structures and are no longer particles that are pushed in the Newtonian inertial systems. Even if they are seen as problems of kinetic energy,

$$\frac{1}{2}mv^2$$

embodies the work of the pushing force, while

$$\frac{1}{2}m\omega^2 \left(\text{or } \frac{1}{2}m\varsigma^2 \right)$$

represents the transportation of energy of the eddy motion under the solenoidal stir. So, we still cannot confuse wave motions and eddy motions as the same concept of physics, even though the concept of wave motions has been sweepingly generalized or mathematicalized.

So, no matter whether it is the Rossby equation obtained by Rossby's linearization or that produced out of our nondimensionalization and comparison of quantitative magnitude levels, the equation suffers from a common deficiency where the underlying rotationality of the fluids or materials is hidden by the formality of quantities. For each realistic fluid, including the atmosphere and oceans, there is no way for the assumption of evenness to be introduced for the range of several thousand kilometers. It is because if the assumption of pure evenness is imposed, then the even fluid cannot be pushed from inside, either. If it could be pushed from within, then the evenness of the fluid has be to destroyed. So, even if a super being could have a mighty power to apply a push, he can only push the entire even fluid. In terms of meteorological science, one should pay attention to Bjerknes's circulation theory (1898), which indicates that due to the unevenness and the consequent solenoidal effects, the earth's atmosphere and oceans rotate, and resolves the mystery of nonlinearity of mathematical nonlinear dynamic equations. What is a pity is that half a century later, meteorological science, which contains Bjerknes's circulation theorem, embraced the theory of Rossby long waves. At least in terms of meteorology, the historical phenomenon seems to be a paradox. In particular, in today's meteorology, where Rossby long waves are seen as a fundamental theory, people have to ask naturally: Since the circulation theorem can be applied to explain land and ocean breezes, how can air of a range of several thousand kilometers be even? No doubt, the circulation integral of the circulation theorem does not comply with the particlization of classical mechanics and also destroys inertial systems of modern science. So, that is why we once pointed out that as a scientist of the west, V. Bjerknes not only made contributions beyond his time in meteorology, but also was the first person, even ahead of Einstein who advocated for wave motions and Schrödinger, who successfully resolved the mystery of nonlinearity of the latter part of the 20th century, if his work is seen from the angle of theoretical physics. Or in other words, if Einstein and Schrödinger truly knew about the significance of Bjerknes's work, they would have modified what they did.

As for correctly recognizing the difference and effects of eddy motions and wave motions, even if we follow the theory of wave motions, we still have to face the problem of whether or not this theory of wave motions can be applied as the theoretical foundation of prediction for evolution or evolution science. Evidently, wave motions are a morphology that appears after materials have moved, and eddy motions are behaviors that appear at least at the same time as the materials move. That is why wave motions are not the cause for movements in materials. Also, wave motions realize energy transfers and removing forces in the fashion of propagating energies and maintain the invariancy of modern science by preventing materials from being damaged. That in fact has also revealed the secret of modern science. Evidently, any theory of invariant materials cannot constitute evolutions of materials. In particular, wave motions are morphological postevent effects. No matter whether it is their mechanism or essences, they represent signals or information of happened events instead of information prior to the occurrence of the events. So, naturally, wave motions cannot be employed for predicting what has not happened. It is just like the situation of trading stocks: When excessive capital is flowing into the marketplace, a bull market in the form of wave motion appears; when excessive amounts of capital are leaving the marketplace, a bear market in the form of wave motions will occur. Roughly 70% of stock trades are losers, about 10% winners, and about 20% break even. What is interesting is that the percentage (70%) of losers is basically the same as the accuracy rate of current weather forecasts. So, the lagging of wave motions is not only limited to meteorological science but in fact involves the problem of how to understand the properties of the whole system of wave motions. To this end, after suffering through 100-plus years, at the same time when we celebrate the successes of the wave-motion system, we should also notice the problems existing along with the system. In terms of meteorological science, the lagging of pressure systems behind weather phenomena in essence has overthrown the current system of forecasts. However, it is a pity that the relevant professionals in meteorology still cannot accommodate the necessary changes in concepts. Even so, the crudity of the reality does not entertain human emotions or feelings; it is still a fact that only after precipitations are pressure systems seen. The core of our method of digitization is not to make use of pressure systems in the form of inversed order structures so that the valuable lead times for forecasting are obtained and that forecasting weather factors can be done directly without crossing over situation predictions of pressure systems.

The eddy effects of rotational movements possess applicability different from that of wave motions. Other than quantitativeness, eddy motions also have directionality so that different ways of rotation lead to different events; and eddy motions come from the unevenness in materials' structures. So, since 1963, we have been able to analyze the transitionality of evolutions by using the direction of rotations and the digitization of the irregular information of specifics, while foretelling what is forthcoming before it actually happens. With over 40 years of practice, even for those observational stations that are 300 kilometers apart from each other, our accuracy rate of forecasting major disastrous weather conditions has reached over 70%; and this rate of accuracy can still be improved another 15–20% if additional stations are added. (The ideal distribution of these stations for the purpose of forecasting major weather disasters using our method is about 100 kilometers.) Although transitional weather changes or disastrous weather conditions are only about 20–25% of all weather phenomena, they represent the central problem of evolution. This is because disasters or major accidents happen along with transitional changes. Or, speaking more generally, disasters or accidents are caused by dislocational transformations of spinning directions and are indicated by peculiar information that represent states and attributes totally different from those of the present or the past. This discovery also holds true for solids beyond fluids. So, the essence of prediction is to seek out transitionalities that change the present or past states or attributes. The currently popular extrapolation of initial-value systems is about monitoring. Its application in the profession of prediction is not entirely about wave motions. Instead, because the rotationality of eddy motions makes pressure systems that have already occurred move very slowly, the needed time for making extrapolation is earned. However, extrapolation is still about reporting what has already happened; and it can also be accomplished by using our method of digitization. When the structural spinning direction stays unchanged, no transitional change will occur. So, one experiences an extension of the present states and attributes, which is an extrapolation of the present.

What we must emphasize is that the classical mechanic system, developed since the time of Newton, is an extrapolation system without involving evolution. Other than extrapolations without change, this system cannot be applied to predict changes in general and transitional changes in particular. That is why modern science cannot be applied to or be treated as the theory of predicting evolutions.

In short, the Rossby long waves, a fundamental concept of meteorological science, no matter whether we are talking about the modeling or the process of solving the model, suffer from many problems even if we follow the conventions of the system of modern science. In terms of predictions, one has to face the additional problem of how to conceptually understand the lagging of wave motions. Evidently, without resolving the dislocations existing in our epistemological understandings, we will not be able to make substantial progress in the study of prediction problems.

4.5 The Problem of Integrability of the KdV and Burgers' Equations

What needs to be pointed out is that one of the authors of this book graduated years ago from the Department of Meteorology of Nanjing University. Since he did not agree with the theory of Rossby long waves, upon his graduation, he left the area of meteorology. While he was working in the profession of water conservancy, he came across the concept of solitary waves in the evolution of floods and the KdV equation. However, what is different from other scholars who study solitary waves is that by taking advantage of the conditions and opportunities of his work, he observed countless rivers, specifically looking for realistic solitary waves in rivers. After doing this for over 30 years, covering all major rivers in China and some of the rivers in Europe and North America, he later required his students to sit at Baopingkou, Chengdu, to possibly capture the natural phenomena. He did not have the good luck to ever see the cylindrical water column with a smooth exterior surface as seen in 1834 by British ship engineer J. S. Russell; Russell claimed that such water columns were created by moving ships. In theory, the problem of solitary waves was mathematically addressed by Korteweg and de Vries in 1895, when they simplified the equation of fluid movements, which was established in the 17th century, and produced the now well-known KdV equation that contains the solitary wave solution.

In the latter part of the 20th century, pushed by applications of mathematical physics equations in theoretical physics, a heat wave of studying nonlinear mathematical physics equations appeared. That was how the concept of solitary waves was greatly expanded when Zabusky and Kruskal (1965) introduced solitons on the basis of particles from physics, leading to such concepts as bell-shaped solitons with the same asymptotic values, torsion-type solitons with different asymptotic values, enveloped solitons, attractors, boomerons, trappons, and so forth. And, in the second half of the 20th century, solitons, chaos theory, fractional dimensions, and so forth, were seen as representative works of the heat wave of nonlinearity. However, together with our colleagues we hold different opinions, and in 1994, 1998, and 2001, we published our relevant works and results, where we touched upon the problems existing with the solitons of the KdV equation, the shock waves of the Burgers' equation, and Rossby waves, respectively, and so forth. Our works on chaos theory caught the attention of the world of learning. What is important is that at the height of scholarly importance and the epistemological concepts, our works brought forward impacts on some of the fundamental concepts of the wave-motion system.

Mathematically speaking, the solitons of the KdV equation and the shock wave of the Burgers' equation are seen as weak-nonlinear, since these equations contain linear terms of higher order derivatives. That shows the world of mathematics is very accustomed to the system of linearity. For example, the KdV equation contains a linear term with a third-order derivative in it, while the Burgers' equation contains a linear term of second-order derivative. So, in mathematics, equations that contain both linear and nonlinear terms are called weak or quasi-nonlinear. Such terminologies specifically show the intention of expanding the concepts of linearity to the studies of nonlinearity. It can also be said that in our modern times, we still have not truly understood the realistic significance of nonlinear equations and still try to continue the system of linearity. Although the contemporary stream of consciousness in human thinking can be understood, the physical world just does not comply with the human conscious thinking. So, continuing the old customs does not mean that these traditional concepts would be reasonable and would meet the calls of practical problems, because as long as one term is nonlinear, the mathematical properties of the equation will be completely changed. So, in essence, both weak and quasi-nonlinear equations are all nonlinear. In particular, the properties of nonlinearity of nonlinear equations cannot be altered simply by reducing their orders. In terms of mathematics, the fundamental characteristics of nonlinear equations are that they do not satisfy the initial-value existence theorem, which should now have become the definite result of mathematics. In the following, we will respectively discuss the problems with regard to the solitons of the KdV equation and the shock waves of the Burgers' equation.

4.5.1 The Modeling Problem of the KdV Equation

In this book, our focus is on the method of series expansion. Other than this method, there are also several other methods, such as the truncation method of nondimensionalized series expansion. In principle, these methods do not have much difference. Considering the fact that the truncation method of nondimensionalized series expansion has been well discussed in our book *Entering the Era of Irregularity*, in the following, we will not touch on this method at all.

For the flowing water in a river current, its local small-scale flow can be described by using the following two-dimensional system of equations of fluid movement in the Euler language, where the water's density is taken as a constant, ρ = a constant.

$$\begin{cases} u_t + uu_x = -RT_x \\ w_t + uw_x = -RT_z - g \\ u_x + w_z = 0 \\ T_t + uT_x - \gamma w = 0 \\ p = \rho RT \end{cases} \tag{4.98}$$

where u stands for the flow speed in the x-direction, w the flow speed in the vertical direction, R the fluid constant of the flowing water, T the temperature, g the gravity, γ the degree of stability of the fluid's stratification structure in the vertical direction, which is taken to be a constant. Equation (4.98) is a system of equations of small-scale fluids in the x–z directions without the Coriolis term. The subscripts stand for partial derivatives, $p = \rho RT$ the state equation of the fluid, and p the intensity of pressure of the fluid.

Based on the traditional method of traveling waves, let us introduce the phase space variables:

$$\begin{cases} u(x,z,t) = \varsigma(\xi) \\ w(x,z,t) = W(\xi) \\ T(x,z,t) = T(\xi) \\ \xi = kx + \lambda z - \nu t \end{cases} \tag{4.99}$$

Substituting Equation (4.99) into Equation (4.98) produces

$$\begin{cases} (k\varsigma - \nu)\varsigma_\xi = -Rk T_\xi \\ (k\varsigma - \nu)W_\xi = -R\lambda T_\xi - g \\ k\varsigma_\xi + \lambda W_\xi = 0 \\ (k\varsigma - \nu)T_\xi - \gamma w = 0 \end{cases} \tag{4.100}$$

The first, second, and fourth equations, respectively, in Equation (4.100) can be written as follows:

$$\varsigma_\xi = -\frac{Rk}{k\varsigma - \nu} T_\xi \tag{4.101}$$

$$W_\xi = -\frac{g}{k\varsigma - \nu} - \frac{k\lambda}{k\varsigma - \nu} T_\xi \tag{4.102}$$

and

$$\varsigma_\xi = \frac{\gamma}{k\varsigma - v} W \tag{4.103}$$

If the function $[f(x,z,t)]_{z=\varsigma}$ can be expanded into a Taylor series about $\varsigma = 0$,

$$f(x,z,t) = [f]_{\varsigma=0} + \frac{\varsigma}{1!}\left[\frac{\partial f}{\partial z}\right]_{\varsigma=0} + \frac{\varsigma^2}{2!}\left[\frac{\partial f}{\partial z}\right]_{\varsigma=0} + \cdots\cdots$$

where $[f]_{\varsigma=0} = f(x,0,t)$, then by expanding Equations (4.100) through (4.103) simultaneously about $\varsigma = 0$ and taking the first two terms of these expansions, we have

$$\varsigma_\xi = \frac{Rk}{v}T_\xi + \frac{Rk^2}{v^2}T_\xi\varsigma \tag{4.104}$$

$$W_\xi = \left(\frac{g}{v} + \frac{gk}{v^2}\right) + \frac{k\lambda}{v}T_\xi + \frac{k^2\lambda}{v^2}T_\xi\varsigma \tag{4.105}$$

and

$$T_\xi = -\frac{\gamma}{v}W + \frac{\gamma k^2}{v^2}W\varsigma \tag{4.106}$$

Substituting Equation (4.106) into Equations (4.104) and (4.105) and ignoring all the terms of above third-order of the product of W and ς lead to

$$\varsigma_\xi = -\frac{Rk\gamma}{v^2}W + 2\frac{Rk\gamma}{v^3}W\varsigma \tag{4.107}$$

and

$$W_\xi = \left(\frac{g}{v} + \frac{gk}{v^2}\right) - \frac{k\lambda\gamma}{v^2}W + 2\frac{k^2\lambda\gamma}{v^3}W\varsigma$$

$$= g_{kv} - \frac{k\lambda\gamma}{v^2}W + 2\frac{k^2\lambda\gamma}{v^3}W\varsigma \tag{4.108}$$

By taking the integral of the third equation in Equation (4.100) and letting the integration constant be zero, we have

$$W = -\frac{k}{\lambda}\varsigma \qquad (4.109)$$

Substituting Equation (4.109) into Equation (4.108) produces

$$W_\xi = g_{kv} + \frac{k^2\gamma}{v^2}\varsigma - 2\frac{k^3\gamma}{v^3}\varsigma^2 \qquad (4.110)$$

Taking

$$\frac{\partial}{\partial\xi}$$

leads to

$$\varsigma_{\xi\xi} = -\frac{Rk\gamma}{v^2}W_\xi - 2\frac{Rk^2\gamma}{v^3}(W_\xi\varsigma + W\varsigma_\xi) \qquad (4.111)$$

Using Equation (4.109) to eliminate W in Equation (4.107), we have

$$\varsigma_\xi = \frac{Rk^2\gamma}{v^2\lambda}\varsigma - 2\frac{Rk^3\gamma}{v^3\lambda}\varsigma^2 \qquad (4.112)$$

Now, by substituting Equations (4.110) and (4.112) into Equation (4.111) and ignoring the third-order term, we obtain

$$\varsigma_{\xi\xi} + \frac{1}{2}\sigma\varsigma^2 - U\varsigma + G = 0 \qquad (4.113)$$

where

$$\sigma = 2\frac{Rk^3\gamma^2}{v^5}\left[2\frac{Rk^2}{\lambda^2} - (2k+1)\right]$$

$$U = -\frac{Rk^2\gamma}{v^3}\left(2g_{kv} - \frac{k\gamma}{v}\right)$$

and

$$G = \frac{Rk\gamma}{v^2} g_{kv}$$

From

$$\frac{\partial}{\partial \xi} (4.113)$$

it follows that

$$\varsigma_{\xi\xi\xi} + \sigma\varsigma\varsigma_{\xi} - U\varsigma_{\xi} = 0 \qquad (4.114)$$

Take $\eta = \varsigma(\xi)$, $\xi = X - Ut$ and simplify Equation (4.114) into

$$\eta_t + \sigma\eta\eta_X + \eta_{XXX} = 0 \qquad (4.115)$$

That is, Equation (4.115) is the well-known KdV equation in modern science, where we note that the subscript X is in uppercase.

However, by carefully analyzing the previous process of modeling, one can readily discover the following problem. That is, if the integration constant of Equation (4.100)$_3$ is not zero, then the general situation will be

$$W = -\frac{k}{\lambda}\varsigma + H$$

or

$$\varsigma = -\frac{\lambda}{k}W + \frac{\lambda}{k}H \qquad (4.116)$$

Substituting Equation (4.116) into Equation (4.107) produces

$$\varsigma_{\xi} = -\frac{Rk^3\gamma}{\lambda v^2}\varsigma^2 + \frac{Rk^2\gamma}{\lambda v^2}\left(1 + \frac{k}{v}\right)\varsigma - \frac{Rk\gamma}{v^2}H \qquad (4.117)$$

Inserting Equation (4.116) into Equation (4.108) leads to

$$W_{\xi} = -2\frac{k\lambda^2\gamma}{v^3}W^2 + \frac{k\lambda\gamma}{v^2}\left(2\frac{\lambda^2}{v}H - 1\right)W + g_{kv} \qquad (4.118)$$

From Equations (4.117) and (4.118), it follows that the first derivatives of ς, W are all mathematical quadratic forms. Evidently, the powers (exponents) of the second derivative of ς, W should be lowered. That is, it will be feasible only if the derivatives of Equations (4.117) and (4.118) are linear forms. However, after substituting Equations (4.117) and (4.118) into Equation (4.111) and ignoring the third-order terms, we have

$$\varsigma_{\xi\xi} = -\frac{Rk\gamma}{v^2}g_{kv} - \frac{Rk^2\gamma}{v^3}\left(g_{kv} + \frac{k\gamma}{v}\right)\varsigma + 2\frac{R^2k^5\gamma^2}{v^5\lambda^2}\varsigma^2 \qquad (4.119)$$

From Equation (4.119), it can be seen that the second derivative of ς is still a quadratic form. Or in other words, in the previous process of modeling, the assumption is that the kept first-order and second-order derivatives of the series expansions are of quadratic forms. That no doubt leads to a contradiction of mathematical formal logic. That is, it is not feasible mathematically that both first-order and second-order derivatives are of the same exponential powers. In other words, according to the basic formal logic of mathematics, equations of the second derivatives should have the order of 1 less than that of the first-order derivatives. Evidently, if we differentiate Equation (4.112) one more time, we have

$$\varsigma_{\xi\xi} = \frac{Rk^2\gamma}{v^2\lambda}\varsigma_\xi - 4\frac{Rk^2\gamma}{v^3\lambda}\varsigma\varsigma_\xi \qquad (4.120)$$

That is why Equations (4.119) and (4.120) are not equivalent. Or from Equation (4.120), one cannot obtain Equation (4.113), which means that the KdV equation in Equation (4.114) or Equation (4.115) cannot be produced. The essence of the so-called mathematical modeling or simplification of mathematical models is to maintain the nonlinear quadratic form in the expanded formulas by using various methods, since without the quadratic forms, there will not be the second term in Equation (4.115). As for the establishment of the KdV equation by using the truncation method of nondimensionalized series expansion discussed in the book *Entering the Era of Irregularity*, the entire process in form is much more complicated than what is shown above. However, the key problem is still about trying to keep the quadratic form, making the entire process look very artificially forced. Based on the excessive desire of chasing after quantitativeness of the traditional system, this whole thing has been evolved to such a degree that quantities are used for the purpose of quantities without even considering the least amount of logic reasoning of quantities. In fact, the perfection of the formal logic is not the same as the physical reality; and the previous modeling of the KdV equation or simplification of the model did not even follow basic calculus of formal logic. So, the KdV equation can only be a product of human consciousness; and the more perfect the mathematical treatment is in the system of continuity, the more the whole thing becomes a play of the relevant rules. As for whether or not the KdV equation is one of the representative nonlinear equations, or whether or not solitary waves possess the generality, one can obtain the definite answers to these questions from the method of how to solve the KdV equation.

4.5.2 The Problem of Solving the KdV Equation

4.5.2.1 The Generality of the Solution of the KdV Equation

Let us rewrite Equation (4.117) as follows:

$$\varsigma_\xi = a\varsigma^2 + b\varsigma + c \tag{4.121}$$

where

$$a = -\frac{Rk^3\gamma}{\lambda v^2}, \quad b = \frac{Rk^2\gamma}{\lambda v^2}\left(1+\frac{k}{v}\right), \quad c = -\frac{Rk\gamma}{v^2}$$

Evidently, Equation (4.121) is a mathematical quadratic model, and in Section 3.3 we have already studied the basic mathematical characteristics of quadratic models. No doubt, the solution of Equation (4.121) must be the following three cases:

If $b^2 - 4ac = 0$, then Equation (4.121) can be rewritten as

$$\varsigma_\xi = a\left(\varsigma+\frac{1}{2}\frac{b}{a}\right)^2 \tag{4.122}$$

Integrating this expression leads to

$$a\varsigma + \frac{1}{2}b = -\frac{1}{\xi+\xi_0}$$

That is,

$$\varsigma = -\frac{1}{a}\left(\frac{1}{\xi+\xi_0}+\frac{1}{2}b\right) = -\frac{1}{a}\left[\frac{1}{\delta\left(1-\dfrac{t}{t_b}\right)}+\frac{b}{2}\right] \tag{4.123}$$

where $\varsigma = kx + \lambda_\chi - vt$, $\delta = kx + \lambda_\chi + \xi_0$, and

$$t_b = \frac{\delta}{v}$$

Evidently, when $t = t_b$, $\varsigma \to \infty$; when $t > t_b$, ς changes its direction. So, Equations (4.121) and (4.122) are problems of blown-ups of transitional changes; and the corresponding so-called solitary solutions do not exist, or the KdV equation does not have any solitary wave solution.

If $b^2 - 4ac > 0$, then Equation (4.121) becomes

$$\varsigma_\xi = a\left(\varsigma + \frac{1}{2}\frac{b}{a}\right)^2 - \frac{1}{4a}(b^2 - 4ac) \tag{4.124}$$

Integrating this equation leads to

$$\ln\left|\frac{\left[(\varsigma + \frac{1}{2}\frac{b}{a}) - \frac{\sqrt{b^2 - 4ac}}{2a}\right]}{(\varsigma + \frac{1}{2}\frac{b}{a}) + \frac{\sqrt{b^2 - 4ac}}{2a}}\right| = \frac{\sqrt{b^2 - 4ac}}{a}\xi + \xi_0$$

If

$$\left|\varsigma + \frac{1}{2}\frac{b}{a}\right| < \frac{1}{2a}\sqrt{b^2 - 4ac}$$

one has

$$\varsigma = -\frac{1}{2a}\sqrt{b^2 - 4ac}\, th\left(-\frac{1}{2a}\sqrt{b^2 - 4ac}\,\xi - \frac{1}{2}\xi_0\right) - \frac{1}{2}\frac{b}{a} \tag{4.125}$$

Although Equation (4.125) is a smooth solution, it is the result of a local constraint and does not possess the generality. And what makes those scholars of solitary solutions disappointed is that even this continuous, smooth solution is not a solitary wave solution. It is because only after one differentiates Equation (4.125) with respect to ξ once, he obtains the solitary wave solution. So, the well-known solitary wave is only a special situation among very special quadratic equations. And, it can be obtained by differentiating quadratic equations (see Section 3.3). As for the generality, if

$$\left| \varsigma + \frac{1}{2}\frac{b}{a} \right| > \frac{1}{2a} \sqrt{b^2 - 4ac}$$

one has

$$\varsigma = \frac{1}{2a} \sqrt{b^2 - 4ac} \, cth\left(-\frac{1}{2a} \sqrt{b^2 - 4ac} \, \xi - \frac{1}{2}\xi_0 \right) - \frac{1}{2}\frac{b}{a} \qquad (4.126)$$

Equation (4.126) indicates that even under the most general condition, the quadratic model also contains the properties of blown-ups of transitionality.

If $b^2 - 4ac < 0$, then Equation (4.121) becomes

$$\varsigma_\xi = a\left(\varsigma + \frac{1}{2}\frac{b}{a} \right)^2 - \frac{1}{4a}(4ac - b^2)$$

By integrating this equation, we have

$$\varsigma = \frac{1}{2a} \sqrt{4ac - b^2} \, \tan\left(\frac{1}{2a} \sqrt{4ac - b^2} \, \xi + \frac{1}{2}\xi_0 \right) - \frac{1}{2}\frac{b}{a} \qquad (4.127)$$

No doubt, Equation (4.127) clearly shows the properties of periodic blown-ups of the transitionality.

So, the KdV equation and the problem of solitary waves, which has been well-known for over a century, are in fact a special case among very special situations. Maybe what makes people laugh is that famous so-called solitary waves can even be obtained by differentiating quadratic equations, representing only a special case of quadratic equations.

Besides, what has to be illustrated is that the purpose of applying series expansion is simply for eliminating the singularity or nonsmoothness of the original equation. However, the fundamental characteristics of nonlinear equations are exactly the singularities and nonsmoothness of these equations that make them different from linear equations. So, one has to consider how much practical significance the KdV equation, as a simplified equation of the N-S equation, still possesses, no matter whether it is seen as a mathematical problem or a physics problem. Also, whether or not the cylindrical water column with a smooth surface as observed by J. S. Russell in 1834 is a solitary wave needs to be reconsidered, because in the following 100-plus years nobody else has ever had the luck to once again see such a phenomenon. Proposing questions is the duty of all scholars responsible for the well-being of science.

4.5.2.2 Bäcklind Transformation and the Problem of Solving the KdV Equation

If we introduce

$$\eta = p_x \tag{4.128}$$

then Equation (4.115) can be rewritten as

$$p_{xt} + \sigma\, p_x p_{xx} + p_{xxxx} = 0 \tag{4.129}$$

Integrating this equation produces

$$p_t + \frac{\sigma}{2}\, p_x^2 + p_{xxx} = 0$$

Introduce the Bäcklind transformation

$$p_x = \frac{3}{\sigma}\left(U - \frac{1}{4}\,p^2\right) \tag{4.130}$$

Equation (4.130) is clearly a quadratic form in p. As pointed out earlier, the general characteristic of this kind of equation is blown-ups of transitionality with the following local special attribute:

$$p = \frac{3}{\sigma}\,2\sqrt{U}\,th\left[\frac{\sqrt{U}}{2}(X - Ut)\right] \tag{4.131}$$

No doubt, Equations (4.131) and (4.125) have similar forms. For this class of equations, we have provided a detailed discussion in Section 3.3. Evidently, if we differentiate the previous equation with respect to x once, we have

$$\eta = p_x = \frac{3U}{\sigma}\,\mathrm{sec}\,h^2\left[\frac{\sqrt{U}}{2}(X - Ut)\right] \tag{4.132}$$

And, Equation (4.132) is the solitary wave solution, which can also be obtained by differentiating Equation (3.6) or Equation (4.125). The so-called solitary wave is the derivative of a hyperbolic tangent function, which only holds true under special local conditions.

4.5.2.3 Direct Integration and the Solitary Waves of the KdV Equation

For a high-order differential equation to maintain its nonlinearity, its reduced order form has to have an order higher than or equal to two. So, using integration to reduce the order is a simple and straightforward method. Introduce $\eta = \varsigma(\xi)$, $\xi = X - Ut$ so that the order of Equation (4.114) can be reduced as follows:

$$\varsigma_{\xi\xi} = -\frac{1}{6}\sigma\varsigma^2 + U\varsigma - G \qquad (4.133)$$

Integrating Equation (4.133) again leads to

$$\varsigma_\xi = -\frac{1}{6}\sigma\varsigma^3 + \frac{U}{2}\varsigma^2 - G\xi + H \qquad (4.134)$$

Evidently, Equation (4.134) is a nonlinear equation of the cubic form. That is, after the order of Equation (4.133) is reduced, the result is an equation of the cubic form (Equation 4.134). No doubt, out of a nonlinear equation of the cubic form, one cannot obtain a solitary wave solution. In particular, in the process of simplifying the model, cubic or higher-order terms of the ς equation have been ignored; and in the corresponding integration process of solving for the solution, no more cubic terms can be reproduced by the integration procedure, which forms a contradiction with the abandonment used in the modeling step. Or, this contradiction is referred to as the infeasibility of mathematical formal logic. That is, when developing the original model by simplifying equations, cubic and higher-order terms are abandoned, while in the process of solving the simplified model, a cubic term reappears, constituting a mathematical inconsistency. Besides, Equation (4.134) does not contain any solitary wave solution; this fact goes against the wish of pursuing solitary solutions.

The current method of solving the model is to multiply Equation (4.133) by ς_ξ. After taking the constant $G = 0$, integrating the result leads to

$$\frac{1}{2}\varsigma_\xi^2 + \frac{\sigma}{6}\varsigma^3 - \frac{U}{2}\varsigma^2 = 0$$

Notice that in this equation, ς^3 term reappears. That represents a common problem experienced by the method of modeling using series expansions. And, this problem has not yet caught the attention of the community of mathematics. Instead, what is done is to rewrite the previous equation as

$$\varsigma_\xi = \frac{1}{\sqrt{3}}\sqrt{3U\varsigma^2 - \sigma\varsigma^3} \tag{4.135}$$

The difference of Equation (4.135) from Equation (4.134) is that the form of ς_ξ that contains ς^3 in Equation (4.134) is changed to the equation in ς_ξ of the power 3/2 by multiplying Equation (4.133) by ς_ξ. In form, multiplying Equation (4.133) by ς_ξ is a mathematical technique; however, this technique has altered the mathematical properties of the original equation. And, if the integration constant $G \neq 0$, then the resultant equation is no longer the KdV equation. That is, in this case, the equation that is solved is not equivalent to the original equation that needs to be solved. So, there naturally appears such a need to reconsider whether or not the so-called KdV equation indeed has solitary wave solution or not.

Integrating Equation (4.135) produces

$$\varsigma = \frac{3U}{\sigma}\sec h^2\frac{\sqrt{U}}{2}\xi = \frac{3U}{\sigma}\sec h^2\left[\frac{\sqrt{U}}{2}(X - Ut)\right] \tag{4.136}$$

Equation (4.136) is the solitary solution of the mathematical physics KdV equation that has been well-known since the 1960s. Correspondingly, for a period of time, the KdV equation was also the center of discussion in the atmospheric science that follows the scholastic fad. In particular, at the end of the 1980s, the KdV equation and the problem of solitary waves, together with chaos theory, were seen as the representative achievements of the science of nonlinearity.

As part of the effort of chasing after the contemporary fad, the laws of conservation of the KdV equation, a simplified equation of fluid movement, also became a research topic of mathematics and mathematical physics.

4.5.3 Mathematical Properties of the KdV Equation and Its Conservation Laws of Energy

It is well known that the three representative scientific achievements of the 17th century are (1) fluids seen as continua of solids of low resistance; (2) laws of conservation of energy; and (3) the establishment of calculus. Now, fluids seen as continua of solids of low resistance has met with the problem of violating physical reality; calculus has been seen as the second crisis and part of the fourth crisis in the foundations of mathematics (Lin et al., 2008); and laws of conservation of energy are one of the main problems studied in physics; they represent the maturity of physics. That is why people are compelled to their continued research on the problem of energy conservation.

4.5.3.1 Representation of Conservation Laws of Energy in Modern Science

At the start of modern science, there appeared three main laws of conservation, which are the conservation of mass, conservation of energy, and conservation of momentum. Mathematically, each law of conservation means that a certain physical phenomenon or event can be described as the following equation

$$u_t = H(u) = H(u, u_x, u_{xx}, \cdots \cdots) \tag{4.137}$$

where the subscripts stand for partial derivatives. If there are functions $T(u)$ and $X(u)$ of the solution u and its derivative with respect to the space

$$\frac{\partial^p u}{\partial x^p} \, (p = 1, 2, \cdots \cdots)$$

satisfying

$$\frac{\partial T(u)}{\partial t} + \frac{\partial X(u)}{\partial x} = 0 \tag{4.138}$$

then Equation (4.138) is referred to as a conservation law of Equation (4.137). In many cases, $T(u)$ and $X(u)$ are polynomials in u and its derivatives of various orders. The function $T(u)$ is known as the conservation density, and $X(u)$ the flux. When $X(u)$ equals zero on the boundary of the region, integrating Equation (4.138) with respect to x produces

$$\frac{d}{dt} \int_{-\infty}^{\infty} T(u) dx = 0 \quad \text{or} \quad \int_{-\infty}^{\infty} T(u) dx = \overline{T(u)} = C = Const. \tag{4.139}$$

It means that $T(u)$ has nothing to do with time t. In this case, Equation (4.139) is referred to as the conservation integral or integral of motion or the constant of motion.

Laws of conservation are a central topic in physics. The traditional applications of these laws in mathematics are in the area of the well-posedness (that means the existence, uniqueness, and stability of solutions) of differential equations in the form of prior estimates of the solutions. This has become the key problems of mathematical physics equations and is known as the theory of well-posedness. That is where the method of designing stable difference schemes using the conservation of energy is from in the current computations of numerical analysis. As a matter of fact, each conservation law of energy is about transfers of energy or the process of transfer instead of the pursuit of the invariant totals or the pursuit of a certain physical quantity that has nothing to do with time. The invariant result of energy conservation does not mean the invariant energy or no relationship with time in the process of transfer or transformation. So, each true conservation of energy should provide the mechanism of transformation or the methods of transfer for the process of change in energies. If energy conservations are only applied as constraints for well-posedness, the true essence of the energy conservations does not seem to have been materialized. That also stands for a deficit of application of the conservation laws and an indication that the conservation laws of energy in modern science are still not complete. And, they have been only applied to maintain the invariancy of materials. The purpose of application or the direction of research has shown that the objects of service are durable goods without involving problems of changes in materials. The so-called prior estimates are not from the mathematical physics equations themselves; instead they employ conservation laws of energy as the constraints for stability. As for problems of evolution, the effects of the so-called conservation laws are to cut off the energy transformation in the form of controlling quantitative instability.

According to the contemporary terminology, the KdV equation with solitary wave solution is a nonlinear equation of wave motion and has been shown to be a special case of quadratic nonlinear equations under the local condition of continuity instead of the general situation of nonlinear equations. From this end, it can be seen that the mathematical properties of the conservation laws of the KdV equation are some of the constraints for the stability considered in the system of modern mathematics. Since they in essence do not involve the problem of energy transformation, they have surely not addressed the physical process and mechanism of energy conservation. That is why even though the KdV equation has infinitely many concrete relationships of conservation laws, none of them clearly addresses the physical process and mechanism of energy conservation. Even from the form of the equation, it can be seen that solitary waves are the results of balancing the quantitative values of nonlinearity and the effects of dispersion. That is why the KdV equation is often called nonlinear dispersive waves. However, the equation's mechanism of dispersion should not be mistakenly seen as nonlinear dispersion. Or in other words, the concept of nonlinear dispersive waves still needs to be carefully pondered.

4.5.3.2 The KdV Equation and Its Conservation Laws

Whitham (1965) first provided three conservation laws for the KdV equation. For example, the earliest KdV equation

$$u_t - 6uu_x + u_{xxx} = 0 \qquad (4.140)$$

can be rewritten as

$$u_t + (-3u^2 + u_{xx})_x = 0 \qquad (4.141)$$

That is, Equation (4.141) is the first law of conservation of the KdV equation, where the conservation density $T(u) = u$ and the flux $X(u) = -3u^2 + u_{xx}$. Integrating Equation (4.141) with respect to x over $(-\infty,\infty)$ produces

$$\int_{-\infty}^{\infty} u(x,t)\,dx = \int_{-\infty}^{\infty} u(x,0)\,dx = M_0$$

which is the conservation of momentum. Evidently, in this case, one has to assume that when $|x| \to \infty$, u and its various derivatives approach zero.

If Equation (4.141) is multiplied by u, we have

$$\left(\frac{1}{2}u^2\right)_t + \left(-2u^3 + uu_{xx} - \frac{1}{2}u_x^2\right)_x = 0$$

Integrating this equation with respect to x over $(-\infty,\infty)$ leads to

$$\int_{-\infty}^{\infty} \frac{1}{2}(u(x,t))^2\,dx = \int_{-\infty}^{\infty} \frac{1}{2}(u(x,0))^2\,dx = E_0 \qquad (4.142)$$

This is the energy conservation of the second conservation law. Its essence is the conservation of kinetic energy of irrotational speed.

If Equation (4.140) is multiplied by $3u^2$, we have

$$u_t^3 - 18u^3 u_x + 3u^2 u_{xxx} = 0$$

If we differentiate Equation (4.140) with respect to x and then multiply the result by u_x, we have

$$\left(\frac{1}{2}u_x^2\right)_t + u_x(-6uu_x + u_{xxx})_x = 0$$

Adding the previous two equations together and organizing the sum produce

$$\left(u^3 + \frac{1}{2}u_x^2\right)_t + \left(-\frac{9}{2}u^4 + 3u^3 u_{xx} - 6uu_x^2 + u_x u_{xxx} - \frac{1}{2}u_{xx}u^2\right)_x = 0$$

This is the third conservation law of the KdV equation. Then, one by one until the summer of 1966, scholars secured nine conservation laws. After that, Miura introduced the Miura transformation and proved that the KdV equation has infinitely many conservation laws.

Let us rewrite the KdV equation as follows:

$$P_u \equiv u_t - 6uu_x + u_{xxx} = 0 \tag{4.143}$$

This equation and the M-KdV equation that Miura developed

$$Q_v \equiv v_t - 6v^2 v_x + v_{xxx} = 0 \tag{4.144}$$

satisfy the following relationship:

Theorem: If v is a solution of the M-KdV equation, that is, $Q_v = 0$, then

$$u = v^2 + v_x \tag{4.145}$$

satisfies the KdV equation (4.143), that is, $P_u = 0$. Equation (4.145) is known as the Miura transformation.

Proof: In fact,

$$\left(2v + \frac{\partial}{\partial x}\right)Q_v = (v^2 + v_x)_t - 12v^3 v_x + 2vv_{xxx} - 12vv_x^3 - 6v^2 v_{xx} + v_{xxxx} \tag{4.146}$$

Substituting Equation (4.145) into Equation (4.143) and simplifying the result produce

$$P_u = (v^2 + v_x)_t - 12v^3 v_x + 2vv_{xxx} - 12vv_x^2 - 6v^2 v_{xx} + v_{xxxx}) \tag{4.147}$$

Since the right-hand sides of Equations (4.146) and (4.147) are equal, so we have

$$P_u = \left(2v + \frac{\partial}{\partial x}\right)Q_v \tag{4.148}$$

From Equation (4.148), it follows that if $Q_v = 0$, then $P_u = 0$ (note that the inverse may not hold true). Introduce the Calilean variable substitution:

$$t \to t', \quad x \to x' - 6ct, \quad u \to u'(x',t') + c$$
$$t' = t, \quad x' = x + 6ct, \quad u' = u\left(x,t\right) - c \tag{4.149}$$

That is, the KdV equation is invariant under the variable substitution of Equation (4.149), where c is a constant. No doubt, from the variable substitution of Equation (4.149), we have

$$u = u'(x',t') + c$$
$$u_t = u'_{t'} + 6cu'_{x'}, \quad u_x = u'_{x'}, \quad u_{xxx} = u'_{x'x'x'}$$

Substituting these expressions into Equation (4.143) leads to

$$u_t - 6uu_x + u_{xxx} = u'_{t'} - 6u'u'_{x'} + u'_{x'x'x'} = 0 \tag{4.150}$$

To show that the KdV equation has infinitely many conservation laws, in Equation (4.149), take

$$c = \frac{1}{4\varepsilon^2}$$

where ε is a parameter. Then, we have

$$t' = t, \quad x' = x + \frac{3}{2\varepsilon^2}t, \quad u'(x',t') = u(x,t) - \frac{1}{4\varepsilon^2}$$

If we take

$$v(x,t) = \varepsilon w(x',t') + \frac{1}{2\varepsilon} \tag{4.151}$$

from Equations (4.150) and (4.148), it follows that

$$P_{u'}(x',t') = P_u(x,t) = \left(2v + \frac{\partial}{\partial x}\right)Q_v \tag{4.152}$$

Substituting Equation (4.152) into Equation (4.151) produces

$$P_{u'}(x',t') = \left[1 + 2\varepsilon^2 w(x',t') + \varepsilon \frac{\partial}{\partial x'} \right]$$

$$\cdot \left[\frac{\partial}{\partial t} w(x',t') + \frac{\partial}{\partial x'} (3w^2(x',t') - 2\varepsilon^2 w^3(x',t') + w_{x'x'}(x',t')) \right]$$

In this expression, if we let

$$\frac{\partial w(x',t')}{\partial t'} + \frac{\partial}{\partial x'} (-3w^2(x',t') - 2\varepsilon^2 w^3(x',t') + w_{x'x'}(x',t')) = 0$$

or

$$w_{t'} + (-3w^2 - 2\varepsilon^2 w^3 + w_{x'x'})_{x'} = 0 \tag{4.153}$$

then $P_{u'}(x',t') = 0$. Equation (4.153) is known as the Gardner equation. From Equations (4.149) (note that $c = 1/4\varepsilon^2$), (4.145), and (4.148), it follows that

$$u' + \frac{1}{4\varepsilon^2} = u = v^2 v_x = \left(\varepsilon w + \frac{1}{2\varepsilon} \right)^2 + \varepsilon w_{x'} = w + \varepsilon w_{x'} + \varepsilon^2 w^2 + \frac{1}{4\varepsilon^2}$$

By eliminating the sign " $'$ ", we get

$$u(x,t) = w(x,t) + \varepsilon w_x(x,t) + \varepsilon^2 w^2(x,t) \tag{4.154}$$

Since as a solution of Equation (4.153), *w* is related to ε, we can assume

$$w(\varepsilon,x,t) = w_0 + \varepsilon w_1 + \varepsilon^2 w^2 + \cdots\cdots \tag{4.155}$$

where $w_i (i = 1,2,\cdots\cdots)$ is to be determined. Substituting Equation (4.155) into Equation (4.154) leads to

$$u = (w_0 + \varepsilon w_1 + \varepsilon^2 w_2 + \cdots\cdots) + \varepsilon(w_{0x} + \varepsilon w_{1x} + \varepsilon^2 w_{2x} + \cdots\cdots)$$

$$+ \varepsilon^2 (w_0 + \varepsilon w_1 + \varepsilon^2 w_2 + \cdots\cdots)(w_{0x} + \varepsilon w_{1x} + \varepsilon^2 w_{2x} + \cdots\cdots)$$

By combining same power terms of ε, we obtain

$$0 = (w_0 - u)\varepsilon^0 + (w_1 + w_{0x})\varepsilon^1 + (w_2 + w_{1x} + w_0^2)\varepsilon^2 + \cdots\cdots$$

So, term by term, we get

$$
\begin{cases}
w_0 = u \\
w_1 = -w_{0x} = -u_x \\
w_2 = -w_{1x} - w_0^2 = -(u^2 - u_{xx}) \\
\quad\cdots\cdots\cdots
\end{cases}
\tag{4.156}
$$

Substituting Equation (4.156) into Equation (4.155) leads to

$$
w(\varepsilon, x, t) = u(x, t) - \varepsilon u_x(x, t) - \varepsilon^2 (u^2(x, t) - u_{xx}(x, t)) + \cdots\cdots
\tag{4.157}
$$

Notice that Equation (4.153) possesses the form of conservation (remove " ' "). So, substituting Equation (4.157) into Equation (4.153) (remove " ' "), combining the same power terms of ε, and taking the coefficients to be zero lead to infinitely many conservation laws of the KdV equation. Here, the coefficient of each even power of ε provides a nontrivial conservation law, while the coefficient of each odd power offers a trivial conservation law.

As an extended mathematical analysis, there is also the problem of integrability of the KdV equation along with the introduction of functional variations. However, when functional variations are introduced, the condition of continuity becomes stricter, including at least such assumptions as that u has continuous first-order derivative, and so forth. It is well known that it is because of the assumption of continuity that the results of calculus become unrelated to the paths, while physical problems just have to be closely relevant to the actual paths taken. Because of this end, at this junction, we will no longer discuss the calculus of functional variations.

4.5.4 Mathematical Properties and Problems of Physics of the Conservation Laws of the KdV Equation

One has to admit that the previous manipulations can be considered a marvel of mathematical beauty, which is also the historical reason why so many mathematicians have been attracted to the study of the KdV equation. However, the investigation of the KdV equation done in the community of mathematics seems to have overly emphasized the formality of mathematics; although infinitely many conservation laws are obtained, the essence of the conservations has not been provided mathematically, and it has not been noticed that producing solitary waves (solitons) out of the KdV equation is only a special case of nonlinear equations without the desired generality. In particular, the accidental observation of the cylindrical water column with smooth exterior surface, as claimed by J. S. Russell, has not been seen again since its initial description in 1843.

Also, whether or not the later solitary waves of the KdV equation are the same as the cylindrical water column with smooth exterior surface still needs to be addressed. Since the first time we got acquainted with the KdV equation over several decades ago, we have paid special attention to capture any phenomena in flowing waters of various rivers that could possibly look like the claimed solitary water column. However, what we have observed is only spinning or spraying water columns instead of anything remotely analogous to the claimed solitary waves of the form of wave motions. Although the community of atmospheric science has also employed solitary waves indiscriminately to its studies of atmospheric flows, what have been accomplished are in fact horizontal eddy currents rotating in the clockwise direction.

4.5.4.1 Mathematical Properties of the Conservation Laws of the KdV Equation

If the third term of the KdV equation is removed, we have

$$u_t - 6uu_x = 0$$

or

$$u_t + \sigma\, uu_x = 0 \tag{4.158}$$

To this end, the literature has mentioned that the solution of this equation experiences quantitatively instable transitions called blown-ups of transitionality without any conservation law. And, even if we rewrite Equation (4.158) as follows:

$$u_t + u_x^2 = 0$$

it still does not contain any conservation law. If the second term of the KdV equation is removed, and we rewrite the third term in nonlinear form, that is,

$$u_t + u_x u_{xx} + u_x u_{xxx} = 0 \tag{4.159}$$

then by using the method of factoring without expansion, we have

$$\frac{A_t}{A^2} = -\lambda = -(U^{-1}U_x U_{xx} + U_{xxx})$$

Separating the variables, integrating the result and taking $A(0) = A_0$ lead to

$$A = \frac{A_0}{1 + (U^{-1}U_x U_{xx} + U_{xxx})A_0 t} \qquad (4.160)$$

No doubt, this is exactly a problem of blown-ups concerning changes of non-linearity with time. And, it does not contain any conservation law of the previous mathematical form.

4.5.4.2 The Problems of Physics Regarding the KdV Equation

It can be readily seen from Equation (4.139) that the conservation laws of the KdV equation of modern mathematics indicate that time t is only a parameter that has nothing to do with change and the change discussed is exactly a denial of changes. A true conservation of physical significance should describe the process of energy transfers, where under the constant total energy, the process of energy transfers has to be related to time, that is, Equation (4.160) holds true. However, Equation (4.160) represents a problem of essence about nonlinearity; that is why Equation (4.159) is a nonlinear dispersion, which cannot be called a nonlinear wave dispersion. So now, it is not difficult to see that the key of the conservation laws of the KdV equation is the equilibria in the instable disturbances of the linear dispersive term u_{xxx}, the spatial quantitative dispersion, and the nonlinear term u_{xx}, reached under very special conditions without spelling out the physical process and mechanism of the dispersion. It can be shown that nonlinearity stands for the rotationality of solenoidal fields. In terms of consistency of equations, it is not appropriate for the KdV equation to use the linear dispersive term u_{xxx} to balance the nonlinear term. So, this end leads to the problem of whether or not the KdV equation actually holds true. Rotational disturbances of nonlinearity should be constrained by opposite spinning of the rotationality in order to reach equilibria; only so, the consistency of the equation can be obtained and the conservation of energy can be achieved through energy transformations. And the conservation of energy transformations cannot be seen as limiting energy transformation.

Based on this discussion, all the contemporary conservation laws, as constrained by mathematical formulas, are conservation laws of physics without any energy transformation; they pursue the conditions of constraint of heavenly eternity. Evidently, conservation laws without energy transformation still cannot become the complete laws of conservation. This end also indicates that modern science is still not yet mature, can only be seen as a school of thought, and is still staggering at the elementary stage of investigating the movements of invariant materials. That is also the essential reason why modern science experiences difficulties when dealing with problems of change. The world without energy transformations is a static figure, which surely is unable to show the dynamics of the constantly changing world. It is exactly because of the formality of quantities that the energy of speed is irrotational, which has to lead to quantitative instability (including the quantitative infinity), a death zone from which modern science seems to have trouble escaping. If one understands that rotations are the inevitable path for energy transformations, then he will realize that in reality, before quantities become instable or the quantitative infinity, transitional changes have already occurred. In this sense, quantitative instability plays the role of signaling the underlying energy transformations; and the rotationality of nonlinearity and its revealed changes with time have clearly indicated that modern science should have focused more on problems of change. However, instead, modern science tightly holds onto

$$\frac{d}{dt}\int_{-\infty}^{\infty}T(u)dx = 0 \quad \text{or} \quad \int_{-\infty}^{\infty}T(u)dx = \overline{T(u)} = C = Const.$$

without considering the process of change implicitly embedded in conservations of energy and without noticing the irrotationality of kinetic energies of speed. Consequently, modern science stays at the formality of the overall invariancy so that changes in energy and the process of energy transfers are constrained by prior estimates. Scholars have worked hard to limit quantitative instability by using energy conservation to design stable difference schemes. The consequence of their work is nothing but the clones that extend the initial values under the condition of well-posedness. Such outcomes can be seen as the static picture of the future = the present. However, the nature is exactly "the future ≠ the present ≠ the past." Problems of change are different from the design of durable goods and cannot be addressed by using the risk values of the prior estimates. We have to know clearly that the consequence of using designs based on risk values to prevent risks is more risky.

Our discussions above readily show that now is time for the scientific community to alter its mode of thinking. It seems that we should consider to yield to risks or to adopt the thinking logic of helping risks to head to their destinations. This end also represents an important topic of ecological balance and environment protection beyond just air pollution control.

4.5.5 General Properties and Integrability of the Burgers' Equation

4.5.5.1 The General Properties of the Burgers' Equation

In the study of wave motions, one can introduce the density ρ for the unit length and the flux $q(x,t)$ of the unit time, and define the speed of flow

$$v(x,t) = \frac{q}{\rho}$$

For the one-dimensional space, the change in the total amount within a time interval $x_1 < x < x_2$ should be the difference of the influx since the moment x_1 and the outflux until the moment x_2. That can be written as

$$\frac{d}{dt} \int_{x_1}^{x_2} \rho(x,t)\,dx = q(x_1,t) - q(x_2,t) \tag{4.161}$$

When $x_1 \cdot x_2 \approx x$, by using the mean value theorem for integrals, we have

$$\rho(x,t)_t = -q_x$$

which can also be written as

$$\rho_t + q_x = 0 \tag{4.162}$$

Based on experiments, the following holds true:

$$q = Q(\rho) \tag{4.163}$$

Substituting Equation (4.163) into Equation (4.162) leads to

$$\rho_t + C(\rho)\rho_x = 0 \tag{4.164}$$

where $C(\rho) = Q'(\rho)$, and $Q'(\rho)$ stands for the derivative of $Q(\rho)$. Evidently, Equation (4.164) is the mathematical quasi-linear hyperbolic equation, or based on the nonlinearity of the second term, it can be directly named as a nonlinear equation.

If $q = Q(\rho) - v\rho_x$, and $v = const. > 0$, we have

$$q_x = Q'(\rho)\rho_x - v\rho_{xx}$$

Assume

$$Q(\rho) = \frac{\rho^2}{2}$$

Then, Equation (4.162) becomes

$$\rho_t + \rho\rho_x - v\rho_{xx} = 0 \tag{4.165}$$

Rewrite this express as follows:

$$u_t + u u_x - \nu u_{xx} = 0 \tag{4.166}$$

That is, Equations (4.165) and (4.166) are the once well-known Burgers' equation, where the third term is the typical form of dissipation in the current system of linearity. In the following, we will not talk about the problem of feasibility regarding mathematically writing dissipation in linear forms.

By letting $u = u(\xi)$, $\xi = x - ct$, and $c = $ const., substituting them into Equation (4.166), and integrating the result, we obtain

$$u_\xi = \frac{1}{2\nu}(u^2 - 2cu + 2\nu\xi_0) \tag{4.167}$$

where ξ_0 is a constant. No doubt, Equation (4.167) is a typical quadratic equation with the following solutions, respectively:

1. When $c^2 - 2\nu\xi_0 = 0$, Equation (4.167) can be rewritten as follows:

$$u_\xi = \frac{1}{2\nu}(u - c)^2$$

Integrating this equation produces

$$u = c - \frac{2\nu}{\xi + 2\nu A_0}$$

where A_0 is a constant. Introducing $\xi = x - ct$ leads to

$$u = c - \frac{2\nu}{\delta\left(1 - \dfrac{t}{t_b}\right)} \tag{4.168}$$

where $\delta = x + 2\nu A_0$, and

$$t_b = \frac{\delta}{c}$$

Evidently, if $t \to t_b$ or $t \to t_b$, then the Burgers' equation (4.166) or (4.167) stands for a problem of blown-ups of transitional changes. It is because when

$t \to t_b$, $u \to -\infty$ (in curvature spaces, the quantitative infinity stands for point of transition). And when $t \to t_b$, the sign of the second term of Equation (4.168) changes, indicating a change in direction of the form

$$\left| \frac{2v}{\delta(1 - t/t_b)} \right| > c$$

When $c^2 - 2v\xi_0 > 0$, there are two possibilities.

a. If $|u - c| < \sqrt{c^2 - 2v\xi_0}$, we have

$$u = -\sqrt{c^2 - 2v\xi_0} \, th\left(-\frac{1}{2v}\sqrt{c^2 - 2v\xi_0} \, \xi - \frac{1}{2} A_0 \right) + c \qquad (4.169)$$

b. If $|u - c| > \sqrt{c^2 - 2v\xi_0}$, we have

$$u = \sqrt{c^2 - 2v\xi_0} \, cth\left(-\frac{1}{2v}\sqrt{c^2 - 2v\xi_0} \, \xi - \frac{1}{2} A_0 \right) + c \qquad (4.170)$$

2. When $c^2 - 2v\xi_0 < 0$, we have

$$u = -\sqrt{2v\xi_0 - c^2} \, \tan\left(\frac{1}{2v}\sqrt{2v\xi_0 - c^2} \, \xi + \frac{1}{2} A_0 \right) + c \qquad (4.171)$$

where A_0 is the integration constant. From Chapter 3 and the previous discussion of the KdV equation, other than Equation (4.169), as a general property of the Burgers' equation, all these cases reveal the problem of blown-ups of transitionality of quadratic nonlinear models. When the corresponding Equation (4.169) is differentiated once, we obtain the solution in the form of solitary waves. Just as what is discussed earlier, the solitary wave solution of quadratic equations is only a special situation under local smooth conditions without any generality. And, what is more important is that the so-called solitary wave solutions are not only owned by the KdV equation.

   That is one of the reasons why we once pointed out that the mathematical solitary waves might not be the same as the cylindrical water column with smooth surface observed by the British ship engineer in 1834 in a river. Solitary waves do not have to come out of the KdV equation; and solitary waves may not even be the solution of the KdV equation. Even if they are a solution of the KdV equation, the solitary waves of the KdV equation still cannot be treated as a representative of nonlinearity. The general or funda-mental characteristic of nonlinearity is that nonlinear systems betray the initial-value existence theorem, where what is important is that nonlinearity reveals blown-ups of transitional changes that are completely different from the initial-value systems.

4.5.5.2 Integrability of the Burgers' Equation

Just because the continuous smoothness of the Burgers' equation is conditional, the corresponding well-posedness of the Burgers' equation should also be conditional. To this end, let us change the notation of the unknown function and rewrite the Burgers' equation as follows:

$$u_t + f(u)u_x - \nu u_{xx} = 0 \tag{4.172}$$

In mathematics, Equation (4.172) is known as an equation of the Burgers' type. To discuss its integrability problem, we can still employ the method of localized traveling waves and take

$$u(x,t) = B(\xi), \quad \xi = x - v_0 t \tag{4.173}$$

and

$$\text{when } \xi \to -\infty, \ B(\xi) \to C^-, \text{ when } \xi \to +\infty, \ B(\xi) \to C^+ \tag{4.174}$$

where $C^{\pm}$ are fixed constants, v_0 the known speed of the traveling wave. Substituting Equation (4.173) into Equation (4.172) produces

$$\nu \frac{d^2 B}{d\xi^2} = f(B)\frac{dB}{d\xi} - v_0 \frac{dB}{d\xi}$$

Integrating this equation once leads to

$$\nu \frac{dB}{d\xi} = \int_0^B f(\eta)d\eta - v_0 B + C \tag{4.175}$$

where C is the integration constant. By using Equation (4.174), we have

$$\int_0^{C^+} f(\eta)d\eta = v_0 C^+ + C + 0$$

$$\int_0^{C^-} f(\eta)\,d\eta = v_0 C^- + C + 0$$

From these two equations, we obtain

$$v_0 = \frac{\displaystyle\int_{C^-}^{C^+} f(\eta)\,d\eta}{C^+ - C^-} \tag{4.176}$$

So, the integration constant is given by

$$C = \frac{C^- \displaystyle\int_0^{C^+} f(\eta)\,d\eta - C^+ \displaystyle\int_0^{C^-} f(\eta)\,d\eta}{C^+ - C^-} \tag{4.177}$$

Substituting Equations (4.176) and (4.177) into Equation (4.175) produces

$$v\frac{dB}{d\xi} = \frac{C^+ \displaystyle\int_{C^-}^{C^+} f(\eta)d\eta + C^- \displaystyle\int_B^{C^+} f(\eta)d\eta - B \displaystyle\int_{C^-}^{C^+} f(\eta)d\eta}{C^+ - C^-} \tag{4.178}$$

By taking $B(0) = B_0, C^+ \rangle C^-$ and separating variables in the previous expression, we have

$$\xi = \int_{B_0}^{B} \frac{(C^+ - C^-)v\,dB}{C^+ \displaystyle\int_{C^-}^{B} f(\eta)d\eta + C^- \displaystyle\int_B^{C^+} f(\eta)d\eta - B \displaystyle\int_{C^-}^{C^+} f(\eta)d\eta} \tag{4.179}$$

No doubt, from the integral expression of Equation (4.179), it can be seen that the denominator is a function of B; and when $B = C^\pm$, the denominator is zero, making the integral in Equation (4.179) invalid. Or in other words, at the moments of $B = C^\pm$, singularity appears; and for $B \to C^+$ or $B \to C^-$, the corresponding values are different. That is, ξ is not unique and takes multiple values and cannot be integrated due to singularities. So, according to the traditional system of linearity, other than smooth solutions under specific conditions, the Burgers' equation in general is not well posed. And, based on the blown-up system of evolution, the fundamental properties of the nonlinear Burgers' equation are the blown-ups of transitional changes of noninitial values.

4.6 Summary

This chapter focuses on the discussion of the contributions and existing problems of the dynamic system of wave motions. The basic thinking logic and methods of linear quantitative analysis developed since the time when the classical study of string vibrations was initially done are systematically introduced. Due to the successes of applying quantitative analysis to the study of vibration problems in solids, the thinking logic and methods developed for handling solids are expanded to the investigations of fluids, where many problems are discovered. That is, generalizations of linear concepts lead to excessive expansions of the concept of wave motions. The purpose of our discussion is about how to correctly understand the problem of wave motions of the system of modern science.

No doubt, the system of quantitative analysis has made magnificent contributions to the study of wave motions. These contributions can also be seen as representative achievements of modern science. At the same time, problems of wave motions of physics provide a place for quantitative analysis to show its capabilities, leading to the glory of the system of wave motions of the 20th century and making the quantitative analysis of mathematics the only key to open the door of the scientific palace. For example, the mechanics of solids of physics has been expanded to almost all areas of natural sciences and even touched on areas of social sciences, leading to the epistemological beliefs that quantitative comparability is the only standard for scientificality, that scientific problems must contain mathematical proofs, and many important and well-known philosophical principles, such as quantitative changes, bring forward qualitative changes, and so forth.

Wave motions have once been seen as the most common physical phenomena and maintain the invariancy of materials in the transfers and propagations of energies, leading to the establishment of the system of modern science without involving changes in materials. And a nonexisting relationship with time is also shown by the corresponding laws of conservation of energy as a central problem of physics. However, the total amount of energy invariant with time does not mean the energies do not change and do not go through transformations. It must be noted that these quantitative laws are laws of conclusions without providing the process of energy changes. So, when mathematics employs conservations of energy, it constitutes the well-posedness (existence, uniqueness, and stability of the solutions) of differential equations in the form of prior estimates. That is also why in the numerical computational schemes, stable difference schemes are designed on the basis of energy conservation. As a matter of fact, energy conservation is about energy transformation or the process of energy transfers instead of the pursuit of invariant totals or an independence of a certain physical quantity from time. The invariant results of the conservation of energy are not the same as that the energy in the process of transformation or transfer is unchanging or independent of time. So, a true conservation of energy should provide the detailed mechanism of transformation or the method of transfer for the process of change. If conservation of energy is merely employed as a constraint for the well-posedness, then the essential significance of the conservation is still not fulfilled, which also indicates the incompleteness of the conservation of energy in modern science. In essence, as of this writing, the conservation of energy has been applied to maintain the invariancy of materials. The purpose of applying and the direction of research on the conservation of energy is embodied in serving durable goods without involving any problem of changes in materials. The so-called prior estimates are not originated in mathematical physics equations themselves; instead, they employ conservation laws of energy as the constraints of stability. As for problems of evolution, the constraining effects of the conservation laws are reflected in stopped energy transformations in the form of controlling quantitative instability. Therefore, the system of quantitative analysis maintains the invariancy of the system of modern science, which leaves the quantitative analysis to face the challenge of evolution problems involving changing materials.

Other than its formality, quantitative analysis is more about its post-event conclusiveness than being ahead of events. Although wave motions possess the functionality of propagating energies in the form of rise and fall, they cannot transport or move materials. That is why the wave motions of the quantitative analysis system cannot foretell the turning points in the rise and fall. That is, waves are created by movements instead of movements being originated from waves. Wave motions are only one morphological state of materials' movements, which is neither unique nor special; and waves are only consequences instead of causes. This end should become an elementary concept of physics.

Successes of mathematics in the investigation of vibrations of solids is not the same as its achieving successes in the study of problems of rotation. That is why when the concept of wave motions is generalized to fluids, one meets with challenge; that is also the reason why turbulences have become a very difficult problem to resolve. So, in this chapter, before we discussed how linearization has been used to handle nonlinear problems, we first provided, still using the current methods of modern science, the characteristics of nonlinearity out of a simplified equation of fluid mechanics, which led to the theory of evolutionary blown-ups. Only after all these were presented, we introduced the inexplicable unidirectional retrogressive waves — Rossby waves, which were obtained by linearizing nonlinear equations. By doing so, we first pointed out the problems caused by excessively expanding the concept of mathematical waves, and second left the end that nonlinearity stands for rotationality wide open for our discussion in the forthcoming Chapter 5.

Considering the fact that the once popular KdV equation and solitary waves were seen as a representative success of the studies on nonlinearity, the KdV equation, Burgers' equation, and solitary waves represent a problem not only touching on the system of wave motions but also revealing the origin where the so-called solitary waves are from and the fundamental properties of weak nonlinear equations. As for the prevalent solitary waves in the current world of learning, they can also be obtained from Burgers' equation instead of being uniquely owned by the KdV equation. What is more important is that they are special cases of solutions of quadratic nonlinear equations. As for whether or not solitary waves mean the same thing as the "solitary wave" observed by J. S. Russell in 1834, this still requires further investigation.

In Section 4.5 above, we especially emphasized that the reason the conservation of energy came from the kinetic energy defined by using squared speed is that in current engineering project designs or in numerical computations people often need to use the conservation of kinetic energy as a constraint for quantitative instability. Evidently, even if the total amount of kinetic energy stays unchanged, it does not mean mathematical invariance; one still has to think about the properties of the underlying problems instead of using prior estimates as the constraints to ignore the physical effects of the processes of energy transformations and the transfers. The essence of the prior estimates is equivalent to blocking materials and energies and disallowing them to change and to transform. So, another reason we specifically discussed the problem of wave motions is that in the current applications, based on the theory of energy conservation, the method of prior estimates or its expanded version of risk values are still widely employed in the design, manufacture and/or construction of goods or engineering projects. It can be said that quantities have greatly perfected the theory of wave motions. However, since quantities themselves suffer from the problems of only being able to describe events after they have already occurred and of the inability to deal with transitional changes (quantities cannot handle the quantitative infinity), the conservation laws of energy in quantitative forms can only be laws of conclusions without providing processes of change. So, we concluded that the method of prior estimates is not from the conservation laws of energy, or it is a misconception of the conservation laws. Therefore, the method of prior estimates is in fact a helplessness of the quantitative analysis of modern science that cannot deal with transitional changes. The true physical significance and methods of application of the conservation of energy should embody and reveal changes in energy and processes of energy transformations. No doubt, prior estimates are a product created out of the human wish of designing and manufacturing safe and risk-free durable goods.

As for how to represent the natural laws or the related epistemological problem concerning safety and no risk, there is still no clear idea in the system of modern science. As a matter of fact, the true safety and assurance are about how to transfer the risk or resolve risks into advantageous events, which is a problem of evolution science. Or in other words, the applications of evolution science are not limited to the area of prediction science. That is why generality cannot substitute for specifics; and irrotational speed kinetic energy cannot at all constrain the transfer of the nonlinear rotational kinetic energy. That is also the essential reason why the conservation of irrotational kinetic energy cannot guarantee the conservation of energy.

Since the beginning time when modern science started to operate, it has been basically employing risk values to resist risks, leading to the result that the levels of risks increase exponentially. The current ideas of level lakes of high dams and tall levees for preventing floods are some representative examples; and in implementing these ideas, people forget about the fundamental laws of keeping ecological balance, where green should be maintained and waters constantly flowing. In particular, when paying attention to the climate warming caused by atmospheric and water pollutions, the current environmental protection ignores the ecological imbalance and warming effects created by the heat break-offs accompanying the appearance of dried-up rivers. To this end, we see the need to alter our way of thinking regarding risks. That is, the current confrontational thinking logic of using prior estimates to deal with risk values should be changed to that of staying away from or yielding to or arranging homes for dangers. This end touches on the law of conservation of stirring energy, which we will discuss in the following chapters.

Discussions on problems of wave motions should make the physical characteristics of wave motions clear. To this end, our study in this chapter has led to the following results:

- They represent the morphological changes caused by nonunidirectional linear reciprocating movements.
- They can only describe regularized changes of small disturbances.
- They can only propagate or transfer energy without transporting materials so that the invariancy system of modern science is maintained.
- They can only be the posteffects of pushing or elastic pressures; that is why wave motions themselves cannot address the turning points of the reciprocating changes.
- They can only be applied to well-posed quantitative descriptions without the ability to address irregular information and without the capability to deal with irregular information.
- They are good at describing repetitions without touching upon changes in materials.

In principle, this chapter looked at the problems existing with the classical wave motions. As for the modern concepts relevant to wave motions, such as gravitational waves, probabilistic waves, and so forth, and their related problems, the essence is a problem on nonlinearity. About this end, we will discuss in Chapter 5. No doubt, to resolve problems of nonlinearity, one has to first address the physical meanings of nonlinearity. Only after that, he can consider the mathematical properties.

Acknowledgments

The presentation in this chapter is based on Born (1926), English and Feng (1972), Gardner et al. (1967), Gu (1978), Hawking (1988), Heisenberg (1927), Lao Tzu (time unknown), Li and Guo (1985), Lin et al. (2001, 2008), Lin and Qiu (1987), Mei (1985), OuYang (1984, 1992, 1994, 1995, 1998a,b,c, 2000), OuYang et al. (1998, 2001, 2002a), OuYang and Lin (2006a,b, in press), OuYang and Wei (1993), Wang (1982), Wang and Wu (1963), Whitham (1986), and Wu and Lin (2002).

Chapter 5

The Circulation Theorem and Generalization of the Mystery of Nonlinearity

The establishment of calculus is listed as not only one of the three major scientific achievements of the 17th century, but it is also the foundation on which modern natural science, in particular, the mechanical system, can be developed. Even today, 300 years after calculus was initially introduced, calculus plays a role in the tools for the theoretical research of natural science, computations of engineering designs, or validations of dynamic theoretical systems; as well as a basic piece of knowledge that all science and engineering professionals are inculcated with since high school. The original human way of knowing the world from analysis of materials and matters has been changed to that of acquiring knowledge based on looking at quantities. In particular, since the 1970s when the scholastic community became fascinated with mathematical modeling, the tide of establishing mathematical models has been seen in almost all disciplines of learning. High-quality works in economics have to contain mathematical models as their supports, while studies of Chinese traditional medicine also began to include such fashionable elements. So, there has appeared in the scientific community such a latent rule that without mathematical modeling, it is difficult for any scholastic paper to be treated seriously.

There is no doubt that science is about challenging the wisdom of man, technology is about the use value for its ability to save people from calamities, and the development of science is stimulated by the calls of problems. Since the time when calculus was initially established, many so-called mathematical models have appeared. However, if we pursue the essence of the mathematical system of calculus, the classical mechanics or the theory of kinematics, developed in the form of mathematical models of mathematical physics equations, is basically about the effects of elastic pressures of the inertial system and the quantitative analysis of movements that are limited to linear dynamics.

So, this understanding has to touch on the need of carefully thinking about the mathematical operation system of calculus in terms of what problems this system can indeed resolve and in front of what problems it becomes invalid.

What has to be noticed is that the development of computers has provided conveniences for mathematical modeling and quantitative computation; however, it is undeniably that the birth and development of computers are established on *quantities*, more accurately on the basis of the symbols "- -, − " of the *Book of Changes* (Wilhalm and Baynes, 1967). So, when computer technology matures, it has to become an "information handler" that is able to deal with information directly. The current fad of mathematical modeling and numerical computation will become a minor function of the future information handler.

Although we have mentioned many times that modern science is a system that is capable of describing invariant materials and is embodied in extracting invariant information out of events, at this junction, we would like to emphasize that calculus that supports the system of modern science or the entire quantitative analysis system cannot resolve the problem of quantitative infinity. That is why the quantitative computations of all mathematical models have to avoid the quantitative infinity or attempt to constrain the quantitative instability, leading to the requirements of mathematical well-posedness. Because of this, the quantitative operations of calculus cannot resolve transitional changes where at extremes, the matter will evolve in the opposite direction. The main function of these operations is about maintaining the current invariancy of the system of modern science. To this end, it can be seen that mathematical modeling and its operations cannot deal with problems involving changes in materials, and the corresponding computations of calculus can only stay outside the door of evolution science.

As for the quantitative operations themselves, we should also be very clear that they are only postevent formality. No doubt, the quantitative formality cannot tell what the objects are. In particular, the postevent formality of quantities seems to suggest to the beholders that without things there would not be any quantity, or quantities cannot appear prior to the events or things. Although this understanding involves the well-known law of philosophy that quantitative changes lead to qualitative alteration, science is still science where people are allowed to change and update their beliefs and opinions. Also, the modern science developed on the basis of quantitative analysis cannot successfully address all problems. For example, in classical mechanics or the system of dynamics, the following questions, and others, are unsettled: What is force? Where force is from? Do materials move only in the form of linear vibrations? How can there be such an irresistible first push in the material world? What should we do when the first push is no longer irresistible?

Even as early as in the 19th century, scholars had believed the absoluteness of mathematics that is applicable in the study of solids. However, such absoluteness only holds true if the moving materials do not change or suffer from damages. At the same time, the scientific community acknowledges the fact that when applied to the study of fluids, mathematics has met with difficulties. As for why such difficulties arise, no answer has been given. Then, should the development of computers help to successfully apply mathematics in the study of fluids before we fully understand why mathematics met with difficulties in the study of fluids? This question is one of the critical problems about how to perceive the system of modern science.

No doubt, the success of applying mathematics in the study of solids in the 19th century has been achieved by limiting the study to the movements of elastic recoverable solids. At the same time, it has been recognized that break-offs of solids and problems of plastic materials are still unsolvable in the system of modern science. Although the followers of Newton have classified problems that are unsolvable in the classical mechanic system as quasi-problems, breakability of solids and deformation of plastic materials, caused by slow expansion and contraction, are after all the physical reality. So, this end reveals that modern science is still not the medicine that can treat all diseases; the part of which can be successfully employed in the study of fluids is limited to that about elastic recoveries under very special conditions.

In terms of mathematics itself, Newton and his followers did not seem to expect that nonlinearity could cause damages to their system of continuity. Euler was the first person to establish the fundamental equation for fluid motion. He did not expect either that from nonlinearity it could be concluded that continuous media are not a fundamental attribute of fluid movements. Even during the rise of so-called modern physics of the 20th century, the physical significance and function of nonlinearity were still not clearly known; and wave motions were widely employed to address problems that no one seemed to know what they were. And, in publications, scholars had tried to either avoid nonlinearity or eliminate problems of nonlinearity. It can be said that other than linearization, there has appeared no new idea as to how to resolve nonlinearity. That was how Schrödinger presumed that physical quantities were phase variables so that they were taken as wave functions ahead of time, which in form made his equation linear in complex numbers such that problems of wave motions could be discussed with the underlying nonlinearity hidden. However, hidden is not the same as not existing. In essence, Schrödinger's equation is still nonlinear, which will be further discussed in the rest of this chapter. What needs to be pointed out is that the question of what nonlinearity is was known as an unsettled open problem in modern physics.

Newton's first law of mechanics is often known as the law of inertia and an example where force is not directly mentioned. Although the second law shows directly the relationship between the force and the movement of the object, it can only be a relationship within the inertial system without spelling out what the force is, where the force is located, and where it is from. Also, we need to pay attention that the force, denoted *f*, in Newton's second law is invariant; and the corresponding moving object is idealized into a particle and seen as a quantity of the material, written *m*. This object *m* moves in the form of changing speed, known as the acceleration *a*, under the influence of *f*. This leads to the well-known mathematical expression of Newton's second law: $f = ma$.

According to the law of inertia, *f* should have a location of support and can only push *m* to move with an invariant acceleration. So, Newton's second law in essence is such a law on an object's movement of constant acceleration with an invariant acting force and invariant moving object. That is, from the very start of modern science, the system is established on some very strict conditions. However, its idea is very clear: Other than the speed can change, the variables *f*, *m*, and *a* are all invariant so that Einstein later even went further and assumed constant speed, leading to a completely invariant system. That is why H. Bergson (1963) believed that "the change as provided by the classical physics is in fact a denial of change, and time is only a parameter that is not influenced by the change it describes. A stable world figure without any process of evolution is still the core of thoughts of the theoretical physics. The system of universe, constructed since the time of Newton, can answer any question, because all those that are unanswerable in this system have been disregarded in the name of quasi-problems." A. Koyré (1968) also once mentioned that "all the movements described by the dynamic system established since Newton are those that have nothing to do with motion; or speaking more strangely, they are movements in time without time," indicating that the system of modern science is invariant and cannot deal with problems of change. And later, I. Prigogine (1980) also recognized that "the scientific system, completed by Newton and Einstein, pursues after the eternal Spinozism." Based on our earlier discussions, we think that it has been very clear about what kind of science modern science is and what kind of problems it could resolve. Our discussions have also provided explanations for why when one faces a problem of change, that problem will almost without exception become a world-class difficult problem within the system of modern science.

Practical experience indicates that as long as a moving object is involved, its acceleration is most likely variant, leading naturally to problems that do not confirm Newton's second law. For example, even if we assume that *f* is invariant, as long as the acceleration is changing, the mass *m* will have to change too. So, Newton's second law $f = ma$ becomes a not closed equation of mathematics and, at the same time, a nonlinear equation. Since variable acceleration belongs to a noninertial system, problems of variable accelerations naturally betray the inertial system of modern science. At the same time when the acceleration becomes a variable, the mass *m* has to be a variable too. Therefore, variable materials and problems of evolution also betray the inertial system. That is the very reason why in modern science, the scales used in the study of evolutions are either astronomical measures or geographical ages. The significance beyond the spoken language is nothing but a self-explanation for the space modern science needs in order to maintain its traditional invariancy.

Evidently, as a scientific theory, downloading the explanation of such a difficult problem as what force is to God is a reason hard for any scholar to speak loudly on. However, as a scientific theory, without being able to address what force is and where force is from, mechanics suffers from some difficulty of convincing inquiring minds. Although Newton's third law openly points out that the acting and reacting forces are equal, it does not spell out what force is, where the reacting force is from, and in particular, when dealing with specific mutual reactions, even if the acting and reacting forces are equal, do they affect each other in a linear form along a straight line? The essence of Newtonian particlization is about applications of quantities, while physical problems are not the same as bunches of particles. Evidently, nonparticlization has to deal with structures within materials and unevenness in materials' morphologies. There is no doubt that when objects with internal uneven structure and morphology interact with each other, the probability for them to act on each other's center of mass is very small. And interactions not on the centers of masses will naturally lead to problems of rotational motion of curvature spaces. So, before the old problem is resolved, there have already appeared the problems of mutual reacting forces and the way of motion caused by the reacting forces. Evidently, if mutual reactions are not limited to the form of straight line, one has to deal with the problem of motion in curvature spaces.

People are familiar with the Newtonian axiom of universal gravitation. However, this axiom does not tell why the universal gravitation has its universality and what gravitation is. To this end, even Newton himself admitted that the mathematical formula of universal gravitation could not explain the cause of universal gravitation. Although Einstein introduced the concept of curved time and space out of curvature equations, he did not explain why time and space could be curved; and his theory of gravitation leads to gravitational waves by linearizing curvature spaces, causing a huge waste of scientific manpower and investment in the consequent pursuit of gravitons. As a matter of fact, the observation that the path of a falling star looks like a curved page in an old book has already shown what universal gravitation is and how to explore it. Hence, the history of research of universal gravitation already shows the difficulties the quantitative analysis system of modern science meets. It is a well-known fact that the air on the earth is a mixture of different components. However, against the rotationality of materials, the concept of atmospheric Rossby waves, obtained by employing quantitative methods, seems to be produced out of the same kind of thinking logic.

To this end, we conclude that acquiring knowledge by analyzing quantities is different from knowing the world based on understanding materials. In other words, knowing materials by using quantitative methods is not the same as knowing the materials by checking into the materials themselves. This conclusion specifically touches on how to recognize forces; how to understand the general, widely existing, and most fundamental forms of movements of materials under mutual reactions;

and what is the difference between modern science and what is produced out of the most general form of motion under mutual reactions. For example, from Newton's second law, third law, and universal gravitation, it follows one by one that the second law contains the Godly force f, the third law only spells out the equal acting and reacting forces without explicitly mentioning God; if God stands for force, there would be at least two Gods to make up the acting and reacting forces; the universal gravitation $\vec{F}$ is given by

$$\vec{F} = G \frac{m_1 m_2}{R^3} \vec{R}$$

between the mutual reacting m_1 and m_2. So, forces come from materials, and the reacting force is from the interaction of the materials m_1 and m_2.

So, from the observation that the path of a falling star looks like a curved page in an old book, it follows that the gravitation in universal gravitation is not a pulling force, while its universality can be explained as the rotationality of the celestial body. That is, rotation is the general form of motion of interacting materials. It should be clear that there were reasons why Newton's third law and his universal gravitation formula were formulated as they were, even though Newton did not clearly point out the general form of motion under the mutual reaction.

It can also be seen that due to the difficulties met by modern science in its quantitative analysis, in particular, operations of calculus have met with the difficulty of numerical computational instability, or rotations always bring damages to continuity while creating changes in materials, modern science has tried its utmost to avoid rotationality. Speaking more concretely, changes and rotationality in materials stand for mathematical nonlinearity as well as problems of noninertial systems, while the entire modern science, including the magnificent achievements of modern physics, is developed on the invention of physical quantities and the inertial system. So, if one still counts on modern science to resolve his problems of rotation and evolution, he will surely face the problem of being unable to do much.

As an epistemological problem, just as solids are special cases of materials in the universe, wave motions caused by the effects of elastic pressures naturally cannot include all forms of motion of materials. Also, the phenomenon of wave motion and the mechanics of the first push cannot be seen as the ultimate theory of epistemology, because the problems of origin and change in forces still exist.

What is unfortunate is that meteorology and seismology, areas of prediction science, have been treated as branches of classical mechanics and studied in similar formats in the establishment of their respective mechanics and dynamics. For example, meteorological science has its own theory of dynamics, known as atmospheric dynamics; the study of earthquakes leads to its seismological dynamics; and geological science contains its own branch called geomechanics. The essence of these theories is about nothing but atmospheric waves on the basis of the system of wave motions; earthquakes of the earth's crustal vibrations on expansions of the concept of vibrations in solids or linearization of nonlinear problems; the theory of gravitations of cosmic, celestial bodies on the basis of generalizing wave motions; and probabilistic waves of the fundamental particles in the microscopic world. So, some inexplicable theories, such as planar unidirectional retrogressive waves, energies that do not do any work, and gravitational waves of irrotational Riemann (curvature) spaces, are created. Consequently, all kinds of disturbances are seen as waves, and the material universe is believed to have been originated from energies of wave motions without any materials.

However, the scientific community is not a solid block of steel without any seam. Nearly 20 years ahead of Einstein, someone had already established the known circulation theorem by using nonparticle circulation integrations, while proving that uneven densities in the atmosphere and ocean can cause eddy currents of spinning motions. Although its form is quite simple and the method of establishment is straightforward, the result has hit at the crucial point of modern science. And, in essence, this work represents a major revolution in the theory of epistemology and methodology. Because of its grand scale of significance, it has not been fully understood by the contemporary scholars and our historical moment.

5.1 Bjerknes's Circulation Theorem

The modern science developed on the quantitative analysis system is known for its study of the physics of the phenomena of wave motions. And, the deterministic solution problems of the mathematical physics equations, which appeared at roughly the same time as calculus, are also known due to their obtainable wave motion solutions. In order to make mathematical physics equations applicable to all physics problems, wave motions, such as those in fluids, including water waves and atmospheric waves, vibrations in solids, electromagnetic waves, all fundamental particles as probabilistic waves, gravitational waves, and the unidirectional retrogressive Rossby waves that are unique to meteorology, are seen as the most general topic of physics. Limited by the capability of quantitative analysis, not only are the diffusion equations of propagating turbulences identified with heat-conduction equations, but also are solutions of wave motions obtained out of these diffusion equations. Although each area of natural science and various problems of engineering touch on problems of wave motion, as a simple and intuitive problem of physics, wave motions embody energy transfers in the form of reciprocating movements of the material media without any of the materials transported. For this reason, we once pointed out that wave motions prevent the underlying materials from being damaged in the form of transferring energies so that they provide a playground for quantitative analysis. That is how the development of the system of wave motions has been reaching such a level that it is no longer limited to the linearized models of Taylor string vibrations, leading to the method of modeling by solving equations by inserting presumed wave motion solutions into the equations, that of treating physical quantities as phase spaces, presuming wave motions and then substituting the quantities into the corresponding relationship of energy or momentum. Or, simplified nonlinear equations are linearized, and the resultant linear equations are seen as new models.

Just as what we have mentioned in earlier chapters, in order to apply indiscriminately the quantitative analysis system, modern physics purposely redefined wave motions as a general concept so that mathematical methods can be employed to investigate problems of wave motions involving as few constraints as possible. And, it has even been openly recognized that the goal of introducing such a general concept as wave motions is for the convenience of employing mathematical techniques and methods. That is how we face the reality as mentioned above that in modern science, from the cosmic to microscopic levels, various physical phenomena have been abstracted as wave motions, leading to the mathematical classification of wave motions into hyperbolic and dispersive waves. In particular, any disturbances, including rotational disturbances, can now be listed in the generality of wave motions. As a matter of fact, even the inertial system of the first push also recognizes that rotations are not a problem of the inertial system. Although the mathematical quantitative value of a rotational disturbance could be the same as that of a wave disturbance, the physical meanings of these two kinds of disturbances are different. In fact, translation, vibration, and rotation are three forms of movements, with both translation and vibration appearing within rotations so that they can be seen as special cases of rotations. The movement with truly very little limitation or without any constraint is that of rotations. That is why there have appeared in modern science quite some situations where theoretical results do not agree with objective reality. It might be because historically problems of meteorology do not carry enough weight and significance on the overall landscape of natural science that a piece of very important work in the scientific history has been little known and did not get a chance to spread across the entire spectrum of science. Even during the heat wave of nonlinearity of the latter part of the 20th century, scholars in meteorology did not seem to know that as early as at the end of the 19th century, their colleague V. Bjerknes had already resolved the mystery of nonlinearity using his circulation theorem. And 40 years after Bjerknes's work, to follow the fad of the wave motions of the 20th century, Rossby obtained his famous unidirectional retrogressive waves, which had also affected the study of oceanography, while bundling up the circulation theorem, which was the sole important result in science at that time, and placing it on the top shelf within meteorological science. Because of this historical reason and how the scientific community is greatly puzzled by the problem of nonlinearity, we feel obligated to pay a revisit to this circulation theorem established at the end of the 19th century. As it has been, scientific truth does not become obsolete because it is from the past.

V. Bjerknes was a Norwegian meteorologist and became known early in his career by introducing his theory of air masses, in which the concept of air masses questioned the Newtonian theory of particles. In 1898, he established the circulation theory. The mathematical expression of the circulation theorem is relatively simple. By taking an arbitrary closed curve in the fluid of concern and the integral of the flow speed vector, one has

$$\Gamma = \oint_s \vec{V} \cdot \delta \vec{r} \tag{5.1}$$

where Γ is an integral along a closed curve, which represents a circulation of a nonparticle kind of the fluid, s is a closed curve, $\delta \vec{r}$ the vector difference of two neighboring points. Hence, Equation (5.1) is known as the speed circulation. If we consider the instantaneous change of this speed circulation with time or the acceleration circulation, we have

$$\frac{d\Gamma}{dt} = \frac{d}{dt} \oint_s \vec{V} \cdot \delta \vec{r} \tag{5.2}$$

The right-hand side of this equation can be written as follows:

$$\frac{d}{dt} = \oint_s \frac{d\vec{V}}{dt} \cdot \delta \vec{r} + \oint_s \vec{V} \cdot \frac{d}{dt} \delta \vec{r} \tag{5.3}$$

Based on the continuity assumption of calculus, the second term of the right-hand side of this equation can be rewritten as:

$$\frac{d}{dt} \delta \vec{r} = \delta \left(\frac{d\vec{r}}{dt} \right) = \delta \vec{V} \tag{5.4}$$

Substituting Equation (5.4) into the second term of the right-hand side of Equation (5.3) produces

$$\oint_s \vec{V} \cdot \frac{d}{dt} \delta \vec{r} = \oint_s \delta \left(\frac{V^2}{2} \right) = 0 \tag{5.5}$$

Therefore, Equation (5.2) becomes

$$\frac{d\Gamma}{dt} = \oint_s \frac{d\vec{V}}{dt} \cdot \delta \vec{r} \tag{5.6}$$

Equation (5.6) provides us with the acceleration circulation. If we substitute the Euler–Coriolis equation that describes a rotational fluid (that is obtained in meteorology by adding the deviation force caused by the earth's rotation into the Euler equation without direct manipulation)

$$\frac{d\vec{V}}{dt} = -\frac{1}{\rho}\nabla p - 2\vec{\Omega} \times \vec{V} + \vec{g}$$

into Equation (5.6), we have

$$\frac{d\Gamma}{dt} = \oint_s \left(-\frac{1}{\rho}\nabla p\right)\cdot\delta\vec{r} - \oint_s 2\left(\vec{\Omega}\times\vec{V}\right)\cdot\delta\vec{r} + \oint_s \vec{g}\cdot\delta\vec{r} \qquad (5.7)$$

If A stands for the area enclosed by the closed curve S, then the second term of the previous equation can be written as follows:

$$\oint_s 2\left(\vec{\Omega}\times\vec{V}\right)\cdot\delta\vec{r} = 2\vec{\Omega}\cdot\oint_s \vec{V}\times\delta\vec{r} = 2\vec{\Omega}\cdot\oint_s \vec{n}V_r\delta\vec{r}$$

where $\vec{n}$ stands for the unit vector perpendicular to the plane formed by the $\vec{V}$ and $\delta\vec{r}$, V_r the component of $\vec{V}$ in the direction perpendicular to $\delta\vec{r}$. That is,

$$\left|\vec{V}\right|\sin\theta\delta r = V_r\delta r$$

So, we have

$$2\oint_s \vec{\Omega}\times\vec{V}\cdot\delta\vec{r} = 2\vec{\Omega}\cdot\oint_s \vec{n}V_r\delta r = 2\Omega\frac{d\sigma}{dt} \qquad (5.8)$$

where σ stands for the projection of the area A on the equator plane. If $\vec{g}$ is taken to be the gradient of the potential function, as often done in modern science, that is, $\vec{g} = -\nabla\phi$, then we have

$$\oint_s \vec{g}\cdot\delta\vec{r} = -\oint_s \nabla\phi\cdot\delta\vec{r} = \oint_s \delta\phi = 0 \qquad (5.9)$$

Substituting Equations (5.8) and (5.9) into Equation (5.7) leads to

$$\frac{d\Gamma}{dt} = \iint_\sigma \nabla\left(\frac{1}{\rho}\right)\times\left(-\nabla p\right)\cdot\delta\sigma - 2\Omega\frac{d\sigma}{dt} \qquad (5.10)$$

where for the first term, we have applied the Stokes formula of transformation between line and surface integrals. Equation (5.10) is the famous Bjerknes circulation theorem that has been ignored and almost entirely forgotten in recent past in the scientific community, since this result does not follow the particlization of the Newtonian form even though it is derived by using the mathematical method of modern science.

No doubt, what is implied by Equation (5.10) is that when the density of the atmosphere is not a constant, together with pressure, it creates a mathematical non-linear term and a physical solenoidal term under the influence of stirring effects. That is why V. Bjerknes once pointed out that all airs and oceans with uneven densities are rotational eddy currents. What is important is that, as an epoch-making contribution, this theorem reveals that nonlinearity stands for rotational movements and has fundamentally resolved the so-called mystery of nonlinearity. What needs to be pointed out is that simplicity does not imply nonscientific; and whether it is scientific is determined by whether it can explain realistic problems. Unfortunately, not only was this result not widespread in the entire spectrum of science, but also did the meteorological community not realize its significance. That led to later scholars, including Einstein, Schrödinger, and others, to have also missed the opportunity to understand the practical significance and function of nonlinearity so that their works pushed coming generations of scholars further into the hole of difficulty.

5.2 Generalized Meaning of Nonlinearity

5.2.1 The Universal Gravitational Effects of the Circulation Theorem

Evidently, if we treat $\vec{g}$ in Equation (5.9) in the fashion of a falling a star following a path like a curved up page in an old book, then $\vec{g}$ cannot be taken as a potential function as done often in modern science. If we take

$$\vec{g} = -\frac{1}{\rho}\nabla\psi$$

then both ρ and ψ are functions of the structure of the star. In other words, both ρ and ψ are variables. Now, the Stokes formula of transformation between line and surface integrals implies that

$$\oint_s \vec{g}\cdot\delta\vec{r} = \iint_\sigma \nabla\left(\frac{1}{\rho}\right)\times\left(-\nabla\psi\right)\cdot\delta\sigma \tag{5.11}$$

That is, the circulation integral of nonparticles has already indicated that the universal gravitation comes from the rotational stirring forces created out of the uneven structure of the stars. Or speaking differently, the concept of universal gravitation, as a weak force in physics, together with other so-called strong forces, cannot be identified as forces. In other words, gravitation itself is not a pulling force; instead it is a form of motion of carrying materials due to the materials' rotation caused by the uneven structures of the materials.

So, universal gravitation is not the same as pulling force; otherwise, the fall of a star would not be like a curved page in an old book. That is why it is relatively easier to overcome universal gravitation by using rotations or forms of rotations.

5.2.2 Gravitational Effects of the Equation of Fluid Movement

Considering the vertical component of the Euler equation of motion in fluid mechanics, we have

$$\frac{\partial w}{\partial t} = -\frac{1}{\rho} p_z - g \tag{5.12}$$

Since the vertical acceleration

$$\frac{dw}{dt}$$

of the atmosphere is smaller than the horizontal acceleration

$$\frac{d\bar{v}}{dt}$$

by two magnitudes, and the horizontal acceleration is smaller than

$$-\frac{1}{\rho} p_z \quad \text{or} \quad -g$$

on the right-hand side of this equation, we have

$$g \cong -\frac{1}{\rho} p_z \tag{5.13}$$

In meteorology, the density ρ in Equation (5.13) is taken to be a constant; and Equation (5.13) has been referred to as the static approximation formula and employed in the computations of transformation between the barometric pressure field and geopotential field. Evidently, using the method of the circulation theorem and considering the density ρ as a variable, then the right-hand side of Equation (5.13) is exactly the stirring force of a solenoid. This end implies that the universal gravitation is not the attraction force along a straight line of the first push system. So, the atmosphere in which humans and all forms of lives exist must be a huge mixture caused by atmospheric rotations; and Lao Tzu's *you wu hun chen* (mixed with materials) must be from the stirring of rotations.

That is, by using nonparticlization, we can conclude that universal gravitation is originated from eddy sources of stirring movements. If later scholars, such as Einstein, Schrödinger, and others, knew and understood the circulation theorem, such theories as gravitational waves, probabilistic waves, and so forth, might not have been created. On the other hand, since V. Bjerknes continued to employ the concept of potential functions of modern science, he lost a great historical moment to resolve and explain the puzzle of universal gravitation. In other words, by introducing nonparticle circulations, one can verify that "universal gravitation" is not gravitation; instead, it is exactly like the path of a falling star that resembles a curved page in an old book. This end reveals the essence of why quantitative laws are not the same as physical laws. The clear and simple circulation theorem not only resolves mathematical nonlinearity but also shows the lofty realm of how physics can lead to solutions of problems.

5.2.3 The Problem of Terrain and Nonlinearity

5.2.3.1 The Topographic Coordinate System

In meteorological science, $-(1/\rho)\nabla P$ is treated as the gradient force of barometric pressure. In form, it can be seen as a quantitative approximation. However, in terms of physics, this treatment has altered the eddy source that causes rotation to a source of elastic pressure that produces vibrations. So, the appropriate name for the term $-(1/\rho)\nabla P$ should be the stirring force of the density pressure. When dealing with problems regarding the attributes of events, one should not use quantitative magnitudes as the criterion for truncation decisions and could be easily confused with nearly equal quantities. Considering that between terrain and the atmosphere, there is a boundary dividing two different kinds of media, this area can be called the *discontinuous zone* of the boundary. Because of irregularities, terrains bring forward protruding effects on weather phenomena. That is why in the study of meteorological problems, the effect or function of terrains has been an unsolved problem in meteorological science. Among many possible reasons for not being able to solve this problem is mathematical nonlinearity. It can also be said that the circulation theorem has something directly to do with this unsettled problem.

Considering the discontinuity existing between the terrain and the atmosphere, a scholar in meteorology in 1957 designed the topographic coordinate system and named it as σ-coordinate system. That is, take

$$\sigma = \frac{p}{p_s} \tag{5.14}$$

where p stands for the pressure at an arbitrarily chosen altitude and p_s the pressure on the ground level. From Equation (5.14), it follows that

If $p = p_s$, then σ = 1; and
If $p = 0$, then σ = 0,

where since p_s is the ground-level pressure; for any ground level no matter what altitude it is at, its pressure can be written as p_s. So, the σ-coordinate system has already implicitly contained altitudes even though in form it has nothing to do with the terrain's altitude. There is no doubt that this idea has very intelligently provided a method to formally deal with terrains and become fashionable for a period of time, leading to the belief that the σ-coordinate system has resolved the problem of terrains. During that time period, the world of learning believed that the purpose of introducing the σ-coordinate system is about transforming the discontinuity of the ground surface into continuity so that according to the well-accepted convention, continuous surfaces are the same as material surfaces, which originated from Aristotle. However, as a matter of fact, true material surfaces are not continuous. This end explains how deeply the assumption of continuity is rooted in the Western world of learning. Although the concept of material surfaces is no longer seen often in the recent literature, the emphasis on the assumption of continuity is nothing but about the ability to employ calculus and its rules of operation.

So, based on the differentiation of composite functions, one has

$$\left(\frac{\partial F}{\partial s} \right)_p = \left(\frac{\partial F}{\partial s} \right)_\sigma + \left(\frac{\partial F}{\partial \sigma} \right)\left(\frac{\partial \sigma}{\partial s} \right)_p \tag{5.15}$$

where $p = p(x,y,z,t)$, $p_s = p_s(x,y,t)$ and introduce the static approximation formula

$$\frac{\partial \Phi}{\partial \sigma} = p_s \frac{\partial \Phi}{\partial p} = -\alpha\, p_s = -\frac{RT}{p}\, p_s \tag{5.16}$$

where Φ stands for the gravitational potential, and

$$\alpha = \frac{1}{\rho}$$

the special volume. So, by taking F as Φ and S as x in Equation (5.15), then we have

$$\left(\frac{\partial \Phi}{\partial x}\right)_p = \left(\frac{\partial \Phi}{\partial x}\right)_\sigma + RT\left(\frac{\partial \ln p_s}{\partial x}\right)_p \qquad (5.17)$$

where the subscripts p,σ stand for the p coordinate and the σ-coordinate, respectively. Now, Equation (5.17) indicates that by using the σ-coordinate transformation, the original linear gradient of the altitude field becomes a two-term expression in the σ-coordinate system. Even if we do not question the feasibility of introducing the static approximation, the second term on the right-hand side of Equation (5.17) is still a nonlinear term.

At this junction, we have to point out that N. A. Phillips (1957), who initially introduced the concept of σ-coordinate system, could have made use of this opportunity to uncover the essence of nonlinearity and the effects of terrains. However, he was very much buried in the achievements of the material surfaces of the terrain boundaries, believing that the mathematical treatment of terrains, as a world-class difficult problem, was resolved. Unfortunately, it is exactly with this problem that no breakthrough was made; instead, an important historical opportunity to solve the problem of nonlinearity was missed.

As a matter of fact, no matter which mathematical transformation is applied, it cannot alter the fundamental attributes of the physical problem. Although Phillips' σ-coordinate system perfected the material surfaces of the system of continuity, it forgot that within the material surfaces, implicitly different altitudes of the terrains still exist, and these various altitudes do not simply disappear. And, in the form of mathematical nonlinearity, they appear in Equation (5.17).

5.2.3.2 The Nonlinear Effect of Terrains

In the present scientific community, problems are still seen as effects of wave motions, which is not only limited to meteorological science. For instance, in the terminology of the related study of radars, the effects of terrains are also similarly called mixed waves of terrains and objects. So, in the following discussion of this part, let us look at the problem of terrain effects from the angle of wave motions.

Let us rewrite the second term in Equation (5.17) as follows:

$$RT(\ln p_s)_x \qquad (5.18)$$

where R is the gas constant, T the temperature, and the subscripts partial derivatives.

Let $T = \hat{T}(p)$,

$$H = \ln \frac{p}{p_0}$$

and take T as a polynomial in H. So, we have

$$T = \sum_{n}^{N} \mu_n H^n \tag{5.19}$$

At the surface of the terrain of our concern, let us take

$$H_s = \ln \frac{p_s}{p_0}$$

where p_0 stands for the pressure at the sea surface that is taken as a constant. Then we have

$$H = H_s + \ln \sigma \tag{5.20}$$

Let

$$H = \sum_{m=0}^{M} \hat{H} \cos \mu_m x \tag{5.21}$$

Then, we have

$$\frac{\partial H_s}{\partial x} = -\sum_{m}^{M} \mu_m \hat{H} \sin \mu_m x$$

Notice that

$$H_s = \ln \frac{p_s}{p_0}, \quad \ln \sigma = \ln \frac{p}{p_s}$$

So, we have

$$H = H_s + \ln \sigma = \ln \frac{p_s}{p_0} + \ln \frac{p}{p_s} \quad \text{or} \quad H = \ln \left(\frac{p_s}{p_0} \cdot \frac{p}{p_s} \right) = \ln \frac{p}{p_0}$$

Substituting Equation (5.21) into Equation (5.20), and then in turn into Equations (5.19) and (5.18), we can obtain

$$RT \frac{\partial \ln p_s}{\partial x} = -R \sum_n^N \sum_m^M \mu_n \mu_m \left(\sum_m^M \hat{H} \cos \mu_m x + \ln \sigma \right)^n \hat{H} \sin \mu_m x \qquad (5.22)$$

From Equation (5.22), it follows that even if the second term in Equation (5.17) or the nonlinear term in Equation (5.18) is expanded according to the system of wave motions, the temperature field of an arbitrarily chosen altitude has something to do with the products of the wave numbers of the terrain field, and the multiplied wave numbers must be far greater than the product MN. Hence, even if terrains are wave motions, Equation (5.22), which corresponds to the second term in Equation (5.17), has already shown the severe short-wave shocks, which have to cause computational instability leading to the eventual spill. Since the atmosphere at high altitudes changes greatly, employing the σ-coordinate system will surely create severe short-wave shocks in high levels of the atmosphere. It can be said that many more short waves are created by employing the σ-coordinate system than from using the p-coordinate system, indicating that the effects of terrains are like a machine that breaks waves. So, whether or not to apply the σ-coordinate system is no longer a problem about computational accuracy. Instead, its essence is about how to understand nonlinearity.

Now, let us rewrite the second term in Equation (5.17) as the general form of nonlinearity,

$$RT \nabla \ln p_s = \frac{RT}{p_s} \nabla p_s = \frac{1}{p_s / RT} \nabla p_s$$

By taking

$$\rho_{ST} = \frac{p_s}{RT}$$

the previous equation can be rewritten as

$$RT \nabla \ln p_s = \frac{1}{\rho_{ST}} \nabla p_s \qquad (5.23)$$

No doubt, what Equation (5.23) reveals is exactly the same as what Bjerknes's circulation theorem does: It is a stirring force term of the density pressure involving the effects of the terrains. So, the essence of the effects of terrains is an eddy-current problem of triggering the occurrence of rotations. From the fact that the problem of terrains has been seen as terrain waves, including leeward waves, mixed waves of terrain and object, and so forth, it suggests the need for a fundamental change in the basic concepts and related theories; it also indicates that the significance and function of the circulation theorem have not been fully understood. Therefore, the thirst for a definite explanation of nonlinearity should first recognize the essence of quantities' post-eventness; otherwise, human epistemology would be trapped in a chaos regarding the physical substance while facing the chaotic formality.

5.3 Mystery of Nonlinearity

The essential reason the problem of nonlinearity is still not settled as of this writing is that great difficulties are met in studies of mutual reactions, where rotational motions of unevenly structured materials are involved, by employing the slaving assumption of the Newtonian system (or the first push) or the methods of quantitative analysis of nonmutual reactions. No doubt, structures have to be uneven vector problems of nonparticles. Since the probability for the push to directly work on the center of mass is very small, rotational movements in materials have to be created due to moments of forces. Now, the problem is, the circulation theorem has provided the stirring force, which consequently implies spinning motions; so, is the nonlinearity in the form of movement the same as a problem of rotation? In terms of nonlinear forms of motion, we have the following simple discussion based on elementary vector operations.

From the vector analysis of mathematics, it follows that the following scalar multiplication formula holds true:

$$\nabla(\vec{a}\cdot\vec{b}) = \vec{a}\times(\nabla\times\vec{b}) + \vec{b}\times(\nabla\times\vec{a}) + \vec{b}\cdot\nabla\vec{a} + \vec{a}\cdot\nabla\vec{b} \tag{5.24}$$

where the symbol $\rightarrow$ stands for a vector. By taking $\vec{a} = \vec{b} = \vec{v}$, we have

$$\nabla(\vec{v}\cdot\vec{v}) = \nabla v^2 = 2\,\vec{v}\times(\nabla\times\vec{v}) + 2\vec{v}\cdot\nabla\vec{v}$$

which can be rewritten as

$$\nabla\frac{v^2}{2} = \vec{v}\times(\nabla\times\vec{v}) + \vec{v}\cdot\nabla\vec{v} \tag{5.25}$$

When

$$\frac{v^2}{2}$$

on the left-hand side of this equation is multiplied by m, we obtain the irrotational kinetic energy. By taking the vorticity, we obtain

$$\nabla \times \left(\nabla \frac{v^2}{2} \right) = 0$$

From Equation (5.25), it follows that

$$\nabla \times (\bar{v} \cdot \nabla \bar{v}) - \nabla \times \left[(\nabla \times \bar{v}) \times \bar{v} \right] = 0 \tag{5.26}$$

For the convenience of intuition, let us take the vector expression for the horizontal spinning movement only. So, we have

$$\bar{v} = \bar{k} \times \nabla_b \psi \tag{5.27}$$

where $\bar{k}$ stands for the unit vector along the vertical direction and ψ the flow function. So, the following holds true:

$$\nabla_b \times \bar{v} = \bar{k} \varsigma \tag{5.28}$$

where

$$\varsigma = \frac{\partial v}{\partial x} - \frac{\partial u}{\partial y}$$

represents the vorticity along the vertical direction. Substituting Equations (5.27) and (5.28) into Equation (5.26) and taking the horizontal vector expression in [ ] of the second term lead to

$$(\nabla_b \times \bar{v}) \times \bar{v} = \varsigma (\bar{k} \times \bar{k} \times \nabla_b \psi)$$

Since $\bar{a} \times \bar{b} \times \bar{c} = (\bar{a} \cdot \bar{c}) \bar{b} - (\bar{a} \cdot \bar{b}) \bar{c}$, the right-hand side of the previous equation becomes

$$\varsigma\left(\vec{k}\times\vec{k}\times\nabla_{b}\psi\right)=-\varsigma\,\nabla_{b}\psi$$

Once again from the analysis of vector, it follows that

$$\nabla\times(f\,\vec{a})=\nabla f\times\vec{a}+f\,\nabla\times\vec{a}$$

So, we have

$$\nabla_{b}\times\left[\left(\nabla_{b}\times\vec{v}\right)\times\vec{v}\right]=-\nabla_{b}\varsigma\times\nabla_{b}\psi \tag{5.29}$$

Taking the horizontal vector of Equation (5.26) and substituting it into Equation (5.29) produce

$$\nabla_{b}\times\left(\vec{v}\cdot\nabla_{b}\,\vec{v}\right)=\nabla_{b}\psi\times\nabla_{b}\varsigma \tag{5.30}$$

Notice that the term $\vec{v}\cdot\nabla_{b}\,\vec{v}$ in ( ) on the left-hand side of Equation (5.30) is a familiar nonlinear term (the same holds true even with the subscripts removed), and the right-hand side is exactly the rotational solenoidal field under the stirring force of a solenoid. This end has very clearly and formally declared that nonlinearity is about the problem of rotationality of solenoidal fields, while intelligently resolving the mystery of nonlinearity. This fact implies that the mystery of nonlinearity, created by the fluid motion equation as introduced by L. Euler over 200 years ago, should have been resolved at the very moment when the equation was initially established by revealing the fact that nonlinearity is a problem about rotationality. It is because at that time, mathematical vector analysis had already been developed into a set of quite mature tools of operation. However, because of the delay in human recognition and the unwillingness to recognize such a reality that goes against the system of wave motions, this ultimate understanding of nonlinearity has been delayed until the present day. This delayed understanding of nonlinearity has also led to the pursuit of linearization, generalization of wave motions, and so forth, in the study of nonlinear problems in mathematical physics, creating many great difficulties. The reason turbulences have been a stronghold in modern science still not possible to be overtaken is that turbulences are not waves; even under the concept of generalized wave motions, they still represent impossibilities in science.

No doubt, the one-dimensional quantitative form of the vector $\vec{v}\cdot\nabla_{b}\,\vec{v}$ is the familiar

$$u\frac{\partial u}{\partial x}$$

which can be rewritten as

$$\frac{1}{2}\frac{\partial u^{2}}{\partial x}$$

What needs to be clear is that quantities are after all formal measurements that cannot tell us the meaning of the substance hidden behind the formality such that the essence of the vorticity

$$\varsigma = \frac{\partial v}{\partial x} - \frac{\partial u}{\partial y}$$

is the angular speed ω, while the quantitative ω represents the uneven distribution of the linear speed and, with its structure of rotation, is different from the irrotationality of any straight-line speed. In particular, it cannot be confused with irrotational quantities. For example, the nonlinear expression

$$\frac{1}{2} \frac{\partial u^2}{\partial x}$$

is a quantity with rotationality. By using calculus, an analytic solution can be obtained, and the solution stands for an asymmetric and irreversible transformation from convergence to divergence. And, from two-dimensional nonlinear equations we have already obtained that inwardly spinning whirlpools can be transformed to outwardly spinning pools and this process of transformation is reversible. Even if we employ tensors to describe nonlinearity, the rotationality of the nonlinearity should not be altered. Correspondingly, from Equation (5.26), it follows that three-dimensional solenoidal spinning fields are well revealed by Bjerknes's circulation theorem and have been well manifested by realistic geo-atmospheric flows and movements of the oceans.

When a current follows downward in a river, the water travels in the form of rolling downward, because the current speed near the surface is greater than that near of bottom of the river. When the speed distribution in the vertical direction is uneven, rolling movements have to be created. At this junction, let us generalize this observation. Living on the earth is like living on the river-bottom of the atmosphere; similarly, since the air at high altitude flows at speeds far greater than those at the ground level, the movements of the atmosphere have to contain the form of rolling currents. No doubt, rolling is also rotation. So, terrains are embodied in nonlinearity, while triggering rotationality. That is the problem of essence we should be looking at when N. A. Phillips (1957) introduced the σ-coordinate system. That is why we consider that Phillips missed his historical opportunity to resolve the mystery of nonlinearity, while giving us the chance to pick up the "gold."

Although the circulation theorem was obtained by using the mathematical methods of modern science, its central idea is the circulation integral of nonparticles. In terms of mathematics, Equation (5.10) is quite simple; however, at the height of epistemology and in theory, what is reflected in the content is epoch making.

When the density ρ and the pressure P, as physical quantities, are uneven variables, $-\nabla(1/\rho) \times \nabla P$ reveals a stirring force of a non-first-push system. Therefore, the simple circulation theorem has very well betrayed the first push system of the Newtonian kingdom, while definitely declaring the existence of stirring forces in mutual reactions. That is, the circulation theorem betrays the wave-motion system of modern science using modern science. Unfortunately, the scientific history immediately after the development of this theorem did not quite understand its significance; otherwise, starting at the end of the 19th century, modern science would have traveled along a different path. More specifically, the significance of this theorem is as follows:

Although the circulation theorem did not directly explain what a force is, from the components of its solenoidal term $-\nabla(1/\rho) \times \nabla P$, it is not hard to see that forces come from structural unevenness of materials. So, it indirectly reveals the fact that forces are not from any supernatural being. That is why Saltzman had seen this result as a major betrayal of the modern scientific system.

The reason the circulation theorem betrays the modern scientific system is that mutual reactions are not any first push. So, this result betrays the inertial system in the form of spinning motions of mutual reactions.

Since, $-\nabla(1/\rho) \times \nabla P$ reflects the mutual reaction of material structures, forces in the quantitative forms are also from material structures. This end completely overturns the belief of the ancient Greek Pythagorean school that numbers are the origin of all things.

The solenoidal term $-\nabla(1/\rho) \times \nabla P$ can reveal different spinning directions when the distributions of ρ and P are different even if quantitatively they might stay constant. So, the nonlinearity of mutual reactions has transformed linear quantities into problems of physical structures and can directly reflect differences in spinning directions. This end also shows that nonlinear problems cannot be linearized and the essential reason why nonlinear equations cannot be analytically solved using quantities.

In meteorological science, the term $-\nabla(1/\rho) \times \nabla P$ has always been referred to as the gradient force of barometric pressure. No doubt, this term was chosen on the basis of the traditional linear thinking, where the atmospheric density ρ is assumed to be constant. If ρ is not a constant, the consequent mathematical properties will be different and the physical significance can be considered revolutionary. Just as what is repeatedly emphasized earlier, even in terms of kinematics, $-\nabla(1/\rho) \times \nabla P$ represents the effect of a stirring eddy source. Evidently, when ρ is a constant, $-\nabla(1/\rho) \times \nabla P$ stands for the effect of an elastic pressure of the first push and belongs to the inertial system. When ρ is not a constant, it means the stirring effect of mutual reactions and belongs to a noninertial system. Therefore, in studies of the atmosphere, ocean, and cosmic nebulas, ρ cannot be taken as a constant. Also, we have to recognize that whether or not ρ is a constant is not simply a problem of approximation. So, $-\nabla(1/\rho) \times \nabla P$ should not continue to be named as a gradient force of barometric pressure (or intensity of pressure); instead, it should be named as the stirring force of density (intensity). And generally speaking, only when the object being acted upon is not a particle, will $-\nabla(1/\rho) \times \nabla P$ be a mutual reaction and a stirring force. Since in the material world, a first push that goes through every situation does not exist, the position of modern science becomes quite clear, and studies of mutual reactions cannot rely on the system of modern science.

As a problem of mathematics, even if we still have doubts about nonlinearity in the form of motion (in the following we will provide further results in the language of mathematics), $-\nabla(1/\rho) \times \nabla P$ has already presented a mathematical expression for mutual reactions and clearly illustrates the physical meaning of nonlinearity. Therefore, for mathematical problems, nonlinear equations cannot be handled by using linearization. Because of this conclusion, many theories and methods established in the 20th century on the basis of employing the method of linearization to deal with nonlinear problems have to be reconsidered or modified.

Although nonlinearity is a term of mathematics, it touches on problems of physics, philosophy, and epistemology. As a problem of physics, in the name of mutual reactions of unevenness, the circulation theorem has walked out of the first push system of modern science and substantially betrayed the inertial system. As a philosophical problem concerning recognition, the circulation theorem reveals the relationship between forces and materials' structures using statements of modern language. At the same time when it deepens our understanding of the material world, the theorem also unifies the concepts of objects and forces into one concept due to the absorption of forces into materials' structurality. No doubt, this will greatly trouble the system of dualism with separate objects and forces developed since the time of Aristotle and is a system-wide reform of Western philosophical history. It is because over 100 years ago in 1898, forces were out of the control of God and melted into materials, constituting a major betrayal to the classical mechanic system. This new understanding agrees well with the ancient Chinese philosophical point of view of objects and figures and the ancient Chinese epistemology of combined heaven and man of monism.

Other than directly telling that the fluids' movements in the atmosphere and oceans are in the form of spinning eddy currents and that forces are germinated within the structure of materials, the historical significance of the circulation theorem is about how it guided mankind into the mysterious door of universal gravitation. At this junction, we should notice the real pity that due to constraints of potential functions of his time, V. Bjerknes personally missed the opportunity to resolve the mystery of universal gravitation.

The problem of terrains that has been bothering meteorological science for over 100 years has been successfully resolved. The effects of terrains on fluids are shown to be mixed eddies (instead of waves) of objects and terrains when expressed in the form of nonlinearity.

Nonlinearity is embodied in the rotationality of solenoidal fields, which settles the more than 200-year-old mystery of nonlinearity. So, it is time for us to clean up many of the theories established on various misunderstandings of the properties of nonlinearity. To correctly comprehend the physical meanings of nonlinearity, in the following we will look at several nonlinearity-related problems, including Einstein's general relativity, probabilistic waves, and so forth.

As for the problem of existence of the circulation theorem, it involves the problem of whether or not circulations are closed, since realistic spinning eddy currents are not closed. Even with this problem, we still should not overlook the historical significance of the circulation theorem. And unfortunately, meteorological science, in which this important theorem was initially established, has bundled up this result and stored it on a high shelf so that it has greatly delayed the discipline to make true and beneficial progress. In other words, meteorology could have become a leader in the development of modern science; however, due to misguidance of concepts, in the past 100-some years, studies in weather and related disciplines did not make much substantial progress. Should meteorological science have evolved along the thinking logic of the circulation theorem since the end of the 19th century, the present meteorology would look completely different.

5.4 Einstein's General Relativity Theory and the Problem of Gravitational Waves

Einstein's mass–energy formula, which is also known as the conservation of mass and energy, and quantum mechanics, developed at the start of the 20th century, have been seen in the community of physicists as two magnificent achievements of modern physics, which are different from the classical system of mechanics. However, if we think carefully against realistic physical problems, especially since we have produced Equation (5.30), on the basis of the structurality and rotationality of materials, we surely feel the existence of the problem of not being self-contained even if we reason using formal logic. For example, Einstein treated light as material. So, a contradiction of formal logic exists with respect to Einstein's mass–energy formula $E = mc^2$ between the mass m and the speed c of light in vacuum. Since we talk about vacuum, m should be 0 so that Einstein's mass–energy formula becomes $E = 0$. Next, modern physics points out that photons do not have rest mass. So, m should also be 0, leading once again to $E = 0$. Now, from Newton's second law

$$f = m\,a = m\frac{dv}{dt}$$

even if a is a constant, the speed v should still be a variable. Einstein (1976) once said: "Modern physics is established on the invention of the inertial system, mass and force." So, Einstein's assumption of constant speed of light leads to another contradiction of formal logic. In particular, if the speed of light is constant, we must have

$$\frac{dc}{dt} = 0$$

so that the corresponding acting force f has to be 0. However, energy is the work done by a force. So, if f is 0, no force actually exists so that no work would be done and no energy would exist. So, Einstein's mass–energy formula is still 0. On the surface, what we have just shown seems to be a game of formal logic played backward. However, in essence, it involves how to understand the energy of high-speed flows. This end touches on an essential problem about the energy-momentum tensors of Einstein's general relativity theory.

Since Newton honestly admitted that he could not explain the reason for the existence of universal gravitation, many physicists of the following generations were motivated to explore the possible answer. Einstein introduced gravitational waves by using irrotational Riemann (curvature) spaces in his attempt to resolve the problem of universal gravitation left open by Newton, leading to his theory of gravitation. This theory has been seen as a breakthrough in physics different from classical mechanics and is applied to theoretically explain the precession of Mercury. However, the precession of Mercury can be explained realistically or experimentally by using refraction of lights. After nearly 100 years of searching for gravitational waves with huge amounts of money and valuable scientific labor, still no definite conclusion has been produced. And presently the gravitational anomaly is still a mystery of science. Surely theories without any practical verification can only be called hypotheses. To this end, in the following, we will cast some problems in terms of nonlinearity related to Einstein's mass–energy formula, Newtonian kinetic energy, gravitation, gravitational waves, and so forth.

5.4.1 The Law of Governance of the Slaving Energy of the Newtonian First Push and the Mass–Energy Formula of Mutual Reactions

5.4.1.1 The Law of Governance of the Slaving Energy of the Newtonian First Push

Although the traditional speed kinetic energy (that will be referred to as irrotational kinetic energy in the following discussions) is perfect in terms of its formality, it does not tell the process of energy transformation and can only show the cancellation of energy in the form of equal positive and negative values without providing any guarantee for kinetic energy conservation. However, an interesting problem can be revealed out of the equation of fluid movement. That is

$$\dot{\vec{v}} + 2\vec{\Omega} \times \vec{v} = \vec{g} - \alpha\nabla P + \vec{F} \tag{5.31}$$

where

$$\dot{\vec{v}} = \frac{d\vec{v}}{dt}$$

represents the acceleration vector, $\vec{\Omega}$ the spinning speed vector of the earth, $\vec{g}$ the acceleration vector of gravity, P the pressure of the fluid, $\alpha = 1/\rho$ the special volume, ρ the density, the mass in the unit volume, and $\vec{F}$ the friction.

To help with the intuition in our discussion of the concept of kinetic energy, let us use horizontal irrotational kinetic energy as our example and ignore the friction. So, we have

$$\dot{\vec{v}}_h + 2\vec{\Omega} \times \vec{v}_h = -\alpha \nabla_h P \tag{5.32}$$

where the subscript h stands for the horizontal direction. Dot multiplying Equation (5.32) by $\vec{v}_h$, which is also known as the scalar multiplication of $\vec{v}_h$ and Equation (5.32), leads to

$$\frac{d}{dt}\left(\frac{v_h^2}{2}\right) = -\vec{v}_h \cdot \left(\alpha \nabla_h P\right) \tag{5.33}$$

where according to the rules of operation of vectors, we have $\vec{v}_h \cdot \left(\vec{\Omega} \times \vec{v}_h\right) = \vec{\Omega} \cdot \left(\vec{v}_h \times \vec{v}_h\right) = 0$. Evidently, the meaning of the conservation of irrotational kinetic energy says that its value is a constant. That is, it should be

$$\frac{d}{dt}\left(\frac{v_h^2}{2}\right) = 0 \quad \text{or} \quad v_h^2 = C \text{ (a constant)} \tag{5.34}$$

So, in order to satisfy Equation (5.34), which is called the conservation of irrotational kinetic energy, we have to require $\vec{v}_h \cdot (\alpha \nabla_h P) = 0$ for the right-hand side of Equation (5.33). Evidently, $\vec{v}_h \cdot (\alpha \nabla_h P) = 0$ holds true only if $\nabla_h P > 0$ and $\nabla_h P < 0$ exactly cancel each other. No doubt, Equation (5.34) in its mere quantitative form cannot vividly describe the process of transformation or fashion of change between $\nabla_h P > 0$ and $\nabla_h P < 0$. In short, the conservation law of irrotational kinetic energy cannot guarantee the conservationality. We will return to this in the following discussions.

The term $\vec{v}_h \cdot (\alpha \nabla_h P)$ on the right-hand side of Equation (5.33) is the slaving control term of the kinetic energy of the Newtonian first push. In order to intuitively express its physical meaning, we can replace the special volume by mass so that we produce (even with the subscripts removed, the physical meaning stays the same)

$$\frac{d}{dt}\left(\frac{v^2}{2}\right) = -\vec{v}\cdot\left(\alpha\nabla P\right) = -\frac{1}{m}\vec{v}\cdot\nabla P = \frac{1}{m}\vec{v}\cdot\vec{f} \qquad (5.35)$$

Notice that the dot product of vectors is a quantity, that is $\left|\vec{v}\cdot\vec{f}\right| = \left|f\,v\right|$ and

$$v = \frac{ds}{dt}$$

where s stands for the distance traveled. So, $\vec{v}\cdot\vec{f}$ represents the change in the work done by the Newtonian pushing force, called power, while the consultant distance traveled under the influence of the force is the work of the force, which is energy. So, the physical meaning of Equation (5.35) is that changes in irrotational kinetic energy are exactly the power for the first push to do work, or called the controlling power of the slaving energy of the Newtonian first push.

5.4.1.2 Mutual Reactions and Einstein's Mass–Energy Formula

If we take

$$\frac{de}{dt} = \vec{v}\cdot\vec{f}$$

then from Equation (5.35), we obtain

$$e = \frac{1}{2}mv^2 \qquad (5.36)$$

This equation is exactly the expression of the irrotational kinetic energy under Newtonian push. By introducing the third law of the Newtonian mechanics, that is, the acting and reacting forces are quantitatively equal, we then produce the speed-kind kinetic energy for the irrotational movement of the mutual reaction:

$$E = 2e = m\,v^2 \qquad (5.37)$$

It might be because Newton was too fond of his first push that he did not derive Equation (5.37). Even so, it is still Newton's third law that points out the quantitative identity of the acting and reacting forces. Therefore, although Newton did not directly and clearly spell out Equation (5.37) during his lifetime, we still refer this equation as to a piece of work of the Newtonian system. However, based on this equation, what is very interesting and strange is that if the speed v of the

Newtonian classical mechanics is replaced by Einstein's speed of light c, then we can obtain Einstein's mass–energy formula immediately from Equation (5.37):

$$E = m c^2 \qquad (5.38)$$

What needs to be clearly seen and noticed is that Einstein's assumption of $c =$ const. has to lead to

$$a = \frac{dc}{dt} = 0 \quad \text{and} \quad f = ma = 0$$

So, the conservation of mass and energy, which is a so-called great achievement in modern science that has gone beyond classical physics, does not agree with the common knowledge of classical physics. What Newton's second law states is the equivalence between force and acceleration and energy comes from the work of force. So, without acceleration there will be no force. Evidently, Equation (5.37) agrees with the equivalence between force and acceleration, while due to 0 acceleration, Equation (5.38) becomes a kinetic energy where no force exists to do any work. To reconfirm our work, we checked through all papers contained in Einstein's collections without being able to locate where and when Einstein ever claimed that Newton's second law is incorrect. Evidently, according to the formal logic, if Newton's second law is correct, then it must be the mass–energy formula that is a mistake. Conversely, if the mass–energy formula holds true, then Newton's second law together with the entire system of modern science has to be thrown into the trash.

5.4.2 General Relativity Theory and Problems with Irrotational Curvature Spaces

It is well known that Einstein's general relativity was developed on the basis of Minkowski's four-dimensional time–space. Even though at this junction, we will not mention any of the problems with respect to this four-dimensional time–space, one should still notice that time should be different from space, since time does not occupy any material dimension (more details will be provided in Chapter 7). So, there is a problem with the statement that "time is equivalent to space" of the four-dimensional time–space. Of course, the conclusion that time does not occupy any material dimension has essentially denied the concept of the four-dimensional time–space, and this end covers a major part of modern science, including the general relativity theory. Those interested readers can consult Chapter 7 in our book *Entering the Era of Irregularity* for more details. Here, let us continue to employ Einstein's original expression, that is,

$$R_{\mu\upsilon} + \frac{1}{2} q_{\mu\upsilon} R = 8\pi G \frac{1}{c^4} T_{\mu\upsilon} \qquad (5.39)$$

where $R_{\mu\upsilon}$ stands for the Ricci (a second-order curvature) tensor, R a curvature tensor of order zero, $q_{\mu\upsilon}$ a metric (a metric tensor), G the gravitational constant, c the constant speed of light (even going along with Einstein's concepts, time and space are curved, so the speed of light should also be a function of the time and space $c = c(x,y,z;t)$, and $T_{\mu\upsilon}$ on the right-hand side of Equation (5.39) is the momentum tensor of the material's energy.

Evidently, if the speed of light is a constant, then $T_{\mu\upsilon}$ in Equation (5.39) has to contain such energy that does not have any force to do work. Surely, Equation (5.39) is a nonlinear equation about changes in curvature and contains a second-order derivative. In general, the first-order derivative of the curvature stands for changes in the curvature and is known as torsion. The second-order derivative of the curvature is the change in torsion. So, Equation (5.39) must be a problem of a Riemann space about changes in curvature. The corresponding introduction of irrotational Riemann geometry itself by Einstein suffers from misunderstandings of some basic concepts. Besides, the physical properties of nonlinearity involve rotationality (see Section 5.3 for more details).

5.4.3 Problems with Energy-Momentum Tensors

According to the popular belief, Einstein broke through the constraint of classical mechanics using theoretical achievements of high-speed flows. A typical representation of high-speed flows is to replace Newton's speed v of movement by the speed c of light. So, accordingly, the energy and momentum of high-speed flows have to deal with mc^2 and mc without any acceleration. Correspondingly, $T_{\mu\upsilon}$ of Equation (5.39) must implicitly contain energy without acceleration. Let us call such energy as one that has no acting force to do any work. Since it is energy without any acting force to do any work, it has to represent a theory of physics of absolute invariancy so that there is no need to talk about transformation and transfer of energies. And if natural science does not have the ability to reveal the transformation and transfer of energy, it will not be seen as a complete system of theories. Since the time when modern science was born in the 17th century, the law of conservation of energy has been seen as one of the signature achievements. However, as of the present day, modern science is still very immature and still has not figured out what it means to be complete.

5.4.4 Irrotational Kinetic Energy and Problems with Energy Transformations

No matter whether it is the push in Equation (5.36) or the mutual reaction in Equation (5.37), all these stand for irrotational kinetic energy, including Einstein's high-speed flows' kinetic energy that do not do any work. It is because Equation (5.34) or (5.35), where we only need to replace v with c, cannot provide the process of transformation of energies. That is why the conservation of irrotational kinetic energy cannot guarantee the claimed conservation. So, even with the conservations of energy, which signal the maturity of the scientific system, modern science still cannot be seen as complete, since modern science still did not make the laws of conservation of energy complete.

Although Einstein introduced the second-order curvature tensor $R_{\mu\nu}$, creating a description different from Newton's quantitative expression of the Euclidean space, and proposed the concept of curved time–space, he did not derive the expressions of momentum and kinetic energy of curvature spaces. In particular, each curvature space of the Riemann geometry is of rotationality, and each rotation has to involve angular speed. However, Einstein's expression of kinetic energy of constant speed did not really provide any new content when compared with what Newton had done, while leaving behind the problem of what kind of energy is the energy that does not do any work. Of course, this problem has to touch on such a fundamental concept of physics as energy coming from the work of forces. For this reason, we refer to his conservation formula of mass–energy as an energy that does not do any work.

Considering that in modern science, the square of speed is seen as the kinetic energy, there naturally appears the problem of what the square of angular speed is in a curvature space. Since light travels along a curved path, the speed of light in the universe not only changes, but also experience variances in angular speed. So, what needs to be noted is that the physical meaning, as delivered by Equations (5.37) and (5.38), is only about the straight-line movements of mutual pushings on the centers of masses of each other. Hence, the left-hand side of Equation (5.39) describes changes in curvature, and the right-hand side the mutual pushings along a straight line, containing the tensor of the energy that does not do any work. That is, Equation (5.39) represents a problem with inconsistent physical meanings on the left- and right-hand sides. The situation can be depicted as follows: Either the linear movements of the classical mechanics and vibrations or stirring motions and rotational movements make consistent equations; however, pushing and rotation, or stirring and linear movements or vibrations lead to inconsistent equations. Therefore, Equation (5.39) stands for an equation with inconsistent physical meanings. In particular, in terms of high-speed flows, taking constant speed means taking the momentum tensor of nonworking energies. Without any work done, there would not be any movement that took place. Without any movement, how could there be any high-speed flow? No doubt, deriving nonworking energy from constant speed has to create problems for the relativity theory at the foundations' level without even mentioning the possible existence of discontinuity in the gravitational field. So, Einstein's general relativity theory together with his theory of gravitation becomes a building floating in the air without much foundation.

5.4.5 *Irrotational Riemann Geometry and the Problem of Linearization*

Evidently, if the speed of light c is taken as a constant, then in form Equation (5.39) becomes

$$R_{\mu\upsilon} + \frac{1}{2} q_{\mu\upsilon} R = K T_{\mu\upsilon} \qquad (5.40)$$

where

$$K = \frac{8 \pi G}{c^4}$$

is a constant. Although tensors are members of a special class of quantities, they are still quantities. So, the right- and left-hand sides of Equation (5.40) with constant speed of light c, even if it is seen as an equation in the quantitative form, are still not consistent, because the left-hand side represents a problem of nonlinearity of the mutual reactions, while the right-hand side the momentum of an energy that does not produce any work. To make the two sides consistent, we at least need to rewrite the speed of light as $c = c(x, y, z; t)$. Consequently, Equation (5.40) becomes

$$R_{\mu\upsilon} + \frac{1}{2} q_{\mu\upsilon} R = \frac{K_1}{\left(c\left(x, y, z; t \right) \right)^4} T'_{\mu\upsilon} \qquad (5.41)$$

where $K_1 = 8 \pi G$ is a constant and in $T'_{\mu\upsilon}$ $c = c(x, y, z; t)$.

Besides, in terms of Equation (5.40), Einstein once applied the symmetry of the first two indexes of the Christoffel symbol, that is $\Gamma^k_{ij} = \Gamma^k_{ji}$. We should notice that the physical meaning of the symmetry $\Gamma^k_{ij} = \Gamma^k_{ji}$ stands for irrotationality. Or in other words, the Riemann geometry Einstein applied is irrotational. By employing the curvature of the Riemann geometry, one should introduce the torsion tensor $S^k_{ij} = \Gamma^k_{ij} - \Gamma^k_{ji}$, where S stands for spinning and can be replaced by T from torsion, and the superscripts and subscripts are dummy variables. It should be clear that Einstein himself knew that he should employ the torsion tensor; otherwise, he would not later collaborate with Cartan to study the problem of gravitation of rotational fields on the basis of the torsion tensor $S^k_{ij} = \Gamma^k_{ij} - \Gamma^k_{ji}$. Unfortunately, he did not make much progress. What is more unfortunate is that Einstein himself did not provide any clear explanation. Science is about exploration; whatever difficulty is met, it is always nice to leave an explanation. Even when no success is made in the exploration, telling the newcomers what happened or what was observed is also important and valuable knowledge.

To this end, let us look at how Einstein derived his gravitational waves in order for us to understand potential problems existing in his work. Since the left-hand side of Equation (5.40) is nonlinear, Einstein introduced a small disturbance to the Lorentz metric tensor

$$g_{\mu\nu} = \delta_{\mu\nu} + h_{\mu\nu} \tag{5.42}$$

where h_μ^λ and h are defined as follows:

$$h_\mu^\lambda = \delta^{\lambda\alpha} h_{\mu\alpha} \tag{5.43}$$

and

$$h = h_\alpha^\alpha = \delta^{\alpha\lambda} h_{\mu\alpha} \tag{5.44}$$

The curvature tensor $R_{\mu\nu}$ can be written as follows by using the Christoffel symbol:

$$R_{\mu\nu} = \Gamma^\beta_{\mu\nu,\beta} + \Gamma^\beta_{\mu\nu}\Gamma^\alpha_{\beta\alpha} - \Gamma^\beta_{\mu\beta,\nu} - \Gamma^\alpha_{\mu\beta}\Gamma^\beta_{\nu\alpha} \tag{5.45}$$

where

$$\Gamma^\beta_{\mu\nu} = \frac{1}{2} g^{\beta\alpha}\left[\frac{\partial g_{\alpha\mu}}{\partial x^\nu} + \frac{\partial g_{\alpha\nu}}{\partial x^\mu} + \frac{\partial g_{\mu\nu}}{\partial x^\alpha} \right]; \quad \Gamma_{\mu\nu,\beta} = \frac{\partial}{\partial x^\beta}\Gamma^\beta_{\mu\beta}; \quad \Gamma^\beta_{\mu\beta,\nu} = \frac{\partial}{\partial x^\nu}\Gamma^\beta_{\mu\beta}$$

Substituting Equations (5.42) through (5.44) into Equation (5.45), taking the first-order approximation, and performing some computational cleaning lead to

$$R_{\mu\nu} = -\frac{1}{2}\delta^{\alpha\lambda} h_{\mu\nu,\alpha\lambda} - \frac{1}{2}\left(\frac{1}{2}\delta^\beta_\mu\, h - h^\beta_\mu \right)_{,\beta\nu} + \frac{1}{2}\left(\frac{1}{2}\delta^\beta_\nu\, h - h^\beta_\nu \right)_{,\beta\nu} \tag{5.46}$$

By taking an appropriate coordinate system with the subscripts changed accordingly, we can make

$$\left(\frac{1}{2}\delta^\beta_\mu\, h - h^\beta_\mu \right)_{,\beta\nu} = 0 \tag{5.47}$$

So, we obtain the following second-order curvature tensor:

$$R_{\mu\nu} = -\frac{1}{2}\delta^{\alpha\lambda}h_{\mu\nu,\alpha\lambda} \tag{5.48}$$

Substituting this equation into Equation (5.40) produces

$$-\frac{1}{2}\delta^{\alpha\lambda}h_{\mu\nu,\alpha\lambda} + g_{\mu\nu}\left(\frac{1}{4}\delta^{\sigma\lambda}h_{,\sigma\lambda}\right) = KT_{\mu\nu} \tag{5.49}$$

By defining

$$L_{\mu}^{\nu} = h_{\mu}^{\nu} - \frac{1}{2}\delta_{\mu}^{\nu}h \tag{5.50}$$

and from the condition in Equation (5.47), it follows that

$$\Box L_{\mu}^{\nu}, \nu = 0 \tag{5.51}$$

By lifting the subscript ν in Equation (5.40), Equation (5.50) implies

$$\Box L_{\mu}^{\nu} = -2KT_{\mu}^{\nu} \tag{5.52}$$

where □ stands for the D'Alembert operator. If it is taken to be in a free space, we have

$$\Box L_{\mu}^{\nu} = 0 \tag{5.53}$$

So, the gravitational field equation (5.40) is transformed by Einstein into the propagation equation (5.53) of gravitational waves.

Even if the reader is not familiar with tensor operations, he can still see that in the entire process of reasoning leading to his theory of gravitation, Einstein followed the conscientious behavior of modern science of the 20th century: Within the irrotationality of the system of linearity, take small disturbances and the first-order approximation, choose an appropriate coordinate system and the corresponding conditions, and then derive the eventual results of wave motions in the free space.

From the end of the 19th century to the first half of the 20th century, the Norwegian scientist V. Bjerknes was the only scholar who betrayed the stream of consciousness of his time and revealed in the form of circulation integrals of non-particles that nonlinearity is about rotational movements under stirring forces. So, it can be said that the so-called mystery of nonlinearity had been resolved at the end of the 19th century. Unfortunately, V. Bjerknes himself and others of the following generations did not recognize the significance and function of the circulation theorem. That is why during the latter part of the 20th century, the scientific community chased after the fad of nonlinearity.

It seems that Einstein never saw V. Bjerknes's work. That explains why he continued to employ the basic scientific operating procedure of his time with the difference that he met with nonlinearity in the operations of tensors. Even so, Einstein still did not realize that nonlinearity does not have any substantial connection with the true theory of gravitation. He presumed linear manifolds, introduced the ordinary methods of linearization, such as irrotationality, small disturbances, first-order approximation, and so forth, and obtained the great epoch-making achievement in Equation (5.53). His entire deduction vividly showed that without assumptions there would be no science, which became the dominant consciousness of modern science.

However, even if scientific research could not exist without assumptions, it still would not mean that one could introduce unlimited assumptions. Besides, in terms of problems of physics, mathematics is only a formal tool and cannot be used to alter the properties of the underlying problems, even though tensors are involved. If the assumed formal logic of mathematics is not related or appropriately associated with the problem of physics of concern, then what is left is only games of mathematics. So, for problems of physics, we have to consider the questions in the following paragraphs.

Einstein named Equation (5.40) a gravitational field equation without directly explaining where the concept of gravitation was from. As for the meaning delivered by this equation, it is still the same as the mutual reaction of particles in the quantitative form $m_1 m_2$ of Newton's universal gravitation (see Equation 5.54), while it was Newton who had first registered the name of gravitation in the scientific community. However, mutual reactions of the particle kind are not limited to attractions and should also include repulsions. Even from Equation (5.53), one can only see waves in a free space. And, it is exactly because free space is employed that no one seems to be able to tell what wave is a free wave. In such a puzzling situation, how can such a "wave" be named as gravitation? And, based on Equation (5.53), the energy–momentum tensors have to be about the mutual reactions of the particle kind so that one has to question how to answer gravitation from T_μ^ν. Even if we do not pursue Einstein's energy that does not do any work, his gravitational waves did not answer what gravitation in the universal gravitation is and where it is from, where Newton at least had expressed his universal gravitation as the mutual reaction of particles.

As a form of mathematics, no matter how complicated and general the energy–momentum tensor T_μ^ν can be, it is after all a formal quantity. And under the assumption that the speed c of light is constant, the left-hand side of Equation (5.40) is nonlinear; so T_μ^ν on the right-hand side should at least be a solenoidal term (which Einstein limited to be an irrotational Riemann geometry), constituting an equation of inconsistent mathematical properties. According to Bjerknes's circulation theorem, nonlinearity stands for eddy motion, and linearity, pushing or vibration along a straight line. If we take the speed of light to be a variable $c = c(x,y,z;t)$, then Equation (5.41) becomes consistent, which leads to a motion with rotationality under stirring effects instead of a wave.

Although irrotational Riemann geometry and linearization might satisfy the formal logic of mathematics, it does not mean that they agree with reality. In particular, Einstein himself knew about this problem by emphasizing that "pure logical thinking cannot provide people with any knowledge about the objective world." However, in his theory of general relativity, he did not clearly explain this problem. Also, his entire deduction contains not only holes of mathematical formal logic but also inexplicable concepts of basic physics.

Science always suffers from the limitations of time. That is why great scientific masters are those who are able to hold on to the core of problems and to foresee the future. For example, Einstein's associative thinking seemed to be stronger than his formal logical deduction; it is in his associative thinking that he had shown what a great master he was. In particular, after taking the speed in Newton's first push system as constant, Einstein was able to introduce the energy of mutual reactions, although he did not notice that by doing so, he lost energies that actually do work; from curvature equations, Einstein thought associatively of curved time–space and equated time with space, leading to the important open problem of time; from the speed of light in vacuum, Einstein predicted torn masses; however, without anything existing in the vacuum, how could there be any torn mass? Einstein deducted that energies could be propagated into the void space, in which sits his energy that does not do any work. However, in a void, how could energy propagate? Einstein foresaw the future and left behind many interesting and puzzling problems.

5.4.6 *Rotationality and Universal Gravitation*

The mathematical vector analysis in Section 5.3 shows that nonlinearity is a problem of rotationality of solenoidal fields (Equation 5.30). Since the term

$$\frac{1}{2} q_{\mu\upsilon} R$$

in Einstein's Equations (5.40) or (5.41) is undoubtedly nonlinear, its mathematical properties cannot be altered even tough tensors are employed. If Einstein were to give up his constant speed of light while taking up Equation (5.41), he would then have created a consistent gravitation field equation. Even though an analytic solution of the equation might not be possible, he could still declare that in terms of rotational solenoidal fields, the general relativity theory enters the exploration of the mechanism of universal gravitation. In this case, he would not need to employ his mass–energy formula of kinetic energy that did not do any work. Therefore, all materials in the universe would form a gigantic field of spinning currents satisfying the principles of solenoidal spin fields. The essential characteristics of the fields of spinning currents are as follows: At the same time when materials are transported, energies are transferred into Einstein's void space, while avoiding wave motions that cannot transport materials. It is exactly because of these spinning current fields' capability of delivering materials' energy that inwardly spinning currents could cause collapses and outwardly spinning currents could lead to expansions. However, red shifts do not seem to be a substantial evidence for the universe's expansion, because wave propagations do not really deliver materials. Instead, eddy motions actually make deliveries. Besides, the conventional concept of explosion is a linear concept of the particle form. As a matter of fact, the mushroom clouds of nuclear explosions are rotational, and the collisions in accelerators are created by using rotations. So, only by employing rotations has one obtained high-speed flows.

To this end, let us rewrite Newton's universal gravitation formula as follows:

$$\vec{F} = G \frac{\rho_1\left(x, y, z; t\right)\rho_2\left(x, y, z; t\right)}{R^3} \vec{R} \tag{5.54}$$

where $\rightarrow$ is the symbol for vector, $R = (r + r_d)$, r the distance between the surfaces of ρ_1, ρ_2, r_d the sum of the radii of ρ_1, ρ_2, and G a constant. Taking the vorticity of Equation (5.54) leads to

$$\nabla \times \vec{F} = \frac{G}{R^3}\left[\rho_1\rho_2 \nabla \times \vec{R} + \left(\rho_1 \nabla \rho_2 + \rho_2 \nabla \rho_1\right) \times \vec{R}\right] \tag{5.55}$$

Evidently, as a mathematical quantity, $\rho_1(x, y, z; t)\rho_2(x, y, z; t)$ in Equation (5.54) is nonlinear, leading to the rotationality of a solenoidal field for universal gravitation. That reveals the physical implications that Newton's quantitative forms did not deliver. So, this observation illustrates that structures can alter mathematical properties. And V. Bjerknes accordingly obtained the twisting rotationality of solenoidal fields by using circulation integrals on the basis of nonparticles; his result has been well validated by practice. Therefore, nonlinearity implicitly contains the physical rotationality of solenoidal fields.

Now, let us employ Einstein's associative thinking similar to that of Equation (5.55): The so-called universal gravitation is not an attraction and pull along a straight line, which, as an essential concept of physics, cannot be completely compared with the so-called strong forces. Instead, it can be thought of as the carry-in force of the celestial body's inward rotation. And, accordingly, there should also be a carryout force corresponding to the outward rotation of the celestial body. The carry-in of the celestial body's inward spinning has been validated by the fact that the path of a falling star is like a curved page in an old book. Each high-speed carry-in is a black hole, while from spins high-speed flows can really be obtained.

As for the carryout of outward rotations, it should be a white hole (white source); and the speed of carrying out could reach such a high level at least near the exit of the white hole that we might not be able to imagine today. So, the universal gravitation should be named uniformly as the universal force existing along with the carrying capability of celestial bodies' rotation. As for whether it is the carry-in of an inward spinning or the carryout of an outward rotation, it will be determined by the attributes and spinning speed of the celestial body. What needs to be clear is that the described is only a conjecture based on equilibrium of inward and outward rotations of solenoidal spinning fields.

Also, from Equation (5.33), it can be seen that the condition for the conservation of irrotational kinetic energy is very strict, because only when the inward and the outward rotations of the solenoidal field cancel each other perfectly, the so-called conservation can be achieved. However, in this case, the kinetic energy is not irrotational; instead it already becomes a rotational stirring kinetic energy. Since in reality the probability for the perfect cancellation to occur is nearly zero, the physical reason why the conservation of the (irrotational) kinetic energy of modern science cannot guarantee its conservationality should be very clear.

Although the rotationality characteristic of nonlinearity is produced out of a very simple vector analysis, other mathematical methods, including quantities, vectors, and tensors (quantities are zero-order tensors and vectors are first-order tensors), should not alter the mathematical properties of nonlinearity either. Therefore, the so-called mystery of nonlinearity, as of this point in time, is completely resolved. And the reason why nonlinear equations do not satisfy the initial-value existence theorem of the linear system is also clearly laid out.

Considering the potential value of our work, let us organize and summarize as follows.

At the same time that the speed of light is assumed to be a constant, Einstein's mass–energy formula loses the underlying material and is led to an energy that does not do any work.

It seems that the theory of gravitation established on irrotational curvature space could not explain why the path of a falling star looks like a curved page in an old book, causing the physical reality separate from Einstein's theory.

Through vector calculus of the mathematical system itself, nonlinearity is shown to be the rotationality of solenoidal fields. So, there is no longer the need to list nonlinearity as a world-level difficult problem. This end of the mystery of nonlinearity is not dependent on employing any inexplicable theory to explain an existing fundamental reality. Also, we hope that the over 100-year-old Bjerknes's circulation will no longer be bundled up and stored on a high shelf. Many scientific workers, including Newton, Einstein, and some of their followers, seem to have forgotten that quantitative equations are merely formal expressions without the ability to answer *whys*. That explains the reason Newton could not address what universal gravitation is, nor did Einstein provide a definite answer to this question out of his theory of gravitational waves.

No doubt, due to dualities of rotations, inward and outward spins coexist. So, each rotational field of flows itself is a discontinuous initial-value field. That is, materials' structures dictate the rotationality of materials and the discontinuity of the initial current fields and well hit the weakness of the system of continuity of modern science. This end explains the reason why modern science has faced constant challenges from various problems.

5.5 Probabilistic Waves of the Schrödinger Equation and Transmutation of High-Speed Flows

Quantum mechanics has been seen as one of the signature achievements of modern physics. Together with Einstein's relativity theory, it used to be known as one of the two black clouds, and then later became one of the two magnificent flowers. The concept of indeterminacy, left behind from quantum mechanics, has been one of the problems of debate of the scientific community and faces many difficulties in resolving practical problems. That is why when a complex problem is considered, a complex answer is often provided; the probabilisticity for an event to occur has been evolved into the randomness of the event itself, leading to the chaos doctrine of the 20th century and conducting the point of view of probabilistic universe.

Since the time when we initially discovered blown-ups in the transitional changes of materials' evolutions, we have vividly sensed the determinacy behind problems of randomness; that is about how to correctly treat irregular events and how to extract information of change out of irregular events. Modern science has been about how to extract eternally invariant information from events. That is the reason why modern science is not fond of irregular information that in fact is the information about changes occurring in the materials or events. Changes in materials or events that tend to destroy the past not only are a physical reality, but also can lead to methods of dealing with irregular information that modern science cannot handle with much success. The transmutation of high-speed flows has similar meanings as the concept of blown-ups we introduced with an emphasis on changes in properties. That is how we have reconsidered some of the fundamental problems studied in modern physics and related to the wave–particle duality. Bohr's probabilistic explanation of wave functions and Bohr's followers have renamed Heisenberg's principle of inaccurate measurement as the principle of indeterminacy (S. W. Hawking, 1988). On this basis, they not only criticized Newton but also called Einstein an old diehard who is against accidentality since he pointed out that quantum mechanics is incomplete. That laid down the historical foundation for the chaos doctrine to appear and to become a heat wave in the scientific community in the 20th century, where restating the principle of inaccurate measurement as that of indeterminacy is for the preparation of posting that God does play dice. Behind all these, the very central problem is how to understand irregular information and how to employ irregular information. Modern science has gone through the stages of not understanding irregular information as the information about evolution, to feeling annoyed by irregular information, and to the current attempt to eliminate information of change, or denying the variancy of the world for the pursuit of regularity.

In the complex number form, the Schrödinger equation is a linear equation with variable coefficients of the unknown wave function as its variable. However, the appearance of complex numbers conceals the implicit nonlinearity of the real azimuth angle and the amplitude. The nonlinearity of fluids involves the evolutionary mutual reactions of the flows and waves and transformations of eddy currents and can appear to be irregular information, quantitative instability, or a problem of mathematical singularity. That reveals the deterministic relationship between rotations and evolutions.

5.5.1 Flow–Wave Duality of Microscopic Material Flows and Quantumization

The flow–wave duality can be obtained from the equation of fluid motion, revealing that waves dwell on flows and involving the directional structure of the fluid. The corresponding set of microscopic materials is also a structure, or the direction of a particle flow can also constitute the structurality of disturbing irregularities. Even based on what Einstein (1976) said, "no matter what, what is described by wave functions cannot be the state of a single system; statistically speaking, it an ensemble, …" an uneven wave of the quantitative time–space can be written as

$$\varphi(x, y, x, t) = a(x, y, z, t) \cos \xi(x, y, z, t) \tag{5.56}$$

If we define the local wave number $\vec{K}$ and the local frequency ω, we then have

$$\vec{K}(x, y, z, t) = \nabla \xi, \quad \omega(x, y, z, t) = -\frac{\partial \xi}{\partial t} \tag{5.57}$$

where ∇ is the gradient operator, and the relationship between the wave frequency and number is

$$\omega = W(\vec{K}) \tag{5.58}$$

From Equation (5.57), it follows that

$$\vec{K}_t + \nabla \omega = 0 \tag{5.59}$$

where

$$\vec{K}_t = \frac{\partial \vec{K}}{\partial t}$$

If in Equation (5.58), ω is a nonlinear function in $\vec{K}$, then Equation (5.59) is a nonlinear equation in $\vec{K}$. Its solution stands for a blown-up problem of break-off transformation of the wave number. In earlier chapters, we have studied that the essence of blown-ups is about transformations of eddy currents in higher-dimensional curvature spaces instead of the quantitative infinity. Even on the opinion of indivisibility of fundamental particles, if the wave number is infinitely large, the wavelength has to be infinitely small, producing an irregular flow of discontinuous particles that destroy the wave of the continuous media. No doubt, *wave* is a collective concept of the particle media, while *particle* is an individual, noncollective concept. That is why Einstein pointed out that wave functions are ensembles and the corresponding quantum mechanics should be a mechanics of quantum states.

For a microscopic uneven flow, we can introduce the following uneven "wave" function:

$$\psi = A(x,y,z,t)e^{i\xi(x,y,z,t)/\hbar} \tag{5.60}$$

and define the local wave number and frequency similarly as

$$R(x,y,z,t) = \nabla\xi / \hbar, \quad \omega(x,y,z,t) = -\frac{\partial \xi}{\partial t} / \hbar \tag{5.61}$$

According to the condition of quantumization, the quantumizations of the energy E of the ensemble and the momentum $\vec{P}$ are given below:

$$\begin{cases} E(x,y,z,t) = \hbar\omega = -\dfrac{\partial \xi}{\partial t} \\ \vec{P}(x,y,z,t) = \hbar\vec{K} = \nabla\xi \end{cases} \tag{5.62}$$

where

$$\hbar = \frac{h}{2\pi}$$

h is the Planck constant. From the relationship between energy and momentum, we have

$$\vec{P}_t + \nabla E = 0 \tag{5.63}$$

Since

$$E = H(\vec{P}) \tag{5.64}$$

Equation (5.63) can be rewritten as

$$\vec{P}_t + \nabla H(\vec{P}) = 0 \tag{5.65}$$

Evidently, if E is a nonlinear function in $\vec{P}$, then Equation (5.63) also stands for a blown-up of nonlinearity or a problem of eddy current transformation. What is shown here is about momentum transformation, which corresponds to break-ups of waves. From the point of view of the theory of blown-ups, it follows that the quantitative infinity does not represent any physical reality; instead it is a description or symbol for transitionality. For nonsmall effective values, Equation (5.63) can grow and cause the momentum to approach infinity due to increases in the nonlinear quantitative products. This end indicates that the momentum experiences a transformation problem of transitional changes.

5.5.2 Nonprobabilistic Annotation of Uneven Wave Functions

First, what needs to be pointed out is that Schrödinger himself derived his Schrödinger equation using a method different from the traditional procedure of modeling using mathematical physics equations. He employed the physical quantities of energy and momentum as parameters on the presumption of even planar waves of free particles. In other words, he presumed the state of the quantum ensemble of concern as a wave function

$$\psi(x, y, z, t) = A_0 e^{\frac{i}{\hbar}(\vec{P}\cdot\vec{r} - Et)} \tag{5.66}$$

where $\vec{r} = x\vec{i} + y\vec{j} + z\vec{k}$, and A_0 is a constant. Then, employing the relationship between the energy and momentum of the free particles

$$E = \vec{P}^2 / 2m \tag{5.67}$$

where m stands for the mass of the particles. So, the corresponding Schrödinger equation with imaginary number ($i = \sqrt{-1}$) coefficient is given as

$$i\hbar \frac{\partial \psi}{\partial t} = -\frac{\hbar^2}{2m} \nabla^2 \psi \tag{5.68}$$

where

$$\nabla^2 = \frac{\partial^2}{\partial x^2} + \frac{\partial^2}{\partial y^2} + \frac{\partial^2}{\partial z^2}$$

If the particles are located in a potential field $U(x, y, z)$, then the relationship between the particles energy and momentum is

$$E = \frac{\vec{p}^2}{2m} + U(\vec{r}) \tag{5.69}$$

So, the Schrödinger equation (5.68) for a planar even wave of free particles is given as follows:

$$i\hbar \frac{\partial}{\partial t} \psi = -\frac{\hbar^2}{2m} \nabla^2 \psi + U(\vec{r})\psi \tag{5.70}$$

If the uneven wave in Equation (5.60) is introduced, will the corresponding Schrödinger equation still hold true? Let us look at this problem specifically in the following.

First, let us compute the following differentiation of Equation (5.60):

$$i\hbar \frac{\partial \psi}{\partial t} = e^{i\xi/\hbar} \left(i\hbar \frac{\partial}{\partial t} - \frac{\partial \xi}{\partial t} \right) A \tag{5.71}$$

From

$$\frac{\hbar}{i} \frac{\partial \Psi}{\partial x} = e^{i\xi/\hbar} \left(\frac{\hbar}{i} \frac{\partial}{\partial x} + \frac{\partial \xi}{\partial x} \right) A - \hbar^2 \frac{\partial^2 \Psi}{\partial x^2} = \frac{\hbar}{i} \frac{\partial}{\partial x} \left[e^{i\xi/\hbar} \left(\frac{\hbar}{i} \frac{\partial}{\partial x} + \frac{\partial \xi}{\partial x} \right) A \right]$$

$$= e^{i\xi/\hbar} \left[-\hbar^2 \left(\frac{\partial^2 A}{\partial x^2} \right) + \frac{\hbar}{i} A \frac{\partial^2 \xi}{\partial x^2} + 2 \frac{\hbar}{i} \frac{\partial A}{\partial x} \frac{\partial \xi}{\partial x} + \hbar^2 \left(\frac{\partial \xi}{\partial x} \right)^2 A \right]$$

it follows that

$$-\hbar^2 \nabla^2 \Psi = e^{i/\hbar \xi} \left[(-\hbar^2 \nabla^2 A + \frac{\hbar}{i} A \nabla^2 \xi + 2 \frac{\hbar}{i} \nabla A \cdot \nabla \xi + \hbar^2 A (\nabla \xi)^2 \right] \tag{5.72}$$

By substituting Equations (5.71) and (5.72) into the Schrödinger equation (Equation 5.70) and letting the real part = real part and imaginary part − imaginary part, we have

$$\frac{\partial A}{\partial t} = -\frac{1}{2\rho}(A\nabla^2\xi + 2\nabla A \cdot \nabla\xi) \tag{5.73}$$

and

$$\frac{\partial \xi}{\partial t} = -\left[\frac{\nabla\xi}{2\rho} + U - \frac{\hbar^2}{2\rho}\frac{\nabla^2 A}{A}\right] \tag{5.74}$$

Here, the mass m of particles has been replaced by the flow's density ρ. Evidently, Equations (5.73) and (5.74) are completely equivalent to the Schrödinger equation. In other words, when the amplitude A and the phase angle ξ of the uneven wave function (5.60) satisfy Equations (5.73) and (5.74), the Schrödinger equation (5.70) holds true. From the conservation of probabilistic density of the Schrödinger equation, that is

$$\frac{\partial R}{\partial t} + \nabla \bar{J} = 0 \tag{5.75}$$

it can be seen that the essence for Schrödinger to introduce the probabilistic density is to treat the movement of fundamental particles as irregular probabilistic. That is where the concept of probabilistic waves is from. As for how to understand irregularity and where irregularity is from, no specific explanation is given (we will discuss this in this section). And, the probabilistic density R and the probabilistic flow density $\bar{J}$ are respectively given by

$$R = |\psi|^2 \tag{5.76}$$

and

$$\bar{J} = \frac{i\hbar}{2\rho}(\psi\nabla\psi^* - \psi^*\nabla\psi) \tag{5.77}$$

which represents essentially a nonprobabilistic treatment of the probabilistic problem, where ψ^* is the complex conjugate of ψ^*. Assume ρ = a constant. Substituting Equation (5.60) into Equations (5.76) and (5.77) leads to

$$R = A^2 \tag{5.78}$$

$$\vec{J} = \frac{1}{2\rho}\left[\left(\Psi^* \frac{\hbar}{i} \nabla \Psi\right) + C.\ C.\right] = \frac{1}{2\rho}\left[A\left(\frac{\hbar}{i} \nabla A + A \nabla \xi\right) + C.\ C.\right] = \frac{A^2}{\rho} \nabla \xi \tag{5.79}$$

where *C. C.* stands for the conjugate of the previous term. From Equation (5.73), it follows that

$$\frac{\partial R}{\partial t} + \nabla \cdot \vec{J} = 2A \frac{\partial A}{\partial t} + \frac{\partial A}{\partial t} + \nabla \cdot \left(\frac{A^2}{\rho} \nabla \xi\right)$$

$$= 2A \frac{\partial A}{\partial t} + \frac{2A}{\rho} \nabla A \cdot \nabla \xi + \frac{A^2}{\rho} \nabla^2 \xi = 0 \tag{5.80}$$

Since $\vec{V} = \vec{J}\,/\,R = \nabla \xi\,/\,\rho$, Equation (5.74) can be simplified to

$$\frac{\partial \xi}{\partial t} + \frac{1}{2}\rho \vec{V}^2 + U = \frac{\hbar^2}{2\rho} \nabla\left(\frac{\nabla^2 R^{1/2}}{R^{1/2}}\right) \tag{5.81}$$

which can be rewritten as

$$\frac{\partial}{\partial t}\vec{V} + \vec{V} \cdot \nabla \vec{V} = -\frac{1}{\rho} \nabla U + \frac{\hbar^2}{2\rho} \nabla\left(\frac{\nabla^2 R^{1/2}}{R^{1/2}}\right) \tag{5.82}$$

This is exactly the form of the equation of fluid motion in Euler language except that the second term on the right-hand side stands for quantum effects and is a mathematical third-order derivative of a nonlinear variable. Evidently, the equation of fluid motion without the second term on the right-hand side of Equation (5.82) cannot be stochastic; the third-order derivative of the quantum effects must be deterministic, since the third-order derivative represents the deterministicity of the torsion of the shear stress force. Since the third-order derivative in Equation (5.82) is of the form of nonlinearity and ρ = a constant is assumed in the process of deduction, Equation (5.82) stands for a both convergent and divergent, irrotational movement.

If $\rho \neq$ a constant, then the continuous Equation (5.80) no longer holds true. From the second equation in Equation (5.62), it follows that $\vec{V} = \vec{P}\,/\,\rho = \nabla \xi\,/\,\rho$, $\rho = \rho(x, y, z)$, and Equation (5.74) becomes

$$\frac{\partial \xi}{\partial t} = -\frac{1}{2}\rho \vec{V}^2 - U + \frac{\hbar^2}{2\rho}\frac{\nabla^2 A}{A} \tag{5.83}$$

Taking the gradients of both sides produces

$$\frac{\partial \nabla \xi}{\partial t} = -\rho(\vec{V}\nabla)\vec{V} - \frac{1}{2}\vec{V}^2\nabla\rho - \nabla U + \frac{\hbar}{2}\nabla\left(\frac{\nabla^2 A}{\rho A}\right) \qquad (5.84)$$

An equivalent equation of this line is

$$\frac{\partial \vec{V}}{\partial t} + (\vec{V}\nabla)\,\vec{V} = -\frac{1}{\rho}\nabla U - \frac{1}{2\rho}\vec{V}^2\nabla\rho + \frac{\hbar^2}{2\rho}\nabla\left(\frac{\nabla^2 A}{\rho A}\right) \qquad (5.85)$$

This fact indicates that the probabilistic waves of the Schrödinger equation of the complex number form are in fact the quasi-regular eddy currents of particles or the quasi-turbulences of fluid mechanics (see the numerical experiments in Section 5.6) under the mutual reactions of the quantum effects and the potential field instead of waves under the influence of pushes; and, they are analogous to the turbulences of classical fluid mechanic equations in the form of multiple irregular eddy currents under stirring effects except that the underlying mutual reactions are shown more completely than the case of equations of classical fluid mechanics. It is because according to the Bjerkness's circulation theorem (1898) for when $\rho \neq$ a constant, it can be shown that the three terms on the right-hand side of Equation (5.85) are solenoidal stirring terms of different scales and the quantum effect term, the third term on the right-hand side of Equation (5.85), is a twist-up — a third order derivative. So, the so-called probabilistic waves of the Schrödinger equation are neither waves nor a problem of indeterminate probabilisticity. In other words, Equation (5.85) has very clearly denied the claim that God does play dice. It is because the vorticity of uneven vectors has to be rotational; and it can be shown that the stirring effects of a solenoidal term, also known as the nonlinear mutual reactions, are eddy sources. So, the corresponding movements have to be rotations.

No doubt, even if quantum effects are included, Equation (5.85) is still a nonlinear equation under mutual reactions; the corresponding traditional fluid mechanic equations are only special cases of Equation (5.85). The so-called quantum effect is a third-order derivative mathematically, and is expressed in mathematical nonlinear form as a term of change in curvature. When seen from the angle of mathematics, Equation (5.85) is a typical ill-posed problem of mathematical physics equations, which does not satisfy the initial-value existence theorem. In other words, Equation (5.85) or the Schrödinger equation cannot be solved using methods developed in the inertial system.

Evidently, not satisfying the initial-value existence theorem is not the same as being indeterminate. Just as what we pointed out earlier, mathematical nonlinearity is a rotationality problem of solenoidal fields. It experiences transitional changes away from the initial-value system and touches on the problem of discontinuity of the initial-value system. So, based on the mathematical properties of the Schrödinger equation or Equation (5.85) where $\rho \neq$ a constant, the quantum mechanics explanation of probabilisticity is not the essence of probabilistic waves; in other words, the mathematicality of the nonlinearity of Equation (5.85) has already ended the mathematical probabilisticity. Since fundamental particles are not the same as states of particles, quantum mechanics does not tell that fundamental particles are waves. The once-popular translation of achievements of modern physics into the philosophical proposition that God does play dice, and the claim that quantum mechanics tells us that all fundamental particles are waves, indicate that the problem of whether or not probabilistic waves are in fact waves is not clearly addressed; and the debate between determinacy and indeterminacy, as triggered by the Schrödinger equation, is no longer meaningful. Nonlinearity and randomness cannot be identified with each other; or in other words, the popular randomness has nothing to do with nonlinearity.

We must point out that mutual reactions are the fundamental form of materials effects. In other words, there does not exist in the material world any first push that experiences no obstacle; and the resistance against movements increases exponentially with the change in speed. So, at a high speed, transmutations — changes in materials' attributes — will surely occur. As for modern science's invariancy in materials of either low-speed flows or high-speed flows, they can only be our one-sided human wish. Evidently, the introduction of the concepts of blown-ups and transmutations naturally leads to the need to address what time is. In other words, if the invariant material, quantitative time, is continually employed in the study of problems of high-speed flows, science will lose its sense of direction. Hopefully, the reader can understand what we intend to say.

As for the end of the Newtonian initial-value determination, it can be accomplished without introducing the concept of probabilisticity. This is because the Newtonian mechanics system is about variable speeds under pushes without touching on changes in materials. In terms of problems that do not involve changes, there is no longer the need to talk about whether they are deterministic. The practical significance of whether deterministic or not is about whether the underlying material changes, and whether we can foretell the states or attributes after changes based on those from before the changes. Only after the states and attributes are made definite, what is left are mathematical problems. No doubt, if materials stay eternally the same, what will be the need for us to explore whether they are deterministic or not?

Quantum mechanics more directly shows the stirring motion of mutual reactions, and constructs the non-initial-value automorphic, transitional evolutions through using transformations between energies and momentums. Evolutions, even with blown-ups, are about macroscopic state changes in materials, while transmutations involve changes in materials' microscopic attributes. The phenomena of birth, age, illness, and death of materials' evolutions do not seem to be problems of probabilisticity, and cannot be seen as the end of determinacy. Evidently, if the phenomena of birth, age, illness, and death were seen as probabilisticity or the end of determinacy, then it would automatically admit that the essence of the scientific pursuit of determinacy is in fact about the eternal invariancy of materials. Therefore, Einstein was indeed innocent when he was accused of being an old diehard for not leaving any space for accidentality. However, at the same time, quantum mechanics should not be seen as incomplete, either, because problems of quantum mechanics pose challenges to the modern scientific system; the essence of the probabilistic waves is the quasi-irregular flows made up of quantum irregular flows and strong potential field pushes under mutual reactions.

5.5.3 Non-Initial-Value Transformation of Energy and Momentum and Transmutation of High-Speed Flows

From the one-dimensional relationship between the energy and momentum of uneven wave functions in Equation (5.63), it follows that the momentum transformation of irrotational uneven flows of particles is

$$P_t + E_x = 0 \qquad (5.86)$$

where

$$E_x = \frac{\partial E}{\partial x}$$

From $E = E(p)$, it follows that

$$P_t + \frac{dE}{dp} p_x = 0 \qquad (5.87)$$

So, we can respectively discuss the blown-ups of low-speed flows and transmutations of high-speed flows.

5.5.3.1 The Problem of the Classical Low-Speed Flows of Particles

The classical relationship between particles' energy and momentum is

$$E = P^2 / 2\rho \qquad (5.88)$$

Let us write the momentum in terms of one-dimensional density and speed,

$$p(= m\upsilon) = \rho u \tag{5.89}$$

Let ρ be the density of a single particle, and it is assumed to be a constant. Then, substituting Equations (5.88) and (5.89) into Equation (5.87) produces

$$u_t + u u_x = 0 \tag{5.90}$$

where the subscripts stand for partial derivatives. No doubt, Equation (5.90) stands for a typical blown-up problem discussed in earlier chapters. This fact implies that even under the condition of low-speed flows, the initial-value's convergence can be transformed into non-initial-value's divergence, constituting a morphological change of the underlying material. This end has been well validated by water flows in oceans and weather evolutions of the atmosphere. Or in other words, convergence would not terminate from becoming quantitative infinity due to assembling at the singular point; instead, it is transformed into divergence in order to achieve equilibrium of the whole system. Therefore, high-speed flows of the general relativity theory should not be an exception.

5.5.3.2 High-Speed Flows of Particles of Relativity Theory

For Einstein's mass–energy formula, even if we take the density of a single photon as constant, we still have $E(= m c^2) = \rho c^2$. So, the corresponding formula of momentum becomes

$$p = \rho c \tag{5.91}$$

So, the relationship between the energy and momentum looks like

$$E = p^2 / \rho \tag{5.92}$$

where c stands for the speed of light. Since the movement takes place in a curvature space, we have to take $c = c(x,y,z,t) \neq$ a constant. So, the difference between Equations (5.92) and (5.88) is only a constant of ½, indicating the mutual reaction of particle flows. Substituting Equations (5.91) and (5.92) into Equation (5.87) while replacing u by $c(x,y,z,t)$ produces

$$c(x,y,z,t)_t + 2c(x,y,z,t)c(x,y,z,t)_x = 0 \tag{5.93}$$

Evidently, the only difference between Equations (5.93) and (5.90) is that of non-linear rotational low-speed flows and high-speed flows under mutual reactions. When the convergence of a high-speed flow is transformed to divergence, the characteristics are different from those of low-speed flows, because light comes from transitions of electrons and indicates atomic structural changes — transmutations. However, what needs to be clear is that the speed of light cannot be assumed to be constant; otherwise, one would obtain Einstein's energy that does not do any work; or, at the same time when Equation (5.93) does not hold true, it also makes the movement of high-speed flows a problem of invariant materials. That explains the reason why fundamental particles do not experience either fission or fusion. To give an expression to transmutations of fundamental particles, the speed of light should not be constant.

No doubt, the rotationality of the transmutations of Equation (5.93) is more intense when compared to the rotationality of the blown-ups of Equation (5.93). The former could be as drastic as "instantly changing the sky," which explains why the black holes of celestial bodies do not end in the form of collapse due to overcrowding; instead, they become sprays in a more intense form. If ρ stands for the structural unevenness of quantum ensembles, then the movement must be rotational with the unification of opposites of inward and outward rotations. Overcrowded inward spin of high-speed flows has to cause changes in materials' attributes at the same time the spin is transformed into a spray of an outward rotation. Similar to the morphological structural changes of low-speed flows, a non-initial-value automorphic evolution is completed, which is not random or indeterminate.

What is interesting is that we once did numerical experiments based on the Schrödinger equation in terms of the transitional changes in curvature spaces (see Section 5.6), and produced the irregular flows of pure quantum effects with twist-ups of the third-order nonlinear derivatives. These experiments not only reveal transitional changes in evolutions that correspond to natural disasters or major accidents, but also specifically represent such damaging events as disastrous weathers or earthquakes. Only when quantum effects and an intense potential field interact with each other can one obtain quasi-regular flows that are analogous to waves but do not belong to the category of probabilisticity. Our results indicate that no matter whether it is the "incompleteness of the quantum mechanics" or that "quantum mechanics implies that all fundamental particles are waves," it is neither the accidentality of the dice played by God nor the indeterminacy due to the turbulences of particle flows. The concept of wave–particle duality itself is a conceptual confusion. Or in other words, the so-called probabilistic waves of the Schrödinger equation itself indicate that the mathematical and physical properties of the Schrödinger equation were not fully understood. So, other than the turbulencality of particle flows, Equation (5.85) still possesses other significant and important properties, involving the third-order derivatives' twist-ups of the quantum effects that are no longer limited to the area of quantum mechanics itself. This end also touches on some of the fundamental understandings on the damaging effects of changes in materials.

This work is a by-product of our investigation of the applications and generalizations of the blown-up principle. We accidentally discussed some not simple but quite sensitive problems of the scientific community in the form of some relatively simple mathematical expressions. By chance, we discovered that the so-called probabilistic waves of the Schrödinger equation do not seem to be quite up to the theoretical perfection. Following are some of the main results along this line of investigation.

In essence, "move without being seen" is not only a proposition of physics but also one of philosophy. There is a need to reconsider the current alteration of the quantitative measurement inaccuracy to the "principle of uncertainty." "Move without being seen" is a natural consequence of materials' mutual reactions in a noninertial system and is not indeterminate. It is just like what ancient Chinese knew: *yi wu ce wu, bu fen wu wu* (when an object is used to measure another object, one faces a measurement inaccuracy caused by mutual interference), which was well-known way before Heisenberg's principle of measurement inaccuracy was introduced. That is why the Chinese would not refer to the phenomenon of measurement inaccuracy between equipment as "indeterminacy."

The establishment of the Schrödinger equation under the assumption of wave functions was a natural consequence of the special moment in time when people treated the irregularity observed in the movements of fundamental particles as probabilistic so that they employed equations in quantum ensembles or quantum states in their investigations of quantum problems or fundamental particles. Since these equations are similar to Equation (5.85), it implies that the so-called probabilistic waves are not the indeterminacy of the dice played by God and illustrate that quantum mechanics does not actually indicate that all fundamental particles are waves.

Material movements in curvature spaces have to be rotational. The speed of light cannot be assumed to be a constant; instead, it should be considered as a variable of high-speed rotations. With the help of rotations, one can retain high speeds and evolution of changing materials. One should always respect that in the material world no first push can exist without any obstacle. So-called free particles in essence all betray the first push in the form of rotation under mutual reactions.

Since along with the assumption of constant speed of light there appears such energy that does not do any work, this assumption denies one of the elementary laws of modern science. It is just like what Einstein once said, "The classical mechanics, quantum mechanics, general relativity theory, and the basic mathematical physics equations cannot provide the difference between the past, the present, and the future" (Einstein, 1976, volume 3). That has already very frankly admitted that modern science and modern physics are still theories in the inertial system about nonevolutionary materials. Since we have reconsidered irregularity, the theory of quantum mechanics, and the transmutations of high-speed flows, our work has essentially entered the realm of evolution science. Whether or not deterministic, God plays dice, leave space for accidentality, probabilistic universe, and so forth are not the messages quantum mechanics truly delivers.

5.6 Numerical Experiments on Probabilistic Waves and Torsion of Quantum Effects

Phenomena of change, such as flowing water, twist-ups of cloud systems, and so forth, can be observed in rivers and cloud maps, where one can see curvatures and changes in curvatures. In terms of mathematics, first-order derivatives are slopes; second-order derivatives are curvatures; so changes in curvatures involve third-order derivatives. And, third-order derivatives are also related to shear stress forces. Based on the custom of quantification, we name third-order derivatives torsions or spinnings. Unfortunately, all the well-known mathematical physics equations contain mostly first- or second-order derivatives except the altered Schrödinger equation that contains a third-order derivative of a nonlinear variable. However, neither Schrödinger himself nor his followers provided any specific explanation for the quantum effects of third-order derivatives. To this end, in this section, we will use this equation as our example to conduct our numerical experiments on the basis of quantitative instabilities.

What needs to be said is that according to the well-posedness theory of the traditional mathematical physics equations, quantitative nonlinear growth is constrained by well-posedness so that the solution satisfies the existence, uniqueness, and stability. However, since nonlinear equations can grow explosively with nonsmall values due to nonlinear products, these equations do not satisfy the theory of well-posedness and the initial value existence theorem, causing distortions in solving these equations. In order to give an expression to nonlinearity and transformations of torsions, based on the fact that quantitative infinity (∞) in Euclidean spaces stands for transitional changes in curvature spaces, we propose the method of not limiting the quantitative growth, while using structure spaces of reciprocating functions to emphasize irregular changes so that the basic characteristics of the evolutionary structural morphology can be analyzed.

Even by using the current quantitative analysis system, there are also different opinions about the concept of probabilistic waves. The probabilisticity (in fact, it is quantitative irregularity) of probabilistic waves was evolved into the epistemological concept of indeterminacy, believing that the probabilisticity of quantum mechanics posed challenges to the Newtonian system, which in the 1980s reached the height of forming the epistemological concept of a probabilistic universe. As mentioned in Section 5.4, Einstein did not believe that God would play dice, so he was labeled as an old diehard against indeterminacy by the scientific community of the 20th century. In fact, the two opposite opinions, God plays dice and God does not play dice, have not touched on the problem of essence of the Schrödinger equation itself, so that none of them has delivered the exact physical significance.

To this end, considering the aforementioned disagreements of the scientific community, our numerical experiments are conducted with three cases:

1. The function of quantum effects;
2. The combined impacts of the intensity of the potential field and quantum effects; and
3. Adjusted unit volume density and intensity of the potential field.

What needs to be explained is that when Schrödinger discussed probabilistic waves, he mimicked the Newtonian system with separate objects and forces and assumed a constant unit volume density, leading to the name of quantum mechanics. If the unit volume density is not a constant, then we have a nonlinear problem. To this end, we first follow what Schrödinger did to reveal the physical essence of probabilistic waves. After that, we adjust the unit volume density and the intensity of the potential field in order to analyze the corresponding structural properties.

5.6.1 The Fundamental Equation

5.6.1.1 The Original Schrödinger Equation

Schrödinger (1926) established an equation in ψ, a wave function, which is the amplitude for a particle to have a given position at any given time, for the movement of microscopic ions, named later as the Schrödinger equation, on the basis of the superposition principle of de Broglie's relations. The original Schrödinger equation looks like the following:

$$i\hbar\psi_t = \left(-\frac{\hbar^2}{2m}\nabla^2 + U\right)\psi \tag{5.94}$$

where

$$\hbar = \frac{h}{2\pi}$$

$h = 6.626 \times 10^{-34}$ is the Planck constant, ψ the wave function, m the mass of an unit volume of particles, and U the intensity of the potential field. The subscript t stands for the partial derivative of ψ with respect to t, and ∇^2 the Laplace operator.

5.6.1.2 An Altered But Equivalent Schrödinger Equation

Considering the complexity of Equation (5.94), we can separate the norm and azimuth angle of the wave function ψ and look at the altered equation that is equivalent to the Schrödinger equation below; see Equation (5.85) for more details:

$$\vec{v}_t + \vec{v} \cdot \nabla \vec{v} = -\nabla \left(\frac{1}{\rho} U \right) + \hbar^2 \nabla \left[\frac{1}{2\rho^2} \left(\frac{\nabla^2 P^{1/2}}{P^{1/2}} \right) \right] \tag{5.95}$$

where $\vec{v}$ is a vector, ρ the density of a unit volume of particles, U the intensity of the potential field, P the probabilistic density, $\hbar = h / 2\pi$, and h the Planck constant. The first term on the right-hand side of Equation (5.95)

$$\nabla \left(\frac{1}{\rho} U \right) = \left(\frac{1}{\rho} \right) \nabla U + U \nabla \left(\frac{1}{\rho} \right) \tag{5.96}$$

stands for the solenoidal term of the mutual reaction between the potential field's intensity and unit volume density of the particles. If ρ is a constant, then it is only the push of the potential field. However, the second term on the right-hand side of Equation (5.95)

$$\hbar^2 \nabla \left[\frac{1}{2\rho^2} \left(\frac{\nabla^2 P^{1/2}}{P^{1/2}} \right) \right] = \hbar^2 \left[\frac{1}{2\rho^2} \nabla \left(\frac{\nabla^2 P^{1/2}}{P^{1/2}} \right) + \frac{\nabla^2 P^{1/2}}{P^{1/2}} \nabla \left(\frac{1}{2\rho^2} \right) \right] \tag{5.97}$$

stands for the quantum effect, that is the quantum effect term of the torsion (a third-order derivative) under the mutual effect of the unit volume density and the probabilistic density.

5.6.1.3 The Difference Equation of the Altered Schrödinger Equation

Here, we apply the method of direct differencing. That is, we apply two-dimensional differences in the $u - v$ space to Equation (5.95). The resultant difference equations are given below:

$$u_{(n+1)} = u_{(n)} + \left(-u_{(n)} \frac{\partial u}{\partial x} - v_{(n)} \frac{\partial u}{\partial y} - \frac{1}{\rho} \frac{\partial U}{\partial x} + \frac{U}{\rho^2} \frac{\partial \rho}{\partial x} + \hbar^2 \frac{\partial}{\partial x} \left[\frac{1}{2\rho^2} \frac{\nabla^2 P^{1/2}(x)}{P^{1/2}(x)} \right] \right) \Delta t$$

$$v_{(n+1)} = v_{(n)} + \left(-u_{(n)} \frac{\partial v}{\partial x} - v_{(n)} \frac{\partial v}{\partial y} - \frac{1}{\rho} \frac{\partial U}{\partial y} + \frac{U}{\rho^2} \frac{\partial \rho}{\partial y} + \hbar^2 \frac{\partial}{\partial y} \left[\frac{1}{2\rho^2} \frac{\nabla^2 P^{1/2}(y)}{P^{1/2}(y)} \right] \right) \Delta t \tag{5.98}$$

What needs to be explained is that since this discussion is about nonlinear variables, the thinking logic of the time is to treat irregularity as random, and nonlinearity does not follow the linearized regularity, we take the probabilistic density P as a nonlinear function with the following expressions:

$$P^{1/2}(x) = \sin(x^4)$$
$$P^{1/2}(y) = \sin(y^4)$$

(5.99)

Let us assume that the spatial distributions of the physical quantities take the harmonic function form, that is,

$$\partial u / \partial x = \sin u, \; \partial u / \partial y = \cos u; \; \partial w / \partial x = \sin w, \; \partial w / \partial y = \cos w; \partial \rho / \partial x = \sin \rho,$$

$$\partial \rho / \partial y = \cos \rho; \; \partial u / \partial x = \sin U, \partial U / \partial y = \cos U$$

Substituting these expressions into Equation (5.97) produces the nonlinear difference equations that we will use to discuss the mutual reactions between flows and waves:

$$u_{(n+1)} = u_{(n)} + \left(-u_{(n)} \sin(u_{(n)}) - v_{(n)} \cos(u_{(n)}) + \frac{\hbar^2}{\rho^2} \left(12x \cot(x^4) - 24x^5 (\csc(x^4))^2 - \right.\right.$$

$$\left.\left. 48x^5 \right) - \frac{\hbar^2}{\rho^3} (12x^2 \cot(x^4) - 16x^6) \sin(\rho) \right) \Delta t$$

$$v_{(n+1)} = v_{(n)} + \left(-u_{(n)} \sin(v_{(n)}) - v_{(n)} \cos(v_{(n)}) + \frac{\hbar^2}{\rho^2} \left(12y \cot(y^4) - 24y^5 (\csc(y^4))^2 - \right.\right.$$

$$\left.\left. 48y^5 \right) - \frac{\hbar^2}{\rho^3} (12y^2 \cot(y^4) - 16y^6) \sin(\rho) \right) \Delta t$$

(5.100)

Other than the symmetry, the purpose of requiring the probabilistic density P is considering the physical concept of third-order derivatives and that the interaction between currents and waves is also an unsettled problem in modern science. Mathematical third-order derivatives express shear stress forces of physics, which are also known as changes in curvature, or torsions (or spinnings) or twist-ups.

The left-hand side of Equation (5.95) together with the first term from the right-hand side constitutes an equation that in form is analogous to a nonlinear equation of fluid motion. According to the interaction of currents and waves of nonlinear equations of fluid motion and other physical properties, such as transformations of rotational currents, from the previous discussions it follows that wave motions are only local and limited special cases under the condition of continuity of the equations. Therefore, if we treat the second term on the left-hand side as an interaction between microscopic current and wave, where the spatial distribution of the speed is seen as a wave and $\bar{v}$ itself a current, then we obtain a scientific proposition about interaction between currents and waves that is still an unsettled problem in the traditional system of wave motions. However, the wave motion existing in the current-wave interaction eventually disappears in the rotational current, which has been well observed in the objective world or laboratory experiments.

Our purpose here is to study the twist-up of third-order derivatives and its effects under the influence of a strong potential field. We will involve the Schrödinger equation with the probabilistic density P a deterministic function of the form as in Equation (5.99) without any probabilisticity. As a problem of fluids, whether from the rotational fluids experiment (the dishpan experiment), realistic observations of flowing currents in rivers, or satellite cloud charts, what is often observed are curvatures of whirlpools and phenomena of variant twist-ups, which seem to have a direct connection with disastrous earthquakes and weather conditions. In the following discussion, we employ the Schrödinger equation as our example because its altered version contains a third-order derivative, which no doubt will bring forward new and deepened understanding about movements of materials.

5.6.2 The Numerical Experiments

For the convenience and organization of our discussion, the following presentation is organized into two cases.

5.6.2.1 The Function of the Quantum Effects

The initial value is positive with small absolute value, and the conditional value is negative with small absolute value. In this case, we take $u(1) = 2$, $v(1) = 3$, $\rho = 0.000002$, $x = -1$, $y = -2$, and $n = 50,000$.

The initial value is positive with relatively large absolute value, and the conditional value is negative and close to zero. In this case, we take $u(1) = 20$, $v(1) = 30$, $\rho = 0.000002$, $x = -1$, $y = -2$ and $n = 50,000$.

5.6.2.2 The Combined Impact of the Intensity Pushing of Potential Field and Quantum Effects

The initial value is negative with relatively large absolute value, and the conditional value is relatively large and positive. In this case, we take $u(1) = -20$, $v(1) = -30$, $\rho = 0.000002$, $U = 100$, $x = 50$, $y = 20$, and $n = 50,000$.

The initial value is negative and close to zero, and the conditional value is negative and further away from zero. In this case, we take $u(1) = -2$, $v(1) = -3$, $\rho = 0.000002$, $U = -100$, $x = -50$, $y = -20$, and $n = 50,000$.

Both the initial and conditional values are negative and close to zero. In this case, we take $u(1) = -2$, $v(1) = -3$, $\rho = 0.000002$, $U = -1$, $x = -1$, $y = -2$, and $n = 25,000$.

Both the initial and conditional values are positive and close to zero.

The numerical experiment with Chiang unit volume density. In this case, we take $u(1) = 2$, $v(1) = 3$, $\rho = 0.000002$, $U = 1$, $x = 1$, $y = 2$, and $n = 25,000$.

What needs to be clear is that this entire experiment consists of 16 difference scenarios. In order not to make this presentation too long, we will in the following only list the experimental outcomes of six scenarios with main figures presented.

5.6.2.3 Numerical Experiments on the Impact of Quantum Effects

The initial value is positive and close to zero, and the conditional value is negative and also close to zero. This experiment shows the pathological quantitative computations (chaos) of large quantities with infinitesimal differences and the process of how the computation becomes instable. The figures are omitted here.

The initial value is positive and relatively large, and the conditional value is negative and close to zero. This case very clearly shows the phenomenon of instable twisting. In other words, pure quantum effects should be rotational eddy currents, revealing irregularities. When $\Delta t = 0.1549689812$, Figure 5.1b shows that when the line segment walks out of the (chaos) concentrated area, it only shows "aging" and will not "die;" that is, the right-hand side of Figure 5.1a is not completely blackened.

When $\Delta t = 0.1549689822$, from Figure 5.1d, it can be seen that when the instability is concentrated in the negative-value area, the line segment walks to "death"; the corresponding Figure 5.1c is completely blackened (died). When the time step is increased to $\Delta t = 0.1549689823$, from Figure 5.1f, it can be seen that the line segments are gathered in the positive-value area; Figure 5.1e shows a region with scattered segments at the right upper corner (a region of rebirth).

When the time step $\Delta t = 0.1549689833$, from Figure 5.1h, it can be seen that the line segment walks out of chaos and moves to the positive-valued region. The corresponding Figure 5.1g shows a shrinking black area. When the time step $\Delta t = 0.1549689843$, the situation is similar to that of when $\Delta t = 0.1549689812$, the segment walks out of the (chaos) concentrated positively valued region, and the functional phase space graph only shows "aging" without any indication of "death." When $\Delta t = 0.1549689848$, the line segments concentrate irregularly in the positively valued area (Figure 5.1j), while the corresponding Figure 5.1i is completely blackened (death).

What deserves our attention is that this experiment indicates that the torsion of the quantum effects functions as an irregular flow under the influence of twisting and spinning. And the darkened block at the quasi-(0,0) region of the phase space is

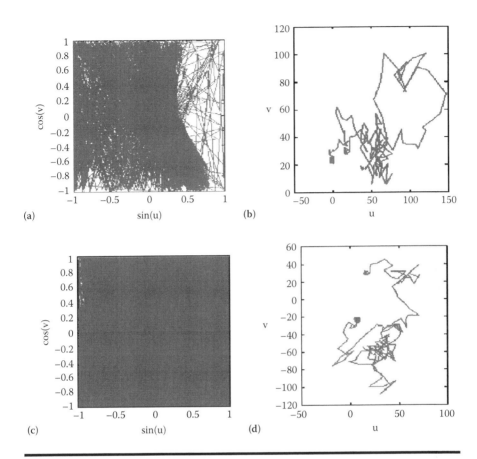

Figure 5.1 **The case involving positive and relatively large initial values with a nearly zero, negative conditional value.**

the so-called chaos caused by large quantities with infinitesimal differences. What has to be clear is that irregular flows should not be comprehended as probabilisticity and are originated in the twist-up motions in the form of the third-order derivatives of the nonlinear function. So, there are causes for irregular flows to appear. Evidently, events with causes should not be seen as random.

If we speak in more ordinary language, the so-called quantum effects of free particles are rotational irregular flows, which can create twist-ups due to rotationality at high speeds, leading to transitional and disastrous changes. And, "true freedoms" are also a reason for changes to occur, since in the physical world no long-lasting state of motion exists without blockage to appear sooner or later.

Or, speaking differently, identifying irregular disorderliness with indeterminacy is an epistemological problem. Even if it is called a disorder, it is still an objective existence and should not be treated as indeterminate. In particular, disorder is about changes, which have already warned about the variancy of the underlying events.

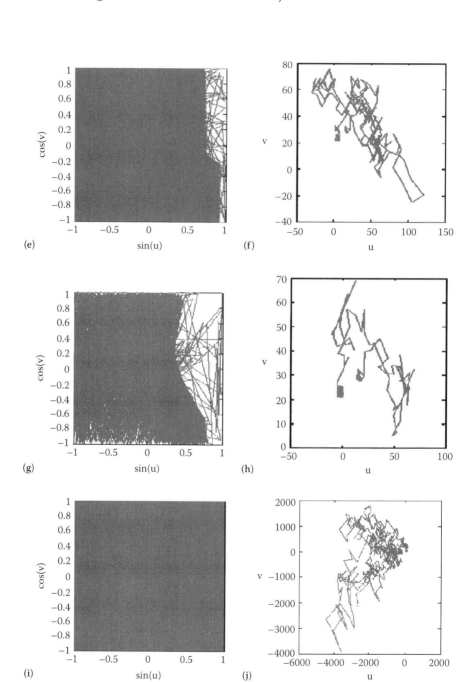

Figure 5.1 (*Continued.*)

The central problem of the so-called determinacy and indeterminacy is whether or not causalities exist in the process of changes in events instead of whether or not we, the humans, sense the regularity or irregularity. Even if it is an irregular or complex event, as long as the cause of change and the results can be clearly expressed, it is determinant. In essence, this is the true meaning of the Schrödinger equation and the problem this equation intends to reveal.

5.6.2.4 Pushing of Potential Field Intensity and Combined Quantum Effects

According to the form of the Schrödinger equation, let us take the unit volume's density ρ as a constant. That is, the first term on the right-hand side of Equation (5.95) stands for the gradient pushing force of the intensity of the potential field. If the unit volume density ρ is not a constant, we have to give ρ a function form, which makes Equation (5.95) a nonclosed equation. Considering the essence of the probabilistic waves from Schrödinger equation (5.94), the unit volume density can be seen as a constant so that we can analyze its combined function with the quantum effect term.

In this case, assume the initial value is negative and relatively far away from zero, and the conditional value is positive and relatively large. Since the time step Δt is relatively small, the corresponding growth in the numerical integral is not major either. When $\Delta t = 0.000000001$, the computational values quickly approach the quasi-region at $(0,0)$ as a straightforward drop (Figure 5.2b). The corresponding Figure 5.2a still shows some light areas without being completely blackened (death). When $\Delta t = 0.0000000016$, the computational results are similar to those in Figure 5.2, except the initial part of the computed values is enlarged when compared with Figure 5.2b; see Figure 5.2c,d. When $\Delta t = 0.0000000016$, there appear in Figure 5.2f quasi-regular density flows analogous to probabilistic waves, and the corresponding Figure 5.2e is completely blackened, indicating a death. When the time step is increased to $\Delta t = 0.000000004$, although the phase space Figure 5.2h still shows a quasi-regular density flow that is analogous to a probabilistic wave, this graph is relatively denser than that in Figure 5.2f and approaches irregularity, where Figure 5.2g is similar to the situation in Figure 5.2e. As a matter of fact, as the computation reaches the step $n = 25,000$, it has already been completely blackened (death) (in our computations, to keep the uniformity, we maintain $n = 50,000$ for different time steps). So, our work indicates that it is the instability caused by irregular flows that speeds up the "death." Since this conclusion is very different from what is conjectured before and touches on a necessary change of the conventional epistemological concepts, we carefully and rigorously repeated our experimental design and the details of the computations in order to verify that the results shown in the graph series in Figure 5.2 are not accidental.

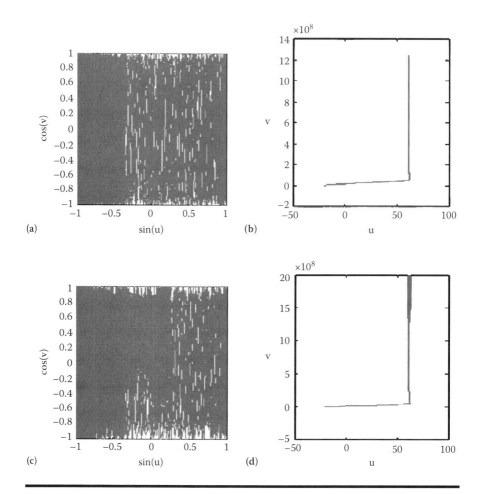

Figure 5.2 **The case where the initial value is negative and far from zero and the conditional value is >0, and relatively large.**

Assume both the initial and conditional values are negative, with the former close to zero and the latter further away from zero. In this case, when $\Delta t = 0.0000000016$, we obtain Figure 5.3b that is opposite of Figure 5.2d, and so Figure 5.3a is also opposite. When $\Delta t = 0.000000004$, Figure 5.3d also shows a quasi-regular density flow that is analogous to a probabilistic wave, and Figure 5.3c is entirely blackened due to instability. When $\Delta t = 0.000000001$, Figure 5.3f becomes denser and deformed and approaches an irregular flow, and the corresponding Figure 5.3e is completely blacked out (death). When the time step is increased to $\Delta t = 0.000000003$, Figure 5.3h is similar to the previous situation in Figure 5.2h, with a thicker density flow and Figure 5.3g is the same as Figure 5.3e. In short, when compared with the previous scenario, in this case, the evolution has been sped up.

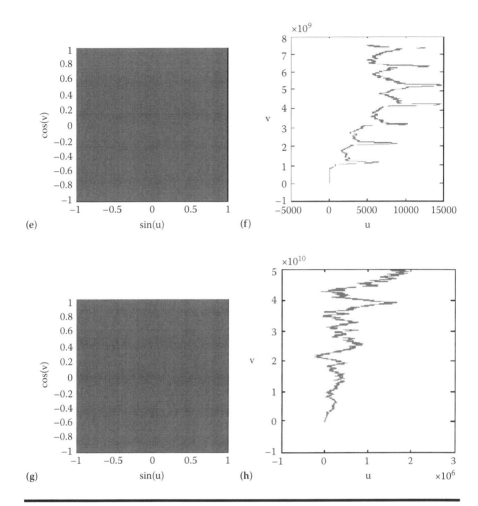

Figure 5.2 (*Continued.*)

Both the initial and conditional values are negative and close to zero. In this case, we half the total number of computation steps to $n = 25,000$. It is not hard to see that under this condition of relatively weak initial value parameters, it becomes easier to produce density flows (Figure 5.4b). And when the time step is relatively large, say, $\Delta t = 0.0000013$, we can still see the situation of thicker density flow in Figure 5.4h. All other figures are similar to those in the Figure 5.3 series, and are omitted.

Both the initial and conditional values are positive and close to zero ($n = 25,000$). In this case, when the time step Δt is relatively small, combined with relatively small initial and conditional values, the evolution values, there will not appear any form close to a probabilistic wave. Such form of quasi-regular density flow only appears when the time step is relatively large. All the figures are omitted.

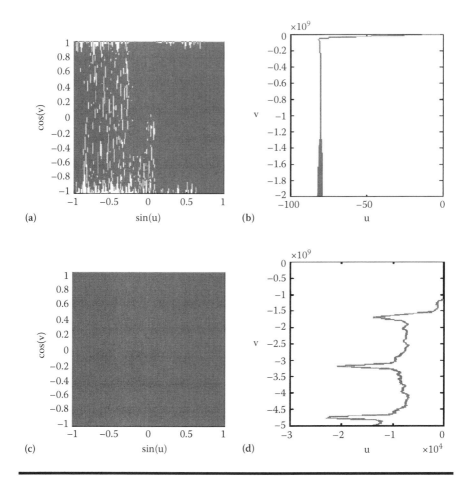

Figure 5.3   The case with negative initial and conditional values such that the former is near zero and the latter is away from zero.

What is presented above is our numerical experiment on the altered equivalent Schrödinger equation. What is shown is that the so-called probabilistic waves are products under certain conditions. Even by taking the unit volume density ρ as a constant, as what Schrödinger did, probabilistic waves could relatively easily appear only when the intensity of the initial and conditional values are relatively large. However, what needs to be pointed out is that under certain conditions, although the quasi-regular density flows in form are analogous waves, their physical essence is irregular flows caused by the twists of the quantum effects. Considering the quasi-regular density flows, created by the mutual reactions of the quantum effects and the push of the intensity of the potential field, it can also be said that the essence of probabilistic waves stands for a form of motion caused by the quantum torsion effects when interfered by the potential field and has nothing to do with the

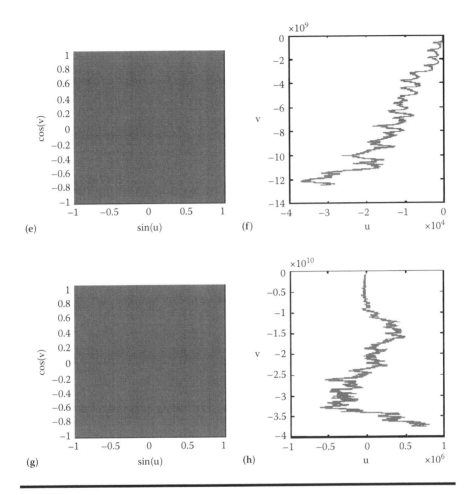

Figure 5.3 *(Continued.)*

so-called probabilisticity. With this understanding in place, "probabilistic waves" are neither waves nor a problem of indeterminate probability. What is more important is that they are products under special conditions without much generality.

No doubt, from our repeated computations with many different scenarios, it can be seen that the quasi-regular density flows created out of the irregular flows of the quantum effects under the pushing influence of the potential field intensity, as a great achievement of modern physics at the start of the 20th century, cannot be used as the evidence of a probabilistic universe, a product of the belief that God does play dice. And, Einstein no longer needs to carry the name of an old diehard who did not give any space for accidentality. Or in other words, when people are often talking about the probabilistic universe, the true achievements of quantum mechanics did not actually deliver such information.

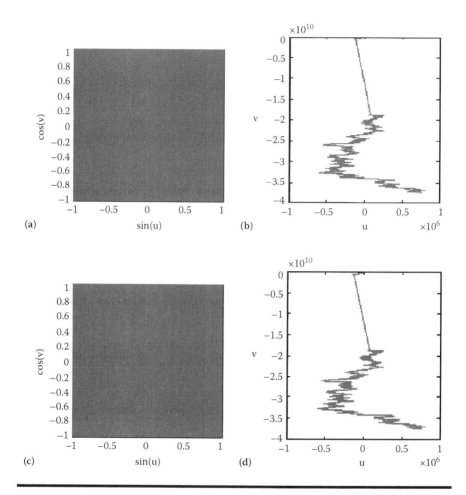

Figure 5.4 **The case for positive and near-zero initial and conditional values.**

5.6.2.5 *Numerical Experiments with Changing Unit Volume Density*

Based on what is obtained above, this work considers the numerical experiments of variable unit volume density. The results follow.

When $u(1) = -20$, $v(1) = -30$, $x = 50$, $y = 20$, $\Delta t = 0.0000004$, and $n = 50,000$, we adjust both ρ and U and produced the following results.

When $U = 100$ and $\rho = 0.00000002$, Figure 5.5b shows the strengthening irregularity in the density flow, making the density flow become irregular; the situation is similar to that in Figure 5.2h. The corresponding Figure 5.5a is completely blackened (death). When $U = 1$ and $\rho = 0.000002$, the computational results are similar to those of when $U = 100$ and $\rho = 0.00000002$, where a density flow with strengthening irregularity in the (u-v) space also appears and approaches an irregular flow.

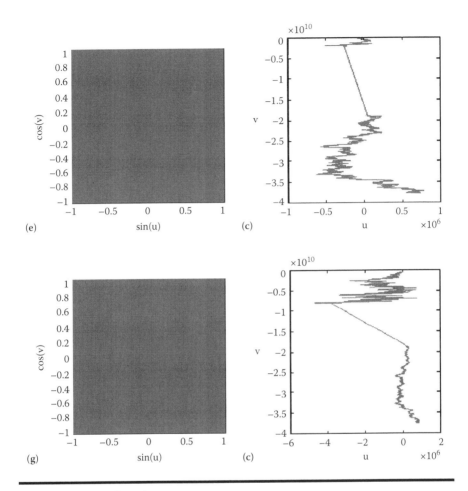

Figure 5.4 (*Continued.*)

The corresponding (sin*u*-cos*v*) space is completely blackened. When $U = 100$ and ρ = 0.0002, in Figure 5.5f there appears a quasi-regular density flow that is analogous in form to a probabilistic wave. And the corresponding Figure 5.5e is completely blackened. This end indicates that the "probabilistic wave" — the quasi-regular density flow — is both conditional and nonaccidental. Its definite physical meaning is the outcome of the mutual reactions between the potential field intensity and the quantum effects.

Based on an altered version of the Schrödinger equation, we discussed the twist-ups of third-order derivatives of nonlinear variables and its structural characteristics of its interaction with the potential field using numerical experiments. Our conclusions can be summarized as follows.

Under the impact of quantum effects (torsions), our experiments show the relatively common phenomenon of irregular flows, leading to the problem of how to

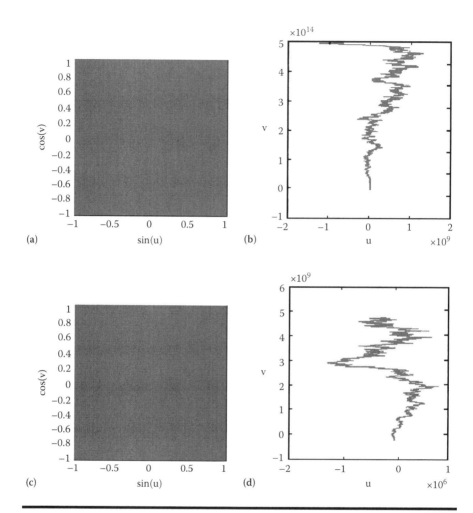

Figure 5.5   Experiments of variable unit volume density.

comprehend irregularity. When the numerical values are relatively small, the pathological computations of large quantities with infinitesimal differences can appear, touching on the problem about the chaos doctrine. This phenomenon reflects exactly the problem of computational inaccuracy of quasi-equal quantities and cannot be seen as a characteristic of instable nonlinearity. It is neither a problem of physics nor the philosophical chaos and cannot be treated as a probabilisticity.

Considering the pushing effect of a strong potential field or the combined impact of the potential field intensity (a strong pushing force) and quantum effect (a twisting force), quasi-regular density flows — probabilistic waves — can appear. In form, they look like waves. However, their physical essence is neither a wave nor probabilistic. The reason why quantum mechanics is different from classical

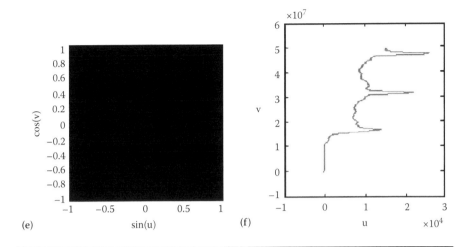

Figure 5.5 (*Continued.*)

mechanics, instead of its probabilisticity, is the attribute of rotationality of nonlinearity and the irregularity of flows — turbulencality, or the determinacy of the interaction between the rotational irregularity and the pushing of the potential field. The essence of probabilistic waves is the combined effort of the linear pushing of the potential field and irregular quantum flows.

Changing the intensity of the potential field and the unit volume density can all lead to irregular flows, while evolving toward quantum effects (twisting motions). The entire experiment indicates that irregularity flows have the generality and the probabilistic waves of quasi-regular flows will eventually be taken up by irregular flows in the evolution. Rotational materials have to suffer from injuries and damages due to different rotation directions so that irregularity is a must. We cannot treat probabilisticity as indeterminate just because the current quantitative analysis system cannot handle it. Concepts of physics should not be named on the basis of whether or not the mathematical tool can handle it or not.

The reason we once again analyzed the Schrödinger equation and its origin and consequences and conducted the relevant numerical experiments is that nonlinear equations depict transitional changes using rotationality and the special physical meanings third-order derivatives represent. Our work reveals that quantum mechanics did not tell us that all fundamental particles are waves, that the claim that God plays dice has nothing to do with quantum mechanics, and that the so-called wave-particle duality is a problem that needs to be revisited.

Speaking frankly, our interest in the Schrödinger equation is more in its equivalent form, since it contains a nonlinear term with a third-order derivative, which is not so in other well-studied mathematical physics equations. This interest has a direct connection with our 1963 discovery of the ultralow temperature phenomenon existing at the top of the troposphere. It is because twist-ups in airstreams corresponding to the ultralow temperature phenomenon are a piece of indicative information for the occurrence of disastrous weather conditions. Later, this discovery was employed in the studies of earthquakes, disastrous accidents, conflicts, and so forth. Disasters become a challenge to the scientificality of the modern scientific system because, among other reasons, modern science did not uncover and illustrate the physical significance of the mathematical third-order derivatives. That explains why we conducted our numerical experiments on quantum effects, leading to our irregular quantum flows and the investigation on twist-ups. As a matter of fact, the specific effects of nonlinear third-order derivatives are not limited to microscopic quantum problems, producing gradually the recognitions that irregular events are about information of change, that problems of evolution do not involve probabilisticity, and so forth. Beyond our initial expectations, this work also reveals as by-products that probabilistic waves are not truly waves and that quantum mechanics did not really tell us whether or not God plays dice. So, it seems that the significance of the Schrödinger equation is still waiting to be further excavated. What should be noted is that the current equations of fluid mechanics do not contain any nonlinear third-order derivatives. Because of this fact, the wish of depending on these traditional equations to address problems of disasters has essentially been broken. That is why we have turned our attention to the study of irregular information, the physical meaning of events themselves, and practical applications.

5.7 Summary

No doubt, our focus of this chapter is on the problem of nonlinearity with emphasis placed on how to uncover the physical essence of events and nonlinearity and whether or not nonlinearity can be transformed into linearity. After collecting as much as possible the previously published works, and on the basis of our own work, we explored the theory and methods of nonlinearity. Our discussion can be summarized into the following few groups.

The problem of nonlinearity should have helped science enter into the era of evolution science. However, due to the historical inertia of holding on to the glory of the wave motion system, a contemporary stream of consciousness has been formed. It was exactly because the scientific community of the 20th century believed in the generality and commonality of wave motions that nonlinear problems were popularly linearized. Even after the circulation theorem, established in the system of modern science, pointed out that the physical significance of nonlinearity is about the rotationality under stirring effects, no major attention was caught so that in the ending part of the 20th century, the scientific community was still chasing after the fad of wave motions. Einstein's theory of gravitations, Schrödinger's probabilistic waves and others further helped to expand and continue the tide of thoughts of wave motions.

In particular, in terms of experiments and engineering, nuclear physics has entered the system of transmutations of microscopic particles, while the works of Schrödinger and his followers are still lagging behind in the system of wave motions. It is not the case that the system of wave motions and its theory come from the concept of probabilisticity; instead, it was Schrödinger and others who first introduced the probabilisticity for quantum densities. Or speaking differently, they presumably treated the irregularity of fundamental particles with probabilisticity in their recognitions. No doubt, for the original Schrödinger equation, during the entire mathematical modeling process from the introduction of physical quantities to the construction of the equation, he well followed the quantitative formal logic of the Newtonian system. What he did differently from the ordinary procedure of the system of wave motions is that he first assumed that the energy and momentum were wave functions of physical parameters before he established the mathematical model. So, the probabilisticity of probabilistic waves came from the predetermined belief and thoughts, which is different from the situation that the classical string vibration is obtained as the solution of a linear equation. What Schrödinger did provided the first incidence for others to treat nonlinear equations of fluid motions in a similar fashion, leading to such a concept as the inexplicable unidirectional retrogressive waves.

Therefore, the probabilisticity of probabilistic waves is not a product of the Schrödinger equation; instead, it came from how to understand the irregularity of information and how to deal with irregular events.

The Bjerkness's circulation theorem first appeared in meteorology. By employing circulation integrals of nonparticles and still using the methods of modern science, Bjerknes had already walked out of the first push system. However, 40 years after the Bjerkness's circulation theorem was initially published, the nonexisting unidirectional retrogressive waves were treated as a fundamental scientific theory. Let us present what we have obtained along this line in a formal language: V. Bjerkness's circulation theorem has already resolved the nonlinearity problem of interactions; at the same time, it transforms the Newtonian first-push system into a second stir so that science has walked out of the inertial system; as for the nonlinearity of the form of motion of the second stir system, it can be shown by using elementary vector calculus that it is the rotational movement of the rotationality of solenoidal fields. So, the 300-plus-years-old mystery of nonlinearity is completely resolved. Or speaking more intelligently, quantitative nonlinearity together with the irregularity of events is more about walking out of the quantitative analysis system while entering the structurality of events themselves.

So, the key for any scientific revolution is a reform in concepts; and each reform in concepts is originated from some pressuring needs or challenge of solving practical problems. Since we discovered the connection between twist-ups of ultralow temperatures existing at the upper layer of the troposphere and disastrous phenomena, we are fortunate to have grabbed the opportunity to reconsider some of the problems existing within modern science with the focus placed on how to understand nonlinearity. As a matter of fact, by employing the vector analysis of the modern scientific system, we have shown very clearly that nonlinearity stands for the rotationality of solenoidal fields, which naturally destroys the continuity of the basic field so that we easily entered the stirring rotations and the area of evolution science. It might be because the scientific community cannot bear to part with the system of continuity that it missed many great opportunities so that problems of materials' rotations are still stubbornly treated as those of wave motions, leading to many jokes. To this end, the worries of Einstein (1976) in his old age cannot be thought as probabilistic: "God plays dice. I believe it is very possible that physics is not established on the concept of fields, that is, not on the system of continuity. If it is so, then all my castles in the air … including the theory of gravitation … together with other modern physics will be all gone."

Considering the 20th century, due to the need of applying mathematical techniques, the concept of mathematical waves once appeared, leading to a whole series of problems concerning its treatment of all kinds of disturbances as waves. So, at this junction, we purposely list the physical characteristics of eddy motions as follows:

- Each eddy motion is a unidirectional curvilinear flow with the irrecoverability and transformability of its structural morphology and attributes.
- There exist nonsmall disturbances and irregularities of disturbances in each eddy motion.
- Each eddy motion can transport both energy and material at the same time; and due to the duality of eddy rotations, during transportations of energy and materials, the morphology and attributes of the underlying materials can undergo changes.
- Eddy motions and stirring motions occur simultaneously with rotation direction appearing ahead of quantities.

Because of the duality in eddy motions, different spinning directions will surely create discontinuities, leading to injuries and damages in the underlying materials. That explains why irregular and small probability events occur, evolve, and disappear, representing physical processes.

With rolling movements of eddy motions, different events and different information appear. That leads to timely differences of events. Problems of eddy motions touch on the problem of what time is, which is still an unsettled open problem in modern science, reveal events' past ≠ the present ≠ the future, and walk out of the inertial system while entering evolution science.

What needs to be emphasized is that how to correctly understand time is a fundamental problem of the fundamental science of all sciences. It can be said that being unable to answer what time is is equivalent to the fact that mankind has not entered the door of science. That is the reason why we have listed the fundamental physical characteristics of eddy motions. No doubt, scientific development is based on the progress in our epistemological concepts. Correspondingly, what time is, as a proposition, is not only about its epistemological, conceptual subvertibility, but also about its important practicability.

Acknowledgments

The presentation of this chapter is based on Akimov and Shypov (1997), Bergson (1963), Born (1926), de Broglie (1925), Einstein (1976), Gardner et al. (1967), Gu (1978), Hawking (1988), Heisenberg (1927), Koyré (1968), Li and Guo (1985), Lin et al. (2001), Mei (1985), OuYang (1984, 1986, 1992, 1994, 1995, 1998b, 2000, in press), OuYang et al. (2001, 2002a,b, 2005a,b, in press), OuYang and Lin (2006b, in press), OuYang and Wei (1993), Prigogine (1980), Schrödinger (1926), Wang (1982), Wang and Wu (1963), Whitham (1999), and Wu and Lin (2002).

Chapter 6

Nonlinear Computations and Experiments

Initially, computers were connected with electronic technology through increased speed of computation. As for the underlying principles of computation, they are based on those of quantities. More specifically, computers' capabilities of transfer and handling numerals come from an application of the concept of yin and yang "- -, —" of the *Book of Changes* (Wilhalm and Baynes, 1967). Since quantitative operations are only one of the functions of "- -, —," or in other words, the capabilities of computers are far beyond quantitative computations, even in terms of quantitative capabilities, computers should be seen as information handlers or an "electronic brain," as known in Chinese, which seem to be more appropriate considering what computers can do.

However, the initial motivation for inventing computers was mainly aimed at solving problems of quantitative computations. Because of their high speed, in principle, there is not any problem where quantitative computation works. That is why corresponding computational mathematics is considered a branch of mathematics. The present problem is whether or not there is an incomputable problem in computation. Or speaking more precisely, what problems are incomputable even if computers are employed?

From previous discussions, it can be seen that people used to believe that problems written in mathematical equations should all be computable, leading to conviction that equations are eternal. Evidently, the eternity of equations should be embodied in computability. If an equation suffers from incomputability or cannot be computed, then this equation becomes meaningless. No doubt, science and technology are about their use value. Equations are not about their "artistic value."

So, the value of existence of equations is not only about their computability but also the desired results one can produce out of equations. However, in the present science and technology, not only incomputable equations exist, but also the phenomena of investing huge amounts of manpower and capital in the computation of incomputable equations. Equations involving large quantities with infinitesimal differences or differences of quasi-equal quantities represent one class of equations scholars are still calculating. Speaking in ordinary language, each equation of large quantities with infinitesimal differences can be seen as one involving $x - y = z$ when $x \approx y$, which stands for a meaningless equation or an incomputable equation. It is an incomputable equation because its practical significance is equivalent to the pound of meat Venice merchants could never provide, as written in Shakespeare's plays. Such a problem of computational inaccuracy of quasi-equal quantities evolved into the chaos theory in the latter part of the 20th century. It should be made clear that the chaos theory was developed to coincide with the claim of modern physics that God plays dice by treating ill-posed quantitative instable problems of mathematical physics equations and irregular events as random. Consequently, large quantities with infinitesimal differences or problems of inaccurate computations of quasi-equal quantities were seen as irregular events without touching on the irregular mapping events underlying the quantitative instability, because in the computations, people ignored the quantitative infinity problem that modern methods could not really avoid. This is why in previous discussions, when we talked about the problem of rotationality, we focused on quantitative instability and the meaning of the quantitative infinity.

Next, based on the terminology of mathematics, ill-posed equations are also incomputable. Of course, the concept of mathematical ill-posedness is not just about problems of quantitative computations. Other than computational stability, mathematical ill-posedness also contains the problems of the solutions' existence and uniqueness. The essence of the stability of quantitative computation is to satisfy the initial-value existence theorem, which involves problems of calculus through differential equations. It is well known that the fundamental requirement for calculus to hold true is the assumption of continuity. However, in the past 300-plus years, people seemed to have forgotten one important problem. At the same time, when people celebrate that the results of calculus have nothing to do with paths, they forget that because of continuity, great difficulties for people to practically apply calculus are created. It is because between any two points, infinitely many continuous points exist; even if one integrates a point each day starting on one point, he will never be able to reach the other point. So, in applications, calculus has been employed in the fashion of big steps. Or in other words, the central problem of practical calculus is caused by the invariancy of the original function, which is the essential reason why results of calculus have nothing to do with the path taken. This is not only embodied in the well-posedness of calculus of satisfying the initial-value existence theorem but also maintains the nonevolutionality of system of modern science.

In ill-posed mathematical physics equations, one experiences the problem that nonlinear mathematical physics equations face with the problem of stability in solutions. The corresponding difference equations, developed for taking advantage of computers, also meet with the problem of solutions' stability, constituting the central problem of numerical computations. It can also be said that the root of the stability problem of quantitative computations is a continuation of the stability problem of the deterministic solutions of mathematical physics equations; and the key of the problem is not with the quantitative computations themselves. So, in terms of the computations of quantitative schemes, we must resolve epistemologically the essence of the so-called stability problem. Or, we should unveil the basic meaning delivered by the concept of stability.

No doubt, other than the existence and uniqueness of the solutions, the core of the well-posedness theory of mathematical physics equations involves the stability under initial-value automorphisms, where when the initial values change slightly, the corresponding solutions are also allowed to alter slightly. In the numerical implementation of the computational stability of difference equations, this principle has to be followed. To this end, a quite good amount of contents in computational mathematics is about how to design computationally stable difference schemes. For example, beyond discussing the agreement and convergence between difference equations and the original differential equations, the central problem is about computational stability other than spatially forward difference, central difference, and backward difference schemes. Many computational schemes have been developed, such as dissipation (Lax–Wendroff), repeated substitution, implicit/semi-implicit schemes, smoothing, filtering, backward time difference, central time difference, energy conservation, reciprocating substitution, simplified equations, initialization and inversed integrations, and innumerable others. In order to achieve stability, deterministic solution problems of mathematical physics equations were even changed to that of solving for the desirable initial values by inverse integration based on the known circumstances as the solutions. That is, the purpose of doing so is to explore what kind of stable initial values should be needed for a given solution, which in essence is about how to fabricate the initial value according to the realistic situation. Based on our earlier discussions, the truly meaningful research should be about the cause behind the computational instability and then how to deal with it.

What has to be pointed out is that quantitative instability collectively appears with nonlinear mathematical models, and computational instability of nonlinearity has nothing to do with truncation and rounding errors. This problem should be an important research topic in computational mathematics. However, in the area of computational mathematics, we have not seen any work remotely related to this class of problems. Because of this, we have spent a good amount of time looking for the reason behind instabilities of nonlinear problems and the fundamental characteristics of numerical computations of nonlinear problems.

After several years of repeated numerical experiments, we discovered that the instabilities appearing in nonlinear quantitative computations originated from products of nonsmall quantities, which is totally different from the results of products of small values in nonlinear problems. Here, products of nonsmall effective values can lead to explosive quantitative growth, which can quickly force the working computer to spill, while products of decimal effective values create an explosive decrease, approaching quickly to zero. So, nonlinear computational instability is different from that of linear equations. This end has been addressed in Chapter 3.

Next, nonlinear equations do not satisfy the initial-value existence theorem and experience the problem of transitionality due to quantitative instabilities. For example, let us look at the simple and standard mathematical model of the quadratic form as studied in Chapter 3 and Figure 3.4 as follows:

$$\dot{u} = u^2 + pu + q \tag{3.4}$$

When $p^2 - 4q = 0$, we have

$$u = \left(\frac{1}{t+c} + \frac{p}{2} \right) \tag{3.5}$$

According to $c > 0$ or $c < 0$, we have Figure 3.4 from Chapter 3, where both I and III are hyperbolas and can experience a spatial transformation of the hyperboloid of the tire as shown in Figure 3.5 (see the relevant discussion in Chapter 3). No doubt, this end has very clearly shown that the quantitative evolutionary infinity of nonlinear models stands exactly for problems of transitional changes in curved spaces, since between the hyperbolas I and III there are not only quantitative changes that are different from the initial values but also discontinuities, which are referred to as blown-ups of transitional changes. So, even in terms of mathematical properties, nonlinear mathematical models no longer satisfy the linear initial-value existence theorem. Thus, the proposition of well-posedness in the form of linear equations itself no longer agrees with the formal logic of mathematics. As a matter of fact, after the fact is pointed out that the deterministic solution problems of quantitative nonlinear equations contain quantitative instability, and the essence that through products of nonsmall quantities the computations can quickly evolve into quantitative unboundedness, if one continues to apply dissipation (Lax–Wendroff), repeated substitution, implicit/semi-implicit schemes, smoothing, filtering, backward time difference, central time difference, energy conservation, reciprocating substitution, simplified equations, initialization and inversed integrations, and innumerable other methods developed for linear systems to deal with nonlinear problems, one is no longer attacking the original problem of his concern. Evidently, nonlinear equations implicitly contain transitional changes in curved spaces; continually applying the concept of quantitative computational stability to limit quantitative instability will constrain or eliminate transitional changes. So, the essential difference between nonlinear and linear problems is very clear.

Due to this understanding, the essence of the current study on the stability of deterministic problems of mathematical physics, the determination of initial values using the method of inverse problems of mathematical physics equations, and other related studies betray the transitionality of nonlinear problems and constrain changes in variables by using the stability concept of linear problems. Although modern science believes in the philosophical principle that quantitative changes lead to qualitative changes, no quantitative changes are allowed in practical applications. It might sound ironic; however, it is indeed the true face of modern science that modern science essentially maintains the realm of eternity.

What is worthy of explaining is that it was British scholar R. F. Richardson who first attacked the quantitative instability of mathematical nonlinearity. During 1916 to 1922, he conducted a series of audacious numerical experiments on equations of fluid mechanics. His results indicated that within 6 hours, changes in barometric pressure reached as much as 145hPa. In terms of numerical computations, his work could not be seen as erroneous; instead, he had revealed the problem of essence regarding nonlinearity. However, scholars in the meteorological science of that time did not pursue what caused the quantitative instability of nonlinearity and what irregular information was. In particular, Charney, an American meteorologist, later badly criticized Richardson in the name of not understanding what meteorological noise (irregular meteorological information) was, which in fact not only revealed how little Charney knew about meteorological noise but also indicated that the meteorological community of the time indeed did not pay much attention to the question of what irregular information was, leading to the abandoning (as of this writing) of irregular records existing in sounding data in the name of "noise." It should be noted that only a few scholars as of today recognize that irregular information represents information of change.

Note that traditionally in meteorology, pressure systems have been considered to be the cause of weather phenomena, and since the 1940s, each pressure system has been expressed as a system of wave motions. In fact, such a comprehension of the meteorological community is still stagnated at the early stages of scientific history, because the essential reason for atmospheric movements is uneven distributions of and changes in the distribution of heat, leading to the later uneven distributions of and changes in pressure densities. Furthermore, studies of meteorological science assume that the atmospheric density is constant, which explains why atmospheric pressures always lag behind weather phenomena. This is why experienced front-line meteorologists know that only after rains can one see pressure systems. This is analogous to trading stocks, where the rise and fall of the market are determined by the amount of capital flowing into or out of the market, which explains that "vibrations" of the pressure systems are the consequences instead of the causes. We have used the stock market many times as our example to illustrate that the current method of weather forecasting, using pressure systems, has mistakenly ordered the causality relationships in the wrong direction, leading to the inability of pinpointing major, disastrous weather conditions. It can be said that the current commercial weather forecasts have treated living reports and monitoring as predictions. To provide true forecasting, the front-line meteorologists must first understand the correct causalities. Evidently, without seeing the essential difference between monitoring and predicting or without being able to correctly understand the causal relationships, there will not be any substantial progress made in the science of prediction.

6.1 Mathematical Properties and Numerical Computability of Nonlinearity

Earlier it was made clear that computations of nonlinear difference equations possess special characteristics different from those of linear equations. However, such an important fact is not well illustrated in the corresponding methods of computations. This is why in quantitative computations, people are still employing the methods and concepts developed for linear problems. In particular, the concept of computational stability itself is from the well-posedness of linear mathematical physics equations, although the well-posedness of mathematical physics equations is meant to be the existence, uniqueness, and stability of the solutions. However, if the solution of an equation does not exist, there will not be any need to talk about the solution's stability. That is why stability is not the central problem of mathematical physics equations. The corresponding concept of the deterministic problem of difference equations of computational mathematics is different from that of mathematical physics equations and implies consistency, convergence, and stability. If the corresponding difference equation is not consistent with the original equation or its solution is not convergent, then there will be no sense in talking about the stability of the solution. So, the problem of stability is not central in the study of difference equations either. Therefore, after all, the concept of stability in computational mathematics is nothing but a continuation of that of linear mathematical physics equations. Because of this realization, there is a need to carefully consider the characteristics of instability of nonlinear equations, which might shed light on the problem of stability of nonlinearity.

To understand the quantitative instability of the solutions of nonlinear equations, let us look at the nonlinear mathematical model of the Newtonian system:

$$\frac{du}{dt} = F = ku^2 \tag{6.1}$$

We will analyze its essence of stability using different initial values. Rewriting Equation (6.1) as a difference equation, we have

$$u^{(n+1)} = u^{(n)} + k(u^{(n)})^2 \Delt \tag{6.2}$$

where the subscript n stands for time step. Now, the problem of stability can be analyzed in three cases.

6.1.1 Computational Instability of Nonlinearity

Aiming for practical applicability, let us take the effective value of the initial value to the first digit after the decimal point. For example, if we take

$$u^{(0)} = 10.1 \text{ m/s}; \quad k = 1 \text{ m}^{-1}; \quad \Delta t = 10 \text{ s}$$

where m stands for meter, s a second, then substituting these values into Equation (6.2) produces:

When $n = 1$, we have

$$u^{(1)} > 1000 \text{ m/s}$$

When $n = 2$, we have

$$u^{(2)} > 10^7 \text{ m/s}$$

This end has already indicated that under the condition of nonsmall effective quantities, only with two time steps, the quantitative computation of the nonlinearity produces the result in Equation (6.2), which is far greater than the first cosmic speed; that is, a particle of the Newtonian mechanics system has already left the earth and flown into outer space. That is the quantitative computational instability of nonlinearity. It can be seen that the computations are correct so that years ago Richardson should not have been blamed for any computational errors and his results had nothing to do with the so-called meteorological noise. The problem comes from the acceleration in Equation (6.1) that is variant. That is a mathematical property of nonlinear equations we mentioned earlier and a difference between nonlinear and linear equations. In other words, if nonlinearity is introduced into dynamic equations, then problems of constant accelerations will be changed to those of variable accelerations, while variable accelerations belong to noninertial systems. So, nonlinearity in general stands for problems of noninertial systems. This is equivalent to saying that in inertial systems, we cannot introduce nonlinearity; or that nonlinearity cannot be linearized. The present problem is after introducing nonlinear models, the quantitative instabilities of nonlinearity are constrained by using the well-posedness theory of linear systems, which naturally creates an abuse of mathematics within mathematics itself or is referred to as mathematical chaos.

So, when studying nonlinear problems, we have to first be sure what the mathematical properties of nonlinearity are, and second what the corresponding physical significance of the nonlinear mathematical problems is. In terms of mathematical properties, nonlinear quantitative evolutions can experience explosive growth within effective quantities, approaching quantitative infinity and escaping the continuity of the calculus system. What should be noted is that the realistic physical space is a curvature space, while the quantitative infinity of the Euclidean spaces corresponds exactly to transitional points of the curvature space. So, nonlinear problems are problems of non-Euclidean spaces; the corresponding quantitative infinity should no longer be treated as a concept of mathematical ill-posedness or singularity.

It can be shown that nonlinearity is the problem of physical rotationality (Chapter 5), which has already betrayed the Euclidean spaces. Even as mathematical methods, nonlinear problems should not be continually treated with mathematics of continuity of modern science. What should be very clear is that even as mathematics, the quantitative computational instability of nonlinearity should not be seen as meaningless, just because of the appearance of quantitative instability or singularity. Also, problems of non-Euclidean spaces should not be transformed into those of Euclidean spaces. Instead, it is through quantitative instable singularity that at the same time when nonlinearity walks out of Euclidean spaces, it enters a much wider spectrum of non-Euclidean spaces and evolutions of materials in the form of rotations. If speaking more in layman's terms or more intuitively, the reason why the system or the classical mechanics Newton and his followers completed has become a classic is that the mathematical tools of the Euclidean spaces have described the existent (nonevolutionary) physics. However, these mathematical games have left behind many disagreements with the reality without expecting that nonlinear mathematics cannot resolve transitional changes, which correspondingly terminates all of the mathematical games. Through digitization of information, nonlinear mathematics has materially entered noninertial systems. Because people are still behind this call of our modern time and are still accustomed to the traditional concepts of inertial systems and the well-posedness of linear systems, the jump into noninertial systems has been delayed.

6.1.2 Errors of Initial Values

One must understand the problem of numerical computations of nonlinearity. In practical computations, one has to face the situation of decimal effective values in the initial value and/or parameters. To this end, let us look at the case with one effective digit after the decimal point. If we take

$$u^{(0)} = -0.1 \text{ m/s}; \quad k = 1 \text{ m}^{-1}; \quad \Delta t = 6 \text{ s}$$

substituting these values into Equation (6.2) leads to, respectively:

$$u^{(1)} = -0.04 \text{ m/s}$$

$$u^{(2)} = -0.0304 \text{ m/s}$$

$$u^{(3)} = -0.0250 \text{ m/s}$$

At this junction, we should note that the effective value is limited to the first digit after the decimal point; so the precious computational value has been trapped in the errors of the initial values. Since the effective value is limited to the first digit after the decimal point, then all values smaller than the first digit after the decimal point no longer exist or are called ignorable. That is where quantities can be employed to compute nonexisting matters more accurately than those that actually exist. This has been well described by the statement that "not all things numbers can describe" (Zhan Yin, from the time of Warring States of ancient China). What is more essential is that events are not quantities, where using quantities to substitute for events was a revolution that appeared between historical eras. As a new revolution, it seems to be time for us to consider how to understand events themselves and how to directly employ events in applications. Limited by the length of this book, we will not go into details along this line.

The familiar quantitative computations of large numbers with infinitesimal differences all experience the problem of computational inaccuracy of effective values. However, what should be pointed out is that this phenomenon should not be seen as a particular characteristic of nonlinearity, since linear or general equations in the difference of quasi-equal quantities also experience this phenomenon. The chaos theory, which was popular during the latter part of the 20th century, was developed out of this problem of computational inaccuracy involving the difference of quasi-equal quantities. For example, in Equation (6.2), if $u^{(1)} > 0$ and $k < 0$, then even for nonsmall effective values, it can also fall into the calculations of error values. That is analogous to the quantitative indistinguishability problem of Shakespeare's Venice merchants who could never cut off a pound of meat. That is to say, even though the 15th century Englishman knew that numbers could not describe all things, it did not mean that the scientific community truly comprehended the problem.

6.1.3 Infinitesimal Difference of Large Quantities

If we rewrite Equation (6.1) as follows:

$$\frac{du}{dt} = ku^2 + c \tag{6.3}$$

then, its difference equation is

$$u^{(n+1)} = u^{(n)} + k(u^{(n)})^2 \Delta t + c \Delta t \tag{6.4}$$

Taking $u^{(0)} = 5$ m/s, $k = 0.1^{-1}$, $c = -2.5$, $\Delta t = 0.1$ s, and substituting these values into Equation (6.4), we can respectively compute

$$u^{(1)} = 5.0000 \text{ m/s}$$

$$u^{(2)} = 5.0000 \text{ m/s}$$

$$u^{(3)} = 5.0000 \text{ m/s}$$

It is interesting that since c is negative, Equation (6.4) becomes an "infinitesimal difference of large quantities," resulting in an "attractor" of the chaos theory. At this point, the reader can easily tell the essence of the so-called chaos theory. That is, nonlinear forms of small quantities and "infinitesimal differences of large numbers" can make nonlinear explosive growth ineffective. If Equation (6.4) just happens to involve irrational numbers or repeating nonterminating decimal numbers, then it will surely become more of a problem of computational inaccuracy or a typical case where not all things can numbers describe.

So, in nonlinear computations, one sooner or later has to experience particular situations of Equations (6.2) and (6.4). Even when the computations appear to be stable, it does not mean that these results agree with the reality. What "not all things can numbers describe" implies are the shortcomings and weaknesses in the description ability of quantitative calculations of specific matters and events. Evidently, we should not treat such incapability and indistinguishability of quantities as a kind of theory or doctrine. Even when linear computations of quantities are employed, we still can face problems other than differences in computational speed. Therefore, the problem of quantitative indistinguishability is not unique to nonlinearity; only the situation in Equation (6.3) is truly special and uniquely true for nonlinearity and is not shared by linearity. Also, this characteristic is enough to make the stability of nonlinear quantitative computations different from that of linear computations.

By now, it should be clear that nonlinear explosive growth of nonsmall effective values have to destroy the system of continuity and consequently make the traditional Riemann integration scheme ineffective. It should be pointed out that this mathematical property of nonlinearity is a problem of computational mathematics. And the analytic solutions of the corresponding differential equations also behave the same. Hence, the computational mathematics has also met with the reality that the deterministic problems of nonlinear equations cannot be resolved by employing the initial-value existence theorem and then expanding the local solutions to the overall solutions by letting $t \to \infty$. In other words, the whole solutions of the deterministic problems of nonlinear equations do not satisfy the well-posedness theory of the stability and do not depend on the initial parameters and boundary conditions. That is, in essence, nonlinearity has met with the scenario that even with all the initial values known, one still cannot foretell everything of the future, which well demonstrates that nonlinearity has walked out of inertial systems and entered noninertial systems of evolution science. This is, the so-called initialization of numerical schemes or the series of stable numerical integral schemes, developed out of the inverse problem of mathematical physics equations, all lose their validity in front of nonlinear problems. The corresponding conventional quantitative indeterminacies, ill-posedness, and other relevant concepts need to be reconsidered. In particular, nonlinearity should not be seen as indeterminate or random just because of the associated quantitative indeterminacy and ill-posedness.

The quantitative instability of the explosive form of nonlinearity discussed here is that with quantities approaching infinity, there is a problem of quantitative unboundedness of the Euclidean spaces and in essence stands for transitionalities of non-Euclidean spaces. In other words, the nonlinear problems corresponding to the rotationality of nonlinearity do not suffer from the indeterminacy due to indefinite growths or ill-posedness. In short, quantitatively instable evolutions of nonlinearity represent transitional changes and are problems of our blown-ups. Even when seen as a problem of epistemology, nonlinearity uncovers the fundamental characteristics of transitional changes through quantitative instabilities. So, the quantitative infinity can no longer be seen as an ill-posed problem or indeterminacy of linear systems. In other words, as a methodology, the traditional quantitative methods have met with problems of instability unsolvable by using quantities. That is, quantitative analysis cannot be employed as the methodology to handle transitional changes.

6.2 Computational Stability Analysis of Nonconservative and Conservative Schemes of Nonlinear Fluid Equations

In order to specially comprehend the numerical instability of nonlinearity, let us look at a particular mathematical model. The following is a model for one layer of the two-dimensional atmospheric fluid:

$$\begin{cases} \bar{u}_t^t = -(u\bar{u}_x^x + v\bar{u}_y^y + g\bar{z}_x^x) + fv \\ \bar{v}_t^t = -m(u\bar{v}_x^x + v\bar{v}_y^y + g\bar{z}_y^y) - fu \\ \bar{z}_t^t = -[u\bar{z}_x^x + v\bar{z}_y^y + z(\bar{u}_x^x + \bar{v}_y^y)] \end{cases} \tag{6.5}$$

where

$$\bar{A}_n^n = \frac{1}{2}(A_{i+1,j} - A_{i-1,j})$$

A can be an arbitrary variable in Equation (6.5); *m* a magnifying factor of the map plane, $f = 2\Omega \sin \varphi$, Ω the average revolution of the earth, φ the geolatitude, fu, fv are respectively the deviation force along the y, x directions. Since modern science treats the speed Ω of revolution of the earth and the gravitational acceleration g as constant, the corresponding $g\bar{z}_x^x, g\bar{z}_y^y$ have been in essence modified artificially to linear gradient forces. So, Equation (6.5) has been artificially altered into a mathematical model with mixed linearity and nonlinearity, becoming a nonlinear model of the first push. Since this model is inconsistent in terms of the form of motion and the acting force, it surely revealed the fact that at the historical moment when this was done, people did not fully understand that nonlinear movements and pushing forces are inconsistent mathematical models. For our purpose, let us for the moment ignore this problem of feasibility and just look at the problem of quantitative stability of this model from the angle of quantities.

For comparison, Equation (6.5) is known as a direct difference scheme. From computational mathematics, it follows that Equation (6.5) is an instable difference scheme.

Due to the epistemological concepts of the time, conservation of energy is shown in the quantitative form of averages so that we will do our comparison using the conservative scheme of the so-called energy conservation. So, we have

$$
\begin{cases}
\bar{u}_t^t = -[(\bar{u}^x\bar{u}^x)_x + (\bar{v}^y\bar{u}^y)_y - u(\bar{u}_x^x + \bar{v}_y^y) + g\bar{z}_x^x) + fv] \\
\bar{v}_t^t = -[(\bar{u}^x u\bar{v}^x)_x + (\bar{v}^y\bar{v}^y)_y - v(\bar{u}_x^x + \bar{v}_y^y) + g\bar{z}_y^y) - fu \\
\bar{z}_t^t = -[(\bar{u}^x\bar{z}^x)_x + (\bar{v}^y\bar{z}^y)_y] + g\bar{z}\bar{m}_y^y
\end{cases}
\tag{6.6}
$$

where $\bar{A}^x$ stands for the average of the variable A along the x direction. From Equation (6.6), it can be seen that in practical applications, conservative schemes go through more smoothing treatments than nonconservative schemes. In other words, the purpose for the system of modern science to employ energy conservation as a constraining condition of stability is nothing but to limit the quantitative growths without explaining the specific mechanism of the transformation process of the energy conservation. No doubt, any energy conservation without a mechanism for energy transformation cannot embody or materialize the desired conservation. Figure 6.1 shows the computational results of the nonconservative scheme of the direct differences, and Figure 6.2 the outcomes of the conservative scheme. The unit of the altitude field is 10-geopotential meters, denoted dagpm in meteorological science. From these figures, the following can be seen.

6.2.1 Desired Quantitative Stability

The conservative scheme did not materialize the desired quantitative stability; and at the same time, it did not realize the conservation of energy, either, except that in comparison to the nonconservative scheme, it showed a little extra stability. For the

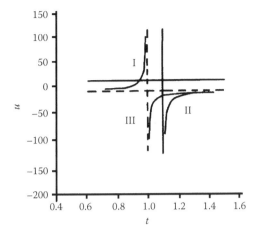

Figure 6.1   Quadratic form and hyperbola.

nonconservative scheme, after step number 582, the computational values experience explosive fall (for the convenience of making the graph, in Figure 6.1, all steps before step number 576 were truncated).

6.2.2 Energy Conservation and Quantitative Growth

The conservative scheme cannot guarantee stability, either. It only slows down the quantitative growth slightly when compared with the nonconservative scheme. At the time step of 4237, the conservative scheme also experiences an explosive fall (Figure 6.2). This end shows, based on realistic computations, that the irrotational kinetic energy of speed of modern science embodies energy conservation in the quantitative form of Euclidean spaces; however, when implemented practically, the Euclidean spaces do not provide any space to go around other than giving an expression to energy conservation by limiting quantitative growths. That explains

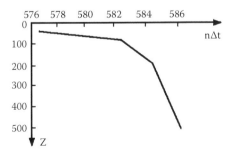

Figure 6.2   Computational results of a nonconservative scheme.

why in the study of mathematical physics problems, energy conservation is pursued as an invariance of integrations and chased after the so-called upper limit estimation prior to experience. That constitutes the overall thinking logic and method of modern science — resist risks using risk values.

Any realistic conservation of energy in nature completes energy transformation and transfer through a transformation mechanism instead of using the condition of any prior-to experience estimation to limit or eliminate changes in energy. So, it reveals that in terms of the conservation of energy, modern science suffers from some epistemological nonclarity, leading naturally to the specific calculation methods, developed by limiting quantitative growths, and the difference schemes of energy conservation, which do not actually possess the capability of constraining energy transfers. What is more important is that when quantitative instabilities are controlled artificially by using computational schemes, it does not mean that nature would follow the man-made computational schemes to pause or terminate its transformations and transfers of energies. That is why in Chapter 5 we specifically employed vector analysis to show that nonlinearity stands for problems of rotationality of solenoidal fields. If we employ this rotationality in our present discussion of conservation of energy, then we see that at the same time when material transformations are realized through rotations, nonlinearity completes energy transformations and transfers also through rotationality. In other words, the conservation of energy of the kinematics accomplishes energy transfers through the mechanism of rotation without applying any of the so-called prior-to-experience estimations. Because of this, problems of these kinds must walk out of the quantitative analysis system of the Euclidean spaces and enter the conservation of stirring energy of curvature spaces.

That is why we once pointed out that modern science is only a school of thought, since it has not perfected the conservation of energy and understood the situation completely. Energies are conserved in the form of transformations; limiting changes in energy is not the same as controlling energy transfers. In terms of practical applications, nonlinear models currently suffer from the problem that computers stopped working due to explosive quantitative growths, which reveals that the quantitative tools meet with the difficulty of dealing with transitionalities. In other words, the current quantitative tools cannot handle the phenomena that at extremes, matters evolve in the opposite direction; such inability is a weakness of the quantitative tools and had been clearly pointed out by Zhan Yin that "not all things can be described by numbers" of the time period of the Warring States of ancient China.

In terms of computational mathematics, from Figure 6.2, one meaningful question is: How can the conservative scheme prolong the time of integration (stability) when compared with the nonconservative scheme? And, in what fashion does the conservative scheme prolong the time of integration (stability)?

6.3 The Form of Computational Stability of the Conservative Scheme

No doubt, scientific research is about questioning at any time and space and even along a newly made progress when an unknown appears. So, posing questions should be encouraged in the development of science. For example, what is found is that conservative schemes can help prolong the time of integration (stability); then in what form is the stability realized? To this end, we are forced to look at our specific computations in order to analyze the processes of change in the computational values. To do this, we shrink the time steps to 1/10 of those used in Figures 6.1 and 6.2 and display the computational outcomes respectively in Figures 6.3, 6.4, and 6.5. In these figures, the black solid curves represent the computational processes of the conservative scheme, and the dotted curves the processes of the nonconservative scheme. From Figure 6.3, it is not difficult to see that the refined changes in quantities lead to opposite phase evolutions out of the conservative and nonconservative schemes. That is, if the nonconservative scheme shows a growth, then the conservative scheme can change the values of the nonconservative scheme to the opposite direction so that the quantitative growth is naturally constrained. In other words, the reason the conservative scheme is more stable than the nonconservative scheme is that the former changes the direction of growth of the quantitative values of the latter.

If we only analyze the changes in the computational values without considering the directions of change, then quantitative changes of the conservative scheme are actually greater than those of the nonconservative scheme (Figure 6.4). At first sight, this does not sound reasonable. However, a careful analysis indicates that due to the smoothing effect, the conservative scheme has strengthened its base. (For example, if the initial value for both schemes is 1 and the results of the first and second steps are respectively 2 and 3, then the actual starting step of the nonconservative scheme is still 1, while the starting step of the conservative scheme is the average 2 of 1, 2, and 3 as its new initial value. That explains the further strengthened base of the conservative scheme than the initial value of the nonconservative

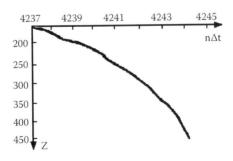

Figure 6.3 Computational results of a conservative scheme.

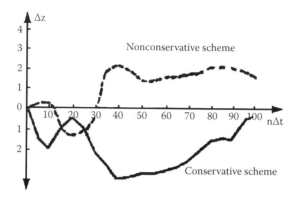

Figure 6.4   Phase difference between conservative and nonconservative schemes.

scheme.) From a greater base, the computational values are naturally greater, leading to the results as shown in Figure 6.4.

To help the reader better understand the quantitatively computational characteristics of nonlinear equations, let us compare computational schemes by looking at the following example of nonlinear equation of the quadratic form, whose analytic solution can be explicitly obtained.

$$\frac{du}{dt} = u^2 + pu + q \tag{6.7}$$

This kind of equation has been mentioned many times earlier. And, it has been shown that when $p^2 - 4q \geq 0$ or $p^2 - 4q \leq 0$, this equation stands for an evolutionary problem of blown-ups. So, let us look at the numerical experiments for the situation

$$p^2 - 4q = 0 \tag{6.8}$$

where all the parameters take the same values as in the analytic solution.

Although the direct difference scheme is an instable scheme, from a comparison analysis of the computational values shown in Figure 6.5, it can be seen that the results of the direct difference scheme are closer to the analytic solution so that the quantitative instability of the underlying nonlinearity is vividly shown. Considering the fact that computational values cannot handle quantitative unboundedness, and automatically jump over the point of the nonlinear blown-up at $t = t_b$, the values shown in Figure 6.5 are those computed for $t < t_b$. To this end, we tested a smooth scheme (where the values of $t > t_b$ do not enter the smoothing), and plotted the results in Figure 6.5 along with the results of the direct difference scheme and the analytic solution, where all the parameters of the numerical schemes are taken to be the same as those of the analytic solution. The left-most dotted solid curve in

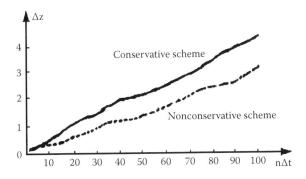

Figure 6.5 **Growths of conservative and nonconservative schemes.**

the figure stands for the results of the smoothing Runge–Kutta scheme; the middle solid curve the analytic solution; and the right-most dotted curve the direct difference scheme. Since the three curves almost entirely coincide with each other for $t < 0.72$, the computational values for before $t = 0.72$ are truncated when constructing Figure 6.5 with only the values with obvious differences remained. Evidently, from Figure 6.5, it can be seen that when compared to the analytic solution, the computational results of the Runge–Kutta scheme shows positive errors; and when getting near the quantitatively instable singular area, the instable growth is faster than both the analytic solution and the direct difference scheme. It is because the base numbers increase in the process of smoothing. In particular, when the quantities increase monotonically, the growth of the base number becomes more obvious. For example, with a monotonic increase, we have $u_2 > u_1$ and so

$$\frac{u_2 + u_1}{2} > u_1$$

Therefore, the starting value used in the initial step is greater than the original base value. And, when the variable decreases monotonically, the base number is also decreased.

For the direct difference scheme, when $t < 0.95$, the computational results show positive errors when compared with the analytic solution; and when $t > 0.95$, negative errors are shown. When the quantitative instability approaches the singularity, the growth rate becomes slower than the instability of the analytic solution. This end indicates that stable schemes of computational mathematics do not have the effect of stability on nonlinear quantitative instability, as well as that nonlinear quantitative computations are not a problem of computational schemes; instead they reveal the problem of incomputability by using quantitative calculations. As a matter of fact, Figure 6.5 has clearly shown that any method or computational scheme that is developed to eliminate quantitative instabilities of nonlinear equations is going to eliminate transitional changes existing in the underlying nonlinear problems.

Smoothing schemes, including all stable computations of computational mathematics, do not have the capability to describe true changes. The essence of the so-called quantitative computations is to limit changes in quantities. As for nonlinear problems, two important problems are revealed: One is about monotonic increases in quantities, where smoothing schemes have the effect of fabricating the starting points or initial values; the other is the changes in the phases of the quantitative values in the computation. In particular, in terms of the singular values of nonlinear transitional changes, false results can be produced due to the effects of changed phases; and the existing transitional changes could be completely eliminated, leading to evolutions of no rise and no fall.

As a problem of epistemology, the so-called stable or smoothing schemes alter the mathematical properties of nonlinear problems, creating misunderstandings of the underlying physical problems. The corresponding computations of nonlinear problems using smoothing schemes have essentially treated the nonlinear problems as linear ones. That is also the reason why we have provided the computational results of the conservative scheme with altered phases. In short, in the study of analytic solutions of nonlinear equations, these equations cannot be simplified by linearization. Similarly, in numerical computations, nonlinear equations cannot be calculated by using quantitatively smoothing or stable schemes of linear systems. Nonlinear equations reveal evolutionary transitional changes through quantitative instabilities. Speaking more accurately, computational mathematics meets with the problem of quantitative incomputability of nonlinear equations.

At this junction, the reader could think on his own about whether or not the problems revealed in the quantitative computations of the currently popular mathematical models actually agree with reality, and why in implementations the initial values have to be "corrected" constantly by using the realistic situations. No doubt, the method of constantly correcting the computed values does not comply with the initial-value existence theorem of the traditional mathematical physics equations, while the operators at the same time do not have the guts to declare that their works are a challenge to or a new development of modern science. In essence, the current numerical forecasting is to apply live report or implicit live report to replace the true forecasting. That is why we once pointed out that quantitative computations can calculate nonexistent matters more accurately than those that actually exist.

Our results can be summarized as follows:

1. Other than the aforementioned systematic dislocations of opposite phases, the conservative scheme also experiences systematic sequential dislocations and nonuniform quantitative changes at individual grids, causing the whole system to lag behind.
2. The spatial smoothing of the conservative scheme makes the computational field shrink and become smaller.

3. A direct consequence of the systematic dislocations is to cause long waves to shrink. The situation worsens with the Euler backward difference scheme and spatial smoothing. It can be said that the Euler backward difference scheme quickly slows down the system by using opposite phases. So, in numerical simulations or quantitative computations, one should not apply the Euler backward difference scheme.

4. For bounded fields, differences of the computed values of the grid points neighboring the boundary might increase. That is the reason why numerical computations often spill and cause the computers to stop working due to accumulations of the differences near the boundary.

5. The amount of computation increases, especially when smoothing schemes such as five-point smoothing or nine-point smoothing are applied.

6. Since conservative schemes raise the base values in the form of averages, within short periods of time (less than 24 hours), these schemes are not as stable as nonconservative schemes.

7. In terms of computational stability, even if smooth initial fields are applied, the stability can only be maintained for about 25 to 30 days.

8. The only achievement accomplished by the result that conservative schemes cannot guarantee stability is that the conservation law of energy of modern science does not give any expression to the mechanism for energy transformation so that the claimed conservation cannot be realized. The design of conservative schemes is complicated, leading to the problem of the inability to actually design conservative schemes.

9. As a problem of epistemology, we should value the transitional changes as revealed by nonlinear problems and their rotationalities.

In particular, the idea for designing conservative schemes is to limit and block energy, which is materialized by using averages to spread energy. Evidently, neither limit and block nor spread is the same as transformations of energy. Just as inextinguishable materials are embodied in changes, energy is also an attribute of materials. So, inextinguishable materials imply that energy is also inextinguishable and can only be transformed. Limiting and blocking energy have to cause accumulations of energy. The quantitative computations of conservative schemes are nothing but to average out the limited and blocked energy. However, averaged allocation is not the same as energy transformation and still leads to accumulation of energy, creating an overall raised level of energy in the computational fields. A realistic conservation of energy has to provide the information about how energy transforms. That is also why we pointed out that the law of conservation of energy of modern science is still not complete and that modern science is still a school of thought.

Our listed problems on quantitative computations specifically give an expression to the fact that in computations, energies are not transformed. And, in the computations of energy-conservative schemes, quantitative instabilities correspond exactly to the concentrated accumulation of energy. So, instabilities existing in quantitative computations have posed a challenge to the theoretical system of modern science. That is, limiting energy conservations to such a law that says nothing about the mechanism on how energy transforms is the same as stating an incomplete law of conservation.

Besides, quantitative computations also implicitly contain the error calculations of large quantities with infinitesimal differences, which can appear to be formally stable. However, such stability does not represent the realistic values due to errors and is meaningless. As for this problem, we will discuss it in the following section on spectral expansions.

The original purpose for us to compute the stability of various numerical schemes is to find out how to understand quantitative instability, what the irregular information that underlies quantitative instabilities is, and how to deal with irregular information. That is how we derived the epistemological understanding that irregular information is that of change in materials' evolutions, and started to make use of the method of digitization to deal with irregular events and information of change.

The problems, as revealed by the various computational schemes of limiting or blocking energies used in quantitative computations, are not originated from within the computations themselves; instead, they are rooted in the lack of a mechanism for energy transformations in conservation laws of energy of modern science. So, our work here has provided a great opportunity to complete the conservation law of energy, or at least it points out that the thinking logic of limiting and blocking energies is problematic. As a matter of fact, the thinking logic of the so-called limiting and blocking energies had been shown useless over 4,000 years ago during the Xia Dynasty in ancient China. That is also the reason why we proposed the concept of second stir, the conservation of stirring energy, and the law of conservation of energy transformations.

The fact that computational mathematics continually employs the well-posedness theory of mathematical physics equations itself reflects that the root problem is at the level of epistemological concepts. That explains why the development of computer technology has not brought forth a revolution in mathematical computation. If speedy computations that are accomplished by computers are employed to limit the quantitative instabilities of nonlinearity, then the invention of computers is the same as nuclear physics being employed to make nuclear bombs so that man can destroy himself more easily. So, we should truly understand the essence of quantitative instabilities of nonlinearity and that the revolutionariness of computers is about the digital handling of information and directly dealing with events so that there will be no need to convert events or information into quantities first and then analyze the events or information through the quantities using formal logic. Since modern science believes in replacing events by quantities and quantitative comparability as the only standard for scientificality, it has not only misguided the quantitative computations, but also established a huge obstacle in the path of scientific development. So, the central problem for the next scientific revolution to come is how to understand that events instead of numbers are the essence.

6.4 Principal Problems in the Quantitative Computations of Harmonic Wave Analysis of Spectral Expansions

What is currently fashionable in modern science is to express problems of physics and engineering as mathematical nonlinear problems; however, mathematical nonlinear problems are in turn problems of quantitative instability and do not satisfy the initial-value existence theorem. Because of this, a widely applied method in almost all areas of studies is to employ the method of spectral expansions in the form of harmonic waves to constrain the quantitative instabilities. The stability theory of differential equations is not only the standard for the analytic analysis of deterministic problems of mathematical physics equations, but also the theoretical basis for numerical computations. However, nonlinear quantitative variables can evolve into quantitative instabilities due to explosive growths of the products of nonsmall effective values, and can also fall into ineffective error values due to drastic decreases of the products of small ineffective values — that is the problem of computational inaccuracy of large quantities with infinitesimal differences. Evidently, these problems should also be reflected in the computations of superpositioned harmonic wave expansions. To this end, in the following, we will conduct numerical experiments on harmonic wave superpositions, where false information is clearly fabricated artificially.

Quantitative instability is not only a characteristic of nonlinearity but also touches on the problem of what the quantitative infinity is. Even from the angle of Riemann geometry, it is not hard to see that quantities are formal measurements in zero-curvature (Euclidean) spaces so that quantitative instability and the quantitative infinity are products of zero-curvature spaces. As for problems of transition of curvature spaces, they have been resolved by using implicit transformations between high- and low-dimensional spaces. (All the details are omitted here.) So, quantitative instabilities and the problem of the quantitative infinity can be dealt with using phase space transformations so that the evolutionary transitional changes, such as those that at extremes, the matters will evolve in the opposite direction, can be successfully analyzed. However, in realistic quantitative computations, one still has to face the problem of large quantities with infinitesimal differences or computational inaccuracy of quasi-equal quantities.

For the sake of convenience of our presentation, let us present our numerical experiments in three cases:

1. The numerical experiment with quasi-equal quantities.

2. The numerical experiments with adjusted time steps.
3. The numerical experiment on the impacts of the initial value and each parameter.

6.4.1 The Basic Computational Formula

$$\frac{dx}{dt} = A\sin x - B\cos x \tag{6.9}$$

When A and B are constant, Equation (6.9) becomes a simple trigonometric equation. When B is negative, we will have an addition of waves; and when both A and B are variables, this equation stands for a nonlinearity of the interaction of waves and flows.

The traditional difference schemes include smoothing schemes, Runge–Kutta schemes, conservative schemes, and so forth, developed wishfully to achieve quantitative stability. The essence of these computational schemes is nothing but extensions of the initial values in the Newtonian system, which in fact constrains quantitative changes. From the fact that the quantitative infinity represents problems of transitions in curvature spaces, we can avoid quantitative instabilities in our numerical experiments by using transformed phase spaces in the form of digitization (see Chapter 9).

The corresponding direct difference scheme of Equation (6.9) is given by

$$x = x_0 + (A\sin x_0 - B\cos x_0)\Delta t \tag{6.10}$$

According to the method (Chapter 9) of transforming quantitative instabilities into reciprocations, or called the method of digitization of quantitative instabilities, we will transform quantitative instabilities into frequencies of changes. Doing so can not only reveal problems existing in quantitative computations, but also reflect the process of quantitative changes and the characteristics of transitional changes. However, what needs to be noted at this junction is that the feasibility of this method is from the theoretical conclusion that time does not occupy any material dimension, which we will study later in sections on what time is.

6.4.2 Numerical Experiments

For our purpose, we designed the following several cases for our experiments.

The quantities involved are quasi-equal. This situation cannot be avoided in superpositions of harmonic waves. In this case, we take the initial value $x_0 = 45^0 = \pi/4$ (radians), $A = B$, and let Δt be the time step.

The situation with adjusted time steps.

The initial value is relatively small, *A* is relatively small and *B* relatively large, and *X* = 0.0009, *A* = 1, and *B* = 100,000;

The initial value is relatively large, *A* is relatively large and *B* is relatively small, and *X* = 1.5, *A* = 10,000, and *B* = 1; and

The initial value is relatively small, *A* = *B* are relatively large, and *X* = 0.0009, *A* = *B* = 12,900.

The impacts of the initial value and the individual parameters of the equation.

X is relatively large, *B* relatively small, the time step relatively small, *A* varies, and *X* = 1.5, *B* = 1, and Δ*t* = 0.00066;

X is relatively small, *B* relatively small, the time step relatively small, and X = 0.0009, *B* = 1, and Δ*t* = 0.00035; and

The parameters stay constant, the time step invariant, the initial value varies, and *A* = 10, *B* = 100, and Δ*t* = 0.03.

Some explanations: All figures on the left-hand side are those of the functional phase space (sin*x* − cos*x*), and the figures on the right-hand side the (x–t) phase space graphs of the corresponding moments. In order not to make this presentation too long, we will only provide three case situations of our experiments.

6.4.2.1 Quasi-Equal Quantities

This situation corresponds to when the initial value approaches the radian value of 45°, where the numerical values cannot be made accurate. Since *A* = *B*, a situation of large quantities with infinitesimal differences is created. The computations are about the error tails of the effective values. Widened time steps only lead to with denser "sawtooth" waves (Figure 6.7). In this case, even if we do not limit the changes in the quantities, the computations are also stable and can be carried on forever. The sawtooth waves are also one of the problems of the chaos phenomenon in harmonic wave expansions. "Death" in the functional phases space indicates the death of chaos, where error value computations are mistakenly seen as stability, leading to useless calculations.

6.4.2.2 Invariant Initial Value and Parameters with Adjusted Time Steps

The numerical experimental process and results for the case of relatively small initial value, relatively small *A*, and relatively large *B*, are shown in Figure 6.7. This situation is equivalent to that of *A*sin*x* << *B*cos*x*. That is, the dominant term in the computation is the second term in Equation (6.9), while the first term can be ignored. However, we can still see the sawtooth wave in Figure 6.7a and the corresponding death (Figure 6.7g). However, this sawtooth wave and the death are different from the situation of large quantities with infinitesimal differences in Figure 6.6; if we

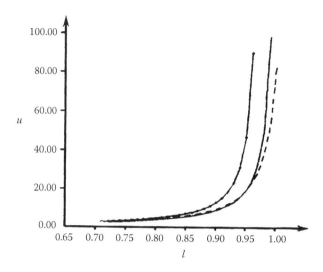

Figure 6.6 Comparison of three schemes.

did not employ reciprocations to deal with changes in the quantities, the computers would be stopped due to quantitative instabilities. Now, a problem appears: In the traditional harmonic wave analysis, although the quantitative instabilities are eliminated by using orthogonality or such computational methods as smoothing and so forth, the error value computations of large quantities with infinitesimal differences cannot be removed so that the computations are trapped in fabricating false information and the quantitative instabilities, reflecting transitional changes, are also artificially eliminated. The consequence is the maintenance of the invariant system of modern science.

Besides, what is very meaningful is that Figure 6.7 depicts the process of quantitative change. That is, the sawtooth waves and deaths in Figure 6.8 are transformed into new quantitative instabilities and "rebirth" in Figure 6.8. This process of transformation reveals exactly the transitional change in the underlying evolution, which is a problem the traditional harmonic wave analysis cannot reveal. Evidently, eliminating instabilities has to remove the processes of changes in quantities.

In this experiment, we also looked at other situations, such as that when the initial value is relatively large, A relatively large, B relatively small, and that when the initial value is relatively small, $A = B$, and so forth. As quantitative instabilities, each of these cases shows the instable sawtooth waves and deaths and the phenomenon of transformation into new instabilities and rebirths. The only difference is in changes of the processes and positions. And, quantitative stabilities and instabilities can all be expressed in the form of sawtooth waves, where the sawtooth waves of quantitative stabilities are meaningless computations, while the quantitatively instable sawtooth waves stand for transitional changes. All the details are omitted.

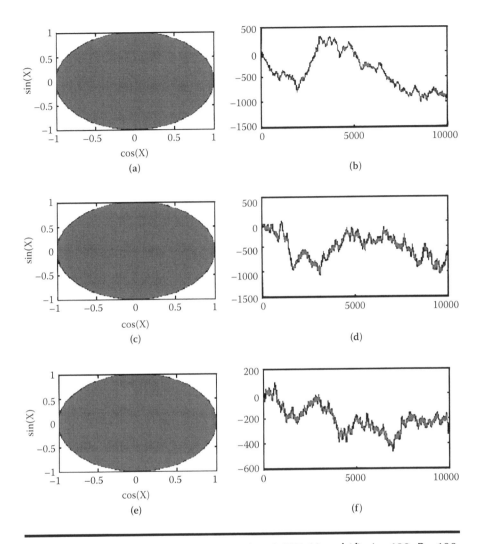

Figure 6.7   (a) and (b): $A = 100$, $B = 100$, $\Delta t = 0.035$; (c) and (d): $A = 100$, $B = 100$, $\Delta t = 0.065$; (e) and (f): $A = 100$, $B = 100$, $\Delta t = 0.115$.

6.4.2.3 Impacts of the Initial Value and Parameters

This experiment also considered the individual cases when X is relatively large, B relatively small, the time step is relatively small, and A varies and is relatively large; when X is relatively small, B relatively small, the time step relatively small, and A varies and is relatively large; when the parameters stay constant, the time step is fixed, and the initial value varies. Limited by the length of our presentation, we will only look at one of the cases considered, where the first coefficient A varies relatively largely, which is equivalent to the computations of varying parameters and $A\sin x$

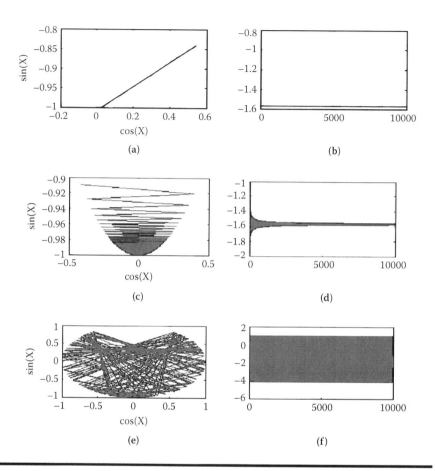

Figure 6.8 Numerical experiments with quasi-equal quantities. (a) and (b): X = 0.0009, A = 1, B = 10,0000, Δt = 0.00001; **(c) and (d):** X = 0.0009, A = 1, B = 100,000, Δt = 0.00002; **(e) and (f):** X = 0.0009, A = 1, B = 100,000, Δt = 0.000040; **(g) and (h):** X = 0.0009, A = 1, B = 100,000, Δt = 0.00005; **(i) and (j):** X = 0.0009, A = 1, B = 100000, Δt = 0.00022.

>> $B\cos x$. The process of evolution is shown in Figure 6.8. Since the initial value is relatively large, the quantitative change starts off as an instable sawtooth wave or the transitionality of death. As A increases, a rebirth appears (Figure 6.8c,d,e,f). After that, along with the quantitative instability, the evolution once again enters into a sawtooth wave or death (Figure 6.8g,h); this evolutionary pattern continues indefinitely. Similar to the situation of nonlinear equations, the computations of the equations with variable coefficients can also represent instable quantitative growths.

In essence, when the coefficients A and B vary, Equation (6.9) is indeed a nonlinear equation. Although Equation (6.9) is a very simple mathematical model, it reveals some of the basic problems existing in superpositions of harmonic waves. To

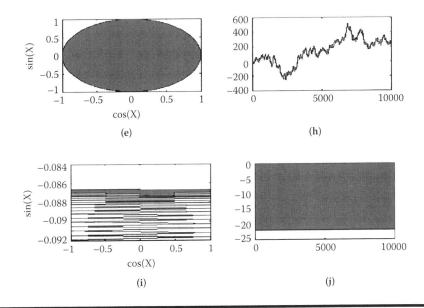

Figure 6.8 (*Continued.*)

this end, our careful analysis indicates that the traditional harmonic wave expansions suffer from the problem of artificially fabricating information, while they do not reflect the processes of quantitative changes.

Our presentation above made use of the quantities of zero-curvature spaces and transitional changes in curvature spaces to analyze the stabilities of the quantitative computational inaccuracy of the pathological equations of large quantities with infinitesimal differences and the transitional changes of the quantitative instabilities of varying coefficients based on a simple harmonic wave equation. Our results indicate the following.

Large quantities with infinitesimal differences lead to pathological equations, whose numerical computations are stable. However, the computations are inaccurate and produce nonrealistic information (false information). The stable sawtooth waves stand for meaningless computations of the evolution, which not only fabricate the computational results but also waste the computational time. Therefore, in superpositions of harmonic waves, one will surely meet the situation of differences of quasi-equal quantities. Even if the quantitative instabilities are constrained in the analysis of harmonic waves, false information will still be introduced due to the infinitesimal differences of large quantities. That is a series problem existing in spectral expansions. The essence is to mistakenly treat error value computations as quantitative stabilities.

The death of the quantitatively instable sawtooth waves is about transitional changes, while the traditional, stable harmonic wave analysis fabricates major false information — invariant quantities, causing the analysis to eliminate all transitional changes. In other words, the method of removing singularities of the spectral expansions is to eliminate transitionalities. So, the harmonic wave analysis of the spectral expansion is not a tool of analysis for nonlinear problems.

The quantitative infinity is traditionally seen as a mathematical singularity. However, it represents a piece of key information about transitional changes in curvature spaces. The core of investigating problems of evolution is about information on peculiarities, which led us to publish our book *Entering the Era of Irregularity* (OuYang et al., 2002a). What is more important is that the quantitative computations using spectral expansions are not a logical deduction of events and create false, artificial events. It can also be said that the created false events are exactly the consequence of using quantities to substitute events, which reveals the unreliability of quantitative manipulations. In particular, quantities cannot handle peculiar events, which indicates that peculiar events are information of change. This end also signals that mathematical models and their solutions based on spectral expansions are not methods for predictions, because methods for predictions have to be able to resolve transitional changes, while each transitional change makes quantitative computations suffer from the avoidable problem of the quantitative infinity.

6.5 Lorenz's Chaos Doctrine and Related Computational Schemes

We look at the chaos theory as a related computational scheme because this theory was developed on the basis of misunderstandings of nonlinear problems. What can be said from our published papers and the book *Entering the Era of Irregularity* is that the problem regarding to the so-called chaos theory has been clearly resolved. Besides, we include our discussions of the chaos theory in this section because the central concepts of the theory came exactly from the computational schemes of large quantities with infinitesimal differences or the problem of computational inaccuracy of quasi-equal quantities. What is worth mentioning is that the chaos theory was originally introduced by Lorenz from studies of meteorological problems, so to reveal problems of the theory, we also have to look back to the profession of weather forecasting and related predictions of disastrous weathers. Also, in particular, our colleagues in meteorological science have expressed their interest to see such a discussion in order to clarify the fact that the Lorenz's chaos theory has nothing to do with weather forecasts.

6.5.1 Problems with Fundamental Concepts

The word *chaos* originated from the Greek word ΧΑΟΣ. In Greek mythology, the word ΧΑΟΣ appeared very early, which was used to express the origin and evolution of the world. In modern English, the word *chaos* means extremely chaotic, order-less, no organization, and so forth. In short, *chaos* has been used as a synonym for *order-less* in modern epistemology. It can be said that by well choosing the word *chaos*, Lorenz has created a shocking effect for several tens of years.

The ancient Chinese believed that before the earth and heaven were formed, the universe was in a fuzzy and unclear state that was described as chaotic. This meaning of chaos is similar to that of ancient Greeks. According to ancient Chinese legends, there is the concept of

盘古开天地的浑沌初开

meaning that at the very start of the earth and heaven, it was all chaotic. And, in Zhuang Zi's *Ying Di Wang*, chaos was once mentioned as the central empire. In other words, no matter whether it was the very start of the earth and heaven or the central empire, chaos, either seen as a proposition or an event, stands for the ancient past that has been long gone. The meaning that is passed on to us, modern man, is that chaos should be identified with such a state that it is fuzzy, not clear, or indistinguishable. Whether or not such a chaos indeed existed at the very start of the earth and heaven would be another matter of concern. What is important is that the ancient people of both the East and the West shared the same understanding.

As an earlier philosopher of ancient China, Lao Tzu never directly talked about the problem of chaos. *Tao Te Ching* (Chapter 25) mentioned that before the formation of the heaven and earth, all materials were mixed, calm and spatial, independently invariant, and rotated without stop. While Lao Tzu talked about uniformly mixed and invariantly ordered states that existed before the formation of heaven and earth. So, the meaning of *chaos* seems to be quite clear: It stands for such a state that it is spatial, calm, and indistinguishable (no difference), and can be well mixed without any sound, instead of any distinguishable (with difference), irregular, or noisy state of no order. Similarly, the meaning Lorenz, the founder of the chaos theory, wanted to deliver was the later state of no order.

Evidently, the message we would like to deliver here is that whether or not there once existed a period of chaos in the evolution of the universe should be a topic of research in the science of celestial bodies. However, on the other hand, disorderliness is indeed an objective existence even in today's world. It is a realization of irregular events and makes the system of quantitative analysis face its unsolvable problems. Modern science is used to referring to *chaos* as disorderliness. If we also employ the word of disorderliness, then we are talking about the problem of "walking into the era of irregularity." In other words, disorderliness makes the quantitative analysis suffer from difficulties. However, it does not mean that problems of disorderliness are random or cannot be successfully dealt with. Besides, why is there disorderliness? What is its physical significance? And what is its function? These are problems worthy of our deepened and careful investigation.

Another reason why we talk about the chaos theory in this chapter is our desire to find out why this theory had created a shocking effect for a period of time, leading to the epistemological concept of indeterminacy. In the community of physicists before the 20th century, there had appeared an epistemological tendency of treating irregular events as indeterminate due to the fact that quantities could not handle these events. Since Schrödinger introduced the concept of probabilistic waves, which went along well with the then-popular Heisenberg's principle of inaccurate measurement, the concept of probabilisticity became an openly accepted theory. Because of the existing classification within the scientific community of the classical mechanics as determinant due to its emphasis on materials' invariancy, there appeared a debate for nearly 100 years between determinacy and indeterminacy. However, irregular events are objectively existent. Their attributes of existence cannot be simply altered just because classical mechanics system cannot deal with them. So, for now, what is very clear is that classical mechanics is a scientific system that maintains materials as invariant. No doubt, without any change in events, there will not be any reason to talk about determinacy and indeterminacy. The corresponding so-called indeterminacy is originated from the fact that the system of quantitative analysis cannot deal with irregular events so that no "deterministic" causalities for changes in events can be provided, leading to the term of indeterminacy.

Similarly, irregular events that cannot be dealt with by using quantities still have their reason for existence and their processes of change. That is, there would be methods beyond the quantities that could be employed to handle irregular events.

Just as any other objectively existing event, there must be a reason for the chaos theory to generate its shocking effect in the scientific community in the past years. The overwhelming dominance of the system of classical mechanics is not weakened because of the appearance of the concept of randomness. The chaos theory was developed on the basis of the nonlinear equations of classical mechanics; and classical mechanics was first labeled as deterministic. So, by creating indeterminacy out of the deterministic system, the chaos theory generated such a feeling that a fire had started in the backyard of classical mechanics so that the Newtonian kingdom could be completely overthrown and that a realm of probabilistic universe could be consequently established.

In short, even if we continue to apply the popular terminology of the modern scientific system, the essence of the debate between determinacy and indeterminacy is that the origin and significance of disorderly events are not successfully addressed and the question of what the governing laws of variable movements are is not resolved.

6.5.2 Problems with Lorenz's Model

Lorenz (1963) employed the Saltzman model (1962), which was developed for atmospheric convections:

$$
\left\{
\begin{array}{l}
u_t + u\,u_x + w\,u_z = -\rho_0^{-1} p_x + \nu \Delta u \\[2mm]
w_t + u\,w_x + w\,w_z = -\rho_0^{-1} p_z - g\pi + \nu \Delta w \\[2mm]
\pi_t + u\,\pi_x + w\,\pi_z = \dfrac{N^2}{g} w + K \Delta \pi
\end{array}
\right.
\tag{6.11}
$$

where u, w are respectively the components along the x, z directions of the speed,

$$
\pi = \frac{\rho}{\rho_0}, \quad \rho_0 = \text{const.}, \quad N^2 = \frac{g}{\theta}\frac{\partial^2 \theta}{\partial x^2} + \frac{\partial^2 \theta}{\partial z^2}
$$

named as the parameter of stratification, ν, K respectively the viscosity and heat-conduction constants. The subscripts stand for the corresponding partial derivatives, and the symbol

$$
\Delta = \frac{\partial^2}{\partial x^2} + \frac{\partial^2}{\partial z^2}
$$

stands for the Laplace operator.

Evidently, Equation (6.11) is a mathematical equation of the first push system of classical mechanics. Since $\rho_0 = $ const., the movement, described by the equation, has been modified to be under a pushing force. That implies that although Saltzmann once praised Bjerknes's circulation theorem as a major betrayal of the Newtonian system, his model (1962) developed for atmospheric convections did not comply with the stirring effects of the solenoidal term of the circulation theorem. That is, the left-hand side of Equation (6.11) describes the nonlinear form of the movement, while the right-hand side maintains a pushing force so that the acting force and the form of movement do not agree with each other. So, Equation (6.11) should be seen as an irrational mathematical model. Besides, along the vertical direction, the atmospheric density changes drastically. Even if such density is written as a quantity, the change in the quantity should also be dramatic. Since π in the third term of Equation (6.11) is already nondimensional, any knowledge on the change in π has to be shown as a change in quantities. So, it can be concluded that Saltzmann did not seriously consider the vertical atmospheric structure, since the dramatic quantitative changes of the atmospheric density implicitly contain drastic changes in the overall atmospheric structure. No doubt, if problems of weather are seen as problems of drastic changes, then vertical changes in the atmosphere can be seen as the major part of dramatic atmospheric changes. Because of this understanding, in our investigations of weather forecasts, we have chosen the vertical direction as our entering point for analyzing atmospheric structures. That is also the reason why we first introduced the figures of digitized vertical structures of information of the atmosphere and how we have paid close attention to the aspect of changes in events when we analyze problems of change.

As is well known, strong convections have been a difficult problem in meteorology for the front-line weather forecasters. And, it is along the vertical direction that strong convections reveal their important characteristics. So, as a problem of epistemology, this understanding plays the key role of how to pick and choose the appropriate methods in order to resolve the problem of concern successfully. The symbol N^2 in Equation (6.11) involves the atmospheric stratification along the vertical direction. For convective weathers, N^2 cannot be seen as a constant; and its structural characteristics along the vertical direction are more directly recognizable than the structurality of the atmospheric density. It can be said that to resolve the problem of convective weathers, one has to start his investigation along the vertical direction of the atmosphere. This not only is our opinion, but also has been validated time and again over the past several decades of real-life practice.

However, according to the Newtonian first push system, or why the Newtonian first push systems could have been materialized, it is the methods of nondimensionalization and treating parameters as constants that quantification has been greatly pushed for. These methods are also indispensable for practically employing the mechanics of the Newtonian system. Just as what was mentioned in Section 6.4 earlier, when coefficients and parameters are variables, they play similar roles as nonlinearity in quantitative computations.

No doubt, the purpose of scientific research is to express the physical reality or to reveal the principles behind practical problems. However, the mathematical model itself in Equation (6.11) contains irrationality of modeling. So, it would be questionable for anyone to count on Equation (6.11) to provide meaningful explanations for the practical problems of concern. Although Lorenz referred to Equation (6.11) as a nonlinear model, in terms of the basic concepts of nonlinear models, Equation (6.11) stands for an irrational nonlinear model. Even if we continue to use the conventional terminology, it is only a weak nonlinear model. Considering that nonlinear mathematical models stand for problems of rotationality, Lorenz's work can only be seen as a mathematical game.

Now, let us introduce the following flow functions:

$$u = \Psi_z, \ w = -\Psi_x \tag{6.12}$$

Substituting Equation (6.12) into Equation (6.11) leads to

$$\begin{cases} (\Delta\Psi)_t + J(\Psi, \Delta\Psi) = -g\pi_x + \nu\Delta\Psi \\ \pi_t + J(\Psi, \pi) = \dfrac{N^2}{g} + K\pi \end{cases} \tag{6.13}$$

where $J(,)$ stands for the Jacobi operator, and all other symbols are the same as those in Equation (6.11).

Although this kind of treatment can be seen widely in traditional kinematics, it does not mean that it is correct. And, based on the basic concepts of the Bjerkness's circulation theorem, rotational movements are originated in eddy sources; or only under stirring effects will there be spinning motions. That is, rotations and eddy sources are equivalent. However, the left-hand side of the first equation in Equation (6.13) stands for the movements and changes of rotational vorticity, and the right-hand side the pushing of a gradient force. In terms of incompressible fluids, it is common knowledge that water cannot be moved by pushing. And, even if it is pushed, only local eddy currents are created. So, the realistic effects of pushing on a body of water are local stirs. In this sense, Equation (6.13) has its difference from Equation (6.11), because even if we do not understand nonlinearity as a problem of rotation, and pushes can be reflected in movements of linear speed, then the effect of pushes should not correspond to rotational movements. No doubt, pushes can correspond to elastic pressure effects, but not the stirring effects. So, in the long period of studying mechanics, there has been the formality of quantifications, which confuses rotationality of materials with nonrotational movements. That is how one can see that the quantitative analysis of the Newtonian system can deal with any problem; however, at the end, the analysis does not tell you anything specific. Although in form this statement seems to be a joke, it does point to the central problem existing with modern science. Although the greatest advantage of the quantitative analysis is its ability to unite all matters and events under its formality, this advantage also reveals its greatest weakness: The united quantities cannot answer what things are talked about.

Due to the application of quantification, modern science suffers from some of the problems at the most fundamental level. For example, the generalized concept of wave motions has encompassed all kinds of disturbances as waves so that disturbances of stirrings can no longer be distinguished from those of vibrations. As pure quantities, the amplitude of a stirring disturbance can be identical to that of a vibration disturbance; however, their physical properties and functions, in particular their forms of motion or principles of transformation, are totally different. So, quantities have their formality and have to be constrained by their formality. Applications of the corresponding system of quantitative analysis will have to suffer from the corresponding constraints. Since the time when Newton published his *Mathematical Principles of Natural Philosophy*, one of the well-accepted laws of behavior by the scientific community is that quantitative changes lead to qualitative changes; and generalized from this principle has appeared the belief that only through quantities can one understand the world and alter the world, and later that quantitative comparability is the only standard for scientificality, and so forth. All these beliefs and opinions reveal that the system of modern science is established on the concept of numbers and the belief that with numbers the natural world will be known. That is completely opposite in principle to the Chinese belief that through matters and events, the natural world will be known.

In particular, when predictions are seen as a branch of modern science, one has to meet with the quantitative formality that is postevent. So, how can postevent quantities be employed to foretell the specifics of forthcoming events? In its originality, this should be a common knowledge on the basis of knowing the world through matters and events. However, after the opinion of knowing the world merely through numbers became popular, modern science has evolved away from the proposition of knowing ahead of time what is coming up. That is why after over 300 years of development, modern science has not yet produced a true and reliable prophet. In other words, the root reason why we still cannot reliably predict earthquakes and disastrous weather conditions has something to do with the exaggerated capabilities of quantities. Besides, quantitative analysis itself does not have the ability to deal with transitionalities, while in applications, changes in quantities are constrained by the stability condition of the well-posedness theory.

The currently popular spectral method is one of the stable methods of quantitative analysis. In other words, using the terminology of modern mathematics, as a system of weak nonlinear equations, Equation (6.13) also suffers from computational instabilities. That was why Lorenz applied spectral expansions to the flow function Ψ and the π density. That is,

$$\begin{cases} \Psi = \sqrt{2}\,\dfrac{K(n^2+k^2)}{kn}\,X\sin kx\sin nz \\[2mm] \pi = \sqrt{2}\,\dfrac{(n^2+k^2)}{gk^2n}\,Y\cos kx\sin nz - Z\,\dfrac{(n^2+k^2)^3}{gk^2n}\sin 2nz \end{cases} \tag{6.14}$$

where k,n are respectively the wave numbers along the x,z directions, and X,Y,Z are the coefficients of the spectral expansion, respectively denoting the intensity of motion, intensity of the linear stratification, and the intensity of the nonlinear stratification. No doubt, as a quantitative analysis of mathematics, one has to deal with the conditions for applying the spectral expansions, because for the spectral expansions to be valid, the functions to be expanded must have continuous first-order derivatives and piecewise continuous second-order derivatives. Since the left-hand side of Equation (6.13) stands for the rotationality of nonlinearity, the duality of rotations makes the original functions discontinuous in between inward and outward rotations. So, no one can guarantee the continuity of the first- and the second-order derivatives.

Next, the central problem of the method of spectral expansions is that the orthogonality of the Fourier analysis can be employed to eliminate quantitative singularities. Evidently, if singularities are mathematical or physical properties of Equation (6.13), then eliminating singularities will have to alter the mathematical or physical properties of the original model. In particular, problems of change are different from problems of product design. In this sense, Equation (6.14) cannot represent Equation (6.13) and instead becomes a fabricated problem. That is, Equations (6.13) and (6.14) are not related at all so that the consequent works become meaningless and useless.

Third, even as a problem of mathematics or physics, Lorenz's substitution of the stratification parameter is very strange and inexplicable. It specifically touches on the following problems.

In Equations (6.11) and (6.13), the physical quantity N^2, the parameter for the stratification, is a constant. However, in Lorenz's expansions, this constant becomes a variable and violates the original mathematical model.

This original single parameter of stratification is split into two physical quantities Y,Z. So, based on the tradition of treatment, this end is not rational both physically and mathematically. In order to distinguish linearity and nonlinearity, one can surely take the first power and second power terms of a certain physical quantity without the need to take two physical quantities. In terms of physics, applying two physical quantities to describe the same event seems to make Lorenz the first and last person to make his quantification for the purpose of quantification.

Evidently, manipulations of mathematical physics equations need nondimensionalization of physical quantities. However, one should note that the variable π in Equation $(6.14)_2$ is nondimensional. So, the right-hand side should also be nondimensional. However, g has its dimension, and the corresponding physical quantities Y,Z for the stratification must have their own dimensions. No doubt, the dimensions of the same kinds of variables are the same. However, since the stratification parameters Y,Z represent respectively linearity and nonlinearity, their dimensions must be different due to different exponents. So, in the process of nondimensionalization, when the linear variable becomes nondimensional, it does not mean that the nonlinear variable can also be nondimensionalized. So, there exists a contradiction in the formal logic of nondimensionalization. Evidently, splitting one physical quantity into two to represent linear and nonlinear stratifications helps to construct the system of equations, on which the chaos theory was developed. However, in the logical manipulation of the model building, many problems of violating the formal logic have been created.

In particular, Lorenz is a scholar in meteorology. So, he should have known the Bjerknes circulation theorem of 1898 in his specialty area ahead of scholars from other scientific areas and that nonlinearity stands for solenoidal effects, and in particular, both the atmosphere and oceans are rotational eddy currents, as pointed out by the circulation theorem. What is ironic is that the original model (Equation 6.11) Lorenz applied was established by Saltzman in 1962, while Saltzman himself had recognized and highly praised that the circulation theorem was the greatest betrayal to the classical mechanics system since the time of Newton in the past 200 years. However, neither Saltzman himself nor later Lorenz dealt with nonlinear problems based on the circulation theorem, and neither provided any explanation for what they did.

Now, looking back for why the circulation theorem was not further developed and applied in meteorological science, the only reason we can find is that the circulation theorem does not fit for quantitative analysis. That explains why even within the community of meteorologists, this theorem has met with the treatment of being tied up and stored on a high shelf.

Out of the consciousness of our time of the system of quantitative analysis, the standard for being curved or straight has been laid out by checking whether or not it fits quantitative computations instead of looking at whether it agrees with physical reality. It seems that after over 300 years, people did not make one thing clear: Quantities can only be computed after the appearance of events instead of before the events. So, the original forecasting of the future is changed to tracing after appeared events, indicating that people have given up their pursuit and exploration of how to foretell what is forthcoming. For example, treating problems of changes as initial value problems of mathematical physics equations itself comes from extrapolations of the initial values instead of the thinking logic of foretelling the future. This example clearly indicates why the prediction of natural disasters has been a scientific stronghold that still cannot be taken successfully. Speaking in layman's terms, as for the problem of foretelling the future, since the very start, modern science has treated it as extensions of the initial values without truly considering making predictions. And even after the Bjerknes circulation theorem was established, the scientific community still did not fully understand what mathematical nonlinearity is. Otherwise, the chaos theory would not have been in existence.

Lorenz substituted the quantitative equation (Equation 6.14) into Equation (6.11) to produce

$$\tau = \pi^2 b^{-2}(1+\alpha^2)K\,t$$

By eliminating t, Lorenz obtained his chaos model:

$$\begin{cases} \dot{X} = -\sigma X + \sigma Y \\ \dot{Y} = rX - Y - XZ \\ \dot{Z} = XY - bZ \end{cases} \tag{6.15}$$

where

$$\dot{A} = \frac{dA}{d\tau}(A = X, Y, Z)$$

$\sigma = v k^{-1}$ is the Prandtl constant,

$$r = \frac{R_\alpha}{R_c}$$

the Rayleigh constant, $R_\alpha = g\alpha h^3 \Delta T \nu^{-1} K^{-1}$, $R_c = \pi^4 \alpha^{-2}(1+\alpha^2)^3$, the critical value is

$$R_{cm} = \frac{27\pi^4}{4}$$

$b = 4(1+\alpha^2)^{-1}$. Here, α stands for the heat expansion coefficient, h the plane distance, ΔT the temperature difference, and all other symbols are the same as before. In computations of his chaos theory, Lorenz took

$$b = \frac{8}{3}$$

$\sigma = 10$, and $r \approx 1 \rightarrow 233.5$ so that r values could be modified in computations. Equation (6.15) is the first chaos model in Lorenz's chaos theory.

As problems of mathematical modeling, our discussion above can be summarized as follows.

No doubt, Saltzmann's model (1962) of the atmospheric convections is a model of the Euler or N-S equation in the Euler language. And, the linear dissipation term in the N-S equation suffers from the problem of whether or not it agrees with reality, which we will omit in this book. The only thing we would like to point out is that a realistic dissipation process stands for the effects of rotational vorticity so that the corresponding mathematical description has to be nonlinear. That is, there is a problem of reconsidering the mathematical models established after the traditional N-S equation in the mechanics of fluids.

Evidently, by assuming that the atmospheric density is a constant ($\rho = \rho_0$) in Equation (6.11), the mathematical properties of the N-S equations have to be changed, which at the same time distorts the physical meaning of the external forcing term. That is, by introducing $\rho = \rho_0$, the resultant equation cannot be seen as being simplified, instead, the original nonlinear stirring source is altered to be a pushing source. That is why both Saltzmann's Equation (6.11) and Lorenz's Equation (6.13) continued the custom of the traditional mechanics of fluids without seriously understanding the essential difference between nonlinearity and linearity, which uncovers an epistemological problem of the modern scientific world. No doubt, under such an epistemological understandings, the scientific community does not realize that mathematical modeling itself has to follow a certain set of elementary principles, where the properties of the acting forces should be consistent with the forms of motion. In particular, we should no longer continue the tradition that the external forcing effects in the form of pushing correspond to the nonlinear forms of motion. This end also uncovers that the traditional method of quantitative magnitude comparisons suffers from the principal weakness of mixing problems of different properties.

For example, before fully understanding what characteristics prediction science possesses, in the past 100-some years, modern science has been employed indiscriminately, leading to various inexplicable theories. At this junction, we would like to use this opportunity to emphasize once again that even within the traditional system of mechanics, mathematical modeling should be carried out on the basis of the principle that the acting force and the form of motion need to be consistent. For example, laymen all know that weather changes constantly, while scholars in meteorology still employ theories of invariance to investigate weathers.

It is well known that the spectral method achieves the goal of quantitative stability by eliminating singularities (due to the orthogonality of Fourier expansions). However, quantitative instabilities and singularities are exactly some of the fundamental characteristics of nonlinearity.

In terms of quantitative analysis, possible changes are

$$\frac{dx}{dt} = 0, \frac{dx}{dt} \to \infty, \frac{dx}{dt} \to 0$$

where x stands for a quantity and t the time. Here we still temporarily use the Newtonian quantitative time. As for the problem of whether or not time is only a Newtonian quantity, we will address that in detail later. In these three possibilities,

1. $\dfrac{dx}{dt} = 0$

   stands for invariant quantity under the concept of stability, which summarizes the essential problem of modern science. It is the initial value stability of the requirements of the well-posedness of modern mathematical physics equations or the problem of initial value automorphism. Because of this, Bergson (1963), Koyré (1968), and Prigogine (1980) once named the Newtonian system as one without evolution. That is the essence of how classical mechanics denies changes in materials.

2. For the second case $\dfrac{dx}{dt} \to \infty$

   it is called a quantitative instability, quantitative unboundedness, or a quantitative singularity. However, singularity in Euclidean spaces corresponds exactly to transitionality in curvature spaces. So, eliminating quantitative singularities is removing transitionalities in evolutions. In other words, it is exactly the singularities in which the mathematical properties and the transitionalities in the corresponding changes of events are revealed.

3. The third case $\dfrac{dx}{dt} \to 0$

> in essence is the same as the first case except that in this case, the feature of processes is vividly shown. This situation is similar to those of large quantities with infinitesimal differences and computational inaccuracy of quasi-equal quantities studied earlier. The essence is the meaningless and useless error value computations. However, what is interesting and ironic is that such a common knowledge as "Venice merchants can never cut off a pound of meat" that all nonscientific people can understand and accept was mistakenly seen as the quantitative instability of the second case above in the chaos theory as well as that quantitative instabilities were seen as problems of randomness.

In short, the system of quantitative analysis will always experience the problems of being unable to describe all things or matters using numbers, unable to count exactly, unable to compute accurately, the incomputability of quantitative unboundedness, and others.

The numerical computations Lorenz did on Equation (6.15) did not escape the constraint of stability of the Newtonian system. And, since his calculations were trapped in the computational inaccuracy of quasi-equal quantities, he mistakenly labeled his results as characteristics of nonlinearity.

6.5.3 *Computational Schemes and Lorenz's Chaos*

From what is discussed on computational schemes based on comparisons in Section 6.3 above, it follows that the quantitative instabilities of nonlinearity are not produced by the computational schemes employed, but instead they are natural consequences of explosive growth in quantities, and that the instable direct difference scheme provides solutions closer to the analytic solutions than other schemes. Because of these conclusions, let us first rewrite Equation (6.15) as the following difference equations using the direct difference scheme:

$$
\begin{cases}
X_{n+1} = X_n + (-\sigma X_n + \sigma Y_n)\Delta\tau \\
Y_{n+1} = Y_n + (rX_n + Y_n - X_n Z_n)\Delta\tau \\
Z_{n+1} = Z_n(X_n Y_n - bZ_n)\Delta\tau
\end{cases}
\tag{6.16}
$$

where $\Delta\tau$ stands for the time step of integration over time, the subscripts $n, n+1$ respectively represent the values of the variables at the time moments of $n, n+1$.

Evidently, the mathematical structure of Equation (6.15) or Equation (6.16) is a system of mathematical equations in three unknown functions involving linear and nonlinear terms. Those readers who are familiar with quantitative analysis can see directly from Equation (6.16) that what is described is the trajectory of a mutually reciprocating double spiral in the phase space.

Since Lorenz was especially interested in sensitivities to the initial values, he purposely referred to Equation (6.16) as a problem of sensitivity to the initial value. This no doubt is a statement created by inheriting the concept of well-posedness of traditional mathematical physics equations. As a matter of fact, computations involving products of quantitative variables are not only sensitive to the initial values, but also sensitive to any of the variables as long as products of nonsmall effective values are concerned, including sensitivities to time steps and parameters.

If $\tau = 0$, let us take the initial values of Equation (6.16) as follows

$$X = 3, \quad Y = 2, \quad Z = 4$$

then we have the following.

6.5.3.1 The Results of Integration with Relatively Small Time Steps Using a Nonsmoothing Scheme

If $\tau = 0.02$, then the computational results of the direct difference scheme are shown in Figure 6.9. What needs to be noted is that the direct difference scheme is an instable scheme in the traditional computational mathematics. Since we applied small time steps in our computations, we could have obtained similar results as Lorenz did using a smoothing scheme.

In order to make our points clear conveniently, we repeated our computations as above by using the smoothing Runge–Kutta scheme while keeping the same initial values and time steps. The results are shown in Figure 6.10.

Our computations show that both of these schemes can be continued without causing the computers to stop due to spills except that the trajectories would become denser and denser. By comparing Figures 6.9 and 6.10, it is not hard to see that these two computational schemes produce basically the same results except that Figure 6.10 seems to be smoother and the left spiral leaf a little smaller. Other than these two minor exceptions, these figures do not have any substantial difference. What should be explained at this junction is that the reason we choose to employ the direct difference scheme is based on how the nonlinear instability should be understood. And, in the numerical experiments of the general nonlinearity in Section 6.3 above, we applied this scheme the first time in the numerical computations of systems of equations. So, when the entire integration is carried out to 10 million steps, along with small time steps, no quantitative instability is ever signaled throughout the entire experiment. In fact, the process of integration can still be continued. Considering that the computational results are getting very small (close to zero) without any tendency to increase again, we eventually decided to terminate the computation at the 10 millionth step.

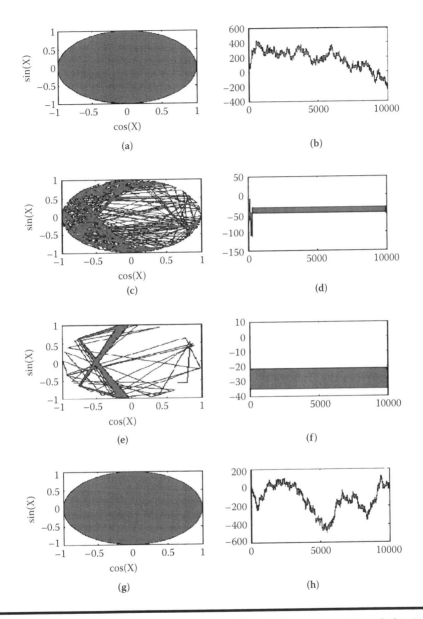

Figure 6.9   Numerical experiments with adjusted time steps. (a) and (b): X = 1.5, A = 12879, B = 1, Δt = 0.00066; (c) and (d): X = 1.5, A = 12,883, B = 1, Δt = 0.00066; (e) and (f): X = 1.5, A = 12,907, B = 1, Δt = 0.00066; (g) and (h): X = 1.5, A = 12,910, B = 1, Δt = 0.00066.

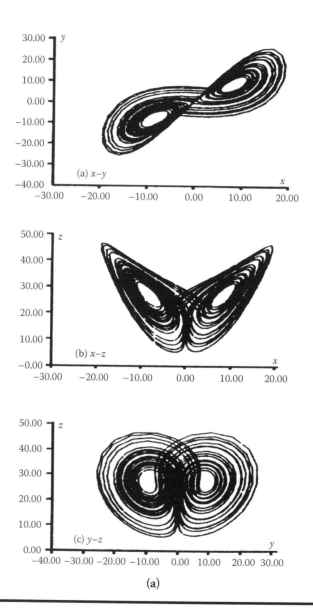

Figure 6.10 Numerical experiments with changing initial values and parameters. (a): Nonsmoothing scheme of the Lorenz model with small time steps.

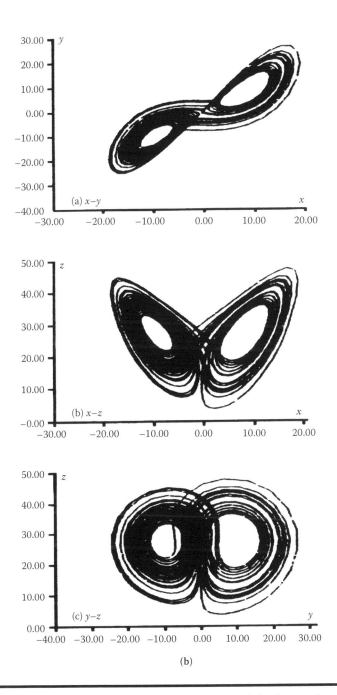

Figure 6.10 Numerical experiments with changing initial values and parameters.
(b): Smoothing scheme of the Lorenz model with small time steps.

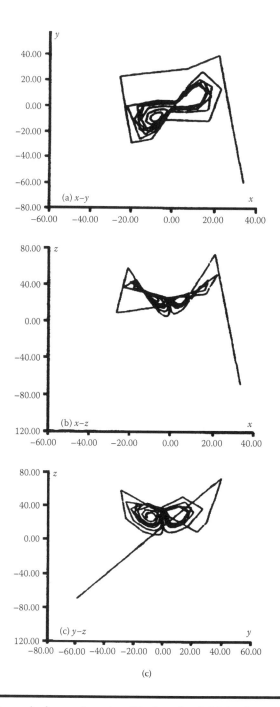

(c)

Figure 6.10 Numerical experiments with changing initial values and parameters. (c): Nonsmoothing scheme with increased time steps.

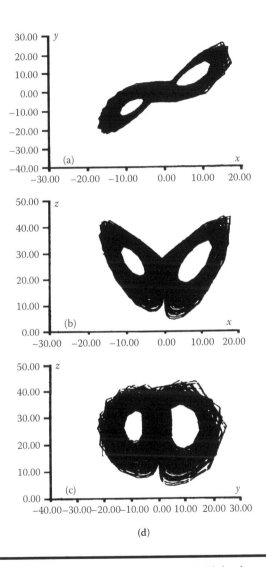

**Figure 6.10   Numerical experiments with changing initial values and parameters.
(d): Smoothing scheme with increased time steps.**

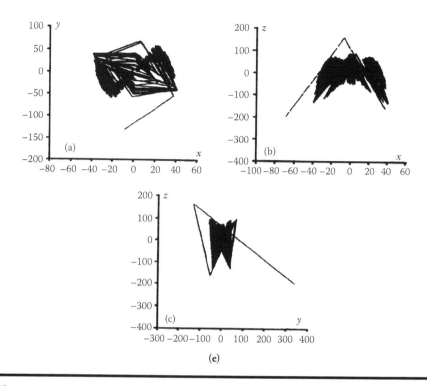

Figure 6.10   Numerical experiments with changing initial values and parameters. (e): Nonsmoothing computational scheme of the Lorenz model with increased parameters.

Evidently, this experiment tells us that when the time step is relatively small; smoothing and nonsmoothing schemes do not make much of a difference. So, a natural question arises: The direct difference scheme has always been believed to be quantitatively instable, so how can it produce roughly the same results as the smoothing scheme?

To address this problem, we traced and analyzed the computational processes on the computer screen. Our results indicate that computational stability has nothing to do with the computational scheme applied; with our setup, it is caused by the computational values trapped in the calculations of large quantities with infinitesimal differences (differences between quantities of equivalent magnitudes). In other words, the actual values participating in the calculations are error values, leading to the trajectories in Figures 6.9 and 6.10 of double spirals. And this double spiral looks analogous to the so-called butterfly, which is completely caused by reciprocating remnant values of the error values in the phase space. It has nothing to do with the "butterfly effect" that a tiny deviation in Australia would cause hurricanes in North America as Lorenz declared. The "butterfly" here is the result of useless and meaningless error value computations and has nothing to do with chaos, randomness, or disorderliness.

If we rephrase what is obtained above in the ordinary language, then the so-called butterfly effect of the chaos theory is Shakespeare's merchants of Venice who could never cut off a pound of meat exactly, or what the ancient Chinese Zhan Yin said: Not all things can be described by numbers.

No doubt, those equations that involve large quantities with infinitesimal differences or the computational inaccuracy due to quasi-equal quantities become incomputable because of their useless or meaningless calculations. At the same time, this end also reveals that equations are not eternal.

6.5.3.2 The Computational Results with Increased Time Steps

Increasing the time step and altering the initial value or parameters can all help the quantitative computations either avoid or enter quasi-equal large quantities with infinitesimal differences. In the following, we will show the computational results in the form of increased time steps. Here, let us recall from Section 6.1 the fundamental characteristics of nonlinear quantitative computations that products of small quantities lead to explosive falls, while products of nonsmall values lead to explosive growth; and only the later situation can constitute the quantitative instability of nonlinearity.

If we take the time step $\Delta\tau = 0.06513178$, then the results computed out of the direct difference scheme are given in Figure 6.11. One obvious character is that the trajectory of evolution — the double spiral, is no longer smooth, leaving behind the signs of forced direction changes, and the computations in all three phase planes spill in the negative Y and Z directions so that the computations have to be terminated. In other words, the trajectory values of the evolution can only be calculated to a certain step (in this example, the computation reached $n = 180$ steps) before they exit Lorenz's chaos (the "butterfly effect" of the double spiral). For the sake of convenience of drawing the graph, Figure 6.11 shows the computational results with the first 20 steps truncated.

Evidently, for nonlinear (singular) problems, the so-called quantitative instability is exactly one of the fundamental characteristics of nonlinearity. Or, we say that nonlinear mathematical models have the characteristics of escaping the continuity of Euclidean spaces and entering the transitionality of non-Euclidean spaces.

At this junction, we would like to mention that if we take $\Delta\tau = 0.06513177$, then if the direct difference scheme is employed, the computational results still enter an "error-value spiral" (or Lorenz's butterfly effect) analogous to that in Figure 6.9. In form, the quantitative difference is only 0.00000001, which constitutes sensitivity to the time steps. So, what needs to be clear is that the essence of this sensitivity is whether or not the computational values are trapped in the error-value spiral of large quantities with infinitesimal differences.

Figure 6.12 shows the computational results using the Runge–Kutta smoothing scheme under the same conditions as above. Its characteristics are similar to the butterfly effect of Figures 6.9 and 6.10. If we apply the smoothing scheme of "one step forward and three steps backward," we also produce similar results. However,

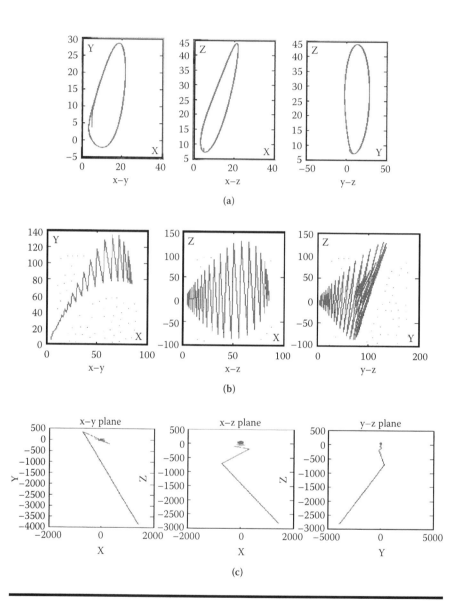

Figure 6.11 (a): σ = 10, *r* = 28, *b* = 7.26853374 – 198, Δτ = 0.008, *X* = 5, *y* = 3, *Z* = 9. (b): σ = 10, *r* = 28, *b* = 198 – 241, Δτ = 0.008, *X* = 5, *y* = 3, *Z* = 9. (c): σ = 10, *r* = 28, *b* > 249.2309892, Δτ = 0.008, *X* = 5, *y* = 3, *Z* = 9, *n* = 69,751.

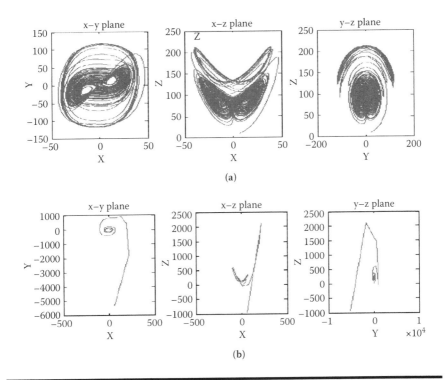

Figure 6.12 (a): $r = 86.72178883513$, $b = 28$, $\sigma = 10$, $n = 80,000$, $\Delta\tau = 0.008$, $X = 5$, $Y = 3$, $Z = 9$. (b): $r = 86.72178883514$, $b = 28$, $\sigma = 10$, $n = 792$, $\Delta\tau = 0.008$, $X = 5$, $Y = 3$, $Z = 9$.

from Figure 6.12, forced trajectory or evolutionary phase changes due to smoothing can be clearly observed.

Our screen tracing shows that the computational results as shown in Figure 6.11 evolve explosively in the negative Z direction, leading to spills and terminated computer calculation. So, the evolution escapes the butterfly effect of large quantities with infinitesimal differences. In comparison, the computational results of Figure 6.12 never enter the negative Z-zone and always stay in the situation of large quantities with infinitesimal differences. This outcome can be seen directly from the third equation in Equation (6.15).

If $Z > 0$, we have

$$\dot{Z} = XY - bZ \tag{6.17}$$

If $Z < 0$, we have

$$\dot{Z} = XY + b|Z| \tag{6.18}$$

Evidently, when the quantitative magnitudes of $|XY|$ and $|bZ|$ are equivalent, Equation (6.17) has to fall into the error-value computations of large quantities with infinitesimal differences. Since the mathematical structure of Lorenz's model (Equation 6.15) is a trajectory of reciprocating computational values, the trajectory of the error-values looks like the pattern of a butterfly. However, it is absolutely not the same as that "when the wings of a butterfly ..., the hurricane will be created in North America." So, all the results shown in Figures 6.9, 6.10, and 6.12 stand for meaningless computations without entering any instability of nonlinearity. What needs to be pointed out is that nonlinear computations possess the characteristics of both extreme instability (products of nonsmall quantities) and extreme stability (products of small quantities).

Besides, we also organized manpower to numerically experiment out various time steps, initial values, and parameters, including such experiments as with variable time steps. The calculations show that when the time step increases from a small value, the computational values can walk out of the so-called chaos or our error-value spiral, leading to spill and terminated calculation. And, if the time step decreases from a large value, then the computational values can enter the error-value spiral. If we change the scheme to smoothing or shrink the parameters right before the computation spills, the computational results can once again return to the error-value spiral. If we employ weak smoothing schemes, such as the second order Runge–Kutta smoothing scheme, or strong smoothing schemes, such as large time steps, large parameters, large initial values, and so forth, we can all make the computation either escape or enter the error-value spiral.

6.5.3.3 The Computational Results with Negative Initial Values

When $\tau = 0$, let us take the initial values

$$X = -3, \quad Y = -2, \quad Z = -4$$

and the parameters

$$b = \frac{8}{3}$$

When $\Delta\tau \leq 0.057927096$, the computational results enter the error-value spiral, the figure is analogous to Figure 6.9 and is omitted. When $\Delta\tau \leq 0.057927097$, which is only 0.000000001 greater than the previous case, the computational results show that as soon as they enter the error-value spiral, they immediately jump out of the spiral and then quickly become instable and force the computer to stop working due to spill (the figure is omitted). However, if before the spill we start to employ a smoothing scheme, then the computation will once again enter the error-value spiral (the figure is omitted).

6.5.3.4 The Computational Results with an Adjusted Parameter

Case 1: For the sake of convenience of our analysis, let us take the same initial values as in Section 6.5.3.3 and $\Delta\tau = 0.05$. That is, this time step $\Delta\tau$ is smaller than those used in Section 6.5.3.3 where the computation enters the error-value spiral. Our purpose of doing so is to analyze the situation of the time step in comparison to the computational error-value spiral in Section 6.5.3.3 with varying a parameter. In other words, when the parameter changes, we do not want to see quantitative instability because of a time step that is too large.

For the varying parameter b, let us look at two situations:

When b is between 8/3~331849, the computational results fall in the error-value spiral. However, in this case, the morphology of the error-value spiral is completely different from those in Figures 6.9, 6.10, or 6.12 (see the graph in Figure 6.13 right before the spill).

When $b \geq 331850$, at step 97,914 the numerical computation suddenly escapes from the error-value spiral and spills in the $Z < 0$ direction, as shown in Figure 6.13. What is interesting is that the trajectory graph in the X-Z phase space looks like a flying bird with both wings widely stretched out instead of a butterfly. Unfortunately, Lorenz did not look at the sensitivity with regard to the parameter; otherwise, it would be a "bird flapping its wings" instead of a "butterfly." What needs to be noted is that in order to show the difference between falling into the error-value computations and escaping from the error-value trap, we only kept the computational results for the steps 96,000~97,914 in Figure 6.13. Later, our student Professor Gongyi Chen used a later-generation computer and did similar computations with a varying parameter. That is how we have the following case.

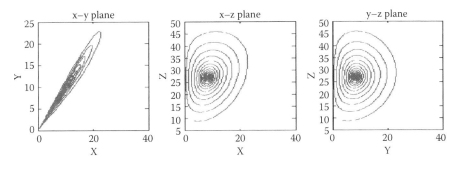

Figure 6.13 $b = 28$, $\sigma = 61.80311085 - 233.1$, $r = 28$, $\Delta\tau = 0.008$, $X = 5$, $Y = 3$, $Z = 9$.

Case 2: Also, for the convenience of comparison, assume the initial value is positive, and take $X = 5$, $Y = 3$, $Z = 9$, $\Delta\tau = 0.008$. That is, in the previous calculations, this time step can make the results fall into the error-value spiral. Similar to the previous case, changes in the parameter b are considered in two situations.

When b is between 8/3–249.23, the computation also falls into the error-value spiral and the graph is similar to Figure 6.9 (omitted). Also, further analysis indicates that when b is between 8/3–7.26853373, the graph is like a butterfly.

When b is between 7.26853374–198, the computational results are depicted as curves as shown in Figure 6.11a. When b is between 198–249.23, the computational results look like a flat or slightly rolled up tree leaf as shown in Figure 6.11b. When $b = 249.2309892$, the computation escapes from the error-value spiral at step 69,750 and spills in the $Z < 0$ direction, as shown in Figure 6.11c. For the convenience of creating the figure, Figure 6.11c only contains the results of up to 69,751 steps.

6.5.3.5 Varying Parameter r

When r is between 28 and 86.72178883513, the computation is trapped in the error-value spiral of Figure 6.12a. When $r = 86.72178883514$, the computation suddenly escapes the error-value spiral at the 792nd step and spills in the $Z < 0$ direction (Figure 6.12b). So, it can be seen that Lorenz's model is very sensitive to parameter r, reaching 1 over 10 billion.

6.5.3.6 Varying Parameter σ

When σ is between 10 and 61.80311084, the computation falls into the error-value spiral, where the shape is like a feather pen. When σ is between 61.80311085 and 233.1, the computation looks as in Figure 6.13. When $\sigma > 233.2$, the computation spills in the $Z < 0$ direction (the figure is omitted).

From the results of the previous series of numerical experiments, it can be seen that Lorenz's model, even if we continue to use the terminology of sensitivity, is not only limited to the sensitivity of the initial values. It is sensitive to the time step, and each of the parameters. And, the sensitivity with regard to the parameters and time steps can be so severe that with the digit located at the eighth place after the decimal point altered, very different results will be produced.

However, this sensitivity is no longer the same as that as introduced by Lorenz. In other words, this sensitivity cannot be understood as that of nonlinear problems, where "a millimeter off at the gunpoint leads to thousands of miles away from the target." Instead, this sensitivity is about whether or not the computation would enter the realm of large quantities with infinitesimal differences. When entering the situation of large quantities with infinitesimal differences, the computation will be absolutely stable, while walking out of the situation of large quantities with infinitesimal differences implies that quantitative instability and explosive growth.

6.5.3.7 The Truncated Spectral Energy Analysis of Lorenz's Model

This problem has been carefully discussed in our book *Entering the Era of Irregularity*. However, to help the reader gain a complete and substantial understanding of the chaos theory, we repeat the materials here.

Multiplying the three equations in Equation (6.15) by X, Y, and Z respectively and then adding the results produce

$$\frac{1}{2}\frac{d}{d\tau}(X^2 + Y^2 + Z^2) = -(\sigma X^2 + Y^2 + bZ^2) - b(\sigma + \gamma)Z \qquad (6.19)$$

Letting

$$E = \frac{1}{2}(X^2 + Y^2 + Z^2)$$

leads to that when

$$E = \text{const.} \quad \text{or} \quad \frac{dE}{d\tau} = 0$$

that is, the truncated spectral energy stays constant, since $b, \sigma, r \neq 0$, we must have $X, Y, Z = 0$; when

$$\frac{dE}{d\tau} > 0$$

the truncated spectral energy increases, so we must have

$$\sigma X^2 + Y^2 + bZ^2 + b(\sigma + \gamma)Z < 0 \qquad (6.20)$$

Since b, $\sigma > 0$, if $\sigma > |r|$, then $Z < 0$ and

$$\left| b(\sigma + \gamma)Z \right| > \left| \sigma X^2 + Y^2 + bZ^2 \right|$$

or

$$|Z| > \left| \frac{\sigma X^2 + Y^2 + bZ^2}{b(\sigma + \gamma)} \right| \qquad (6.21)$$

So, when $Z < 0$ and Equation (6.21) is satisfied, then the truncated spectral energy increases, which exactly reflects the situation that the computation spills in the $Z < 0$ direction.

When

$$\frac{dE}{d\tau} < 0$$

the truncated spectral energy decreases. So we have

$$(\sigma X^2 + Y^2 + bZ^2) + b(\sigma + \gamma)Z > 0 \qquad (6.22)$$

Evidently, if

$$\left| b(\sigma + \gamma)Z \right| < \left| \sigma X^2 + Y^2 + bZ^2 \right| \qquad (6.23)$$

then $Z > 0$. Evidently, if $Z > 0$, then the evolutionary values can be trapped in the error-value spiral and lose all practical values due to large quantities with infinitesimal differences.

If $\left| b(\sigma + \gamma)Z \right| < \left| \sigma X^2 + Y^2 + bZ^2 \right|$, then Z can be either greater than 0 or less than 0. Evidently, if $Z > 0$, the situation is similar to what is shown above. That is, the computation is trapped in Lorenz's chaos. When $Z < 0$ and Equation (6.21) is not satisfied, then constrained by

$$\frac{dE}{d\tau} < 0$$

the energy has to decrease. In other words, $Z < 0$ under this condition can still constitute large quantities with infinitesimal differences so that the computation is trapped in the error-value spiral due to different signs of X and Y.

So, it can be said that by using truncated spectral energy, we have essentially illustrated the chaos phenomenon, or we have essentially pointed out the principal problems existing in the chaos theory. Considering the impact of the chaos theory on the scientific community of the time, on the basis of the works presented earlier, we also conducted serial computations on other well-known chaos models, such as the Brussel oscillator, the forced Brussel oscillator, Rössrel, and other models. These computations and related works cost us a great deal of time and energy, leading to over 10,000 figures generated. Interested readers can locate some of these works in relevant publications.

As affected in the second half of the 20th century by the thought wave that God plays dice, we originally planned to employ numerical computations of nonlinear equations to support the concept of disorderliness of the opinion that God plays dice. However, our work instead showed that the error computations of large quantities with infinitesimal differences were mistakenly treated as the disorderliness; otherwise, there would not be the one-time fad of chaos theory. If chaotic phenomena or disorderly events were seen as the quantitative instability of nonlinearity or quantitative singularity so that the concepts of randomness and indeterminacy were introduced, then Lorenz's chaos theory would not have helped any. Even so, what is found is that the tools of quantitative analysis do not have the ability to deal with problems of singularity, "disorderly" peculiar events, or irregular events and information. Or speaking conversely, disorderly or irregular events or information do not comply with the calculus of formal logic. However, that does not mean that disorderly or irregular events do not follow their own laws of physics except that these laws are not written in the form of quantitative formal logic.

6.5.3.8 Discussions on the Phenomenon of Lorenz's Chaos

The phenomenon of Lorenz's chaos is nothing but the fact that not all things can numbers describe or the problem of Shakespeare's merchants of Venice who can never cut off a pound of meat under the condition of quasi-equal quantities. Based on the previous numerical analysis and the truncated spectral energy analysis on Lorenz's chaos model, we see the following.

The results of analysis on the changes in the truncated spectral energy agree with the results of the numerical experiments, where the former method can more convincingly reveal the problems of essence. What is important is that even if we follow what Lorenz did, we still cannot obtain the result that chaos is a consequence of increased changes in the truncated spectral energy. Since the computational instability has been constrained, only decreasing energy changes can correspond to his chaos. Evidently, decreasing energy changes ($dE/d\tau < 0$) are a realization of approaching invariant energy, that is, E ≈ a constant, indicating that the underlying movement approximates the invariant initial value. In essence, it is a realization of the Newtonian quantitative stability. In other words, nonlinearity is different from linearity because of its fundamental characteristics of approaching quantitative instability, leading to chaos or disorderliness. However, Lorenz mistakenly treated the problem of computational inaccuracy of stable error-value computations as the quantitative instable chaos of nonlinearity.

Each realistic chaotic or disorderly phenomenon stands for a problem of $dE/d\tau > 0$. That should be problems of singularity and explosive growth (that does not include the explosive fall of the products of small quantities) caused by quantitative instability existing in the computations of nonlinear problems using nonsmoothing schemes. The so-called sensitivity to the initial values is not Lorenz's butterfly flapping its wings of the initial values. And the true sensitivity comes from the quantitative instability or singularity of nonlinear mathematical models. So, the phenomenon of Lorenz's chaos is not any realistic, disorderly, irregular chaos; and his chaos cannot be seen as a characteristic of nonlinearity, either.

All realistic phenomena that can reflect nonlinear chaos and all singularities of mathematical models that cause computational values to grow or fall explosively are all eliminated by Lorenz's computational schemes. Nonlinear quantitative instabilities are exactly the key characteristic of transitionalities. Or in other words, blown-ups or "deaths" of the original initial value systems appearing in the evolutionary values of the underlying evolutions are the essential characteristics of nonlinearity, which indicates that nonlinear models possess the attribute of destroying their own initial value (or parametric) systems. So, mathematical nonlinear evolution models are closer than linear models to the objective and realistic evolutions that chaos implies change, and at extremes, matters evolve in the opposite direction. The phenomenon of Lorenz's chaos is a realization of error value computations of large quantities with infinitesimal differences of the stability problem of quantitative calculations. It has nothing to do with problems of evolution. Even out of the computations of large quantities with infinitesimal differences of linear equations, one can also produce the phenomenon of Lorenz's chaos except that the form or state of expression is not as severe as that of nonlinear equations.

Out of our huge amount of computations, we not only gained understanding of the essence of quantitative computations and verified the fact that the system of quantitative analysis cannot handle irregular information, but also were pushed to enter the era of irregularity so that we increased our speed to investigate irregular information by employing digitization.

Acknowledgments

The presentation of this chapter is based on Lin et al. (2001), OuYang (1984, in press), OuYang and Lin (2006a,b), OuYang et al. (2002a,b), OuYang and Wei (1993), and OuYang and Wei (2005).

Chapter 7

Evolution Science

Many times earlier we have mentioned that modern science treats m as a constant by particlizing materials, which leads to the concept of inertial system and Newton's second law. What this statement means is that modern science is about movements without any change in materials. That implies that evolution science should be the investigations on changes in materials so that the materialistic m can no longer be particlized. No doubt, as long as the material of concern is not a particle, one has to represent its structurality. The materialistic structurality has to contain forces due to its gradients (since forces exist within materials' structures, there is no need to count on the help of an external "God"); so moments of forces are formed by arms of forces of the structures. Since the probability of interacting exactly on the centers of mass is extremely small, the common form of motion under interactions should be that of rotation. In the physical world, rotations naturally cause collisions, leading to injuries and damages in materials. No doubt, such injuries and damages have to cause changes in the structures and morphologies of materials. That explains why each evolution is closely related to spinning motions.

Modern science has been in operation for over 300 years. One of the first problems people encounter is that based on modern physics, no one can really tell why and how materials age. The essence of why the concept of evolution has been limited in modern science to those of geologic ages and astronomical times is that modern science is such a "science" that does not deal with changes in materials. However, neither Newton nor his followers ever clearly explained this problem, while hinting that modern science could resolve any problem, which even caused Einstein and others to believe that the future was a continuation of the initial value. This end is very well illustrated by Laplace's self-declaration (Kline, 1972): "From knowing the initial value, I can foretell everything about the future," in his promotion that the future is the same as the present, where the present is evidently his initial value so that statistics is nothing but replacing "the future is the same as the initial value" by "the future is the same as the past." At the same time that Laplace's statement becomes a celebrated dictum, it also very well provides a positioning for modern science. That is the reason why modern science has pushed evolution away to the geological vicissitudes of geoscience and the problems of heavenly scale disasters of astronomical science.

So, two natural questions of primary importance arise: What kind of science is modern science? Why didn't it resolve the problem of what time is? What time is, how to apply time, and so forth should be seriously considered in physics, and appropriate methods of application should also be provided, instead of just dwelling at the level of epistemological debates.

No doubt, the purpose of our publishing the book *Entering the Era of Irregularity* (OuYang et al., 2002a) is in essence to explore the rules of operation, if any, that govern the changes in materials or events, and to address problems such as the following and others: What kinds of problems do irregular events or information stand for? Do they have anything to do with changes in materials and events? How can they be dealt with? However, limited by historical reasons and streams of consciousness of our time, in principle, our book *Entering the Era of Irregularity* only points to the problems without revealing the underlying reasons. So careful readers have readily sensed that we had more to say beyond what is in that book. They sent us numerous requests expressing their wishes for learning what would follow *Entering the Era of Irregularity*. As urged by the readers and our students, in this book we systematically present our recent works on evolution science and the system of information digitization.

The establishment and development of modern science in essence were originated from Copernicus's concept of heliocentricity over 500 years ago, which triggered the debate of the 15th century that led to the death of G. Bruno and the imprisonment of Galileo. The system of modern science, which was later completed by Newton, Einstein, and their followers, and when specified to physics is characterized by the invention of the concepts of inertial system, force, and mass, becomes such a system of analysis that quantities are employed to represent events. The corresponding problems of physics, at least in terms of problems of wave motions, can be resolved by solving well-posed dynamic equations, leading to glories of modern science. That is why over time, quantitative analysis gradually becomes the method of science, leading to such belief that the scientific system, as completed by Newton and his followers, can address any problem, and those problems it cannot address must be quasi-problems. What is not truly recognized in the system of modern science is that quantitative analysis can be employed to describe wave motions because wave motions can transfer energies in the form of vibrations while maintaining the invariance of materials. Unfortunately, wave motions are not the only form of movement of materials. And, what can be very clear now is that wave motions are basically an approximated motion at the human bodily scale under the condition that no material can be damaged; the more general and fundamental form of material movements is that of rotations in curvature spaces where damages and changes in materials accompany the movements. So, when modern science treats all micro- and macroscopic problems by using methods developed for wave motions, it has to experience difficulties when faced with problems involving changes in materials or events. The reason why turbulences have been an extremely difficult, unsettled problem is because the turbulences are originated in rotational subeddies — the chaotic flows are about irregularity and variability. Since the time when modern science advocates the concept of mathematical waves, there has appeared the confusion between the concepts of eddy motions and wave motions. That has surely constituted a trap for modern science itself.

Besides, due to historical reasons, modern science has been quite slow on addressing a whole series of problems like what time is, what events' irregularities are, why there are irregularities with events, how we can make use of the irregularities, and so forth. As one of our attempts to face this end, we published our book *Entering the Era of Irregularity*.

7.1 Specifics of the Concept of Noninertial Systems

7.1.1 Dualism, Materials, Attributes of Materials, and the Concept of Noninertial Systems

In applications, the system of quantitative analysis, as the methodology of modern science, mainly employs well-posed dynamic equations and the statistical methods of stable series. The essence is either quantitative regularization or stabilization of time series without answering what irregular or small probability events or information are. That is where and how we established the positioning of modern science and the conclusion that information cannot be altered, and published related papers. As a problem of epistemology or philosophy, irregular information touches on such fundamental problems as Lao Tzu's "whatever is described as Tao, it is not the Tao." It is exactly with fundamental problems that modern science deviates from the essentiality, multiplicity, and variability of natural events; instead, it pursues the generality and the prescribed unity by using formal quantities.

Evolution science is the science that investigates material changes that can be specified as movements of variable accelerations and rotations. That is why evolution science will not be a part of modern science and will be able to deal with nonlinearity and problems of noninertial systems. So, it is because of the existing problems unsolvable by using modern science that evolution science is different from modern science. In particular, modern science avoids materials' and events' structures and attributes through the inventions of physical quantities and parametric dimensions; by replacing structural transformations of events with quantitative formal logical calculus, physics is established in modern science.

So, a new problem appears: If the quantitative formal logical calculus cannot be used to replace transformations of events, then what rules do these transformations follow? In particular, the current formal logical calculus of the quantitative system, including dynamic equations and statistical methods, cannot satisfactorily deal with irregular events or small-probability information. So, it is natural to ask: What are the physical meaning and effects of irregular events and small probability information? And, even if the physical meaning and effects are clearly understood, then how can irregular events and small probability information be practically employed?

Newton's second law places *f* outside the object without clearly spelling out where the force is from; so this law naturally cannot tell where the acceleration *a* dwells. Modern science declares the existence of *f* and *a* independently outside the object, naturally making physics lose its eventfulness at the level of foundations. Our earlier analysis indicates that the invention of physical quantities is done at the very foundations of modern science and that at the same time this invention also implicitly brings forward with it difficulties. It is because physical quantities have essentially separated material or events' attributes from the underlying materials and events while making these independent attributes act back on the materials or events. For example, the acceleration *a* of a moving object *m* originally dwells on *m*. However, it is taken artificially out of *m* and is then multiplied back with *m*. Similarly, forces exist within materials. However, they are also separated from the materials in the form of physical quantities and then treated as the "God" that operates on the materials. If events were seen as the essence, then all these symbolic operations would become unthinkable. However, the opposite has been true: These operations are the laws of modern science. That explains why the physics of physical quantities that complies with the formal logic of quantities naturally cannot give an adequate expression to the essence of events. Any material in motion experiences mutual reactions, leading to age and damage. That is why the system of modern science cannot successfully resolve the problem of aging in materials.

Besides, materials stand for the primary problems of physics, since without materials there would not be any physics. The corresponding physical quantities, such as forces, accelerations, and so forth, are only attributes that depend on materials. Without materials, neither *f* nor *a* would exist so that the physics of physical quantities would not exist. Now, looking back through history, the reason why modern science did not enter the field of evolution seems to have something to do with the method and starting point on which problems are attacked.

Therefore, the essence of modern science is the investigation of problems of invariant materials, leading to the invention of inertial systems. For evolution problems with changing materials, they have to walk out of inertial systems due to variable accelerations. So, when employed to study changes in materials, the system of modern science has to face the problem of theory not meeting the challenge of practice. That might be the reason why in the past 300-plus years, no scholars who are able to foretell the future were ever produced out of the system of modern science. According to Aristotle, physics is a science that investigates changes and processes of change in materials, and anyone who is unable to predict the future cannot be known as a physicist. After all, changes in materials are the basic foundation of physics, which is why problems of evolution will appear sooner or later. They are both the need of our modern time and the necessity of scientific development.

When people pursue freedom with prices as high as sacrificing their lives, the science applied is still stagnating in the "slave" era of the slaving first push. This seems to be an inconsistency of our time, since science should start to change ahead of time and call for thinkers and activists in order to bring about changes in time.

7.1.2 Quantitative Formality and Variability of Events — Existence and Evolution

In previous chapters, we have already spent a good amount of time and space discussing problems existing with quantities. To a certain degree, the introduction of numbers is originated in various peoples and their individual cultural backgrounds with differences in recognition and comprehension. As we have pointed out, Pythagoras, an ancient Greek school of learning, believed that numbers are the origin of all things. That is, the ancient Greeks were the first to treat the existence of quantities as prior to that of materials and the most fundamental reasons for all materials and events. The latent influence of this belief has been substantially felt in the entire spectrum of modern science, leading to the opinion of our modern time that time is prior to existence.

The ancient Greeks believed that without comprehending numbers and properties of numbers, one could not fathom all things and their relationships of interactions. They made materials and events second to numbers, while believing that events and their relationships are determined by quantitative logical calculus. Although the later scientific system of the West did not entirely evolve along the direction of Pythagoras, at the epistemological basis or philosophical thinking, this system has definitely inherited the proposition that numbers are prior to materials, leaving a space for "God." Even after the birth of the western materialism, along the written lines of materials being the first, the "heavenly" position of quantities still exists in the root system of beliefs, producing such philosophical laws as "quantitative changes lead to qualitative changes" (where quantitative changes become the cause of material variance), "without mathematical proof, no theory can be seen as scientific," and so forth. For quite a period of time, these beliefs have greatly affected the development of fundamental science and substantially maintained the invariance of the system of modern science. Although people can recognize and take part in societal development and reforms, any truly meaningful and effective societal change has been originated in changes and developments of science and technology. And, in these changes and developments, the scientificality problem of the law that quantitative changes lead to qualitative changes is deeply involved.

There is no doubt that without things, there would not be any quantity. In other words, it is very clear that quantities appear after events. This end definitely spells the primary position of materials and the consequentiality of quantities. As early as in the dynasty of West Zhou in China, people had already recognized that human knowledge comes from the analysis of materials and events instead of from analyzing quantities. And, Zhan Yin, a famous quantity calculator of the Warring States era of ancient China, was the first to point out that not all things can numbers describe. One should note that in the physical world, there does not exist "zero material," while the mathematical world has to have number zero. Without the number zero in mathematics, the entire system of numerical computation would be paralyzed. And, one should also note that the invention of number zero has essentially denied the belief that numbers are the origin of all things. It has been verified by experts that the number zero was an invention (not a discovery) of ancient India. In the 7th century, it was brought to the West. Only after the invention of the number zero and being brought to the West, the calculus system of quantitative formal logic was developed. It is exactly because of the quantitative zero that there appeared operations of numbers, which also makes it inappropriate to substitute the analysis of materials and events by quantitative operations.

Chinese people have always held the belief that events are the substance. In ancient China, the legendary peasant tasted all kinds of grass in order to observe these materials' physical attributes, such as their tastes, smells, and their medical effects. By using the system of knowing the world from analyzing materials, this legendary man and his followers established Chinese herb medicine, which has made its important contributions in the over 5,000-year history of China. This history of course does not mean that Chinese people did not know about numbers; instead they had long ago realized that numbers could not reveal the attributes and functions of materials and the specifics of these attributes and functions. In *Entering the Era of Irregularity*, when we pointed out that all laws of modern science are quantitative laws, no one seemed to object to our observation; however, when we declared that all laws of modern science are laws of results, several readers felt that we tried to secretly set up a trap of formal logic to trick people. Pressured by the huge investments made in the scientific community in the attempt to resolve the problem of foretelling the future, we pointed out that quantities are of the attribute of being postfact and that the future cannot be foretold by using quantities only. No doubt, being postfact means the hindsight of the laws of results; the formal measurements of quantities are about the fact that numbers cannot address the states and attributes of events. We used the example of flipping hand and covering hand, where the former hand indicates clouds and the latter rains, to illustrate that the hand in both situations represents the quantity "1." However, this number "1" cannot tell people the state of either flipping or covering, and of course it is unable to distinguish the properties of the flipping and covering hands. Evidently, some readers have already realized the message we would like to deliver. That is, the purpose of investigating problems of change is about predicting the future so that the postevent formal tools can no longer be employed in these studies.

What is more important is that due to the need of survival and challenging disasters and diseases, since the earliest recorded time in history Chinese people have been more aware of the overall states, functionalities, and attributes of materials, such as properties, tastes, sounds, colors, relative positions, and so forth, and recognized that numbers always suffer from the weakness of being unable to describe clearly. That is why in the *Book of Changes* (Wilhalm and Baynes, 1967), known as a signature of Chinese civilization, the method of eight diagrams, which take characteristics of material structures by looking at the object at a distance and the body itself when nearby, has been employed to know and explore the world. That might have explained why at the root of its civilization, modern science could have appeared in China. That is, at the very start of scientific exploration, the Chinese had fastened their minds by investigating changes in materials. And, variability problems of materials are exactly problems of nonlinearity and noninertial systems, as we discussed earlier, and were referred to as *ge wu zhi zhi* (know the world by analyzing materials and matters) (see the representative work of West Zhou dynasty, named *da xue*, "higher learning"), instead of *ge shu zhi zhi* (know the world by analyzing numbers) of the western world or Pythagoras of ancient Greece. So, due to varied living conditions, the difference between Eastern science and that of the West is originated in the difference of how to know the world. That explains why modern science did not appear in China. However, it does not mean that there is no science in China. In other words, modern science is not the sole representative of scientificaility. From this discussion, it becomes easier to understand why Newton could have written his *Mathematical Principles of Natural Philosophy* and later why such an opinion as that quantitative changes lead to qualitative changes could have become an important philosophical law of the West, leading even further to the current belief that quantitative comparability is the only standard for scientificality. All of these together with other related works of the West have constituted the system of *ge shu* (analyzing numbers) of modern science. What should be clear is that the essence of Newton's *Mathematical Principles of Natural Philosophy* is how to mathematically describe principles of nature with calculus theoretically established on the assumption of continuity that has never been followed through in applications. Or, one main "achievement" of calculus is the independence from "specific paths taken." However, this result does not mean that physical phenomena also have nothing to do with any particular path. Therefore, Newton's *Mathematical Principles of Natural Philosophy* has clearly indicated how modern science employs the philosophy of *ge shu zhi zhi*, and it essentially did not escape the shadow of the Pythagorean belief that numbers are the origin of all things.

It should be noted that mathematics originated from the study of quantitative measurements so that it has become one of the most important components of human knowledge. Due to the need for survival, it gradually evolved into the symbol of human wisdom. No doubt, human wisdom is also a reflection of wealth, while the measurement of wealth cannot be done without quantities. However, in terms of the foundations of human wisdom, quantities cannot exist without the existence of materials. As the foundation of fundamental knowledge, quantities have to depend on materials and events. That leads to the quantities' attributes of being postfact and formal. In other words, numbers exist dependently of *things*, which can be seen in that quantities do not occupy any material dimension. This has essentially indicated that quantities have nothing but formality and cannot appear before events.

Now, it becomes clear why we have questioned the philosophical law that quantitative changes lead to qualitative changes. Besides, when facing the problem of predicting major natural disasters, we are acquainted with irregular events that quantities cannot deal with. So, we realized that the manipulations of the entire system of quantitative analysis, as a natural consequence of the constraint of the mathematical principle of well-posedness, are the quantitative logic of the already-happened initial values and parameters. In other words, the outcomes of quantitative manipulations are based on the initial values and parameters of the past and "clones" of the entire initial-value and parametric system under the restriction of stable initial values and parameters. The formal logical changes in the corresponding quantities themselves do not really satisfy the "wishes" of manipulatibility of quantities. That is why the ultimate results of quantitative manipulations become translations of the initial-value and parametric system without any process of manipulation. Even according to the law that quantitative changes lead to qualitative changes, under the restriction of mathematical well-posedness quantities have lost their potential to change so that there is no longer any need to talk about qualitative changes. Besides, the substantial support of modern science comes from the development of calculus, while the core contribution of calculus is its irrelevance with any specific path and invariant translations of the original functions, and at the same time, the physical world is exactly the opposite where particular paths matter and the original functions do vary. This end constitutes the problem that the formal logical reasoning of quantities cannot reveal the logical transformation of events themselves. In other words, in terms of the relationship between events and the laws of changes of events, it has not been shown mathematically that events can be identified with quantities and satisfy the operations and logic of quantities. To this end, in mathematics, there is a branch of study named the logic of mathematical physics, emphasizing that the individual characteristics of elements of sets are not considered, which in essence has admitted that quantities are incomplete.

The conclusion that quantities are postfact, formal measurements has clearly revealed that without things there would be any quantity, and with invariant things quantities would not change. The statement that quantitative computations can produce more accurate results for nonexisting matters than for those that actually exist is not just a satirization on the quantitative tools but also a realistic function of quantities. It provides a specific expression for the incompleteness of numbers and the fact that not all things can numbers describe. In particular, since quantities are appendages of Euclidean spaces, they become formal measurements on the imaginary numbers lines in Euclidean spaces, leading to the problem of quantitative unboundedness and quantitative infinity. That creates the crippling difficulty for quantities that they cannot even resolve their own problem of infinity. However, each realistic physical space is curved, any growth in such curved spaces experiences transitions before even approaching the quantitative infinity. This fact reveals the limitation of quantitative analysis and is one of the specific differences between the Eastern and Western ways of knowledge exploration.

Cultural differences are created by those existing in the geographical conditions and environments so that people employ different methods to solve their individual problems. Different methods of knowing and exploration naturally create different spheres of livelihoods and regional streams of consciousness. No doubt, in the learning atmosphere that not all things can numbers describe, in the Eastern philosophies, the Chinese would not have made laws of nature to depend solely on quantities. However, not solely depending on quantities does not mean that China has never had schools of learning. By using the states and attributes of materials, such as functionalities, tastes, sounds, colors, and positions, to understand the world, there must involve characteristics and mutual relationships of materials. That is why Chinese established their theory of yin and yang and that of five elements — the theory about states and mutual reactions of materials, which is a system of logical transformations of structural characteristics of materials. When one stands at the level of formal logical manipulations of quantities, he would not be able to fathom logical transformations of structural characteristics of materials. Similarly, based on logical transformations of structural characteristics of materials, he would experience difficulties in comprehending quantitative formal logical manipulations.

For example, the school of quantitative logic criticizes the transformation theory of five elements (gold, wood, water, fire, and earth) for its usage of the meanings of these elements in the logical transformations of mutual constraints and collaborations by labeling this theory as not deductive. On the other hand, the school of five-element logical transformations directly finds faults with the theory of quantitative logical manipulations, where by using numbers, all things can be added together so that no one knows what is meant by the results; without the existence of zero materials, there is the quantity zero, so how can quantities be used to substitute for events? When zero is used to multiply any given number, the result is zero. So, if quantities could be employed to substitute for materials, then zero would melt the entire universe into nothing …. All these aspects of the quantitative analysis have to make people laugh.

Using today's language, the essence of the five-element theory is about problems of noninertial systems. That is why there are *xiang ke* (mutual constraints) and *xiang sheng* (collaborations) and other related words in the Chinese language; these words have been widely used in Chinese daily lives. Some scholars believe that Eastern philosophies focus more on the holistic view. Our earlier discussions indicate that this stereotypical description is too general without specific clarification.

The difference in ways for the East and the West to embark on scientific knowledge is that using the inertial system of the first push, the West entered the realm of science, while advocating separation of disciplines; in the East, China has traditionally investigated natural and social sciences as a whole by using noninertial systems of mutual reactions — comprehending natural laws and designing how to practically deal with matters at the same time. That is, the East and the West have completely different starting points. With this understanding, the *Book of Changes* could be translated into the *Theory of Evolution* in modern terms and essentially studies changes in materials in the form of mutual reactions. That is, the East has long ago entered evolution science. On the other hand, due to the eternity of materials, the West entered modern science ahead of the East in the form of the first push.

No doubt, movements with change and those of invariant materials stand for problems of different calibers. Their epistemologies and methods of treatment are naturally different. Evidently, due to the formality of quantities, modern science has made its important contribution in terms of the general knowledge on the movements of invariant materials and problems of vibrations using quantitative formal logical manipulations. However, it is also because of the formality of quantities that modern science is unable to enter evolution science, although it has been wandering outside the front door of the evolution problems of materials' mutual reactions. However, the Eastern and Western sciences can be combined together to form a more complete system of natural science by using the concept of rotational stirring energy. In short, to trace sources and to locate reasons, which are the tasks of science, considering the fact that the changes of rotational materials do not comply with the formal logical manipulations of quantities, changes in events should not follow the formal logical transformations of quantities, either. In particular, variable events' interactions, which reflect the logical transformations of structural characteristics of materials of the five-element theory, cannot be simply replaced by quantities. That is the origin of ancient Chinese science. From pursuing the durability of commercial goods, modern science developed the formal logical manipulations of quantities; by imposing quantitative stability of wave motions, which does not comply with the logic of structural transformations of the five-element theory, modern science has maintained the invariancy of materials.

Evidently, both the structural transformations of the five-element theory and the formal logical manipulations of the modern science system have their individual purposes and specifics. However, when influenced by a school of thought for a long time, people tend to have emotional troubles in accepting other ways of thinking, historically leading to losses of valuable lives. In particular, in our modern days, it has become difficult for people to accept the logic of transformations of events under mutual reactions. These transformations are not currently expressed in the form of mathematical formulas but digitization of states and properties of changes. In terms of problems of change in movements, the nonquantification of "rotational directions" of materials is emphasized.

What deserves our special attention is that the focus of the structural trans-formations of materials is on the changing materials and related events, while the formal logical manipulations of quantities are designed to handle problems of movements involving invariant materials or events. That is also the reason why quantitative logical manipulations cannot deal with irregular events or small prob-ability information. This end also indicates that irregular events or small probabil-ity information stand for variant events or information, which is a problem modern science in essence has not recognized. In other words, based on logical reasoning, the reason why formal logical manipulations of quantities cannot contain chang-ing information in their discourse is because variable information does not comply with the quantitative calculus of formal logic. That is why the *zhi ji* (minor, small changes) of the *Book of Changes* stand for the central problem of predicting changes in events.

Leibniz, a founder of calculus, recognized that events are the essence. However, his work in principle did not involve variable events. In terms of evolution science, what Leibniz said should be restated as variable events are the essence.

7.1.3 *Rotational Movements and Material Evolutions*

No doubt, people always face problems like why humans go through the stages of birth, age, illness, and death, why vegetation experiences blossoming and wither-ing, and why materials always age. And in terms of basic science, these are also unavoidable problems. The corresponding studies of physics should at least provide an answer as to why materials age. If physics is about the study of changes and pro-cesses of change in materials (Shakespeare), then it should provide the reasons why materials age and the corresponding process of change so that scientists who are able to foretell the future can be produced. So, even according to Shakespeare, the physics of modern science still cannot be seen as adequate or is at least incomplete. Science is about not only renewing the structure of human knowledge, but also chal-lenging the wisdom of mankind. If we say that over 300 years ago using quantities to substitute for events, leading to the development of the quantitative formal logic system, can be seen as a challenge to the human wisdom of that time, then the sci-ence of the probabilistic universe, established 300 years later by continually employ-ing the quantitative system, even if this science is correct, can no longer be seen as a challenge to human wisdom. It is because substituting events with quantities is after all a past idea. Our previous discussions indicate that one of the problems that can truly challenge human wisdom is the study of feasibility of substituting events with quantities and how to extract information of change from events.

We once pointed out that in the 20th century, the scientific community estab-lished the realm of wave motions on the basis of mathematical achievements, which for a period of time made wave motions the most general and common form of movements without pursuing why eddy motions and turbulences have been some of the most difficult open problems in modern science. Instead, these open problems were seen from the angle of indeterminacy without fully understanding uncertainties. In other words, even in terms of uncertainties, one should also first address what uncertainty is.

As a matter of fact, rotations are the fundamental movements of materials and of the most generality, while the so-called waves are just a phenomenon of vibration in rotational motions. Modern science has avoided rotations because it has met with difficulties in using quantities to deal with problems of rotation. However, the exception is that V. Bjerknes (1898), a Norwegian scientist, established his circulation theorem, which has been seen as an outstanding work within modern science by simply employing integrations of circulations of nonparticles. This result was the first in modern science that declares that the fundamental motions in fluids, such as the atmosphere and oceans, are rotations (as a matter of fact, the main form of movements of solids is also rotations). It can be said that this Bjerknes circulation theorem has challenged the human wisdom of the time. Unfortunately, this work is not widely known in the scientific community so that as of this writing, it is still only listed in textbooks of meteorology, where it is briefly mentioned and then is packed up and stored away. If meteorological science was developed along the direction as outlined by the circulation theorem, then we would not see the present difficulties in the forecast of disastrous weathers; and in terms of the whole of natural science, there would not have appeared the series of inexplicable theories produced by linearizing nonlinear problems. Or at least during the end of the 19th century or the beginning of the 20th century, the scientific community would have entered the era of evolution science. This discussion clearly explains why modern science has limited the concept of evolution in the studies of problems of change with extralong time scales, such as geological ages and astronomy, because changes involving extralong time scales cannot avoid rotations of celestial bodies.

Next, based on the thinking logic of the first push, modern science limits kinetic energy only as the square of speed, including Newton's V^2 and Einstein's c^2, so that a natural question arises: What is the square ω^2 of an angular speed, where the angular speed contains the speed and provides an expression for the spatial uneven distribution of the speed? Since the square of angular speed is different from that of speed, we are forced to introduce the concept of stirring energy. In essence, the contemporary scientific community does have the ability to consider this problem. However, as constrained by conventional concepts, modern science has missed the energy of rotations. What is more important is that modern science has too much focused on its thinking in Euclidean spaces so that the laws of physics about the existence and transformations of materials in curvature spaces have been completely ignored.

In fact, the concept of stirring energies can be employed to transform the first push of modern science into changes in interacting materials. It is because as long as nonparticlization is concerned, even if we are still in the first push system, the probability for a force to directly act on the center of mass is nearly zero so that rotation movements are created. In terms of interactions, as long as the point of action is not on the center of mass, a rotation has to be produced. And, due to the complementary duality of rotations, where inward and outward rotations coexist, different directions of movements collide into each other creating subeddies. And, collisions of subeddies lead to sub-subeddies … until irregular eddy currents are formed and transformed into heat energy, which makes up of a new structural form and new state of materials.

So, in terms of the process of evolution, due to the unevenness in material structures, uneven arms of forces combined with gradient forces created out of uneven structures produce rotational movements of materials. That is, uneven structures of materials "produce" rotational movements of materials. In the alternative transformations of the different spinning directions in the duality of rotations, new, uneven materials are "created" through transformations of heat, and so forth. Hence, rotational movements play the central role of transformation and conduct the birth and being-born-again evolution or the process of evolution of all things in the universe. As for the physics that expresses and represents changes and process of changes in materials, it should be able to answer the question and describe the process of aging in materials, while creating scientists who can foretell the future. That is why we once concluded that the true physics should be the physics of processes and that investigating the process of changes in materials should be a particular task of the postmodern science and would be the development direction of modern science.

Mr. D. N. McNeil many times suggested us to change the concept of second stir to that of the "first stir," while renaming the currently known first push to the "second push." Considering the path of scientific history, since the phrase of first push was first employed and the term of second stir was initially introduced in our various publications, in this book, we continued this historical fact without altering anything as suggested by McNeil.

Our suggested process physics is different from modern physics, where variable accelerations and rotations naturally introduce changes in materials — evolution, and the mathematical formula of Newton's second law also becomes a nonlinear equation. So, in simple language, noninertial systems can be expressed as follows: The process physics is concerned with rotational, changing materials or events and mathematical nonlinearity does not belong to the modern science of the inertial system.

It is exactly because stirring energies include traditional kinetic energies of modern science that it can also contain modern science. That is also why we mentioned postmodern science in *Entering the Era of Irregularity* without specifically spelling out the reason. It can be said that the postmodern science of interactions can make modern science a phenomenon of a historical time period through investigations of problems of rotations, while it indirectly illustrates the significance and functions of rotations in curvature spaces. This postmodern science also deals with the problem of what time is, which has been an open problem in modern science, explains how to correctly comprehend the meaning of irregular or small probability information, and truly enters the era of digitizing information.

7.2 What Is Time?

7.2.1 The Problem

What is time? No matter whether it is seen as a proposition or not, this problem of time should correspond to the physical essence involved as a component of the basic knowledge of mankind. The system of modern science even openly admits that from Monroe to Newton to Einstein, including the most recent Prigogine, nobody has resolved this problem of time. After all, the problem of what time is should be one of the most fundamental problems of science. Or speaking more accurately, without understanding what time is, our basic knowledge about nature could only be nearly nothing. Historians often describe ancient ages as those without much true knowledge. However, without knowing what time stands for, in essence, modern man is analogous to those of the ancient past of knowing nothing. So, answering what time is in essence challenges the wisdom of man.

However, when people are still arguing whose work is more important, it seems that only a few scholars noticed that the problem of what time is has been clearly addressed in the *Book of Changes*. To this end, we have to wonder during the past several hundred years or 3,000 years what people have been doing and thinking about. Accompanying the problem of what time is, there are also many meaningful problems, such as: Where is time? What is force? Where is force located? What is energy? Where is energy stored?

Glancing through the *Book of Changes* makes it clear what time is: Variable events are the essence. So, when Leibniz, a contemporary of Newton and a cofounder of calculus, pointed out that events were the essence, while criticizing Newton's quantitative time, should be seen as an understanding with relatively lofty realm. This statement also makes us realize why modern science has placed the problem of time aside and no followers of Newton continued to investigate this problem because in the conscious thoughts of the time, people did not at all consider the basic proposition of variable events. Besides, it is also due to the fact that conceptual development of thoughts has been constrained by quantities so that the resolution of the time problem has been delayed. If at the time Leibniz could have recognized variable events among events, then Leibniz could possibly have altered the path of development of modern science.

Although in *Entering the Era of Irregularity* we introduced the concepts of space and time, which are different from those of modern science, we should make it clear that we dislike focusing our studies on the theoretical right and wrong; instead we first focus on solving practical problems and then propose formally our theory or opinions. That is the reason why it took us a long time before we eventually published such conclusions as "time does not occupy any material dimension and is parasitic on the changes of materials," "time cannot come about prior to existence," and related opinions about variable (or changing) events, since initially we had not practically tested these opinions in solving realistic problems. The gradual appearance of these opinions in our research papers was a consequence of debates with various colleagues. It was also because of these debates that they forced us to empirically apply these opinions on the handling of irregular information so that consequent works appeared over time (see Chapters 8 and 9 in this book).

No doubt, the central value of the concept of time is variable or changing events. If changes in events did not exist, then modern science could have been seen as a representative of true science; or in other words, science would have evolved to its ultimate stage. The core or the problem of essence surrounding variable events is also the core and essence of what time is and can also be seen as particulars of noninertial systems.

The essence for the problem of time to be a proposition is also about walking out of the inertial system. So, it forces us to carefully think about why our forefathers in the scientific community had been that fond of quantities to such a degree as being frenzied. Our analysis seems to suggest that the quantification of time could have led to the establishment of the system of physical quantity analysis so that the functions, attributes, and other aspects of materials could be separated from materials or events while becoming formal problems; by doing so, formal logical manipulations could be employed to substitute for the laws of realistic evolutions of materials or events. However, what is unexpected is that variable or changing materials or events do not comply with manipulations of formal logic, since variable events involve time.

Evidently, time does not occupy any material dimension; the quantification of events naturally leads to the introduction of parametric dimensions. That is why modern science is a quantitative analysis system constructed on the framework of physical quantities and parametric dimensions. For example, our previous problems as to whether or not such physical quantities as forces, energies, light, and so forth actually occupy material dimensions and where these quantities are from are all raised from the quantification of their functions, attributes, and other aspects. However, the evolution science that is made up of variable materials and changing events has to investigate where forces and other physical quantities come from and how they change; otherwise, this science would not be able to materialize the goal of foretelling the future. The essence of comprehending time as the difference in the time order of materials stands for the central problem of variable materials and changing events.

That is where "increase and decrease come in their own times" (*Book of Changes*) and "all matters resolve with time" (Zhuang Zi) come from. Therefore, the essence for time not to occupy any material dimension is that time can no longer be treated as a physical quantity and a parametric dimension, as what has been done in modern science; the corresponding equations with time treated as a quantitative variable or a parameter seem to have formally considered time; however, because of the requirements of well-posedness of mathematical physics equations, any relationship between events and time has been eliminated so that variable events are not considered. That is why some French scholars have noticed that modern science in fact denies evolution and that time has already become a parameter that has nothing to do with changes (Bergson, 1963; Koyré, 1968).

Therefore, in order to correctly understand and apply the concept of time, one has to describe and provide the differences existing in changes in materials that are also called variable events. No doubt, without changes in materials or events, there would not be the past and the future so that the concept of time would become useless. This end also touches on the nearly 100-year-old debate between determinacy and indeterminacy within modern science and makes this debate unnecessary. In other words, events are shown by changes in materials or matters; and this understanding at the same time points out the particularities of how to employ variable events. That is, time tells people that they have to admit to the timely differences of events; otherwise they would not be able to distinguish changes in the event. This end surely involves the problem of evolution processes of materials' aging, illness, and death. In this sense, one can see that evolution science is different from modern science. Or, modern science is only a special portion of evolution science. We should particularly note that other than the existence of events there is also the realisticity of changes in events. In other words, it is exactly because events have their variability that events cannot occur at the same time.

Due to this understanding, the popular practice of employing the theoretical system of modern science as the theory of prediction science should be seen as a major epistemological mistake. In particular, modern science cannot handle irregular or small-probability information. Not only that, it also employs such theories as nonstability, complexity, or randomness to eliminate the existing irregular or small-probability information. This fact shows that the scientific community did not comprehend what time is and in essence did not understand variable events.

7.2.2 About the Concept of Time

If time only had attributes of quantities, then among all physical quantities, time would be the only one that exists everywhere, and no one could ever escape the punishment of time. However, it is indeed also a *thing* that no one can see or feel so that one has to ask the question of *where* time is. Even so, the problem of time seems to be very special, since if no one mentions what time is, people in general would not even think of such a problem. However, as soon as it is raised as a question, people suddenly realize that they do not know how they should start attacking the problem. As if time is seen as a physical quantity in the form of quantities, it only appears in the changes of events. If there were no change in events, time would disappear and could not be traced anywhere.

It can be said that in the past 300-some years, the public has been convinced that the development of science and technology has been so advanced that man could fly high to grab the moon and dive deep in the ocean to capture the sea floor. However, with such a victorious science and technology, mankind still could separate the past, the present, and the future from each other and could not answer why events would not occur at the same time. Not only can these seemingly simple problems not be answered, but also does no one seem to know where to start to locate the answers.

7.2.2.1 Time in China

In the *Book of Changes*, there are at least two places that directly talk about the problem of time. One is hexagram 1 (quian), the creative; and the other hexagram 41 (sun), decrease. In fact, hexagram 41 has already very clearly answered what time is. That is, *sun yi ying xu, yu shi xie xing*, meaning that decrease does not under all circumstances mean something bad; increase and decrease come in their own time (Wihhelm and Baynes, 1967). In today's language, this sentence can be directly translated, "Time is reflected in the state of changes in materials' movements." The clearest implication is that time is not material and does not occupy any material dimension, or time cannot be identified with space. As for indirect mention of time, the first hexagram (quian) of the *Book of Changes* states that *tian xing jian, jun zi zhi qian bu xi* (the movement of heaven is full of power, thus the superior man makes himself strong and unstirring) (Wihhelm and Baynes, 1967). This sentence should have implicitly contained the concept of time. It is because the first three characters *tian xing jian* can be directly translated into "materials do not get destroyed," and the rest into "appear to be vividly alive in the changes of movements so that people can only try their best along with the changes in materials."

Next, although Lao Tzu, the great sage who has influenced people's thinking and intelligence for over 2,000 years, did not directly talk about time, if we attentively read the first part of Chapter 25 of *Tao Te Ching*, which is translated into the following by English and Feng (1972), "Something mysteriously formed, born before heaven and earth. In the silence and void, standing alone and unchanging, ever present and in motion. Perhaps it is the mother of ten thousand things," we can see that the concepts of time and space are essentially included. The way materials can be mixed has to be from stirs, which have to produce rotations. This implies that the universe is not only curved but also rotational. Rotations surely cause differences between before and after in materials and events. However, over 2,000 years later, Einstein still employed irrotational Riemann geometry to study the universe. The corresponding "standing alone and unchanging, ever present and in motion," undoubtedly means implicitly that the commonality of materials' rotations and events' changes — the forever passing of ages and time. That is why we say time is from materials' rotational movements, just as what we mentioned earlier; from the relationship between rotations and materials' evolutions, one has to address the problem of time.

Besides, as of today, people are still using the phrase *universe* (*yu zhou*). This phrase was initially used by Shi Zheng in his book *Shi Zi* written during the time period of Warring States in ancient China. We should be clear that the Chinese phrase *universe* contains different meanings from the English words *cosmos* or *universe*. The Chinese *universe* (*yu zhou*) means space (*yu*) and time (*zhou*) together, while the English *cosmos* and *universe* only mean space. The first character *yu* specifies materials occupying all imaginable fields in all directions; the second character *zhou* stands for precisely "coming from the past and heading into the future," representing changes in materials' rotational motions and the sense of before and after existing in materials' changes, explaining why events cannot occur at the same time moment. However, *yu* comes before *zhou*, and *zhou* follows after *yu*, indicating that space of the fields occupied by the materials is primary, while time does not occupy any material dimension. In the latter time period during the Warring States, there appeared another scholar named Zhuang Zi. He introduced *yu shi ju hua*, where *shi* means time, emphasizing more on materials' changes over time. He specified that no matter or event can stay unchanged forever and all things change through the difference between the before and the after so that the concept of time was introduced. Evidently, if materials did not possess any disparity in location and change in their movements, then time would become meaningless and be no longer necessary for us to treat as a concept or proposition. Later, during the Tang Dynasty, Li Bai, a great scholar and poet, more clearly pointed out that time does not occupy any material dimension, leading to the conclusion that time is a science in the culture and that separating the natural and social sciences may not be scientific.

In short, in China *time* can be summarized as reflecting the disparity existing between the before and the after experienced in the movements of materials or events. However, time does not represent any material; instead, it is a concept that exists in materials.

What needs to be pointed out is that Mo Zi, a contemporary of Shi Zheng, also posed such a concept of *jiu yu*, where *jiu* stands for time. What is different here is that later scholars in China mostly cited Shi Zi's "universe" so that *jiu yu* was washed away in history. As for the words *shi jie*, which are still widely used in modern China, the character *shi* represents time, while *jie* means space. However, the currently accepted meaning of *shi jie* — world — practically contains space only.

7.2.2.2 Time in the West

The concept of time in the West can be traced back to Plato. He believed that the prototype of natural laws comes from absolutely static figures, which were both consciously and unconsciously inherited later by Newton and Einstein, causing modern science to evolve away from the investigation of materials' changes. Galileo's *Discources on Two New Sciences and Dialogues Concerning Two New Sciences* (1638) specified the "physical" time, in which he used "evenness and continuity" that foreshadowed the later development of the concept of time. Thirty years later, Monroe introduced the quantitative time, treating the concept of time as a geometric straight line with only quantitative length. The version of time can be seen as the repeated summation of a series of time moments and also the flow of a time moment. It can also be represented as a circular ring. Newton continued Monroe's quantitative time and expanded it to absolute time and space. Just as what is stated at the start of his *Mathematical Principles of Natural Philosophy*, Newton defined *time* as follows:

"Absolute, realistic mathematical time; in terms of its properties and essence, it flows evenly forever without depending on any external matter." Evidently, in the understandings of time, the East and West are substantially different.

There is no doubt that the quantifications of time and space have something do to with the dualistic system of the Western philosophical recognition. Not only can acting forces exist independently outside materials, but also can time and various physical quantities, introduced later, be floating independently external to materials. What is worth our attention is the purpose of Newton starting his *Mathematical Principles of Natural Philosophy* by first defining quantitative time. It seems that his aim was the consequent physics of physical quantities instead of natural philosophy. In other words, the introduction of the concepts of absolute time and space existing independently outside of materials is to meet the needs of the classical mechanical system. If time and space were not quantitative, Newton would then not be able to establish the mechanic system of particles. It should be clear that in the system of modern science, time is very important. It can be said that if time were not quantitative, modern science would be essentially ruined. So, it can be seen that time is an important proposition of science.

The quantifications of time and space not only confuse time with space in the history of science, but also lead to the concepts of parametric dimensions and physical quantities. Time becomes the fourth dimension, and physics the theory of physical quantities. Speaking more clearly, mathematics was originally a tool for the study of physics; with the introduction of parametric dimensions and physical quantities, mathematical principles dominated all physics laws, as evidenced by the later mathematical equations. In particular, the reason why Einstein identified time with space and treated three-dimensional evolution problems as four-dimensional existences no doubt is all from time being seen as a material dimension in the form of a parametric dimension. So, the modern science of the inertial system constructs classical mechanics on the basis of the quantitativeness of parametric dimensions and physical quantities, which is called the foundation of modern science. Evidently, the central idea is to pull out the physical properties and treat principles of quantities as physics or the laws of myriad things. Because of this, weaknesses and deficiencies of quantitative principles will have to be brought into physics and become weaknesses and deficiencies of physics. Among all the greatest weaknesses is that quantities cannot handle transformations of eddy currents corresponding to quantitative unboundedness. Therefore, we must provide information on the range of validity for modern science so that people can clearly see the boundary of those problems solvable by using modern science.

What should be known is that as soon as Newton's absolute time and space were publicized, he was severely criticized by his contemporary Leibniz, who pointed out that events are the essence. This opinion possesses not only the revolutionality of concepts, but also is ahead of Einstein in terms of their understandings of the problem of time. Even though Einstein also joined those who criticized Newton's absolute time and space with as severe criticism as that of Leibniz, other than proposing curved spaces, he did not really stay out of the quantitativeness of time and did not provide any explanation as to why space is curved. Instead, he treated changes in materials as four-dimensional existences so that time was completely identified by space. On the basis of all these, Einstein (1997) believed that time is a character of human consciousness, becoming an illusion of man. That is, he had traveled farther away from the truth than Newton.

What modern science has not noticed is that even if we record time based on the earth's rotation, time is also a variable of time itself, which changes along with the changes in the earth's rotation. Currently, the 24-hour day on Earth is partitioned into 86,400 portions, called seconds, which is only an approximation of the earth's rotation. Over 600 million years ago, each day on Earth was shorter than 21 current hours. That is, the same time length of a day or a second changes through the ages.

In order to repudiate that time has an "arrow," in his late years Einstein (1997) had to admit that modern science, including his relativity theory, couldn't tell the difference between the past, the present, and the future. This end implies that he also recognized that time in modern science is only a parameter that has nothing to do with time. This end explains why modern science is still standing outside the door of evolution science. Now, it can be seen that in the stream of consciousness at that time, no one seemed to understand what variable events were, so that the 20th century entered the system of indeterminacy.

When the system of modern science extols its glories of the past 300-plus years, it seems that people have forgotten the unsettled problem of time, which makes the foundation of modern science shaky and leaves many problems unsolvable in modern science. For example, is time reversible? Why can different laws of physics use different scales of time? The law of radioactive decay uses the time measured by uranium 238 with half-life = 4.5 billion years. Does this time have the same meaning as those used in Newton's kinetics and the laws of gravitation? As of this writing, no answer has been obtained for this question. Even worse, people do not even know how to secure such an answer. These and many other fundamental problems directly lead to the problem of scientificality of modern science.

Even though in the 1980s, I. Prigogine determined that time is irreversible and that time comes ahead of existence, he still did not correct the separation between physical quantities and materials (Prigogine, 1980).

That is, in the West, time is identified with quantities. Even though Leibniz pointed out that events are the essence, an opinion of a lofty realm, he did not have the power and influence to alter the Western stream of consciousness of the ages and consequent methods.

7.2.3 The Concept of Time

From the history of learning the concept of time, as summarized above, according to the knowledge behavior of modern man, we can see at least the following attributes of time that are different from those of space.

Time comes from the changes existing in the rotational movements of materials. So, it reveals the differences of events' past, present, and future. Time directly shows the variance in properties, states, locations, and functionalities of events in the *time* order. No matter whether the relevant quantity changes or not, people can directly tell or distinguish materials' or events' past, present, and future based on the changes in properties, states, locations, and functionalities. And, such essential problems as the properties, states, and functionalities of materials or events are exactly the areas modern science has not entered.

Time dwells in the rotational movements of materials so that it cannot exist independently outside the materials. Time is an attribute of materials so that it does not occupy any material dimension. In other words, as an independent concept, time is not a material and cannot independently exist or change without being associated with certain material. It is also a function of time itself. Since time comes from the changes in materials' rotational movements, time also varies. That is, both time and materials' movements appear at the same time, which is the most important attribute of materials that modern science has missed. If the materials' attribute of time means that time is parasitic on materials so that as long as there is material, the material will have to move, and as long as the material's movement exists, there will be time, then time cannot exist independently outside materials as described by Newton, Einstein, or Prigogine. Or, we conclude that time cannot exist before existence.

At the same time when time possesses the attribute of direction, it also possesses the attribute of quantitativeness, where the direction points only forward and cannot be reversed. Even when the spinning direction of materials is reversed, time still moves forward together with the reversed materials' rotation so that past events will never reappear again. Even if we took a rocket flying at a speed faster than that of light, we still would not be able to travel through the time tunnel to have breakfast with Einstein since his death in April 1955. Although this attribute of time was employed by Prigogine to introduce the third development period of science, Prigogine only imagined the torn time on top of the system of linearity — there is no negative time — and proposed his linear "arrowhead" using the dynamicality of time choice of space to substitute for Newton's staticality and Einstein's existence. However, Prigogine's time is still existent independently outside materials, leading to his time ahead of existence. Since modern physics, consisting of quantum mechanics, relativity theory, and others, does not possess any substantial difference from Newton's system of classical mechanics, these modern theories still maintain the eternal invariance of materials. But, it seems that now is the time for science to rise to the true second or higher stages of development — the evolution science emphasizing changes in materials' movements. If we use what time is as the mark, then the directionality of time has made time into a vector, constituting a challenge to the entire system of science.

Time possesses the attribute of connecting events and shows the process differences between events' past, present, and future. That is, each material has its origin and destination and its causal relationship and why events cannot occur at the same moment of time. All these realizations and others form the basics of evolution science. Since materials' evolutions do not get directly shown with quantities, materials' aging is exactly embedded in the materials' properties, states, and functionalities. Not only is modern science not deep enough to investigate materials' properties, states, and functionalities, but also it treats the relevant problems as quasi-problems, excluded from the Newtonian system (Bergson, 1963).

In terms of time itself, the concept of time does not possess any physical entity. Together with the multiplicity of materials, time's attributes of direction and quantitativeness also show their multiplicities. That is why earthly time is different from Venus time; Norwegians are annoyed by their own long life spans so that as of recently, they continue the custom of "self-termination"; it might be because they do not like to exist too long; and at a very different time scale, bacteria reproduce rapidly. The corresponding too long life span and too long existence are not on the same time scale. So, time can neither be unified using the speed of the earth's rotation nor follow the wish of man to flow along the quantitative straight line or the geometric circular ring. Because of the differences in movement directions, changes in materials' structures appear, which can be recognized through the before-and-after differences in the structural properties, states, and functionalities of materials. So, the quantitative attribute of time can only be visible after instead of before materials' aging. Hence, quantitative time is merely the postchange results of materials or events.

Just because quantities are postevent formal measurements, Leibniz's belief that events are the essence is from a highly lofty realm of knowledge, which should be the pride of the Western scientific community. If scientific history had gone along in the direction as what was pointed out by Leibniz, evolution science would have appeared over 200 years ago, without having to fall into the current situation where modern science can only describe movements that have nothing to do with time, or speaking more oddly, the movements along the axis of time without time (Koyré, 1968). What should be noticed is that from "events are the essence" to "variable events are the essence" represents a jump in epistemology; it directly shocks the theory of probabilistic universe of the indeterminacy, leading to reforms in the relevant methodologies and representing a transition from the system of quantitative analysis into the analysis system of digitization. So, studies on variable events should constitute a challenge to the wisdom of modern man.

What should be made clear is that we usually like to first test out our theoretical results by actually solving practical problems before publishing the results. Based on our previous understandings, just at the time when we were about to practically work on the conclusion that time does not occupy any material dimension using practical problems, colleagues from different countries invited us to deliver a speech on the problem of time by saying that in the past 100 or so years, they had not seen any Chinese scholar studying this problem. To meet this challenge, combining what we discussed above with our practical applications of our works, let us summarize our results on time as follows.

Time is originated from the rotations of materials' movements. That is why time shows the directionality and orderliness of events. However, time is not a material and cannot occupy any physical dimension. So, time cannot be identified with space. Or, according to the definition of physical dimensions, time cannot be treated as a coordinate dimension of materials or the parametric dimension of physical quantities.

Time possesses the attributes of direction and quantity of rotation. However, these attributes are not an attribute of time itself. Instead, these attributes are from the underlying materials or parasitic on the materials. Together with the changeability and multiplicity of the materials' movements, time also possesses the attributes of changeability and multiplicity.

Each rotational direction presses the underlying materials to age so that the materials' evolution plays out naturally.

By rotational directions, we mean un-uniformities in the directions of materials' rotations, leading to damages or breakups caused by collisions in materials' properties, states, or functionalities, and maybe damages or breakups in the materials' structures. That is why rotational directions press materials to age. Quantities are only formal measurements recorded after aging. Quantities are not the reason for aging, and the essence of aging is not shown in quantities. The purpose for us to emphasize directions is that directions are not quantities but structures and that in the study of evolution problems, directions are more important than quantities. Directionality appears ahead of the development of events and can be observed with less error than quantities. So, it can be a piece of useful information for foretelling what is forthcoming and is the reason why we proposed the prediction method based on digitizing information. In particular, on the basis that time does not occupy any material dimension and that the essence of events or information is not about any quantitative physical quantity and cannot be treated as a parametric dimension, we have employed phase space transformations to analyze the state and structurality of variable information so that variability and invariancy of information can be quite easily recognized.

So, a third method different from the kinematics and statistics of modern science — digitization of irregular time series information (or called digital analysis of information) — is established and practically tested. This reveals that the essence of informational irregular disturbances is not waves; instead it is irregular revolutionary currents (or called changes in revolutionary currents) created out of eddy transformations. That also explains the reason why the traditional system of quantitative analysis cannot deal with irregular or small-probability information. It is because irregular or small-probability events are those that change, while changing events do not comply with the calculus of quantitative formal logic.

The effect of our method has been way beyond our initial expectations, and this method can be employed to make predictions of major natural disasters. It can be said that the system of modern science has not only listed major natural disasters as world-class difficult problems, but also did not have any hope to provide a resolution in the foreseeable future. However, by employing the method of information digitization (or called digital analysis of events), we have provided not only a specific method to predict major disastrous weathers but also a method that is easier than the conventional procedures developed for weather forecasting. The reason is that the irregular or variable information existing ahead of the arrival of major natural disasters is more noticeable and easier to recognize than that of ordinary weather conditions.

7.3 Stirring Motion and Stirring Energy

We once questioned modern science for its kinetic energy and the law of conservation of kinetic energies developed on squares of speed. It is well known that the expressions of energy, such as Newton's

$$e = \frac{1}{2}mv^2$$

Einstein's $E = mc^2$, and others, all employ squares of speed to construct kinetic energy. So, we posed the following question: Can the square of angular speed make up some kind of kinetic energy? As a matter of fact, we have had the habit of questioning. When we were still in college, we had individually asked similar questions. However, no answer was provided for us from our professors. What really motivated us to look into this problem of rotational kinetic energy seriously was after we accidently saw the mechanical device that was initially invented in ancient China (during the Bei Wei Dynasty) for preventing shakes and vibrations, the rightward deviation phenomenon of floodwater, and the solenoidal effects of Bjerknes's circulation theorem. In particular, the essence of the circulation theorem is a betrayal of the system of particle quantitative analysis, where rotations under stirs of solenoids represent problems of nonparticle structures. It might be because we had training in figurative thinking in our earlier ages that we were very interested in the physical significance and conservation of rotational kinetic energies (that is later renamed as stirring energy). In other words, the revolutionariness of the circulation theorem is its transformation of problems of particles into those of structures so that the relevant scientific exploration enters into curvature spaces and returns problems of quantities back to their origins on matters and events.

It should be recognized that all theories and opinions are originated in materials and are to be applied back to materials. This is known in ancient China as *ge wu*, meaning understand and deal with problems based on the structures of matters. Evidently, in the physical world, it is impossible to avoid interactions between materials so that any absolute freedom does not exist and no true first push can reasonably exist. In particular, all matters are nonparticles with zero probability for them to interact exactly on their centers of masses. Therefore, the result of the interactions has to produce rotation movements. With this understanding in place, we will employ the concept of nonparticles to discuss stirring energy and its conservation, which have been missed out by the system of particle quantitative analysis in the past 300-plus years, and some other elementary laws of evolution science.

7.3.1 Rotation and the Problem of Stirring Energy

Rotationality is originated in the unevenness in materials' structures, and rotations create new materials in the form of altering the existing structures. This end reveals the stable existence of structures and the instable evolutionality. Within the system of modern science, other than eliminating instability, we have not seen anyone working on the meaning and effects of instable energies in the research on these energies. Within the system of quantitative analysis, established since the time of Newton, it seems that all theories and methods were developed to deal with the existence of quantitatively stable systems. As for instable energies, scholars have not paid any attention to the problem. No doubt, what have been obtained in modern science are theories and laws that are postexistence of materials without touching on the reason why materials exist and the processes of material existences. Therefore, the evolutionality and related processes of course have not been concerned with. So, at the same time when the problem of foretelling the future challenges modern science, it has also challenged the wisdom of modern man.

Evidently, materials' stable existence and instable evolutionality are unavoidable problems of natural science. Besides, how can any of the conservative schemes designed for calculation of computers realistically limit instable energies existing in nature? Since the time when Newton published his *Mathematical Principles of Natural Philosophy*, all the theories and methods, established on the basis of quantitative analysis, have been seen as high-level (no longer elementary) scientific achievements. Later, Immanuel Kant, the German philosopher who founded the school of systematic metaphysics by manipulating the quantitative analysis based on the knowledge developed on mathematical physics equations, took advantage of the universal formality of quantities and developed his unified normal procedure of thinking. He concluded that without quantities there would not be any theory or method. His work has become the norm of scientific consciousness and behavior in the past 300-some years. What has to be pointed out is that this behavioral norm of the past 300-some years has touched on neither the past nor the future.

No doubt, people could not imagine what infinity is and feel puzzled with infinitesimals. To this end, we once pointed out that the quantitative equivalence of infinity is nothing but an expanded quantitative form within zero-curvature Euclidean spaces and is not a physical reality. Each physical space is either a concavely or convexly curved non-Euclidean space, where the quantitative infinity should correspond to transitions in the physical space. To practically test this result, we have waited for many years in order to come up with some comparative, associative thinking and methods. Eventually, we used the methods of making use of small to explore what is large and making use of large to explore what is small in order to either positively or negatively confirm our theoretical results in practice.

As for the method of making use of small to explore what is large, we analyze the fundamental components of a material structure by decomposing the existing material, because no matter whether what is considered is a cosmic universe at the scale of stars or a microscopic atomic system, the underlying material has its internal structure. Although some people might argue that at present science is still not able to decompose materials into their fundamental levels, what is undeniable is that all collected information about the materials in the universe comes from the effects at the atomic scale of the materials' structures.

The present problem is, can fundamental particles be infinitely divided according to being so small that they no longer have any internal structure of the formal logic of quantities? This end touches on the disapproving capability of formal logic. To this end, let us think in the opposite direction by making use of large to explore what is small. From the observable, relatively stable existence of three-ringed stability of the macroscopic universe, consisting of galaxies, stars, and planets, it can be seen that the stars, as the macroscopic secondary circulation, hold the main concentration of materials and energy of the universe. Correspondingly in the microscopic world, materials are classified in the molecular, atomic, and electronic scales. What is interesting is that the atomic scale, as the microscopic secondary circulation, also contains the main aggregation of energy, where the theory of nuclear energy has been proven and confirmed. These two observations at least explain that the quasistable existence of natural materials needs at least three levels of circulations, where the second-level circulation contains a huge amount of energy and the circulations of the planets' scale of macroscopic systems and those of the atomic scale of microscopic systems represent the special zones with concentrated nuclear energy.

So, from these observations, we can easily expand what we see to the scale of human activities or mesoscale systems, where we can well observe the meteorological three-ringed system that is made up of the frigid zone, temperate zone, and torrid zone. In this meteorological system, it is also the secondary circulation, the temperate zone, that contains the most energy. Each typhoon (or hurricane) is also a three-ringed system that consists of the typhoon eye, the zone of torrential rains and fast winds, and the outer region of subtropical high pressures, where it is once again the area of the secondary circulation that gathers most of the energy of the system.

Based on this analysis, the stable mechanical device invented during the Bei Wei Dynasty of ancient China (AD 386–534) for three-ringed energy transformation, and the observation that all natural rivers in the northern hemisphere tend to have lakes along the right-hand side to help reduce the damaging effects of floods, we recognize that the three-ringed circulation existing in energy transformations has played the role of coordinating and constraining energy transformations within stable existences, that non-three-ringed circulation helps to destroy relative stabilities and conducts instable evolutions, and that without eddy motions there would not be any transformation of kinetic energy. This kind of rise and fall of energy determines the internal heat — the dynamic equilibrium — of the eddy movements.

However, in the past 300-some years, from Newton to Einstein, nobody has truly recognized the significance and effects of rotations. In their works, even though Einstein had concluded that time–space is curved, people still employ the concept of speed of zero-curvature spaces to represent momentum (mv or mc) and kinetic energy (mv^2 or mc^2) of materials. Although the concept of angular momentum has involved that of angular speeds, the definition of angular momentum is still introduced analogously to that of linear speeds ($m\omega r$). In essence, ωr still takes the form of the linear speed v. That is, the unit [m/s] of ωr is the same as that of the linear speed v. To this end, it can be said that in basic physics, Newton, Einstein, and their followers did not provide a mechanism for kinetic energies to transform and transfer, which constitutes a major negligence and pity of the past 300 years. As a matter of fact, the commonly existing and fundamentally important momentum and kinetic energy are those of $m\omega$ and $m\omega^2$ involving angular speed. They are created by the stirs of the structural rotations of materials in curvature (non-Euclidean) spaces. So, quantitative analysis cannot describe rotational directions of materials where directions are made up exactly of problems of structures. In particular, quantitatively stable automorphism of initial values has eliminated quantitative instabilities so that harmonies and conflicts existing along with structural transformations can no longer be revealed.

7.3.2 Conservation of Stirring Energy and Three-Ringed Energy Transformation

The conservation of material stirring energy and the principle of three-ringed energy transformation in fact stand for problems of transformation and transfer of overall energies, where each scale of heat-kinetic energy transformation is completed through three-ringed eddy motions or rotations. The proof for the principle of three-ringed energy transformation of the conservation law of stirring energy is quite simple. In the following, we will look at the initial argument.

The conservation of energy is one of the three most important achievements of modern science; together with the continuity of fluids (that is, fluids are treated as continua of low-resistant solids so that movements of fluids can be discussed as wave motions) and calculus, it constitutes the basics of the theoretical foundation of modern science. Here, kinetic energy is described by using the square of speed, including the speed of light in Einstein's conservation of mass and energy. If we say that in the fundamental theoretical system of the past 300-plus years Newton's achievement was his first push, then Einstein's accomplishment would be the interaction between mass and energy. However, in essence, energy is also the work done by force. So, the problem of kinetic energy of the past 300-some years seems to have been resolved with kinetic energy written as the square of speed. What is important is that no matter whether it is the conservation of kinetic energy or that of the total amount of energy, no one seems to know how energies change and through what process or how energies are transformed or transferred. Just as in Newton's third law, it does not provide any information about the form of motion and the process of interaction; instead it only describes the interaction and the consequence of the movement. According to Aristotle, the task of physics should be the investigation of the process of movement and changes in materials. Besides, when looking at an individual person, the human history, or the development of the universe, everything is about the problem of physical processes and changes. So, laws of physical processes and changes are not only important for practical purposes, but also the true laws of physics.

Even based on the formal logic, one can conclude that if the square of speed is the kinetic energy, then the square of angular speed should also stand for a kind of kinetic energy. In other words, in the study of kinetic energies, the current system of knowledge has neglected the square of angular speed (ω^2), where the square of speed and that of angular speed not only mean different concepts but also possess different functions. Since the unit of speed is [m/s] and that of angular speed [1/s], it indicates that angular speed implicitly contains an uneven distribution of speed. Even as a quantity, angular speed is originated in the measurement of the rotation of materials, while speed is a measurement of straight-line distance traveled by an object within a unit time interval. That is,

$$\vec{\omega} = \vec{i}\,\omega_x + \vec{j}\,\omega_y + \vec{k}\,\omega_z \tag{7.1}$$

Since the angular speed represents an uneven spatial distribution of the speed, we have

$$\vec{\omega} = \begin{vmatrix} \vec{i} & \vec{j} & \vec{k} \\ \dfrac{\partial}{\partial x} & \dfrac{\partial}{\partial y} & \dfrac{\partial}{\partial z} \\ u & v & w \end{vmatrix} \tag{7.2}$$

where the symbol $\rightarrow$ stands for vector, u,v,w respectively the components of $\vec{V}$ along the x-, y-, and z- directions. For a two-dimensional horizontal movement, the angular speed in the vertical direction is given by

$$\omega_z = \frac{\partial v}{\partial x} - \frac{\partial u}{\partial y} \tag{7.3}$$

Introducing the flow function leads to

$$v = \frac{\partial \psi}{\partial x}, \text{ and } u = -\frac{\partial \psi}{\partial y}$$

According to the traditional quantitative analysis, we assume that ψ has continuous second-order derivatives and can be approximated by using simple harmonic disturbances. As for the whole system, let us introduce the following combination of disturbances of different scales:

$$\psi = \sum_n \psi_n$$

So, we have

$$\nabla^2 \psi_n = -\mu_n^2 \psi_n \tag{7.4}$$

Considering the horizontal problem, let us take $V^2 = (\nabla \psi)^2$. From Equations (7.4) and (7.5) and the plane divergence theorem of orthogonality, it follows that

$$\oint\!\!\!\oint_\sigma V_n^2 \, d\sigma = \oint\!\!\!\oint_\sigma \left(\sum_n (\nabla \psi_n)^2 \right) d\sigma = \oint\!\!\!\oint_\sigma \left(\sum_n \nabla \psi_n \cdot \nabla \psi_n \right) d\sigma$$

$$= \oint\!\!\!\oint_\sigma \left(\sum_n \nabla (\psi_n \nabla \psi_n) \right) d\sigma - \oint\!\!\!\oint_\sigma \left(\sum_n \psi_n \nabla^2 \psi_n \right) d\sigma \tag{7.5}$$

$$= \oint\!\!\!\oint_\sigma \left(\sum_n \mu_n^2 \psi_n^2 \right) d\sigma$$

and

$$\oint\!\!\!\oint_\sigma \omega_z^2 \, d\sigma = \oint\!\!\!\oint_\sigma \left(\sum_n (\nabla^2 \psi_n)^2 \right) d\sigma = \oint\!\!\!\oint_\sigma \left(\sum_n (\mu_n^2 \psi_n)^2 \right) d\sigma$$

$$= \oint\!\!\!\oint_\sigma \left(\sum_n \mu_n^2 \cdot (\mu_n^2 \psi_n^2) \right) d\sigma$$

Substituting Equation (7.5) into this last equation produces

$$\oiint_{\sigma} \omega_{\tilde{z}}^2 d\sigma = \oiint_{\sigma} \left(\sum_n \mu_n^2 V_n^2 \right) d\sigma \tag{7.6}$$

Comparing Equations (7.5) and (7.6), it follows that the closed integral of the square of the angular speed contains that of the square of the traditional speed. That indicates that even if we use quantitative methods, the physical meanings of speed and angular speed are different. What is very interesting is that unexpected by both Newton and Einstein is what is revealed in Equations (7.5) and (7.6): Not only does the conservation of stirring energy contain that of the speed kinetic energy, but also do these equations provide information on how kinetic energy transforms and transfers and their processes, which constitutes a major negligence of the traditional conservations of kinetic energy and total energy. In other words, without rotational transformations of eddy motions, energy transformations cannot be revealed and cannot be completed.

At this junction, we would like to mention that considering the difficulty of changing accustomed beliefs and how referees review submitted papers, we, together with our students Gongyi Chen and others, once employed the concept of vorticity from meteorology in the place of angular speed to investigate stirring energy. When we studied the physical meaning of stirring energies, our literature search indicated that in the meteorological community, R. Fjörtoft (1953) had already introduced the kinetic energy of nondivergent flows in the form of such flows, where in form ς^2 and V^2 are similar and $\varsigma^2/2$ is known as the kinetic energy for nondivergent flows. As for the meaning it describes, it should be the kinetic energy of rotational flows. However, in Fjörtoft's original paper, he did not clearly point out the difference in terms of physical significance between his vortical kinetic energy $\varsigma^2/2$ and the speed kinetic energy $v^2/2$. In particular, Fjörtoft applied his kinetic energy for nondivergent flows to the design of stable numerical computational schemes in order to limit instable energy. Evidently, instable energies in nature cannot be truly constrained simply by employing artificially designed computational schemes. Besides, realistic instabilities are the key factors of changes. So, it implies that Fjörtoft did not fully understand the meaning of instable energies. Even so, Fjörtoft's work after all had noticed another kind of kinetic energy beyond that of squared speed. In particular, if we understand the kinetic energy of nondivergent flows as our stirring energy of rotations, then we have to say that it was Fjörtoft who first discovered the stirring energy. Unfortunately, only due to the limitations of his historical time, Fjörtoft did not fully understand the physical significance of stirring kinetic energy.

For us, from the very start we began our work of energy transformations on the recognition that angular speed is different from linear speed so that we naturally have to address the problem of the physical significance of stirring energies, which led to the concept of second stir. So, in our teaching and during our lecture tours, we used to mention the two "half" important achievements in the history of meteorological science. The first half-accomplishment is Bjerknes's circulation theorem, where most realistic rotational eddy currents are not closed and Bjerknes introduced a potential function. That left behind incompleteness and imperfection in this work and constitutes the first half of the regret. The second half-accomplishment is Fjörtoft's introduction of his kinetic energy for nondivergent flows. In this case, not only was his comprehension incomplete, but also was the concept introduced for the purpose of designing conservative numerical computational schemes so that a great historical window for peeking into the idea of stirring energies was lost, leaving behind the second half of regret. If modern science developed along the lines of the circulation theorem and transformations of stirring energies, the current situation of not only prediction science but also the entire natural science would be completely different from the present state of difficulty.

On the other hand, Fjörtoft's work does not reveal the essential meaning of the kinetic energies of rotational flows and is fundamentally different from our concept of stirring energies. In all the studies of the kinetic energies of nondivergent flows, none of the authors seemed to have recognized that wave motions and rotational movements are completely different: Wave motions transfer energies in the form of propagation without transporting materials, but for rotational eddy motions, at the same time when they complete energy transformations, they also complete transportations of materials. In particular, the concept of stirring energies contains that of the traditional kinetic energies as a special case so that our field of study is greatly enlarged. Hence, we believe that the importance of investigating transformations of energy and forms of transfer of energy is about discovering the detailed transformation and transfer processes of materials' energies.

7.3.3 Conservation of Stirring Energy, Process of Energy Transformation, and Nonconservation of Stirring Energy and Evolution

If we consider circulations with only two levels, from the conservation of stirring energy in Equation (7.6), it follows that

$$\mu_1^2 v_1^2 + \mu_2^2 v_2^2 = c_1 = const \tag{7.7}$$

From Equation (7.5), it follows that

$$v_1^2 + v_2^2 = c_2 = const \tag{7.8}$$

Let Δv^2 stand for the change in the speed kinetic energy between two neighboring time moments; to satisfy the conservation of energy, we must have

$$\begin{cases} \mu_1^2 \Delta v_1^2 + \mu_2^2 \Delta v_2^2 = 0 \\ \Delta v_1^2 + \Delta v_2^2 = 0 \end{cases}$$

(7.9)

Eliminating $\Delta v_1^2, \Delta v_2^2$ from Equation (7.9) leads to

$$\left(\mu_2^2 - \mu_1^2\right)\Delta v_2^2 = 0 \quad \text{and} \quad \left(\mu_1^2 - \mu_2^2\right)\Delta v_1^2 = 0$$

(7.10)

Since $\mu_1 \neq \mu_2$, Equation (7.10) implies that both Δv_1^2 and Δv_2^2 have to be zero. This end implies that if the circulation has only two levels, then between these two circulation levels, energy transformation or transfer cannot be completed, leading to blockage of energy and consequently instable evolutions. This fact can be applied to explain why natural rivers located in the northern hemisphere either break through their embankments on the right-hand side or have accompanying lakes formed along the right-hand side. The breakthroughs of the embankments or the naturally formed lakes transfer and store energies and reduce the strengths of the floods from the upper streams (the primary level of the circulation).

If we look at a circulation with three scale levels, from Equations (7.5) and (7.6) we have

$$\begin{cases} v_1^2 + v_2^2 + v_3^2 = c_1 = const \\ \mu_1^2 v_1^2 + \mu_2^2 v_2^2 + \mu_3^2 v_3^2 = c_2 = const \end{cases}$$

(7.11)

where c_1, c_2 are constants. Let Δv^2 stand for the change in the speed kinetic energy between two neighboring time moments. To satisfy the conservation of energy, we must have

$$\begin{cases} \mu_1^2 \Delta v_1^2 + \mu_2^2 \Delta v_2^2 + \mu_3^2 \Delta v_3^2 = 0 \\ \Delta v_1^2 + \Delta v_2^2 + \Delta v_3^2 = 0 \end{cases}$$

(7.12)

Eliminating Δv_1^2, Δv_2^2, and Δv_3^2 respectively from Equation (7.12) produces

$$\begin{cases} \left(\mu_1^2 - \mu_2^2\right)\Delta v_2^2 + \left(\mu_1^2 - \mu_3^2\right)\Delta v_3^2 = \mu_1^2 c_1 - c_2 = const \\ \left(\mu_2^2 - \mu_1^2\right)\Delta v_1^2 + \left(\mu_2^2 - \mu_3^2\right)\Delta v_3^2 = \mu_2^2 c_1 - c_2 = const \\ \left(\mu_1^2 - \mu_3^2\right)\Delta v_1^2 + \left(\mu_2^2 - \mu_3^2\right)\Delta v_2^2 = c_2 - \mu_3^2 c_1 = const \end{cases}$$

(7.13)

If we take $\mu_1 > \mu_2 > \mu_3$ (similar results hold true for the opposite ordering), then the left-hand sides of the first and the third equations in Equation (7.13) are positive, while the left-hand side of the second equation is negative. This fact indicates the following.

From Equations (7.9) and (7.12), it can be seen that the conservation of the pure speed kinetic energy cannot limit the propagation of energy, since other than the speed kinetic energy, there is still stirring energy that continues to be transported.

When Δv_1^2 and Δv_3^2 decrease, Δv_2^2 will increase. Conversely, when Δv_1^2 and Δv_3^2 increase, Δv_2^2 decreases. That is, in a three-leveled (or ringed) circulation, energy transformations and transfers are completed through the second-level circulation so that the transfer and transformation process of energy is specifically given.

Notice that $\mu_2^2 - \mu_3^2 < 0$. So, when Δv_1^2 experience an increase, the corresponding Δv_3^2 also increases. Conversely, when Δv_1^2 experiences a decrease, Δv_3^2 also decreases. This end indicates that the designs of conservative computational schemes using speed kinetic energies cannot really guarantee the conservation of the overall level of kinetic energy; transformations and transfers of stirring energies are completed through the second-level circulation; and between the primary and the third-level circulations, there does not exist any energy transfer and transformation. That end explains not only that the conservation of the traditional speed kinetic energy cannot limit instability, but also that instable energies represent the mechanism of transformation in realistic physical processes. That is, a process physics of evolution is revealed. So, instable energies should not be artificially controlled. The essence of artificially controlling instable energies is to eliminate transformations and changes in energies.

7.3.4 Conservability of Stirring Energy and Physical Meaning of Energy Transformation

The studies on the conservability of stirring energies and physical meanings of energy transformations can be roughly summarized as follows.

1. If the second-level circulation cannot materialize transformation and transfer of energies, energy blockage will have to be created, leading to instabilities in the accumulation of energy and triggering the third-level circulation to destroy the original second-level circulation in order to achieve the system's equilibrium and stability. To this end, both Yellow and Yangtze Rivers in China can be looked at as examples. For instance, the Yellow River is known for its floods. That is why Chinese history can be seen as a history of struggling with floods. When it flows through the area of the east longitude 100°–110°, the Yellow River, located in the northern hemisphere, curves north and forms a great bend area on the left-hand side of the main river course. So, a current flowing in the opposite direction to the deviation force of the earth

movement is created, where the deviation force inevitably pushes the current to flow to the right. That is why along the Yellow River, it is very easy for floods to break through dykes and dams, causing great losses. For example, with a rate of flow at 6,000 m³/s at the flood peak, Huayuan Kao Dam in Zheng Zhou can be overrun. As for the corresponding Yangtze River, only looking to the east of Hubei, we can see such major lakes as Dongting Hu, Boyang Hu, and Tai Hu. Not only are these lakes located on the right-hand side of the Yangtze River, but also was Xuanwu Hu of Nanjing connected to the Yangtze River in the ancient times. That is why it has been believed that even the tributaries of the Yangtze River can easily entertain the rate of flow of several tens of thousands of cubic meters of floods at their peaks. In particular, it is exactly because the huge lakes on the right-hand side of the Yangtze River constitute a second-level circulation to transfer energies that the capability of redirecting floods is greatly enhanced. In comparison, we can tell that from Huayuan Kao downward, there basically does not exist any large-scale lake along the Yellow River. That explains why it has been quite easy for the Yellow River to flood, causing great losses. It is because its main course does not coincide well with nature, where the river flows to and curves to the left against the effect of the earth's deviation force.

2. The transformation and transfer of circulative energies affirm the cause for materials' stable existence through the conservation of the quasi-three-ringed stirring energy. That might provide a theoretical explanation for Lao Tze's claim (time unknown) that out of three, all the myriad things are born, for such empirical beliefs about changes that nothing goes beyond three, and so forth. That is, these and similar beliefs are concepts not only known since ancient times but also established on some well-understood principles of physics (even though these principles were not known in modern science).

3. Since energy transformations are completed through indirect (the second level) circulations, a natural need arises to question some of the methods and concepts used and studied in the traditional theories, including direct damping (or cut-off), limiting instability, forcing dissipation, and so forth. Or, it can be said that modern physics does not recognize the effect of indirect circulations. And not only can it not describe changes in physical processes, but also has it missed an important law on energy transformation — the conservation of stirring kinetic energy. This end indicates that many laws of physics are descriptions of events without touching on causes and processes of evolutions. That is why we claim that modern science did not bring up such scientists who can foretell the future. In other words, the proposition of foretelling the future at least at the present time still challenges our current level of wisdom.

4. Due to the holding capabilities of indirect circulations, between the first-level and the third-level circulations no direct energy transfers exist. The inevitable consequence is that the first-level circulation "moves," while the third-level

circulation "does not move." When the second-level circulation accommo-dates the retransfers of energy from the first-level circulation, it at the same time transfers to both the first- and third-level circulations so that a transfer of energy is completed in the form of dispersion. This process should be that of realistic, physical "dispersion" and the fundamental principle of dealing with water problems used by King Yu, the founder of Xia Dynasty (ca. 21–16th century BC), and the father and son of Li Bing family of ancient China.

5. Each three-leveled circulation transformation constitutes a quasi-closed sys-tem, completing transformations and transfers of energies. The fourth-level, fifth-level, and any other higher-level circulations at most receive or further transfer the remnant energy out of the three-level circulation without being able to constitute a system's quasi-stability. And, the mechanism of physi-cal adaptation process is that instable subcirculations readjust themselves to the primary three-level circulation. And, since remnant energies exist in the transformations of each three-leveled circulation, it explains why some plan-ets have satellites, while others do not. It is because at the planetary level, celestial circulations might be of some inevitable by-products adjusting the remnant energies. All subcirculations, which do not follow the three-ringed energy transformations, are not stable. And, the instability of the fourth or higher level circulations would adjust themselves toward the quasi-stability of the atmospheric three-ringed circulation. This end can be observed in the evolution of the realistic atmospheric circulations and the experiment of spin-ning fluids. As for the stability of the natural elements, made up of micro-scopic materials, and the instability of man-made elements, we have not seen much explanation provided by experimental and/or theoretical physicists. These phenomena might have something to do with the equilibrium problem of redistributing the separable energy transformations. Our idea of three-ringed energy transformation might provide a new way of thinking to attack this problem.

6. Currently, we have seen many three-ringed stabilities, such as (1) the stability in the cosmic scale, consisting of galaxy level (including the Milky Way), star level (including the solar system), and planets (including the earth); (2) the stability in atmospheric fluids, consisting of the frigid zones, temperate zones, and torrid zones; and (3) the stability of typhoons' fields; (4) the stability in the elementary elements of the microscopic world, consisting of molecule lev-els, atomic levels, and electronic levels. As for the bodily activities of animals and humans or optimal designs of mechanical devices, and so forth, it does not seem accidental that they all reflect a quasi-three-ringed energy transfor-mation and material transfer of the conservation of stirring energy.

7. What needs to be pointed out is that stability is relative, and according to the point of view of evolution, there absolutely does not exist any stability. Any stability can be destroyed by the instability of the underlying systems' non-three-ringed circulations. However, because of the conservation of stirring

energies, each instability must adjust itself toward a three-leveled circulation. And the readjusted three-leveled circulations might not be the same as the original system's three-ringed circulation. So, instabilities and conservation of stirring energies not only are the reason for the existence of materials' forms, structures, or attributes, but also reveal the evolutionary process from the start of the development to the end of a change. And, for non-three-ringed circulations, changes in their direction of rotations can occur prior to their process of adjustment, which provides both the theory and methods for prediction practices of materials' evolutions.

8. What should be said is that three-ringed circulations constitute the systemic transformations of materials and energies, shown in the form of structural rotations. Its essence is no longer limited to certain phenomena of physics. Instead it is a problem about laws of physics. It has been only because in the past 300-some years, we have been so confined by the concept of Euclidean spaces that we overlooked the rotational aspect of materials, left out the recognition that the general form of materials' kinetic energy is that of stirring energy, and did not realize that the concept of angular speed contains that of (linear) speed as a special case. If we extract and purify this as a principle of philosophy about nature, we have: The significance of quasi-three-ringed circulative energy transformation is no longer limited to the "triangular" stability of the statics; it has already touched on the stability of materials' motions and the philosophical problem of evolutions where at extremes, matters will evolve in the opposite direction. This is exactly one of the very important problems that modern physics and natural philosophy have failed to address, and opens the door for a new physics on processes and evolution science in the name of conservation of stirring energies.

Since wholeness is a fundamental concept in systems science, it implies that the law of conservation of stirring energy is the existence law of systems stability and that the nonconservation at the same time constitutes the evolution law regarding systemic processes of changes. What is interesting is that the concept of stirring energy can contain the traditional speed energy as a special case. So, we can expect that the concept of stirring energy can transform modern science to the evolution science through its nonconservations.

Considering the fact that other than revealing the incompleteness of the conservation laws of energy of modern science, what is more important is that stirring motions and stirring energies can directly enter into the transformations of materials and events and the physicality of energies. So, we should leave space for the investigation of transformations of energies. To this end, let us summarize the main results obtained so far.

1. The quasi-three-ringed stability principle of the conservation of stirring energies, which can be seen as a realization of the concept of multilevel systems

(Lin, 1989; 1990), reveals the procedural aspects that all materials have places to come from and to move to and all events have a start and end, and the transformation of energies. It not only contains the quantitative laws of modern science but also reflects the laws of naturally existing events' processes, making it clear that events cannot be identified with quantities. This principle can be applied to product designs in engineering (there are such products in circulation already) or ecological environment planning or employed to resolve problems regarding evolutionary predictions. In particular, it is necessary to forecast natural disasters. However, accuracies in forecasting are not naturally the same as our ability to effectively reduce or prevent the losses of the predicted disasters. Practically effective reduction and prevention of losses can be materialized through the use of the accomodability of the second-level (indirect) circulation existing in the transformation or transfer of stirring energies, while sufficiently taking advantage of the resources brought forward by the disasters, such as floods, to benefit the regional economies affected. The currently higher and steadier dykes and dams employed to prevent flood disasters are not only a potentially greater danger for the next flood disaster but also a waste of the flood resources. As a financial investment and a practical safety measure of disasters, what has been obtained in the previous discussion will not only cut down the amount of investment, but also lead to techniques and methods for long-term effective disaster preventions.

This end also involves the problem of ecological balance when the flows of rivers are artificially controlled. These human interferences increase the risk to the existing high dykes and dams as well as cause imbalances in the ecological environments due to dried up rivers. These problems can be more serious than air pollution. So, to achieve true ecological balance, we should always make sure that mountains are constantly green and river waters always flow. Remember that the water quality produced by the natural filtration of vegetation is far better than that obtained through artificial means.

2. As for the problem of evolution prediction, it involves how we understand instable energies and the corresponding irregular events. Irregular information comes from irregular events and has to be reflected in instable energies. So, the essence of limiting instable energies is to eliminate irregular events; and in terms of problems of prediction, it is equivalent to eliminating information of change. So, to truly face the challenge of predicting the future, we have to keep and employ events of change. So, at the same time when the thinking logic and the methodological system of limiting energy flows and transformations in modern science constitute exactly a denial of evolutions, they also declare the fact that they cannot foretell the future.

3. Corresponding to a scientific system, there should be relevant laws. In particular, in terms of the study of evolution problems, laws on processes have to be established. Since at the same time when stirring energies reveal the transformation process of energies they also describe the effects of instable energies,

the significance of stirring energies is shown vividly in terms of the under-lying mechanism of physics. Since stirring energies include the traditional speed energy as special cases, the law of conservation of stirring energies can bridge modern science to evolution science. Besides, the concept of angular speeds enters curvature spaces in terms of the curvatures of the fields of flows. Due to this understanding, Einstein's treatment of the curvature problem by taking the speed of light as a constant has already violated the space–time curvature. Or in other words, in essence, the discussion in the form of speed kinetic energy (mc^2) has limited the problem of focus in Euclidean spaces. What is more important is that the assumption that the speed of light is constant has already made Einstein's mass–energy as such "an energy of not doing any work," leading to the problem of whether or not the mass–energy formula and the general relativity theory hold true.

What deserves our attention is that curvature spaces are not limited to the Riemann geometry of convex curvatures. There is also a problem with regard to concave curvature spaces. The corresponding modern dynamic equations are written in zero-curvature (Euclidean) spaces. So, even if the equations are cor-rect, they still could not be introduced into the realistic physical space. In par-ticular, angular speed implicitly contains changes in (linear) speed. So, as long as the angular speed is a variable, it implicitly means a problem of changing acceleration. Modern science has admitted that any problem involving changing accelerations belongs to a noninertial system. Therefore, the concept of stirring energy and its consequent evolution science must have gone beyond the range of modern science and helped to reveal many problems existing in the foundations of modern science.

7.4 Physical Quantities, Parametric Dimension, and Variable Events

7.4.1 The Physics of Physical Quantities

It can be said that because of the established concepts of physical quantities, mod-ern science constitutes a system. In other words, without physical quantities, there would be no modern science. The essence of the physical quantities is to quantify events. As for whether or not any of the quantifications of events truly represents the events or not or whether or not the relationships and manipulation of quantities can be used to substitute those of events, as of this writing, we have not seen any systematic investigation. Evidently, if quantities and operations of quantities could not be employed to represent events and their relationships, then there would be a serious problem of principle regarding substituting the true physics by using that of physical quantities developed based on quantitative manipulations.

For example, even based on the concept of physical quantities, mass (quantity) means the attribute of a matter or material, indicating whether the attribute is good, bad, superior, or inferior, providing a sense of order. However, modern physics messes up this sense of order of quality and keeps essentially only the quantity. Mass is one of the elementary quantities employed in modern physics, with the "direction" of materials' attributes excluded. This end can more or less explain the purpose of the physics of physical quantities. Even though direction is an important property or factor of materials or events, since it is not a quantity, this factor has to be disregarded in the study of physical quantities. From the very start when physical quantities were initially introduced in physics, they have lost the fundamental function of events with only the formality kept.

It can be claimed that without mass there would not be modern physics. The concept of mass can be seen as the doorway to enter modern physics. Before Galileo, some scholars mentioned the concept of mass. However, limited by the western stream of consciousness of the time, since the concept was not clearly identified with quantities, it was seen as irregular or not possible to unify until the year of 1678, when Newton wrote *Mathematical Principles of Natural Philosophy*. Newton was the first person who clearly stipulated mass as the measurement of a material's density and volume. Newton limited the word *quality* into a specified concept of quantities, creating the origin for the quantified physics of physical quantities, such as events, factors, information, and so forth. This end also paved the way for the later convention that quantities can be seen as parametric dimensions so that time and any parameter of physical quantities can be identified with material dimensions. Therefore, the wisdom of replacing events or information by quantities was shown, tolerated, and accepted in the West. On the other hand, with their root of thinking deeply in the *Book of Changes*, the Chinese were the first to introduce the method of hexagrams to derive the transformations of the states and attributes of events and realize that not all things can be described by numbers (incompleteness of numbers). What is interesting is that over 3,000 years later, K. Gödel (1931), a mathematician of the West, employed exactly the binary system as first shown in the *Book of Changes* to establish his incompleteness theorems of mathematics. This fact in essence has shown why the Chinese people of the East did not establish the physics of physical quantities by employing quantities to substitute for events. In other words, the idea of replacing materials' attributes by quantities could not be accepted by the Chinese due to their structure of knowledge.

No doubt, altering materials' attributes into quantities of the materials is the most basic assumption on which modern science is developed and is also the central difficulty for pushing for quantifications, because quantities after all stand for problems of formality, and formality is naturally different from the attributes and functions of materials. Although transforming *qualities* into *quantities* has been seen as the wisdom of mankind and well accepted by those who are accustomed to quantitative analysis, whether or not this wise behavior can stand the test of time is still unknown.

Evidently, since even the attributes of materials can be replaced by quantities, it can be concluded that there is nothing quantities cannot achieve. This end reminds us of Pythagoras's philosophy that numbers are the origin of all things. Although the current version is that quantities are the reason or law of all things, it is still the same as quantities' determining the attributes of materials; otherwise, the definition that mass stands for the quantity of the material of concern would not hold true. The reason for us to talk about the quantification of mass is that the physics of physical quantities is established on this concept, which explains why this theory suffers from a congenital deficiency and why since the very start, this theory has experienced various challenges, one after another.

For example, density is also a concept of mass, the mass in the unit volume. Modern science is called an *exact* science because of the rigor of formal logic, while excluding other schools of thought from being exact. However, it is exactly in the manipulations of the elementary concepts — there is density in mass and mass in density — that conceptual cycles that are based on mass, mass is derived, are formed. In modern-day textbooks of physics, mass is seen as a (quantitative) measure of the amount of matter contained in a body, indicating at least that modern physics is a quantitative theory.

Second, the fundamental laws of physics established on physical quantities did not point out where the force f and the acceleration a, as physical quantities, come from. Speaking more specifically, the system of modern science does not tell us where the force f and the acceleration a are located. Evidently, the acceleration a has to be with the moving material m. Without the material, the acceleration would not exist. As for the force f, according to ancient Chinese Mo Zi, it is originated from the internal structure of the material. That is, f also exists inside the material. In fact, even if we do not talk about the feasibility of the dualism of separate objects and forces of Western philosophy, the essence of treating the objects and forces separately from each other is the separation of quantities and materials so that quantities, space, time, functions of materials, and so forth can all be established in the form of physical quantities independently outside the materials and can directly control or dominate the materials. Speaking more precisely, human feelings exist independently outside materials, advocating that these feelings dictate materials. That is the main information modern science tries to advocate. Here, the mass m has been employed in conceptual cycles. That is, the laws of modern physics are written in the form of quantities and can dictate materials established on cyclic concepts.

In modern science, established on physical quantities, quantities have been employed universally without paying much attention to the differences in materials or time in terms of their attributes, states, and functions. As for the consequences of existing problems of quantifying events, the reader is advised to consult with Section 6.1. Since realistic accelerations are variant, to satisfy Newton's second law, one has to take the average value of the variant acceleration.

This end in essence has declared what modern science tends to tell us is: Modern science developed on the inertial system is such a system of knowledge in which quantitative averages are applied to express the generality. In other words, since the time when modern science was born, it has stood for a science where all peculiarities are eliminated. Because of the constraint of average accelerations, what are considered are exactly problems of motion of invariant materials. Otherwise, if m is a variable, then the acceleration a has also to be a variable. Evidently, if a is a variable, the underlying problem will have to involve a variant acceleration, which would be against Newton's original intention. To this end, modern science also admits that variant acceleration stands for a problem of noninertial system. So, according to the formal logic reasoning, let us introduce variable acceleration; then the material m has to be a variable too. That is,

$$f = m(t)a(t) \tag{7.14}$$

where the time t is also seen as a quantified physical quantity and further as a generalized parametric dimension. If we do not introduce t as a physical quantity or a parametric dimension, then Equation (7.14) would be degenerated back to Newton's second law. Hence, from this discussion, we can see why the laws of modern science hold true without "time" of time.

So, when pursuing the essence of modern science, it is found that the foundation of modern science is to study problems of movements with invariant materials; this end also reveals the reason why modern science needs the concept of inertial systems. It also shows how modern studies are not about the original states of materials and that the human wisdom of the time when modern science was initially established stayed at the desire of pursuing eternity. Corresponding to inertial systems is the concept of noninertial systems. Our earlier discussions have clearly shown that variable materials and mathematical nonlinearities are all problems of noninertial systems. This end also reveals that the essence of the noninertial systems of nonlinearity, movements of variable accelerations, and the not closed Equation (7.14) is about such problems of variant materials caused by rotations. Studies of such problems are called evolution science.

Besides, since the force f comes from materials' structure and in evolution science materials' structures also change, Equation (7.14) is both a nonlinear equation and a mathematically nonclosed equation. So, what Newton's second law tells us is that the event or information under the average acceleration without any of its own characteristics only describes the movement of unchanging materials, and that evolution science involves specific events or characteristic information, where materials change harmonically without complying with the inertial system. This end also makes it clear that the physics, representative of physical quantities, is indeed a science without dealing with changes due to the generality of quantities where variable events have to be eliminated. What constitutes true fundamental science is exactly evolution science where changes in materials and events are addressed.

Now, we see that physical quantities have made physics, in particular the theoretical physics, through the principle of materials' transformation into a formal logical calculus of quantities. What we need to notice is that materials or events do not in general follow the manipulation of formal logical calculus due to their variability. However, without physical quantities and parametric dimensions, there would not be modern physics, leaving behind the locality of quantitative physics. Or in other words, modern science has not entered the physics of materials' or events. If prediction science is also seen as part of physics, then the resultant theory will be a physics of event processes.

7.4.2 The Physics of Physical Quantities and Nonquantification of Events

If, continuing the formal logic of modern science, the average acceleration describes the generality of events, then variable acceleration will represent specifics. So, if the generality described by the average acceleration reflects the invariance of materials, then the specifics represented by the variable acceleration will stand for the variability of materials. So, events that change correspond to specific information. Or the variations of events with time not only are the objective existence but also reveal the evolution of materials. Hence, investigations on specific information will inevitably lead to evolution science. To this end, we are obligated to reconsider irregular events from the angle of materials changes. That is how we concluded that irregular information is that about changes and that evolution science does not contain stochastics and randomness.

There is no doubt that these results expound greatly on the belief of the indeterminacy of modern science. However, if we carefully ponder that events are not equal to quantities, then these conclusions seem to be basic knowledge, since no matter how small an event's probability is, it is a basic fact about occurrence or existence. In fact, the averages widely employed in modern science without any constraint are such a minor nonreality that it can be ignored more convincingly than small-probability events.

What is more important is that it is exactly because irregular information is information about changes that we need to reconsider whether or not physical quantities can truly describe events and their information.

7.4.2.1 Problems with the Physics of Physical Quantities

Our previous discussion implies that the direct consequence of introducing physical quantities is to change the physics of materials' laws into the quantitative physics that is only about quantities. This quantitative physics inevitably has to dwell on mathematical physics equations. As is well known, the deterministic problem of mathematical physics equations has to be quantitatively well posed. So, the first problem met by quantitative physics is numerical instability. And, in Euclidean spaces, quantities have to face the problem of unboundedness. So, physical quantities run into the difficulty of being unable to handle the problem of ill-posed quantities. Since modern mathematics did not provide any method to deal with ill-posedness, such a lack of methods has to keep the physics of physical quantities at the stage of unchanging materials. Even if we do not question the validity of the so-called laws of physics, the corresponding law of philosophy that quantitative changes lead to qualitative changes is not one people speak about and has never been implemented in modern science. In other words, mathematical principles have not illustrated, as what Newton had hoped for, problems of natural philosophies, and instead revealed the fact that mathematics has limited quantitative changes in the name of well-posedness.

Next, since the physics of physical quantities represents the eternal existence, many laws in modern physics have to be stated as quantitative conclusion laws. For example, Newton's third law, the conservation of momentum, the conservation of mass, the conservation of energy and the later conservation of mass–energy, and so forth, are all written in the quantitative form. None of these laws explains where the force and energy are from or where they are stored; in which form movements and interactions of materials take place; and through what process energies are conserved, except the quantitative conclusions describing the phenomena.

There is no need to deny that the postevent nature of quantities as formal measurements has already explained why the laws of physics, written in physical quantities, not only stay at the level of formality but also appear after events. This end has essentially revealed the reason why the laws of physics of physical quantities can only be laws of conclusions without providing information on the *whys* behind the events and the relevant processes and changes. For example, the law of conservation of energy does not explain through what fashion or process the conservation of energy is materialized. The law of conservation of mass–energy does not, either, provide any information on how mass could be torn. Newton's third law only tells that the quantitative magnitudes of the acting and reacting forces are the same without spelling out clearly how the forces interact with each other and where the acting force is from. The law of universal gravitation does not address the meaning of universality and what the gravitation is. Even if we do not question Einstein's energy that does not do any work, we still have to ask: How can the speed of any light traveling in a curved space be constant? And, how can there be any irrotational Riemann geometry in curvature spaces? As a matter of fact, some simple vector analysis can be constructed to show that nonlinearity stands for the rotationality of solenoidal fields. So, we have to ask: How can linear manifolds be introduced to the nonlinearity of tensors?

What is important is that the consequent theory of gravitational collapse suffers from the problem of inconsistency. According to Newton's theory, when an object collapses to a point — a singular point — with infinite density, one can directly observe from the outside, and the singular point can only be related to materials;

but from Einstein's general relativity theory, the singular point has something to do with the geometric structure of the space–time, the corresponding time needed for the celestial body to collapse equals the quantitative infinity, and it is impossible to detect from the outside the existence of such a singular point. Evidently, even if we do not pursue whose theory is correct, we still have to notice that singular point and infinity infinity come from the quantitative concepts of Euclidean spaces. And none of the theories provides a description on how a celestial body collapses and any details on the process of collapse. They did not answer why the universal gravitation can be universal and if the gravitation here is an attraction or not. In particular, what is the time structure of the space–time? In other words, can time be identified with space? Collapse is a contracting movement of materials and gravitational waves are propagations of energy without any materials' movement.

If 300 years ago the intelligence of mankind was for people to establish the physics of physical quantities on the basis of numbers, it will be the mental deficiency of man if we believe today on the basis of modern physics that the properties, states, and functionalities of materials or events can be identified with numbers. It is because the material world cannot have zero, and physics does not investigate "nothing." However, mathematics has to have zero; otherwise, the quantitative manipulation of the entire mathematics would be paralyzed. The so-called quantitative singularities and infinity infinity are problems of transition of events. Since quantities cannot deal with quantitative unboundedness, the physics of physical quantities inevitably has to experience problems irresolvable by using quantities.

There is no doubt that the physics of physical quantities has to be constrained by quantities. In particular, quantities can be used to compute objectively nonexisting matters more accurately than those that actually exist. Scholars often treat averages as the generality. However, the probability for an average value to appear is smaller than the chance for small-probability events to occur. So, in general, averages do not objectively exist. Besides, variant events do not comply with the logical calculus of quantities, which implies that the physics of physical quantities has to suffer from the problem of not completely agreeing with the reality, including the possibility of pushing physics into the investigation of "nothing" while creating such an inexplicable belief as that time exists before existence.

7.4.2.2 Nonquantification of Variable Events

Over the years, we have studied such questions as these and many others: Where are quantities from? What are the cause and process of changes in quantities? Are quantities the only method for us to know the world? Why are quantities postevent formal measurements? Our affirmative answer to the question of why quantities are postevent formal measurements no doubt indicates that from anything, quantities can be collected, that quantities cannot become the cause of any event, and that since the first day when the laws of the physics of physical quantities were established, they have been laws without any explanation about the causes and processes.

That is why the physics of physical quantities cannot foretell what is forthcoming, or quantities cannot predict. The formality of quantities surely provides the generality of numbers without being able to describe the specific changes and processes of events' properties, states, and functionalities. As soon as our book *Entering the Era of Irregularity* was published, many readers and scholars highly praised our conclusions, such as quantities are postevent formal measurements.

Science itself is a process. Its development is originated in calls of solving problems. In terms of the intention of science, science is about constantly challenging existing knowledge promoting progress in the development of human wisdom. That is why science always has been fascinating to man and accompanied by huge economic values.

The so-called process has to involve the differences between the before and after of events. If events possess only spatial differentialities, then we have the calm colors of a static world. If an event shows its difference along the order from before to after, then one sees the wonders of a dynamic and changing world. Speaking more comprehensively, changes in an event include those in terms of location, state, property or attribute, and so forth. Speaking strictly, changes in location, state, and attributes cannot be separated. Changes in location without involving the event's state, properties, and so forth, can only be human imagination or assumption in order to meet some specific needs. Because of the feeling of order, the problem about time has to appear.

There is no doubt that if no differentiality existed in each event from before to after, then the world would forever be a still picture. For a purely static picture, no matter how colorful it might be, sooner or later it will lose its attractiveness. So, without the past, the present, and the future, time would become a meaningless concept. So, what is more attractive to people is the wonder of change that naturally leads to these and other questions: What is an event? (Or, what is information?) How can events be classified? How can events be described? That is to say that compared with materials, the concept of events should be more general and involve such elements as materials, matters, and their states, attributes, the shown phenomena, and so forth. More particular events can be grouped into two categories: relatively unchanging events and those that are variable. Modern science used to refer to variant events as small-probability events or irregular events. In principle, what modern science investigates are relatively invariant events. Out of the need for treatment using mathematical techniques and the desire of maintaining the invariant system of modern science of quantitatively describing events, such concepts as parametric dimensions, physical quantities, and so forth, are introduced. What needs to be pointed out is that quantities are only an attribute of events and are only limited to the postevent formality. An event can contain at least such properties as color, smell, taste, character (attribute), direction, state, quantitative formality, and so forth. Hence, physical quantities are only one quantified attribute of events. Now, a natural question is, can quantified physical quantities be employed to substitute for the events from which the quantities were initially abstracted? To this end, the ancient Chinese belief that not all things can numbers describe has provided a negative answer. However, in the past 300-plus years since the time of Newton, other than Gödel's (1931) incompleteness theorem of mathematics and our second stir and the incompleteness of the quantitative analysis, there was no detailed explanation or argument; and instead, it has become a convention and custom to employ quantified physical quantities in the place of events. However, this problem reveals indeed a fundamental weakness existing in the foundation of modern science.

There is no denial that physical quantification of events is necessary in order to develop the relevant mathematical tools; and application of these tools indeed provide to a certain degree a description of the movements of invariant materials and the related problem of oscillation. That is, quantities have been important or magnificent contributions to modern science. However, the scientificality of science is not limited to one aspect of quantities but is more about the functionalities and effects of the events' attributes. Its signal of development should be about how to resolve the unsettled problems left behind from the past contributions and glories. In other words, the development of science comes from the calls of problems. Solving open problems is always the number one task of science. Existing problems are always challenges to the magnificent achievements of the past. Or, as long as unsettled problems exist, science has to be evolved further. And, the significance of pushing for further scientific development is to challenge human wisdom and expand human intelligence.

The first problem, which constitutes a direct challenge to modern science, is that about fluids' turbulences, which is relevant to materials' evolutions, as admitted by modern science. It has been a long-lasting solid stronghold still not possible to conquer. In particular, if we face the objective realities, we can say that all fundamental forms of materials' movements can be summarized as turbulences. It is because fluids are the majority of all materials, which constitute almost 99% of the universe. If we attempt to understand materials' flows through movements, then the system of wave motions of modern science does not describe the basic movements. The more general and fundamental forms of movements are rotational motions. That is why to understand turbulences, one has to resolve the problem of rotation. Even though modern science contains fluid mechanics, since the problem of rotation has not been resolved, it still cannot be called the science of fluids. Since fluids are afraid of twisting, as long as there is a twist, an eddy will appear. As a matter of fact, solids are afraid of twisting, too. Twisting is a problem of stirring motions. Each stirring motion has to be rotational. Each rotation indicates variable acceleration, leading to a noninertial system. In other words, the general and fundamental movements of materials are not those that are possible for the inertial system to summarize and are rotational movements that have annoyed the inertial system and made the inertial system helpless.

Considering the wide-ranging existence of noninertial systems, the achievements and magnificence of modern science, in terms of the problem of rotation, are only about the specific movements that keep the underlying materials unchanged. Even in terms of specific problems, modern science can only be seen as a school about wave motions. What is important is that modern science did not resolve rotations. And, its incapability of solving the problem of rotation should not be used as an excuse to treat turbulences as problems of stochastics.

Next, materials' aging, damage or breakage, or called decrepitude or death, are a problem of common objective existence. The events that are described by using quantities are only a realization of regularized events, which does not mean that events can be regularized. And, the idea of regularizing events is exactly abstracted from irregular events. Or in other words, without irregular events, there would not be such a thing as regularizing events. In particular, when dealing with problems of change, the goal is to look for differences instead of the overall similarities. Only when we are looking for differences or are good at finding differences can we truly handle complex problems. In other words, it is exactly because of the differences in specific events that human knowledge and wisdom are challenged and science and technology are further developed. So, specifics and particularities are the mother of generality. Without specifics, there would not be any generality.

So, the essence of irregular events is variable events. The corresponding reason why the system of quantitative analysis cannot deal with irregular events is that quantities cannot handle variable events. Therefore, variable events become an obstacle and problem for modern science. That was why we published our earlier book *Entering the Era of Irregularity* to introduce the problem of variable events. Since addressing unsettled problems is the call for the development of science, at the same time when variable events become an unsettled problem, it also challenges the system of modern science.

In other words, when people try different methods to eliminate irregular events or small-probability information, they should first ask, what is an irregular event? Where are irregular events from? What could be the functions of these events? It might be because of quantities that the conceptual thinking of modern man has been ossified so that man has refused to admit the challenge of irregular events imposed on his intelligence. Or, it might be due to the sophistry of the logical reasoning that problems of irregularity or small probability have first been disregarded as quasi-problems and later labeled as problems of randomness and indeterminacy. However, after more than 100 years of investigating problems of irregularity and small probability, the soul of these problems is still around and still bothers modern science, so that people do not know what to do.

Evidently, if we intelligently accept the challenge of irregularities or small probabilisticity, then it is found that the so-called irregular or small-probability events or information are nothing but variable events or information that the system of quantitative analysis cannot deal with successfully. This realization enriches the structure of human knowledge and helps to bring our level of intelligence to a higher ground. So, irregular events or small-probability information, whether treated as a theory or method for handling epistemological problems, represent a magnificent topic of research, because the essence of why science is referred to as science is that problems challenge the accustomed wisdom of man.

Besides, the investigations of variant events involve such a science — evolution science — that people are more interested in. Evolution science specifically deals with the physics of processes and changes. What is different from modern science is that in evolution science, the future is not the same as the present, and there is a process bridging the present to the future; whereas in modern science, the present existence becomes the eternal existence, where the present invariance guaranteed by the initial value stability can lead to the never-ending time of infinity. For example, Laplace, a follower of Newton, declared that from knowing the initial values, he could tell you everything about the future. And, Einstein's realization that neither classical mechanics, quantum mechanics, relativity theory, nor the basic mathematical physics equations can provide the difference between the past, the present, and the future, has admitted that modern science did not touch on materials or evolution of events. It can be said that modern science merely lists the laws of results using quantitative analysis without dealing with processes or paths. For instance, conservation laws of energy did not tell us what the conservation process of energy is; and, additionally, they did not touch on the mechanism of energy transformations. The so-called entropy (it is in fact a condition to limit an increase in heat) is also a state function that has nothing to do with a specific path. Just like the situation of the prediction that all people are always walking to their eventual death, since there is no doubt about its truthfulness, it becomes useless, because what people care about is how each person walks to his eventual death — the process of dying. However, for why people always walk to their eventual death, modern science did not provide any answer, either. As for statistics, a branch of modern science, it does not have any essential difference from the system of dynamics, except that experts in this branch themselves believe it is different. Even though in form indeterminacy is used, its concrete procedure is still pursuing the regularized determinacy by ignoring small-probability information using stable data series, changing *the future = the present* of the dynamics to *the future = the past* without providing any explanation for the difference between the past, the present, and the future. Regarding criticizing the determinism of Newton's system, it itself did not understand what *determinacy* really means. In the *invariance* theory of Newton's system, since nothing really changes, what is the significance of talking about determinacy or indeterminacy? That is why Einstein was seen as an old diehard who was against chances, leaving behind an ironic and unjustified allegation in scientific history. Irregular information is such information that is about changes; that is why there is no randomness in evolution science. In other words, the concept of randomness is a product of the system of quantitative analysis.

Third, lack of evolution processes is the same as the past = the present = the future. So, prediction becomes meaningless, and it is unnecessary to have the specific science and professionals to concentrate on how to forecast. So, the concept of determinacy has been introduced with respect to problems of changes. And, the invariance system that throws the hat of determinacy to Newton and Einstein should be seen as a misunderstanding of elementary concepts. Or in other words, modern science, established by Newton, Einstein, and others, did not touch on predictions and is a theoretical system useful for the design and manufacture of durable goods. This system is still in use because people have confused monitoring with prediction. That is why predicting transitional changes has become a difficult problem for the profession of forecasting. Therefore, foretelling what is forthcoming is a problem that challenges modern science and reminds people about how modern science should be correctly treated. Since major natural disasters occur during transitional changes, lack of understanding of these changes essentially means that there is no prediction science.

Combining and including rotational motions with what is known is the key for us to reach evolution science. In other words, the central problem to the three aforementioned challenges to modern science is how to understand rotations and to solve problems of rotations. The reason we list physical quantities and time as specific and individual topics in this chapter is that we like to reconsider in light of rotations the concept of physical quantities and parametric dimensions of modern science and the question of whether or not time can be treated as a physical quantity.

We particularly like the phrase *yu shi ju hua* (melt with time) (Zhuang Zi, from the era of Warring States in ancient China). First, it is because *yu shi ju hua* summarizes the home for the material world to return to without specifically emphasizing differences in time scales. Second, it is because our contemporary stream of consciousness has already started to *yu shi ju hua* or at least show the tendency of changing its direction. In particular, people have begun to realize that digitization is not the same as quantification, which should be seen as a major progress of our time and has started to impact the contemporary stream of consciousness. As is well known, quantification always meets with irregular events that are impossible to be dealt with by using quantities. So, it is natural to ask: what is an irregular event? In order to address this problem, we have to reconsider what information is.

If we understand information as symbols of events, then information will be names or labels of the events, where the events are objectively existent, while the information must exist only subjectively. The events represent such a deterministic concept that each matter must have its beginning and ending and every event its causes and consequences. Here, quantum events are a concept beyond quantitative mechanics and naturally not the probabilistic wave of accidental events. Quantities not only appear after events, but can also make the events disappear in the form of quantities. Even as the generality, averages are only approximations (realistic averages are also roving and cannot essentially represent the overall effects).

Although events can be labeled using quantities, it does not mean that events follow the calculus of formal logic. Integrals can be manipulated without paying much attention to the paths. But that does not mean that events have nothing to do with their "paths." The counts of people can take the sequence of 1, 2, 3, ...; however, the corresponding numbers cannot be seen as the individuals being counted.

The counts can be logically manipulated using the operations of addition, subtraction, multiplication, and division. But the specific people do not follow the rules of these operations; when men are added with women, what is obtained means neither men nor women. Quantities cannot handle irregular events, which at the same time implies that irregular events do not comply with the formal logic manipulation of quantities. So, the physics of physical quantities must contain logical contradictions against some elementary properties of events. Events do possess the attribute of quantities. However, more important and essential are their properties, states, and functionalities, which cannot be well revealed by the formal logical manipulations of quantities. If quantitative averages can approximately describe generality, then not many, if any at all, quantities can do so for specifics. Hence, quantities are not events and cannot enter variable events; and the physics of physical quantities cannot substitute for the physical laws of events. Accordingly, we should establish a direct method of analysis for events. Characteristics of events are in structural properties. That is why the analysis method for events is called a "structural method," leading to our structural predictions. However, in this method, time cannot be identified with a parametric dimension of quantities.

7.4.3 Material Dimensions and Problems with Quantitative Parametric Dimensions

The best way to validate a theory is to see whether or not the conclusions of the theory are supported by practical applications. By a material dimension, we mean the spatial decomposition of the field occupied by the materials. Even though each material dimension can be labeled using quantities, materials and events are still different from quantities. On the other hand, for each quantitative parametric dimension, it is first quantitative, which does not stand for any independent, realistic entity. For example, the time variable, one of the earliest parametric dimensions employed in modern science, is not a concept of any realistic entity, but an attribute of some objective materials. It dwells on materials and is not a pure quantity, which makes it different from pure quantitative physical quantities. It travels with materials and has its directionality. Besides, many physical quantities used in modern science are essentially not concepts of objective materials. Einstein identified time with space, which fundamentally confused the difference between material dimensions and parametric dimensions. It can also be said that the dimensions of objective materials in Euclidean spaces are four-dimensional (in non-Euclidean spaces, they are more than four-dimensional), because the existence of any static entity needs support (otherwise they would have to be spinning at high speeds). So, each four-dimensional existence means that in the static state, there is a fourth supporting dimension. However, this supporting dimension is also an objective entity and cannot be identified with time of nonobjective entities. So, our four-dimensional existence is different from Einstein's four dimensions. For example, in classical mechanics, force, acceleration, energy, and so forth, are attributes of materials. Since they exist on top of materials, they do not occupy any material dimension, either. If we see physics as knowledge about the principles of materials' transformations, then these principles of transformations cannot be identified with principles of physical quantities, either. This physics will have to produce physicists who can foretell the future. Or in other words, in such a new era of learning, if one cannot predict what is forthcoming, he cannot be called a physicist. So, the studies of relationships between physical quantities instead of parametric dimensions should be referred to as phase space dimensions, which should be employed separately with variables of material dimensions. The current method of mixing parametric and material dimensions, such as $f = f(x, y, z, t)$, has to experience instabilities due to manipulations of quantities, leading to the constraints of linear systems and having nothing to do with time t. That is the reason for the popular requirements of well-posedness of mathematical physics equations. At its origin, quantitative analysis should investigate quantitative changes. However, in specific implementations, various methods have been developed to limit these changes in quantities. So, the natural consequence is that modern science ultimately involves movements that have nothing to do with the parametric dimension of time.

Although quantities have their own rules of logical manipulation, whether or not events or information follow these rules, modern science did not provide any affirmative or otherwise theoretical proof or experimental testing. On the contrary, quantitative analysis of variable events has essentially declared that modern science is powerless in addressing this problem. Just as what is well known, quantities can inarguably provide the generality without pointing to specifics. So, if events or information do not follow the rules of quantitative manipulations, then the physics of physical quantities cannot be identified with the physics of the principles of materials' transformations.

For example, the chaos doctrine, which was once a hot topic of discussion in the latter half of the 20th century, started from the Lorenz chaos model, which was obtained from expansions of parametric dimensions. This model has had nothing to do with the materials' dimensions of Salzman's convection model. In particular, a same physical quantity about the atmospheric stratification is separated into two parametric dimensions using both linear and nonlinear strata, respectively. This example shows how the physics of physical quantities has employed parametric dimensions to such a degree that it no longer has anything to do with the underlying physical problem, indicating confusion between physical quantities and parameters. Evidently, although the model established in this fashion experiences chaos, it is not the same as the original convection model stands for chaos. Besides, in 20th century science, the problem of computational inaccuracy of quantities was even being seen as a scientific theory. So, it should be the time for us to consider the problem of how to make scientific investigations satisfy the scientificality. In the applications of parametric dimensions, physics has been developed to such an abusive stage that it has led to the study of nonmaterial fabrications, imagined spaces, and nonexistent time. Were these works created by the greatest physicists of history?

That is why we used to name the physics of physical quantities as a school of the physics of the principles of materials' transformations, since material dimensions cannot be mixed up with parametric dimensions. One of the reasons to this end is that quantities cannot handle irregular or variant events, which inevitably leads to the thought that quantitative averages cannot be used to substitute for the generality of events, either. Also, events themselves should have their own direct method different from the quantitative formal logical analysis that should be able to deal with quantitatively unsolvable problems.

The reason we have placed so much emphasis on this problem is the tragic aftermaths of major natural disasters. To practically forecast these major disasters, we have to take risks by trying out different methods and theories. In our practice, what we first discovered is that irregular information cannot be replaced by quantities and does not follow the manipulation rules of quantitative logical calculus. Next, we discovered that quantitative averages are not the same as the generality of events and are only cognitive wishes of man. Exploring along these lines of thinking, we gradually found that the parametric dimensions of modern science suffer from various problems. For example, although the heat of the chaos doctrine is already cooled, all the corresponding quantitative instabilities, complexities, and so forth are originated from the fact that quantities cannot handle irregular events and information and directly or indirectly have something to do with parametric dimensions. For the exploration of knowledge about nature and the essential properties of epistemological problems, in order to gain an upper hand with respect to complex problems, one should start his analysis with specific events. Only by doing so will he truly acquire knowledge from studying materials. In other words, only after fully understanding the essence of specific events has he the potential to truly fathom the overall trend of development or change of the whole event. That is, the message we delivered in our book *Entering the Era of Irregularity* is to first comprehend the properties of specific events.

To satisfy the need for resolving practical problems, during the past several decades we have repeatedly experimented our nonquantitative digitization analysis (event analysis). That is how we realized that irregular information stands for changing events that create information of change, which directly reveals the fact that quantities and physical quantities cannot deal with rotational movements and transformations of materials.

No doubt, if time is not seen as a parametric dimension in the form of quantities, the reader can naturally see the direction of development of science. If the proposition that events are not the same as quantities belongs to the category of problems of foundation, then when people are fighting to get on the shoulders of the giants, they should first check whether or not these giants are standing on the sandpile of physical quantities, because those historical giants whose shoulders rest on the sandpile can no longer hold up or support the new giants of the coming generations!

Acknowledgments

The presentation in this chapter is based on Bergson (1963), Chen et al. (2005a,b,c), Einstein (1976), English and Feng (1972), Fjörtoft (1953), Gödel (1931), Koyré (1968), Lin (1989, 1990), OuYang (1998b), OuYang and Chen (2006), OuYang et al. (2000, 2001a,b, 2002a), Prigogine (1980), and Wilhalm and Baynes (1967).

Chapter 8

Irregular Information and Regional Digitization

Because during his college years he had many disagreements with the results of conventional meteorological science and in terms of the useful value of a scientific technique, and because he and his classmates did not learn any reliable methods for practically predicting weather conditions, Shoucheng OuYang left the profession of weather forecasting, majoring in meteorology, after 5 years of college. However, fate tends to go in the opposite direction of wishes. In the second year after college (1963), a major flood caused by torrential rains occurred in the northern China. Considering the fact that he had majored in meteorology at the top-ranked university, his boss arranged for OuYang to look into this natural event for the purpose of predicting whether or not such a disaster could possibly occur at the upper reaches of Songhua River. If possible, then the dam located at Songhua Lake could potentially collapse. Knowing his true ability to predict zero-probability disastrous weather conditions, OuYang helplessly accepted this assignment by starting to collect relevant data and information.

When OuYang and his colleagues were analyzing the available information, they mistakenly employed the wrong graphing tool and unexpectedly discovered the existence of the ultralow temperature near the top of the atmospheric troposphere. Numerous case studies have found that this phenomenon possesses the triggering effect of causing and transporting such severe weather conditions as convective winds, torrential rains, torrential snows, hailstones, hurricanes, and so forth. This discovery made OuYang ponder the significance and practical implications of this ultralow temperature. Because the phenomenon of ultralow temperatures stands

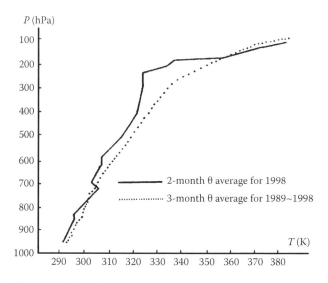

Figure 8.1   Three-month averages of the potential temperature at Chengdu station.

for an irregular event (Figure 8.1), if it possesses some important physical meaning and effects, then any of the observed ultralow temperatures cannot be weakened by using numerical methods. In Figure 8.1, the vertical axis stands for the barometric pressure, denoted P (hPa); and the horizontal axis, the temperature, denoted using the scale (K) of the absolute temperature. The dotted line represents the averages of the potential temperature θ for the months of March, April, and May at the Chengdu station during 1989–1998; the solid curve, the averages of the potential temperature θ for the same months in 1998. This special year, 1998, was a time during which the drainage areas of the Yangtze River suffered from severe torrential rain floods. And, the Chengdu station is one of the observation stations located at the upper reaches of the Yangtze River. Even with this comparison between the averages of the 3-month time period and those of over the 10-year time span, a clear characteristic of the ultralow temperature of about 20 degrees below the 10-year averages is vividly shown (200 to 250 hPa). In terms of the significance of practical applications, it implies that a drop in temperature near the top of the troposphere naturally causes activities in atmospheric convections. In terms of epistemology, first, one has to provide an explanation for why the temperature near the top of the troposphere can drop; and second, the changes in the ultralow temperature can be reflected in the size of the area covered and in the differences in altitude. In terms of what we have observed, changes in the ultralow temperature can horizontally cover an area from several hundred square kilometers to several thousand square kilometers, and vertically from the altitude of 400 to 150 hPa, where it is mostly seen in the range of 300–200 hPa.

The traditional meteorological science only states the fact of low temperatures near the top of the troposphere without explaining the reason for the appearance of such phenomena. After analyzing our data for over 30 years, we once saw such a low temperature as –83°C, and the ultralow temperature above a strong typhoon is generally below –70°C. Evidently, the sudden appearance of such low temperatures at the top of the troposphere cannot be plausibly explained by using the physical processes of plasma states of the earth's long-wave radiations. For this reason, we sensed the importance of this problem, so we looked for help from various sources. After consulting with relevant experts in chemistry, we were provided with the following understanding. This phenomenon of dropping temperature at the top of the troposphere should have something to do with the ionization effects of photochemical reactions. That is, when the adequate amount of water molecules are acted upon by either the ultraviolet light or the Rontgen rays, the water molecules are ionized so that their volumes are increased and the temperature is lowered. The relevant reaction formula of ionization is

$$4H_2O - \text{(ultraviolet radiation)} \rightarrow 4OH{-}\uparrow + 4H^+\uparrow$$

According to our consults, this process of reaction can be entirely carried out in laboratories with the exception that in high altitude, the density of the air is extremely low so that the four resolved hydrogen nuclei cannot as easily obtain electrons from their environment as in a laboratory. So, these resolved hydrogen nuclei would naturally be enlarged in volume, causing sudden drops in temperature. Evidently, the ionization process makes four volumes of molecules resolve into four volumes of negative oxyhydrogen ions and four volumes of hydrogen nuclei, creating a sudden expansion of eight volumes. The amount of heat required for the reaction helps to lower the temperature. Even in a laboratory, the temperature can be lowered as much as 15° to 25°C. It can be said that the ionization of water molecules provides not only a constructive explanation for the formation of the earth's troposphere, but also a basis for the analysis of why precipitation after thunder increases. Actual explorations of the atmosphere have already seen temperature changes of several tens of degrees during thunderstorms (see the area near 500 hPa in Figure 8.2). In the past, these sudden temperature changes had been ignored as misguided cases. As a matter of fact, this phenomenon is a characteristic of thunder rains or thunderstorms. Both this phenomenon and that of ultralow temperatures near the top of the troposphere have something to do with the ionization of the molecules in the atmosphere. Therefore, in order to study problems of disasters due to atmospheric changes, one has to go beyond the physics of molecules. In Sections 5.5 and 5.6, we pointed out that when twist-ups are involved with the third-order derivatives, each atmospheric ultralow temperature stands for a twist-up in fluids that can be seen directly from the relevant cloud charts (Figure 8.3). In particular, Figure 8.3 shows three obvious "twist-ups" in the atmosphere: One is the tropical windstorm twist-up of the west Pacific Ocean

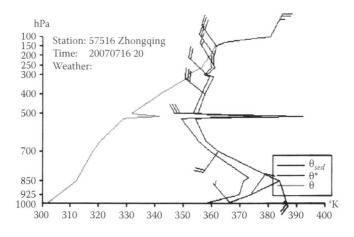

Figure 8.2   Sudden changes in atmospheric temperature for thunder rainstorms.

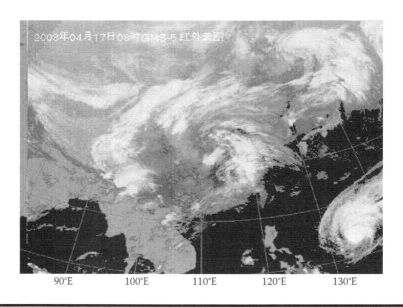

Figure 8.3   The cloud chart at 8th hour on April 17, 2003.

located to the east of 130°E; the second one is the twist-up of the occlusion front located along the same longitude near 50°N; and the third one is the twist-up of the convective cloud form at around 120°E and 40°N. The 6-hour precipitation corresponding to the third twist-up is 129 mm. The reason we have used a concept different from those of traditional meteorological science is the correspondence between atmospheric twist-ups and disastrous weather conditions. To this end, the atmospheric twist-ups (ultralow temperatures) appearing along with clear skies provide 12 to 24 more hours of advance predictability than the relevant cloud charts.

More importantly, all fluid dynamic equations employed to describe the movements of the atmosphere do not at all involve third-order derivatives so that the nonlinearity of third-order derivatives of course has never occurred in related research. No doubt, the practical value of meteorological science is in the area of correctly forecasting disastrous weather conditions; otherwise, there would be no need to have the profession of weather forecasting around. Since disastrous weather conditions correspond to the third-order derivatives of the atmospheric twist-ups, how can one resolve the problem of predicting disastrous weather by using equations without any third-order derivatives? That is why Shoucheng OuYang, a meteorologist by training, and Yi Lin, a mathematician, spent many years of their time and efforts in the study of probabilistic waves and the Schrödinger equation. It is because if the third-order derivative term is removed, then the Schrödinger equation will be degenerated into a ordinary dynamic equation of fluids. And, the Schrödinger equation is also the only nonlinear equation among all the well-known mathematical physics equations of our modern time that contains a third-order derivative term. Our study reveals that the essence of the so-called probabilistic waves is nothing but quasi-regular flows of particles under the mutual influence of the push and the quantum effects (nonlinear third-order derivatives) of potential fields. It is well known that first-order derivatives stand for slopes; second-order derivatives, curvatures; and third-order derivatives, changes in curvatures. As is similarly well known, people can observe the powerful damaging effects of twisting eddy currents in flowing waters. What attracts our attention is that twisting atmospheric flow patterns represent the underlying mechanism for the appearance of disastrous weather conditions. What is interesting and practically meaningful is that earthquakes also represent twisting movements of the "structural knots" in the crustal plates of the earth and appear after suddenly reversed twists.

Evidently, if the physical mechanism of the ultralow temperatures is revealed, not only is it a major breakthrough in the study of meteorological problems and related theories, but also has its significance gone far beyond the need for just forecasting weather phenomena. Each ultralow temperature makes the vertical structure of the atmosphere extremely uneven and touches on the problem of accurately predicting the occurrences and developments of rolling currents existing in the troposphere and the appearances of torrential rains and such severe convective weathers as hailstones, strong winds, tornadoes, sandstorms, and so forth. This phenomenon influences the fundamental principle of how man can possibly influence the weather. In terms of the problem of the essence of physics and philosophy, it provides factual evidence that deepens and enriches the epistemology. That is, the fundamental recognition of the twists associated with ultralow temperatures and vertical rolling movements in the atmosphere confirm exactly that other than their direct connection with the appearance of lives on the earth, under the effect of ultraviolet lights, water vapors also produce the troposphere for the earth near the ground level of the atmosphere, while creating ultralow temperatures. Evidently, the existence of the troposphere near the ground is one of the basic conditions for different forms of lives to appear and to be sustained. So, in accomplishing the task of protecting the ecological environment for the purpose of human existence and evolution, other than utilizing the phenomena of ultralow temperatures, there is also the problem of how to protect the constantly appearing ultralow temperatures in order to maintain the currently existing troposphere.

In terms of applications, colleagues in the prediction professions should modify their ways of thinking and cannot continue to treat irregular information as misguided cases and/or as meaningless randomness. In particular, they should be fully aware that irregular information is significant and usable. At the very least, all collected irregular information should be made directly available to the front-line forecasters. All relevant parties should recognize that information is a resource that should be fully utilized. In terms of weather forecasting, the information of ultralow temperatures existing along with clear skies possesses more advanced indication than that contained in cloud charts. The former has the effect of guiding the movement of convective weathers. Traditionally, air movements on the isobaric surface of the 500 hPa are treated as guiding flows on the basis of the movements of pressure systems. However, pressure systems cannot truthfully represent the movements of weather phenomena (in the following we will point out the problems existing with pressure systems). What is more important is that correctly locating an ultralow temperature also helps to correctly determine the specific region of occurrence of disastrous weather conditions (or predicting the placement, as known professionally). So, in terms of forecasting torrential rains, the current arrangement of sounding stations in China is still too scattered. Stations should be about 100 to 150 kilometers apart from each other. With such an arrangement, the information of ultralow temperatures can be more effectively utilized, and weather forecasts of much greater accuracy can be accomplished. There is also the need to improve the current sounding technology.

At this junction we would like to point out that the formation and development of ultralow temperatures are affected by water vapors; and that evolutions of existing ultralow temperatures are also influenced by water vapors. For instance, in the previously mentioned decomposition photochemical reaction, the decomposed four oxyhydrogen negative ions can also be reabsorbed and return to the formation of water vapors. In this process, additional heat is released and some oxygen is produced. That is, we have

$$4OH^- \uparrow ==== 2H_2O \downarrow + O_2 + 4e$$

This indicates that the oxyhydrogen negative ions floating in the atmosphere also have the effect of giving off heat. We once discovered the phenomena of increasing temperatures at the bottom of the stratosphere, which involves the problem of resolving negative ions. As a problem of practical application, it has been a well-known fact that the temperature at the top layer of the troposphere can increase. Its essence is that there are resolution reactions of ions taking place due to the release of potential heat along with condensations of water vapors during the rise of water molecules after precipitations; and the reason for the appearance of fresh air after precipitation is the negative oxygen ions created by thunder. What is important is that changes in ultralow temperatures have something to do with water vapors and conduct the evolution of the weather. So, among the various sources of irregular information, ultralow temperatures are the key information. That is also why we believe that when solving complex system problems, we have to start with the specifics instead of the generality as commonly seen in modern systems research.

Because the phenomena of ultralow temperatures are closely related to the ionization of the atmosphere, studying the problem of possibly forecasting disastrous weather conditions cannot be only based on the physics of molecular states. At the same time, it also indicates that the problems considered in meteorology are not simply applied problems of general physics, because even if we do not emphasize the fact that weather stands for problems of evolution science, in terms of the problem of weather-related disasters, they involve photochemical processes of change. In terms of the physics of processes, one should pay attention to the relevant problems of change, because changes in solar ultraviolet lights have to cause changes in ultralow temperatures. However, among the factors that affect the strength and the placement of ultraviolet light are also the relative positions of celestial bodies, the earth's mutation, movements of the poles, speed of the rotation, the traveling paths of cold air, and so forth. Hence, meteorological problems are absolutely not limited to the single area of geo-atmosphere. The optimal effect of the comprehensive mutual reactions of various factors can make photochemical processes play special roles. That is, at the same time as the temperature near the ground level increases, the reaction processes force drops in the temperature at the top of the troposphere, creating an extremely uneven temperature or heat structure in the troposphere. The extremely uneven structure has to cause the atmosphere to go through drastic adjustments, leading to disastrous weather conditions in the form of atmospheric rolling currents. In this sense, processes of disastrous weathers play the role of reducing or adjusting the existing uneven atmospheric structures, where the structural unevenness mainly refers to the vertical structure of the atmosphere. The movements that are directly involved in the adjustment are subeddy currents, the secondary circulations in the conservation of stirring energies, which help to carry out transformations and complete exchanges of heat. In terms of problems of prediction, horizontal pressure systems that have already occurred represent the results of the underlying transformations and can no longer be seen as the cause. That is why irregular information, such as ultralow temperatures, indicates uneven vertical atmospheric structures that in turn cause atmospheric transformations.

What we would like to point out is that the blown-up theory of evolution and the digitization of information are an epistemologically different theory and method from those of modern science and its methodology. To this end, let us briefly explain as follows:

1. According to the current, widely employed theories and methods of the profession of meteorology, horizontal pressure systems are employed as the theory and method for weather forecasting. However, in theory, no study has formally pointed out whether or not horizontal pressure systems are results of the heat transformations in the atmosphere; and no one seems to have paid attention to the vertical structural changes of the atmosphere. This end actually explains why horizontal pressure systems lag behind weather phenomena.

   Atmospheric movements are caused by changes and uneven distributions of heat. Changes in heat have to cause changes in density and redistribution of the density. The corresponding changes in and redistribution of horizontal pressure systems have to always follow behind changes in heat. That is why we have employed the regional digitization (the V-3θ graphs) constructed by using the vertical distribution of the potential temperatures. Because the potential temperature is a function of both temperature and pressure, it can be employed to reveal the structural characteristics of the vertical distribution of the atmosphere. At this junction, we would like to mention that to target weather problems, we have designed two methods of digitization. One focuses on the forecast of regional disastrous weather conditions using the information of significant spheres collected at sounding stations; and the other using automatically recorded information. This second method can be employed in any area of learning as long as time series data are available (see Chapter 9 for more details).

2. Because numerical computations encounter the difficulty of nonlinearity, the relationships between geopotential altitudes and pressures and between winds and pressures are being investigated in current meteorological science using static approximation formulas of the linear system, assuming that the involved densities are constant. This end also constitutes the artificial reason for why horizontal pressure systems lag behind the relevant weather phenomena.

   This problem agrees with the discovery of the front-line forecasters of the 1950s that only after precipitations do pressure systems appear. Based on this discovery of the front-line forecasters, we repeatedly verified their claims with case studies and drew our conclusions on why this problem appears. That is the reason why, when we developed our methods of digitization, we employed the reverse information structure without utilizing pressures. In form it seems to be a matter of technique; however, in essence, this modified method has contained a major change in our epistemological understanding. That is why in our various publications, we have repeatedly mentioned that the current weather forecasts are live reports instead of predictions made before the thing predicted actually occurs.

3. The high- and low-pressure systems computed out of weather maps or numerical schemes only represent horizontal (in the vertical direction) vorticities (that is, the ς vorticity). The reason is that the ς vorticity is only a result instead of the cause of a rise or fall of the airflow. That is, after rising, one obtains a positive vorticity; after dropping, one gets a negative vorticity. To truly make valid predictions, one has to employ information that is prior to the appearance of the predicted event. That is why in the method of digitization of the blown-up theory, we utilize vertical vorticities (in the horizontal direction, that is, the ξ,η vorticities). By doing so, we are able to employ the information of rising and dropping air, which provides more valuable lead time for our predictions than using pressure systems. By introducing the concept of rolling currents, where clockwise-rolling currents rise and counterclockwise currents fall, we are able to obtain information that is ahead of the appearance of the traditional information of pressure systems. Therefore, our digitization method is different from the currently employed numerical methods and has modified how information is utilized. That is, the information that is prior to weather phenomena is employed in the place of pressure systems, constituting a system of methodology that first qualifies and then quantifies transitional changes in the factors of the desired predictions. At the same time, this system can be employed to forecast transitional changes in the pressure systems.

4. From analyzing actual data, we discovered the information about ultralow temperatures at the top of the troposphere. This discovery, regarding not only meteorology but also the induced epistemological problems, has called for a change to be made to the current knowledge system, because it has touched upon the central problem — the twist-ups — about transitional changes in materials. In terms of meteorology:

   – First, the information of ultralow temperatures plays the role of stimulating and triggering the occurrence of weather phenomena and affects the movements of convective cloud systems. All of these have altered the traditional concept of treating 500 hPa as the guiding airflows that dictate the movements of pressure systems. That is, the valid prediction of weather conditions is about that of significant factors instead of the prediction of the pressure system.

   – Second, transitional changes stand for the key of the prediction problem. In transitional evolutions, horizontal pressure systems lag behind the relevant weather phenomena. That is why we concluded that to forecast disastrous weathers, we couldn't depend on pressure systems.

- Third, the layer of ultralow temperatures belongs to specific information among all available irregular information that signals forthcoming changes. This layer in general is concentrated at the altitude of 250 to 150 hPa. For extremes, it could also appear at the altitude of 400 to 300 hPa or that between 200 to 100 hPa. It is exactly because of its irregularity and changeability that one faces a problem that cannot be dealt with by using the currently available numerical analysis or computations of numerical schemes. In other words, we say that peculiar events do not comply with the calculus of formal logic of the quantities. That is why we established the methodology system of digitization.
- Fourth, each region within which an ultralow temperature appears is exactly the area of twist-ups where changes in the curvature of the atmosphere occur and disasters appear.

Just as what has been mentioned, twist-ups stand for changes in curvature and are described by nonlinear third-order derivatives. However, the equation (the N-S equation) that has been used to describe the event just does not contain any third-order derivative, revealing the fact that current fluid mechanics suffers from its own innate deficiency. Twist-ups stand for specific events that the system of modern science does not touch on and are also the root cause of various natural disasters that are not limited to only weather-related events. They also represent problems the mathematical equations of the current continuity system cannot deal with. That is also the reason why we introduce the method of digitization to handle irregular information. In other words, twist-ups bring forward challenges to the quantitative analysis system at the very fundamental level. It can also be said that without resolving the problem of twist-ups, there will not be any chance for us to develop an effective method to analyze the essence of disastrous weathers so that we will never be able to effectively predict such specific events as disasters, accidents, and so forth. All such convective weathers as torrential rains, hailstones, strong winds, sandstorms, heavy snows, and so forth are conditions caused by ultralow temperatures. That is why we published our book *Entering the Era of Irregularity* (OuYang et al., 2002a).

8.1 Digitization of Region-Specific Information and Prediction of Disastrous Weathers

At the very beginning when we were forced to establish the method of digitization, we encountered problems that traditional quantitative analysis methods could not handle. However, reality indicated that in order to resolve the problems, we had to employ specific information. In order to include irregular information in our consideration, without much choice we had to apply figurative methods. Because the idea of figurative analysis is widely applied in the *Book of Changes* (Wilhalm and Baynes, 1967), we had an opportunity to look over that classic. What we gained most from studying the *Book of Changes* is that the book has well addressed what time is and helped to guide us into the realm of evolution science. We found the following from studying the *Book of Changes*:

This book has long ago, more than 3,000 years ago to be more specific, documented the human wisdom about how to deal with variable events, while modern science advocates the eternity (invariance) of events. So, the *Book of Changes* and modern science constitute two major sources of knowledge. Using quantities to substitute for events is both human wisdom and a major mistake. Variable events do not comply with the calculus of formal logic of quantities; these events should be investigated by using structures and principles of structural transformations.

With this newfound knowledge, we tested our analysis of transitional changes on the basis of the discovery of ultralow temperatures by using atmospheric rolling currents that have the effects of mixing different materials. Our initial success came from the greatly improved accuracy rate of forecasting regional torrential rains. Over time, we gradually generalized this method to the forecasts of both large area and local area torrential rains, severe convections, strong winds, sandstorms, dense fogs, and high temperatures.

8.1.1 Basic Logic Used in the Design of Digitization of Regional Disastrous Weather Conditions

8.1.1.1 Choice of Heat Analysis

No doubt, uneven distribution of heat and changes in the distribution are the essential cause of atmospheric movements. Because the problem of prediction is different from that of product design, there is a need to pursue the reason and process of events. Fortunately, traditional meteorology also admits that atmospheric movements and changes come from uneven distribution of heat and changes in distribution. For this reason, we select the concept of potential temperature of meteorological science to represent the vertical distribution of heat. That is,

$$\theta = T \left(\frac{P_0}{P} \right)^{R/C_p}$$

where θ stands for the potential temperature, T is the air absolute temperature (in K), P_0 is the air pressure at sea level (hPa), P is the pressure at any altitude, R is the gas constant of air, and C_p is the specific heat capacity at a constant pressure. Because in applying this formula experience is involved, in our computations we employ the formula as standardized by the Chinese Bureau of Meteorology.

Considering the need of advance notice of any valid prediction, we replace the imaginary potential temperature θ_{se} that is traditionally computed using the altitude of condensation by θ_{sed} that is calculated using dew point temperature. The significance is that when water vapors condensate, it is about the time of rainfall, while it is more meaningful for us to analyze the humidity of the atmosphere before the condensation in order to make advance predictions. Also, in order to make a comparison of the distribution of water vapors in the air, we also compute the potential temperature θ^* under the assumed saturated state. And, we notice that any atmospheric precipitation is not a consequence of how much water vapor is contained in the air column but the uneven vertical distribution of the water vapor. For example, if the water vapor in the air column concentrates at the top of the troposphere, no precipitation will occur. In general, we refer to θ, θ_{sed}, and θ^* as three distribution curves, known as 3θ curves.

8.1.1.2 Order of Information and the Reversed Order Structure

Earlier we specifically mentioned that the horizontal pressure information of the atmosphere is a piece of posterior information among many kinds of available information. However, the traditional meteorological science is fond of this factor *pressure*, which is known as a system and has been seen as a cause for weather changes. For instance, as for the reason for the snow and ice disaster that covered southeastern China for more than half a month in January 2008, the common explanation was the stability of the circulation systems. Speaking strictly, a *circulation* system is different from a *pressure* system. However, the professionals in the area of meteorology know well that the so-called circulation systems stand for the pressure systems on the horizontal isobaric surfaces. If we further question why the circulation system could be stable, there will be no answer given.

1. Currently, the traditional weather forecasts are based on the push of pressure gradient forces. The order of information used is pressure, temperature, humidity, and wind, with the pressure system as the core. The tools of analysis for making forecasts have evolved from the isobaric surface charts or contour surface maps of the 1930s to the current numerical forecasting graphs, all of which are constructed on the basis of pressures or isobaric surface charts. That is also the basic reason why the predictions of transitional disastrous weathers have been constantly incorrect. In other words, for transitional weathers, pressure systems cannot be used as a predictor. With this understanding in place, in our digitization of information, we employed the information structure that is the opposite of the tradition order of information. That is, we apply the following order: wind direction, wind speed, humidity, temperature, and pressure (as a matter of fact, the pressure factor is not used at all). And, we specifically divide the wind factor into two

factors: wind direction and wind speed. Because there is less error in the observation of the wind direction, we emphasize the use of wind direction in our information order of prediction. No doubt, the wind direction factor cannot be treated as a quantity, and the uneven distribution of wind speed is also a problem of structures. That is how in our method, we employ the information of two predictors of direct significance.

What needs to be noted is that changes in wind direction represent transitional changes, which provide very valuable information. However, since wind direction is not quantitative, it forces us to adopt the method of figures.

2. The reason why the humidity information is moved to an earlier position than temperature information is that of our initial need of predicting torrential rains. To this end, one has to know the difference between atmospheric structures of general precipitation and torrential rain weather. Later in our practical applications we discovered that the humidity factor is also extremely sensitive to various disastrous weathers, and its sensitivity is not any less than that of wind. Because the problem of water vapor is not entirely quantitative, what is important is how water vapor is distributed structurally and how it changes; water vapor not only conducts torrential rains but also directly affects drought weathers.

At this junction, we would like to point out that the traditional weather forecasts are done on the basis of horizontal cross-sectional maps of the atmosphere, while our tool of prediction is based on vertical cross-sectional charts along with the rotation direction of rolling currents. Intuitively, changes in the significant predictors are far more drastic in the vertical direction than in the horizontal direction. No doubt, knowing the source and origin of water vapor is important in weather forecasting. To this end, the corresponding 24-hour short-term forecasting should trace upward to about 10~15 longitudinal/latitudinal distance from the area to be predicted.

The basic idea of what is discussed above can be summarized as follows: Movements of materials come from the unevenness in their structures (in terms of the atmosphere, it is mainly from uneven distributions of heat). Interactions between rotational movements lead to irregular information and irregular events, where confrontational interactions cause damage to the original structures, leading to natural disasters and human accidents. Considering that the quantitative analysis system cannot handle irregular information, we employ the methodological system of digitizing information. Complex problems are about peculiarity of information instead of overall generality. And, peculiar events are about the properties and attributes of the events instead of the magnitudes of the relevant quantities. That is why we purposely chose the digitization method to emphasize the relevant peculiarities. And, our method of regional digitization is only designed to process meteorological sounding data with the goal of revealing irregular information, such as ultralow temperatures. This is why this method is limited in scope in terms of applications.

The corresponding method of digitization of automatically recorded information does not suffer from this limitation and can be employed in almost all kinds of investigations involving changes.

What is different from the currently accepted epistemology and modern scientific system is that our methods focus on the uneven distribution and change in heat and the belief that the constraints of nonparticle structural rotations are the fundamental principle of change for the worldly materials.

8.1.1.3 The V-3θ Graph of Digitized Regional Information and Explanations

8.1.1.3.1 Wind Velocity Vector V and Atmospheric Rolling Currents

Earlier we stated the significance and purpose for why we choose to reverse the order of information and how our method treats the wind velocity vector V as the most important piece of information. So, to illustrate how to apply our V-3θ graphs, let us first explain the concept of rolling currents made up of wind vector and wind speed, and how to recognize an existing rolling current.

The phrase *rolling current* is a term we adopted from the hydraulics of rivers. A rolling current stands for an eddy current existing in a vertical cross-section of the atmosphere and can be fathomed as a rotational movement just like a moving wheel. Similar to currents in rivers, the airflows in the atmosphere can also travel vertically as rolling currents. In fact, that is how different airs are mixed. From our works, it can be seen that the effects of rolling currents are not only limited to mixing airs, but they also play the role of conducting weather changes.

8.1.1.3.1.1 Rolling Currents of Wind Vectors (Directions) — For the convenience of introducing and understanding the concepts, let us respectively provide typical cases for the Northern Hemisphere in Figures 8.4 and 8.5. The graphs in Figure 8.4 show a clockwise rolling current made up of east winds at the lower sky and west winds at the higher sky (left graph) and south winds at the lower sky and

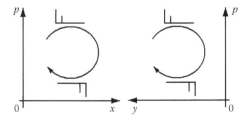

Figure 8.4   Clockwise rolling currents made up of east winds (lower sky) and west winds (higher sky) (left) and those of the south winds (lower sky) and north winds (higher sky) (right).

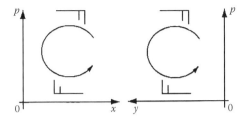

Figure 8.5 Counterclockwise rolling currents made up of west winds (lower altitude) and east winds (higher altitude) (left) and those of the north winds (lower altitude) and south winds (higher altitude) (right).

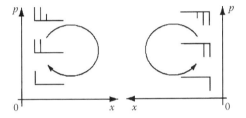

Figure 8.6 Clockwise rolling currents a west wind (left) and counterclockwise currents of an east wind (right).

north winds at the higher sky (right graph). And Figure 8.5 shows the counterclockwise rolling currents, made up of west winds at the lower sky and east winds at the higher sky (left graph), and north winds at the lower sky and south winds at the higher sky (the right graph).

8.1.1.3.1.2 Rolling Currents of Uneven Wind Velocities — By *uneven wind velocity* we mean that when the wind directions are the same, the distribution of the wind speeds in the vertical direction is uneven. Here, we illustrate the basic characteristics of clockwise and counterclockwise rolling currents for the general case where the wind speed increases with altitude. As for other scenarios, the reader can make his own determination using the principle presented here. Figure 8.6 shows a clockwise rolling current (left graph) of a west wind, whose strength increases with altitude, and a counterclockwise rolling current (right graph) of an east wind, whose strength also increases with altitude.

For the rolling currents as shown in Figure 8.7, uniform south and north winds are considered. For readers who are not in the area of meteorology, let us explain the flaglike signs appearing in the figures. These signs stand for wind direction and wind speed, where each horizontal bar represents the wind direction and each vertical bar the wind speed of 4 meters per second; two vertical bars signal a speed of 8 meters per second, and so forth. In the left graph of Figure 8.7, a uniform north

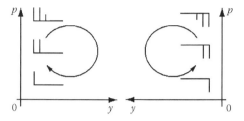

Figure 8.7   Rolling currents with uniform south and north winds.

wind is shown and increases in strength with the altitude. For the professionals, there is no need to explain in Figure 8.7 that when the horizontal bar is on the top, it means a north wind, and when the horizontal bar is on the bottom, it stands for a south wind. To help the reader comprehend how to analyze rolling currents, let us look at Figure 8.8 (inverted). In this case, the left-hand side stands for a south wind and the right-hand side, a north wind. In other words, if the observer stands facing the observation station, the scenario of the north wind represents a counterclockwise rolling current, and with the scenario of the south wind, a clockwise rolling current. What needs to be specifically noted is that almost all clockwise rolling currents of south winds in the Northern Hemisphere appear to the south of the fast airflows of east winds; they have been employed in the analysis of tropical typhoons and tropical windstorms or the strong winds and torrential rain weathers appearing on the southwest sides of northbound traveling typhoons. As for situations in the southern hemisphere, they can be analyzed similarly.

Only some very simplified cases of how to recognize rolling currents have been discussed thus far. When dealing with practical situations, things might suddenly become very complicated. For example, often-seen northwest winds, northeast winds, or northeast winds and southeast winds, and so forth, may not even be uniformly one-directional (e.g., a northwest wind may not be exactly distributed along 45°). However, in terms of the Northern Hemisphere, as long as one learns that northwest winds are representative of cold air, that when mid- to low-atmospheric (below 700 hPa) winds are partially from the south, including southwest and southeast winds, or there is an east wind near the ocean surface, then combined with the upper layer atmosphere (above 500 hPa), one has a clockwise rolling current; otherwise, one has a counterclockwise rolling current. In general, one can relatively easily and quickly learn how to recognize rolling currents from actual cases.

We would like to point out that the discovery of ultralow temperatures and the effects of rolling currents not only caught our initial attention but also drastically changed our way of thinking.

First, the information on ultralow temperatures represents an important piece of information regarding disastrous weather conditions.

Second, the problem of rolling currents is not a problem of the dynamic system of the first push. Instead, it is from confrontational stirring forces and causes rotational movements. The corresponding disastrous weathers only reflect how severe the confrontations are. Because of this, we have to think about such problems as the cause of the confrontations, where the stirring forces are from, and how these stirring forces change. As for the root cause, it is the changes in the structures and moments of force of the confrontational materials that interactions between the materials experience uncertainties. That is why we conclude that by continuing the first push system of particles, we can no longer resolve the problem of prediction of disastrous weathers.

In terms of fluids, it is very easy for them to rotate and to alter their forms of existence under the stirring effect of uneven structures. The current meaning of the word *evolution* is a process of gradually changing or evolving. However, its Latin root stands for *rolling forward*. That is to say, rotations are the cause of changes. In other words, the ancient westerners had already noticed the problem of changes in materials. Although the concept of rolling currents we introduced is from hydraulics of rivers, for the atmosphere, which is in fact a fluid, the form of motion of rolling currents is more general and possesses the essence. Therefore, when we look back in history, if the meteorological profession developed along V. Bjerknes's circulation theorem, it would not be in its current state of confusion. We borrow the term *rolling currents* from hydraulics because we like to cause colleagues in meteorology to ponder its significance and also because atmospheric rolling currents are extremely important in terms of transitional changes in weather conditions. It can even be said that if we can fully understand rolling currents, we then have fundamentally comprehended transitional changes in weathers. What is practically meaningful for predictions is that transformations in the direction of an existing rolling current can foretell changes in weather conditions; and observational errors in wind directions are smaller than those of wind speeds, humidity, temperature, and other factors. That is also the reason why we once pointed out that direction is both an epistemology and a methodology. Combined with the distributional characteristics of water vapor and the ways water vapor is transported in the atmosphere, direction can fundamentally help us deal with the prediction of such major disastrous weather conditions as torrential rains.

As for the understanding of rolling currents, just as the Latin root meaning of the word *evolution*, the problem of rolling forward in principle does not repeat itself even if sometimes by accident quasi or approximate repetitions do appear. The reason is that confrontations of rolling currents possess the changeability of the attributes of events and/or structures. Nonperiodic changes are mostly observed. That is why modern science encounters the challenge of world-class difficulties. In applications, one should note that the layer structure or the vertical difference of the atmosphere is much greater than that along the horizontal direction, which can be more than at least two magnitude levels; and that is why the characteristics of severe weather phenomena are shown more clearly in the vertical direction, and why changes in vertical whirlpools of the form of rolling currents appear ahead of changes in horizontal whirlpools. That is the reason we adopt the concept of rolling currents in the vertical direction instead of the traditional horizontal wave motions. Even if we apply the information of the same time moments, observations on rolling currents still provide more advance information than those of horizontal whirlpools (or the current information of pressure systems). They help distinguish the information on transitional changes in a timely fashion. That is why this method can first forecast weather phenomena and then foretell changes in the pressure system.

No doubt, when analyzing atmospheric rolling currents, one does not need to employ the traditional weather maps of the pressure or altitude systems, because rolling currents have already involved changes in the structure of the entire troposphere.

Because how the current meteorological profession understands irregular information is still very different from ours, most countries and regions either limit their submission of irregular information or modify their collected irregular information using quasi-linear means. For this reason, in the design of our V-3θ graphs using mainly the data of the significant spheres, we try our best to acquire as much irregular information as possible in our computations. In the future, as the needed irregular information is fully supplied, the rate of forecasting accuracy will expectedly be improved.

8.1.1.3.2 Relationship between Atmospheric Rolling Currents and Weather Evolutions

8.1.1.3.2.1 — Rolling currents, combined with an uneven distribution of heat, can cause nonuniform, multiple divergent eddy currents in the atmosphere, constituting a self-contained circulation system in the entire troposphere. This is the necessary condition for a convective system to form and develop. It is exactly because a self-contained circulation system is formed within each convection that relatively independent pressure systems of different scales are created. This end also illustrates why pressure systems lag behind rolling currents. In other words, rolling currents, and in particular the directions of rolling currents, appear ahead of not only pressure systems but also the relevant weather phenomena. What is practically useful is that the general self-contained circulation systems that have a life span of 6 hours and that affect ordinary human activities can all appear in the form of rolling currents that occupy the entire troposphere.

Our practice indicates (the details will be provided later in the following sections) that such disastrous convective weathers as hailstones, severe precipitations, strong winds, sandstorms, torrential rains, and so forth can all be predicted using clockwise rolling currents. Their differences can be completely recognized using our V-3θ digitization of information.

8.1.1.3.2.2 — For the discontinuous zones existing in the flow fields of westerly systems before troughs and after ridges (for easterly systems the inverse directions are used), rolling currents appear before these discontinuous zones in the horizontal flow fields. Here, we try to make use of the traditional terminology; otherwise, we can simply refer to these regions as clockwise rolling current zones, which possess the lead time needed for making predictions. Due to the difference in directions of the sinking and rising airflows in the front or back of a rolling current, the distribution characteristics or structures of horizontal weather-scale systems can change so that relatively strong rolling currents can cause the original weather-scale systems to split or change in their characteristics. What is important is that changes in the direction of a rolling current foretell the transitionality of the changes.

As a problem of epistemology, the strength of a rolling current is determined by the unevenness of the atmospheric structure. That explains why our analysis of digitization of irregular information can predict the direction of movement and form of change ahead of using forces.

Next, in the material world, first push that experiences no obstacles does not exist. For addressing the problem of natural disasters, we must consider confrontational interactions and consequent changes in materials or events.

Third, it is time to reconsider the proposition of treating the problem of prediction as a branch of modern science. What is very important is that the regularization of modern science is no longer adequate for the study of the problem of prediction; extrapolation of pressure systems itself is not about prediction. Instead it is a live report of the events that have already happened.

If we continue epistemologically to dwell on the "prediction and warming" of the form of monitoring, in essence we keep ourselves outside of the door of a prediction science.

8.1.1.3.2.3 — Other than the unevenness in the horizontal heat and mass fields, what is important with regard to the effects of rolling currents is the unevenness of these fields in the vertical direction. In particular, the heat unevenness made up of the existence of an ultralow temperature near the top of the troposphere and a rise of temperature at the ground level constitutes the essential reason for the drastic development of convection. For example, during the middle of the summer in an area located at the mid or low latitudes, the temperature difference between the ultralow temperature near the top of the troposphere and at the ground level can reach about 100°–120°C. This fact does not seem plausible when explained by using the long-wave radiation theory.

8.1.1.3.2.4 — Due to the misguided belief that quantities can be employed to address any problem, irregular information has been regularly ignored. That exactly explains why an opportunity of practically employing rotations has been lost. No doubt, for any positioning system, there are only two forms for rotations or rolling currents: one is to rotate inwardly and the other outwardly. So, no matter how complicated a movement could seem to be, its inward and outward rotations should be relatively easy to distinguish. And this binary distinction can be generalized to two classes of symbols used in our digitization. If we make use of the concepts of yin and yang from the *Book of Changes* from over 3,000 years ago, then the simplicity of the inward and outward rotations becomes the theoretical problem of reconcilability of directions of movement. As a methodology, it is also simpler than the quantitative analysis without being trapped in the irregularity and complexity of quantities.

If ultralow temperatures stand for specific irregular information, then the first task we have to take on is the method of how to deal with and how to apply irregular information. Otherwise, the investigation of changes in weather would inevitably fall into a difficult situation. The current studies of problems of change have been based on the available theories on invariance. Considering how many times the forecasts for major disasters have been incorrect, it is time for us to seriously look at how to deal with irregular information.

As for the epistemology of quantities, we should fully understand the basic characteristics and problems of quantities. The problems involving changing quantities, considered as of now, are nothing but the following three situations: the quantities are invariant, the quantities increase, and the quantities decrease. And, within each of these situations, quantities exist that cannot be dealt with by using quantities themselves.

No doubt, invariant quantities imply keeping the underlying materials or events constant, which becomes a core problem facing the system of modern science. When dealing with increasing quantities, we have to face the problem of quantitative infinity. That is why the condition of well-posedness has to be introduced in modern science, and the consequence of the well-posedness is still about invariant quantities. When dealing with decreasing quantities, the essence is still the same as keeping quantities unchanging, while the relevant computations tend to fabricate events where due to infinitesimal differences, error values are treated as the true events. In essence, modern science does not at all touch upon problems involving variable quantities. So, the currently popular studies on complexities are not about quantitative regularizations themselves; instead they are about how to understand complexities. Differences are the realistic nature of the world and cannot be simply referred to as complexities just because quantities cannot deal with these realistic differences.

In essence, so-called complexities have already indicated that events, in particular, variable events, do not comply with the calculus of quantitative formal logic. If we can look at events through their attributes and properties, it might be much easier for us to understand the events than if we employed quantities. For example, the five to six billion people in the world all look different. However, if we separate the world population into men and women, then only two forms of human beings exist: male and female. Similarly, no matter how complicated the irregularity of an event can be, if we treat the event according to its potential rotational directions, there are only two possibilities. Therefore, for the analysis of rolling currents, we can apply the right-hand spinning rule as follows.

1. In terms of the Northern Hemisphere (for the Southern Hemisphere, it is similar to the tropical weathers to the south of fast east wind flows, just reverse the instructions that follow), if the directions of the airflows at the

upper and lower layers of the troposphere do not agree with each other, then use the four fingers of the right hand to point to the direction of the south or the east wind with the origin of the coordinate system taken to be on the left side (along the latitudinal direction) or the right side (along the longitudinal direction) of the observation station; the constituted rolling current is named a clockwise rolling current (Figure 8.4); inversely, the rolling current is referred to as counterclockwise (Figure 8.5).

A clockwise rolling current signals a transitional change from pleasant weather, such as a clear sky, to gloomy weather, such as strong winds, traveling clouds, rains, and so forth. A counterclockwise rolling current indicates the opposite. That is, it signals a weather change from gloomy weather to the better. Here, the wind vector can be decomposed into components along the four main directions of east, south, west, and north so that the overall analysis is done according to the most dominating components.

In terms of the Northern Hemisphere, when the upper layer contains northwest winds and the lower layer southeast winds, it is an obvious sign for a clear sky to turn to that of strong winds or cloudy weather. These are the mostly seen transitional weather changes in the northern hemisphere. For the specific prediction analysis regarding the severity of the strong wind or the cloudy weather, one needs to pay attention to the distribution of the temperature difference and the structure of water vapor in the entire troposphere, based on which the severity of the relevant weather conditions can be predicted. In the following sections, we will employ practical situations to illustrate how to utilize the V-3θ graphs.

2. When faced with one-directional winds in the troposphere, in general due to ground surface frictions and increased air density at lower altitudes, the wind speed is smaller at lower altitudes and greater at higher altitudes. (However, it should be noted that with practical situations, one might well experience the opposite scenarios, where the wind speeds at higher altitudes can be smaller than those of the lower altitudes; the general reason for such scenarios to occur is that the sounding balloon is trapped in a eddy current or entered a layer of clouds. Front-line weather forecasters need to pay close attention to these scenarios.) In this case, due to the shear effect of the varying wind speed, a rolling current can be produced. That is, when the winds in the troposphere blow in the same direction, a rolling current may still be created. That is also why the troposphere stays constant with well-mixed airs. In this case, the folding right-hand rule consists of having the four fingers of the right hand point to the direction of the greatest wind speed with the origin of the coordinate system always located on the left side of the observation station in the determination of the direction of the rolling current, see Figures 8.6 and 8.7 for more details. This also explains why we have provided the inverted Figure 8.7.

What one needs to be careful of is that when faced with the same directional winds from east/south or from west/north, the east or north upper altitude fast winds create a counterclockwise rolling current, while the south or west upper altitude fast winds produce a clockwise rolling current. These rolling currents might cancel each other. Due to this reason, the forecaster has to pay attention to the speed of the winds and the source of the winds. For example, winds from dry deserts surely have different effects than those from the upper sky of a humid ocean. So, applying wind directions is not just about the magnitudes of quantities.

Besides, in applications, one-directional winds should not be mingled with other kinds of multidirectional winds. And, pay attention to the situation of decreasing wind speed with altitude. For example, when a north wind or east wind weakens with altitude, then it implies a clockwise rolling current; when a west wind or a south wind weakens with the altitude, then it indicates a counterclockwise rolling current.

In short, rolling currents can alter the structure of the atmosphere; and that is how forces can also be changed by rotations, leading to a more even atmospheric structure, which reflects the weakening forces or energy. On the other hand, when the atmospheric structure becomes more uneven, it signals an increasing force or energy. This end reveals how detailed the method of digitization of information can be, where the relevant prediction analysis involves not only forces or energies but also changes in the forces or energies. Therefore, by applying the directions of rolling currents, one can not only resolve the problem of transitional changes, but also make use of the resource of information to a greater extent.

8.2 The Digital Design and Functions of the V-3θ Graphs

The initial idea for our V-3θ graphs began to form in 1963. At that time, satellite cloud charts and radio technology were not in use. Under the requirements for using as much of the irregular information as possible and by emphasizing the attributes and altitudes of clouds, we replaced the logarithm representations, accustomed in meteorology, back into their original curves. That is, we constructed our V-3θ charts to directly represent the information of the significant spheres using the P-T (pressure and temperature) coordinate system using such factors as wind vector, humidity, and temperature. So, the difference between this one and the traditional one is that we make use of ignored irregular information (of the significant spheres) centered on peculiar information. Based on the vertical cross-sectional characteristics of the structure of irregular information, we reveal the direction of the existing rolling current, which indicates how the weather would change. Our primary purpose is to capture the transformations of the existing rolling currents in order to predict the tendency of weather evolution and to specify the kinds of forthcoming disastrous weathers. That is, we analyze:

■ The effects of the existing rolling current in the current flow field.
■ The potential effects of rolling currents.

Here, the meaning for the effect of the rolling current on the current flow field is that the occurrence and maintenance of cloudy and raining weather, especially sustained developing periods, are all clockwise rolling currents. Conversely, the occurrence and maintenance of clear weather are all about counterclockwise rolling currents. The meaning of the potential effect of rolling currents is that about 24 to 36 hours before clear good weather (or cloudy rainy weather) there appears a clockwise (or counterclockwise) rolling current, and then the cloudy rainy (or clear good) weather will turn to clear good (or cloudy rainy) weather. Corresponding to the 3θ structures of the important factors, if the structure of the 3θ field is uneven, while its flow field is a counterclockwise rolling current, or if the structure of the 3θ field is even, while the flow field is a clockwise rolling current, then the counterclockwise (or the clockwise) rolling current will be transformed into a clockwise (or counterclockwise) rolling current. Or, if the 3θ field becomes uneven, then the clockwise rolling current strengthens (or the counterclockwise rolling current weakens) or if the 3θ field becomes even, then the counterclockwise rolling current strengthens (or the clockwise rolling current weakens).

In specific applications, one should at the same time carefully analyze the differences in the wind directions and wind speeds. When there is not a misguided case, the emphasis should be placed on the differences in wind directions.

If we compare our method with the traditional concepts of meteorology, our method emphasizes the baroclinity of the atmosphere and indicates that the baroclinic atmosphere is not a wave. In other words, what is conducted by the baroclinic atmosphere is rolling currents. This constitutes an essential difference from the traditional concepts. Speaking differently, traditional meteorological science extends the positive pressure atmospheric wave theory into the baroclinic atmosphere, which in essence is a mistake created on incorrect conceptual foundations.

The symbol V in V-3θ graphs stands for the wind direction and wind speed directly observed among other sounding data. What is unfortunate is that due to historical reasons, before the front-line forecaster acquires this piece of information, the data collector has treated the data by using quasi-linear means. It should be clear that the problem of observation itself is about observing an object through another object, in which the problem of inaccuracy must exist due to mutual interference of the objects. This is an inevitability of the material world. However, determining which information to use and which information to ignore represents a problem of epistemology. At this junction, we would like to once again emphasize that based on what is known now, as discussed in this book, the available information resource should be utilized as much as possible and no one should be allowed to use any quantitative method of regularization to damage or distort information. Here in the V-3θ graph of an observation station, the wind vectors are still shown using the accustomed symbols and placed along the θ^* curve. The corresponding 3θ are respectively θ, θ_{sed} and θ^*, constituting three curves in the P-T coordinate system, where T is labeled using the absolute temperature K.

What needs to be pointed out is that in traditional meteorological science, the assumed potential temperature is θ_{se}, while in the design of the V-3θ graphs, to meet the need of making predictions with lead time, the original temperature of the condensation altitude is replaced by the dew point temperature. That is why our charting software outputs θ_{sed}. As for θ and θ^*, we have already explained this in detail in Section 8.1; all the details are omitted here. What is worthy of specific mention is that each V-3θ graph covers as much various information from the ground level to 100 hPa as possible, including part of the ultralow temperature information. Due to differences in the current epistemological understanding, there is still quite an amount of observational information that is not being included.

The main functions of the V-3θ graphs are established on the methodological system of uneven atmospheric structures with the core focused on structural unevenness instead of quantifications of particles. The system of methodology is rooted in the occurrence, development, and maintenance of the rolling currents existing in the troposphere along the vertical direction and the relevant transformations. The system is warranted on the basis of the epistemological understanding that rotational movements cause divided flowing motions and on the theory of evolutionary transformations.

The logic of design of the V-3θ graphs is completely about staying away from quantitative analysis systems that cannot deal with irregular information by using figures of materials' structures. In other words, to include irregular information we use the method of digitization. By doing so, we emphasize the transformations of spinning currents, as indicated by the irregular information. Therefore, we can reveal transitional changes. These changes have been seen as a world-class difficult problem in modern science. Also, by using the method of digitization, we study the evolution problem of materials' aging, change, and transitions by focusing on the conservation of stirring energy. What has to be illustrated is that our concept and method of figures are not about the static graphs of pure geometry. They instead directly touch on the laws of transformation of events. Because energy transformations must borrow the process of change of materials so that the necessary time and space are needed, we established the law of conservation of stirring energy (Section 7.3). Here, the basic concept we need to be very clear about is that each accumulation of energy has to involve materials' structural unevenness instead of the magnitudes of quantities. Therefore, the central idea of designing the V-3θ graphs is first about how to represent the structural (or figurative) characteristics of irregular information and then about how to reveal the unevenness of structures.

1. If the 3 θ-curves increase with P while leaning to the right, then it means that the internal structure of the troposphere is even. It is a nonrealistic ideal case. Such a purely even atmospheric state can be rarely seen in reality. So, purely even linear distributions can only be situations created to make easy comparisons with uneven structural states.

2. If the 3 θ-curves decrease linearly with P or barely change with P, then it means that the vertical structure of the troposphere is extremely uneven and has instable energy to be released. In general, extreme unevenness occurs at a certain altitude layer and appears mostly ahead of the cloudy rainy weathers, in which the structures of θ_{sed}, θ^* at the mid and lower altitudes almost always contain such extreme unevenness. When combined with the θ curve that is either left leaning, or approximately perpendicular to the T axis, or forming an angle with the T axis greater than 70°, an existing clockwise rolling current in the entire troposphere sends a definite message of a forthcoming severe weather condition.

3. If the 3 θ-curves increase with P and lean toward the right side of type (1), then the situation can be divided into either an overall state or a regional circumstance. If the overall pattern tilts to the right, then it stands for an absolute stability; if the pattern either leans to the right locally or is quasi-parallel to the T axis regionally, then it mostly shows the characteristics of warm strati, where the θ_{sed}, θ^* curves are close to each other or coincide with each other (Figure 8.9). Also, if the altitudes of ultralow temperatures during the half-year of winter are lowered, the pattern of overall right leaning can also appear. What needs to be noted is that in these kinds of overall extraordinarily right-leaning patterns there are the irregular situations of left leaning at various local layers, which mostly coordinate with warm strati for the weather conditions either before a major snowfall or during a major snowfall (Figure 8.9). This figure shows an overall right-leaning pattern of the 3 θ-curves with P. However, in the region below 900 hPa, the curves show extremely instable left leaning; in the region of 700 to 600 hPa, the curves also lean left with relative instability. In 500 to 400 hPa, an extremely instable ultralow temperature layer leans left; and in the atmospheric layer 925 to 800 hPa the θ_{sed}, θ^* curves indicate the existence of some warm cloud

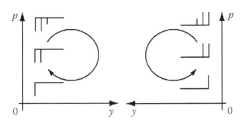

Figure 8.8 Converted version of Figure 8.7

layers of strati, where these two curves are either very close to each other or coincide. In this example, we employ Figure 8.10 to illustrate three problems, indicating a very special individual scenario. Generally speaking, an overall right-leaning pattern stands for stability, even if it is about the disastrous windy snow weathers. So, structurally stable V-3θ graphs are not limited to the stability of clear good weathers. In terms of the quantities with regard to stability, the stability of extraordinary overall right-leaning patterns in general can last for over 15 days. The major difference between the stability of clear good weathers and that of gloomy bad weathers is that for clear good weathers, there does not exist any layer of ultralow temperatures, which is often combined with a local right-leaning phenomenon of increasing temperature at the layer corresponding to the would-be ultralow temperature, see the θ curve at 300 hPa in Figure 8.10, where the folding turn to the right indicates sudden temperature increase.

So, right-leaning overall stabilities may still have different structural states. No doubt, this end is a problem one should pay attention to in his prediction analysis. At this junction we would like to say that Figure 8.10 shows an extraordinary stable structure that is controlled by a subtropical high pressure, stretched westward, of a counterclockwise rolling current of east winds, while Figure 8.10 stands for an overall structural stability controlled by a clockwise rolling current of west winds with regional instable structures. That is, complex problems can be controlled in entirety with their characteristics of the overall stability. However, the overall stabilities do not imply that the specific events are the same. That is the reason why we time and again emphasize the importance of specifics. In short, the stability, as shown by the extraordinarily stable structure in Figure 8.10, is about a clear hot weather,

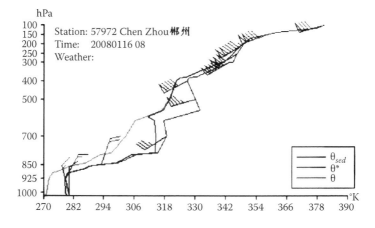

Figure 8.9 Local irregular left tilt and warm strati in an overall extraordinarily right-tilting pattern.

that is a sustainable stability, while the overall upper stable structure with instable regions in Figure 8.9 represents the characteristic of the overall stability of the sustaining icy snow weather.

4. Speaking generally, the 3 θ-curves in the V-3θ graphs at the mid and low altitudes of the troposphere lean slightly toward the left, in particular, the $θ_{sed}$, $θ^*$ curves should show an obvious left-leaning trend, which can even form obtuse angles with the T axis. The reason is that due to the effect of heating from the ground and that of water vapor, slight instability at the lower layers of the troposphere is created. So, the 3 left-leaning θ-curves always stand for an instable atmospheric structure, which becomes more obvious along with a disastrous weather condition. Other than severe disastrous weathers, the overall pattern of the θ curve in general always leans to the right with a slight left-leaning tendency at mid and low altitudes. What one needs to be careful of is that when the 3 θ-curves show an overall right leaning pattern, it stands for a stable structure. However, an overall stability does not always mean clear, good weather, where the stability of gloomy bad weathers might be shown as an overall right-leaning pattern, too (Figure 8.10). That is why for individual situations, the relevant results should be drawn based on the specified conditions. If the 3θ-curves in the V-3θ graphs at the mid and high altitudes are very close or coincide with each other, it means that at the high altitudes there is little water vapor.

From Figure 8.11, it can be seen that at above 400 hPa the $θ_{sed}$ curve first coincides with the θ; and at above 200 hPa all three θ-curves completely coincide. When we look higher, we see a turn, from which location on the θ-curves are quasi-parallel to the T axis. It means that we have entered the stratosphere.

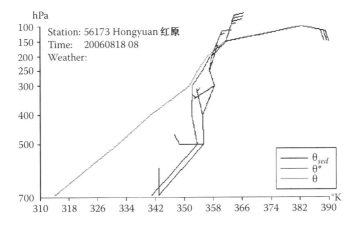

Figure 8.10 The V-3θ graph at the 8th hour, August 18, 2006.

5. In the Northern Hemisphere, when there appear partially southern winds, the 3 θ-curves would lean left; when there appear partially northern winds, the curves would lean right. These combinations of wind directions and the patterns of the 3 θ-curves are referred to as rational or positive combinations. If the opposite combinations appear, they will be referred to as opposite combinations. Each opposite combination indicates a misguided case; the prediction analyst should trace the reason and study the cause.

6. In general, it is all quite normal to see the 3 θ-curves showing a left-leaning tendency with the θ curve agreeing with the $θ_{sed}, θ^*$ curves, or the θ curve leans to the right while both $θ_{sed}, θ^*$ curves lean to the left. If the θ curve leans to the left while $θ_{sed}, θ^*$ lean to the right, one has an irrational combination. The prediction analyst needs to check into potentially misguided cases in the provided information on the basis of the θ curve while looking into potentially misguided cases in the $θ_{sed}$ curve.

7. If the 3 θ curves show an instable structure by leaning to the left (in particular in the mid and low altitudes), then there must exist a potential clockwise rolling current. In other words, if a counterclockwise rolling current exists, then one faces one of two possibilities: One is that the information of the counterclockwise rolling current has a problem; the other is that the information about the counterclockwise rolling current is correct, then within the coming 12 to 24 hours, the present rolling current will change into clockwise. If the 3 θ curves show a stable structure, where both the overall and local structures are stable, by leaning to the right (in particular at the mid and low altitudes), then there is a potential counterclockwise rolling current or no rolling current. In other words, even if there is a current clockwise rolling current, it will soon change into counterclockwise. These facts indicate that structures determine what is next, where rolling currents or directions of movement are only expressions of the underlying structures. In other words, the digitization analysis of information itself touches on not only forces and changes in the forces, but also the origin where the forces are from and how the origin of forces changes.

8. Although currently not all the information of the significant spheres is publicly available, from the constructed V-3θ graphs based on the available information of the significant spheres before disastrous weathers actually occur, it can also be seen that there appears a sudden left leaning, or quasi-perpendicularity to the T axis, or turns in the overall right leaning at near the region of 300 to 100 hPa. All these patterns indicate that at the altitude region an ultralow temperature exists. In this case, one should check on other relevant conditions and be warned of potentially forthcoming disastrous weather. It can even be said that without an ultralow temperature, there will not be a forthcoming disastrous weather of the convective form.

9. Because such disastrous weather conditions as torrential rain and others are closely related to the properties and altitudes of clouds, to reflect the properties and altitudes of clouds in our design of the V-3θ graphs, which took

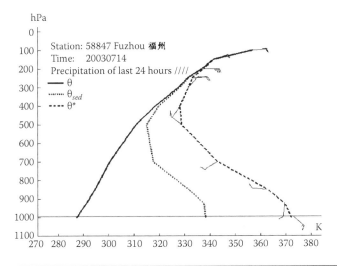

Figure 8.11   The V-3θ graph at the 8th hour, July 14, 2003.

place before satellite cloud maps were available, we introduced θ_{sed} using the dew point temperature. By combining the saturated θ^*, the structural characteristics of the properties and altitudes of clouds are clearly shown, where the relevant altitudes are accurately indicated. If we compare our V-3θ graphs with the widely available cloud charts today, our specific expressions of clouds' properties and altitudes are still better than those shown in the cloud charts in terms of the accuracy of clouds' altitudes, distinguishability of cloud layers, and densities. And, in the V-3θ graphs, one can also separate the strati of warm clouds from those of cold clouds. When the irregular information is sufficiently available, the V-3θ graphs also provide the stratification of convective clouds (Figures 8.12 and 8.14). In Figure 8.12, one can see respectively the altitude of the layer of clouds, the density, and the convective clouds above the cloud layer. Correspondingly in Figure 8.14, the left turn and the tendency for θ_{sed} and θ^* curves to meet at the altitudes of 700 to 600 hPa represent a layer of cold clouds and the convective clouds underneath. Here is some empirical knowledge we acquired from our many years of practical applications: When θ_{sed} and θ^* curves are close to each other at near 3K, if the 3 θ curves present an irregular pattern with multiple turns, it means that there are indigested clouds, where the altitudes of the turns are those of the indigested clouds. If the 3 θ curves coincide either approximately or completely and the right-leaning turns are almost parallel to the T axis, indicating states of strong inversions, then there are layers of warm clouds (Figure 8.12 and Figure 8.9), where the altitudes of the strong inversions are those of the layers of the warms clouds, and the densities are shown by the closeness between the θ_{sed} and θ^* curves.

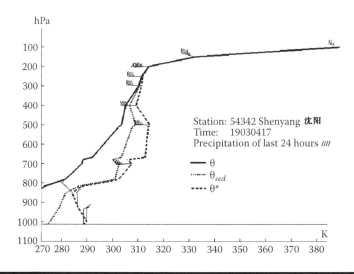

Figure 8.12   Shenyang at the 8th hour, April 17, 2003.

In particular, if in the atmosphere above a layer of clouds there appear several places where the θ_{sed} and θ^* curves are close and make turns, then it means that there are indigested clouds above the layer of clouds. Under such conditions, the precipitation of a torrential rain would be increased to at least another magnitude level.

8.3 Structural Characteristics of Major Disastrous Weathers

The major disastrous weathers considered in this book are those of transitional changes. Their forecasts can include the kinds of monitoring reports using continuous extrapolations, and those of long-term maintenance periods of such severe weathers as nonstop torrential rains, droughts, snow disasters, ice disasters, and so forth. So, our method in essence can be employed in not only short-term forecasts. Here, we will mainly talk about such disastrous weathers as severe convections, hailstones, torrential rains, sand/dust storms, dense fogs, high temperatures (including droughts), and the like.

8.3.1 Severe Convective Weathers

Severe convective weather is traditionally known as suddenly appearing disastrous weather. In essence, it means the phenomena of drastic changes in local weathers, including hailstones, tornadoes, strong winds, the black winds, sandstorms, and heavy precipitations. Different forms of severe convective weathers respectively have their specifically individual structural characteristics as shown in their V-3θ graphs. Although there is always an ultralow temperature appearing ahead of each severe convective weather, different forms of severe convective weathers possess their different, distinguishable characteristics in their ultralow temperatures. Because the relevant figures take a good amount of space, in this book, we will introduce our method by mainly focusing on the often seen severe convective weathers in midlatitudinal areas. As for other types of disastrous weathers, the reader can master the basics of our method by actually going over about 10 typical cases.

8.3.1.1 Hailstone Disastrous Weathers

The characteristics of the ultralow temperatures of the disastrous weathers as hailstones are different from those of heavy rainfalls, tornadoes, heavy rain gushes, and so forth. The difference includes the following: The thickness of the ultralow temperature is comparably greater than other convective weathers with severe cases of dropping temperatures reaching as much as 50 to 100 hPa; the corresponding morphology of the ultralow temperature is mostly seen as steep angles or sudden left-leaning turns, where the thickness of the ultralow temperature is prominently shown; the relevant physical meaning also agrees well with reality, which is after all the necessary depth and thickness of an icy freezing zone needed to form the hailstones. Only with a thick and deep layer of ultralow temperatures could the task of "manufacturing" hailstones be accomplished.

Next, the structural characteristics of constituting hailstones also involve how the water vapor is distributed in the atmosphere and the existing clockwise rolling current in the entire troposphere. In terms of the distribution of water vapor, hailstone weather does not need an abundant amount of water vapor in the mid and low atmosphere. For most cases, it only needs a certain atmospheric layer in which the humidity is relatively high with all other layers relatively dry (Figure 8.15) or the distribution of the water vapor is relatively even (Figure 8.13) (here in the figures, the solid horizontal line stands for the sea level, an earlier way of representation of our earlier version of the software; in the second version, the horizontal sea-level line is no longer needed, where the vertical axis automatically adjusts its length) with a certain layer in the mid or near-ground levels containing a relatively high temperature and humidity (Figure 8.13); in particular, at the mid-atmospheric altitude there is a layer of cold clouds (Figure 8.15) that causes instable convection.

It should be explained that a greater amount of humidity in the atmosphere is good for the prediction of relevant magnitudes. It is because of the relatively greater amount of humidity in the atmospheric layers in Figure 8.13 that the Lijiang area experienced such an extraordinary scale of icy hailstone disaster. In all the hailstones with structures similar to that in Figure 8.15, we can only predict local severe hailstone weathers without much to say about the magnitudes. In short, limited by the current spatial arrangement of the observation stations, their distributional

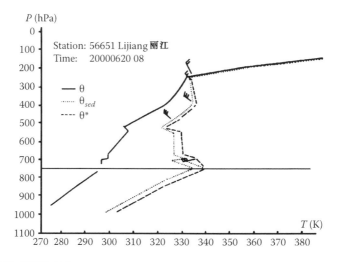

Figure 8.13   The 8th hour on June 20, 2000, Lijiang.

densities, and quality of information delivered, our method provides an improvement regarding whether or not hailstone weather would occur, where the prediction of the relevant magnitudes is still an open problem waiting to be resolved. We believe that even if the spatial density of the observational stations is not increased, as long as the depth of information delivered and the correct locations of ultralow temperatures can be adequately provided, the current level of accuracy in our prediction of magnitudes regarding hailstone weathers can be improved.

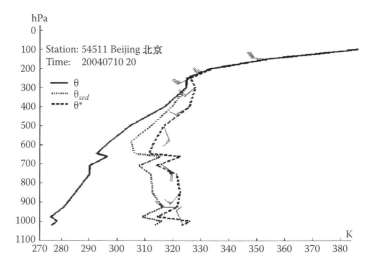

Figure 8.14   Beijing at the 20th hour, July 10, 2004.

As for the forecasting of specific regions suffering from predicted hailstone weather, it is still one of the difficult problems in modern science. In particular, under the current situation of relatively scattered distribution of radiosonde stations, almost no reliable methods exist. However, based on our method, we can make our predictions with increased accuracy based on the location of the ultralow temperature, combined with its wind directions, wind speeds, and their location of concentration. No doubt, if one can coordinate the delivered sounding information with the GPS positioning system, he will be able to improve his forecasting of the specific region of hailstone weather. For the front-line forecasters, they should understand that the location of an ultralow temperature is exactly the location of an atmospheric twist-up, which corresponds to the region in which the disastrous weather will occur. By using our method, the predicted region that will experience hailstones can be shrunk to an area of the size of about 100 to 150 kilometers. That has been the reachable limit based on the current distribution of radiosonde stations of 300 square kilometers apart from each other. To reach this limit, the forecaster has to be extremely competent and patient in his ability to carefully analyze his information. Otherwise, it is very easy to miss the forthcoming disastrous events. Therefore, for regions of historically high occurrence of convective weathers, the forecasters must satisfy the requirement of being able to carefully analyze information. Linkang Gu, one of our students, has worked in Xichang, an area in which severe convective weathers frequently occur; by using our method, he once reached a rate of accuracy in his forecasting of severe convections of over 90%. He summarized his success by directly using his local station's sounding information in a timely manner, he received the original quality and timeliness of the information, which were generally lost if he first submitted the data to the central information station and then waited for the data being wired to be delivered to him again from the central station.

For the prediction of severe convective weathers, one should pay attention to the following aspects:

Other than the uneven structure of the 3 θ curves, what is noticeable is the obtuse angle formed by the T-axis and both of θ_{sed} and θ^*. This formation can distinguish the forthcoming severe convective weather from a torrential rain, constituting the main characteristic of the occurrence of severe convective weathers. That is, although they are both convective weathers, severe convections and torrential rains can be separated.

Ultralow temperature is not a single station or individual regional, local phenomenon. In the winter season, it can appear at the upper skies of the Qinghai and Tibet plateaus or other low latitudinal areas. Under specific circumstances, the west winds carried by ultralow temperatures in the Eastern or Western Hemispheres can reach as deep as about 10°N. In drought areas during the spring season and early summer season or areas located to the south of 50°N, ultralow temperatures can appear. And, during the summer season, ultralow temperatures can appear in areas to the north of 50°N. All these observations indicate that the phenomenon of ultralow temperatures has some direct connection with solar radiation and seems to appear before the sun heats up the surface of the earth by first cooling the top layer of the troposphere and causing severe convections in the troposphere. That is the reason why ultralow temperatures play the role of guiding the evolution of the atmospheric system. Unfortunately, traditional meteorological science only notices the increasing or decreasing temperatures at the bottom cushion and near the ground level of the atmosphere. That explains why meteorologists did not predict the occurrence of the extraordinarily heavy rain gush along Huai River in August 1975 and the snow and ice disasters in southern China in January 2008. Even after the events had already occurred, they still could not provide any plausible reasons for the occurrences. In the current profession of meteorology, the most applicable reason for disastrous weathers has been the evil "siblings": the El Niño and La Niña. However, the provided explanations also make the meteorologists who made the connection seem lacking self-confidence.

So, for the problem of forecasting disastrous weathers, one has to expand his attention beyond just the mid- and low-altitude atmospheres and problems of the bottom cushion. If we use the terminology of modern science, what we try to say is that the mid- and low-altitude atmosphere and the bottom cushion can only provide sufficient conditions for disastrous weathers to form; more important necessary conditions, the changes in heat caused by the sun and the relevant changes in ultralow temperatures, are still needed. Also, in practical applications, one needs to be careful that when analyzing ultralow temperatures, it is not about the magnitudes of the relevant quantitative values; instead it is about the relative difference between the ultralow temperature and the near-ground atmospheric temperature. Although these relative temperature differences are different from one region to another, our practical experience of the past 40-plus years indicates that the temperature difference between the top of the troposphere and the near-ground level can reach 80° to 100°C. When such a temperature difference appears, one should be very cautious about the possibility of a forthcoming severe convection. In specific forecasting practice, one should also pay attention to the direction of the rolling current of the convective atmosphere; as soon as the existing clockwise rolling current and the humidity show drastic changes, the forecaster should consider issuing relevant predictions.

To emphasize the concept that ultralow temperatures cannot be judged by using absolute magnitudes of quantities, let us make it clear that even if the overall temperatures in Figures 8.12 and 8.13 were decreased or increased by 5°C, the relevant hailstone disasters of severe convections would still occur. So, as long as there appears an ultralow temperature, which is adequately combined with water vapor, temperature, and other conditions at the mid- and low-altitude atmosphere, no matter whether it is in the spring season, summer season, or winter season, the phenomenon of hailstones will appear. In other words, when applying our method, do not dwell on the concept of absolute magnitudes of information. In practical applications, ultralow temperatures appear ahead of the relevant weather phenomena. So, as soon as the forecaster discovers the ultralow temperature, he will need to pay close attention to the combinations of humidity and temperature in the mid- and low-altitude atmospheric layers; when the changes agree with a certain type of severe convection, he can then issue his accurate forecast.

Ultralow temperatures above high plateaus or mountainous areas tend to be lower than those above plains. The reason is that above high plateaus or mountainous areas, a bottleneck effect combined with the direct solar radiation due to relatively higher ground levels can more easily cause the temperature at the bottom cushion to increase more than the free space of the same altitude over a plain; so, uneven atmospheric structures over high plateaus or mountainous areas can form more easily than those over plains. This end explains why people experience more severe convective weathers in high plateau and mountainous areas than in areas of plains. In principle, if sufficient information is supplied with no misguided cases, by making use of irregular information and automatically recorded information, the forecasting of severe convective weathers could be substantially improved. However, the current problems in China are that the arrangement of radiosonde stations is too scattered, especially in high plateaus and mountainous areas, and that the collected information has not been sufficiently employed. The key to the resolution of these problems is the way of comprehending irregular information, where at extremes the opinions could be as far apart as opposite of each other.

8.3.2 Local Severe Rainfalls and Thundershowers

8.3.2.1 Local Severe Rainfalls

Local heavy rainfalls are also known as thundershowers with precipitation above that of torrential rains; and within a few hours or 10-plus hours, their precipitation can reach those of heavy rain gushes or extraordinarily heavy rain gushes. Although the time spans tend to be short, the concentrated heavy rainfalls often cause the transportation systems in metropolitan areas to be paralyzed or medium- and small-size dams to overrun. Or, when there are accompanying thunders, the local communication systems may well be interrupted. This kind of weather phenomenon is basically similar to that of hailstone disasters with limited locality and suddenness so that the traditional methods of extrapolation can hardly be useful. It is exactly because of their sudden appearance of transitionalities that the digitization method of information of the blown-up theory finds its superiority. As long as the currently installed network of radiosonde stations can capture the needed marginal information, by using the digitization method of oscillating time series information (Chapter 9), one can basically produce effective forecasts. At this junction, we will only focus on introducing the relevant V-3θ characteristics of the sounding data. As for the oscillating time series information, we will systematically discuss this in Chapter 9.

For the structural characteristics of local heavy rainfalls, we will illustrate using Figures 8.16 and 8.17.

From these figures, it can be readily seen ultralow temperatures exist at the altitude of 300 to 100 hPa. However, their morphologies are different from that of ultralow temperatures of hailstones. They are relatively thinner with simple turns, the θ curve at mid and low altitudes leans to the left slightly, revealing the instable

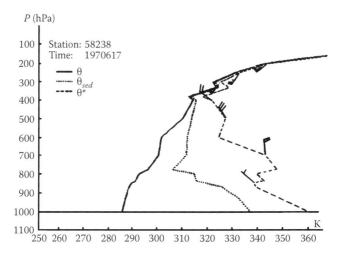

Figure 8.15 The 8th hour on June 17, 1971, Jiangpu (Nanjing).

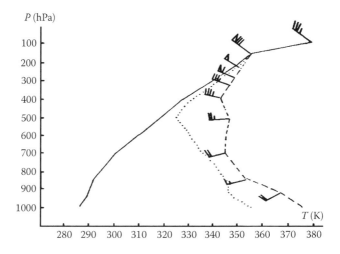

Figure 8.16 The V-3θ graph at the 8th hour, July 20, 1998, Wuhan.

structure of the airs in the troposphere, and the θ_{sed}, θ^* curves at the mid and low altitudes below 500 hPa are quasi-parallel and form obtuse angles with the T axis. This end indicates that the distribution of water vapor in the mid and low atmosphere is adequate for formulating an overall convective state; next, the entire troposphere constitutes a clockwise rolling current consisting of southwest winds at the mid and low altitudes and northwest winds at the high altitudes. Therefore, it is sufficient to forecast that within the future 12 to 24 hours a local heavy rainfall will appear. Here, the southwest winds for Wuhan City (Figure 8.16) had gone beyond 500 hPa so that the forecast for the amount of rainfall should be somewhere between torrential rain and extraordinary torrential rain. The actual record showed 159 mm within 24 hours. This torrential rain almost completely paralyzed the transportation of the entire Wuhan City. The corresponding southwest winds for Beijing City (Figure 8.17) only reached above 700 hPa and the obtuse angles formed by the θ_{sed}, θ^* curves and the T axis are also smaller than those in Figure 8.16. Therefore, one can only forecast the rainfall at the magnitude of heavy rain to torrential rain with exceptions that at some individual localities it might reach the level of heavy rain gush. Rainfalls have relatively strong locality with extremely uneven distribution. This specific rainfall concentrated at the center of the city and the area south of the city. For example, the magnitude of rainfall at Tiananmen was 104 mm; at Mentouguo, 84 mm; at Shijingsan, 74 mm; however, the north districts, Yanan, Miyun, and Pingu, basically did not receive any precipitation. This event of local heavy rainfall caused accumulation of indigested water covering the bottom roads of elevated road systems and the entire city's transportation was paralyzed for over 4 hours. Unfortunately, the local meteorological departments using the traditional methods of forecasting were unable to issue any warning about the heavy rainfall. Afterwards, the missed forecast was excused by using the reason that local disastrous rainfalls are an unsolved world-class difficult problem.

8.3.2.2 Thundershowers

In general, torrential rains with heavy precipitation tend to be mixed with thunders and rains. For thundershowers, they may not be torrential rains and heavy rain gushes with severe precipitation. So, there is the need for us to illustrate the corresponding differences. For thundershowers, sometimes the sounding information can capture the characteristics of thunder blasts (Figure 8.2). However, such chance is quite rare, because the sounding balloons are released at predetermined times, while thunder blasts do not always occur at those predetermined time moments. The sudden change in temperature at the altitude near 500 hPa in Figure 8.2 cannot be seen as an error in the provided information; instead it is caused by the sudden expansion and contraction of the air due to the ionization effect of the thunder blasts. So, in the meteorological problem of thundershower disasters, one has to deal with the physics of molecular states and the effects of ionization. Comparing Figure 8.2 with

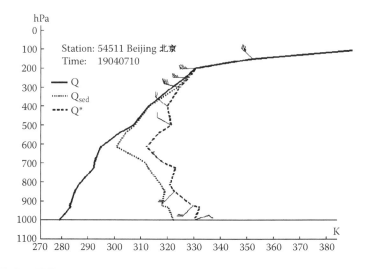

Figure 8.17   The V-3θ graph at the 8th hour, July 10, 2004, Beijing.

Figures 8.16 and 8.17, it can be seen that they have similar structural characteristics and properties. That is how the reader can tell that in mid-July 2007 Chongqing City also experienced a regional thunder rain gush with heavy rainfall. However, if compared with Figure 8.18, it can be seen that although the θ curve does not seem to be much different from those θ curves in Figures 8.16 and 8.17 and there is also a clockwise rolling current, the θ_{sed}, θ^* curves present structurally different properties from those in Figures 8.2, 8.16, and 8.17. First, the θ_{sed}, θ^* curves are not quasi-parallel; instead their distance changes alternatively, revealing insufficient water vapor for a heavy rainfall. Second, the θ_{sed}, θ^* curves do not form any obtuse angle with the T axis, indicating that the degree of instability of the corresponding atmospheric state is relatively weaker than that of a severe rainfall.

At this junction, we would like to mention that for the problem of thundershowers, the alternating changes in the distance between the θ_{sed}, θ^* curves do not have to concentrate at the mid and low altitudes. They can also appear at the middle or slightly higher in the troposphere, constituting a "bee-waist" shape (Figure 8.19). When combined with an existing ultralow temperature and clockwise rolling current, one can always forecast the forthcoming occurrence of thundershowers. For thundershowers, the rainfalls are mostly uneven or scattered, generally without much precipitation except in individual localities where the degree of rainfall could reach several tens of millimeters or even the level of torrential rains (Figure 8.19). Evidently, the bee waist (area of water vapor concentration) in Figure 8.18 is not near the ground level, where the recorded rainfall of 24 hours at the observation station of Jinan City was 80 mm. In principle, based on the currently scattered distribution of radiosonde stations, the location of thundershowers may not happen to be at the exact location of an observation station.

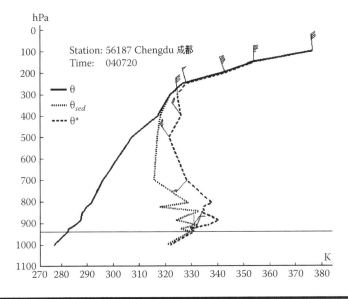

Figure 8.18   The V-3θ graph at the 8th hour, July 20, 2004, Chengdu.

That is why the forecasts of thundershowers and thunderstorms have to rely on the analysis method of structural transformations of automatically recorded information (we will look at how to predict the locations of thunderstorms in Chapter 9). This end will be practically useful for airliners, space-related projects, and military exercises. That is, thundershowers should not be overlooked just because their amounts of rainfall are insignificant (Figure 8.18). Besides, thunders and lightning involve the problem of ionization of the atmosphere and warrant further, deepened investigation.

8.3.3 *Predicting the Amount of Rainfalls*

Modern science has been pursuing quantification. To a degree, such pursuit has exaggerated the importance of quantitativeness. If we understand quantities correctly, then they are forever the concepts of relativity and approximation. In other words, those who advocate quantifications should be clear that there is no absolute quantification because quantities are descriptions of formal measurements; and in applications, quantities reflect different properties and varied requirements from one problem to another. No doubt, the precision in quantification required by modern equipment will mostly likely be more delicate than that of carpenters' work. However, as technology constantly advances with time, the ideal precision in quantification can never be materialized. That is, we are faced with the infinity problem in the precision of quantification. In other words, the ultimate results of quantification are about being "roughly" alright. So, when people religiously treat quantitative comparability as the only standard for scientificality, it means that these people do not truly understand what quantities are.

In terms of the network of radiosonde stations scattered about 300 kilometers apart, the possible precision of quantification can be compared to doing embroidery with a fishing net. It would be difficult to catch any "fish" smaller than 1,000 kilometers. Even if we were to employ our digitization method of automatically recorded information, the only fish that could possibly be caught would have to be bigger than several tens of kilometers or even over 100 kilometers.

China has historically been a country greatly impacted with rains and floods, with floods mainly coming from torrential rains. China's economic prosperities have been created largely through agriculture. That is why the Chinese have paid extreme amounts of attention to problems related to rainfalls. To this end, our work presented in this chapter started with the potential rainfalls related to torrential rains. Because it was an assigned project, the funding was guaranteed. Aiming at the specific needs of agriculture, irrigation works, and hydroelectric projects, special attention was paid to the amounts of rainfalls along the upper reaches of major rivers. That led exactly to our investigation and exploration of regional precipitations. So, speaking frankly, the methods of irregular information and regional digitization, as presented in this chapter, should not be seen as our discovery. Instead they were the consequences of meeting the assigned tasks of forecasting regional torrential rains. In our practices that received no support of any kind from relevant authorities, we discovered that all the disastrous torrential rains that could cause damage to river valleys are basically from the precipitations of large area strati, where combined with the strati, convective clouds help to make the rainfalls heavier. For pure convective clouds, even if they might be severe, the total amounts of rainfalls will be limited and would not cause disasters of the scale of the entire river valleys. What caused us to change our fundamental concepts and beliefs in our works was that the effects of cold strati are not any less than convective clouds, and when combined with convective clouds, their consequent rainfalls have gone way beyond our expectations. The true meaning of learning science is about challenging our conventional wisdom. Our initial helplessness gradually turned into aggressive force to introduce the concept of evolution science.

So now, the reader can understand why in our design of the V-3θ graphs, we give specific representation to the characteristics of strati, and after over 40-some years of development, the modernized cloud charts still cannot label the altitudes, stratification, and densities of cloud layers as clearly as our V-3θ graphs. Next, what needs to be explained is that our method of computing the forecasted amounts of rainfalls is not based on traditional meteorological science; instead, it is developed on our special understanding of the concept of river valleys from the professions of irrigation works and hydroelectric powers. As a simple example, if a torrential rain falls only along a river, the resultant flood will not be much greater than those produced by light to moderate rains covering the entire river valley. In particular, our repeated empirical computations indicate that the amounts of rainfalls are not determined by the amounts of water vapor contained in the atmospheric columns; instead, they are determined by the amounts of water vapor input from the surrounding areas in the mid and low atmospheric levels. In other words, the inflow space of water vapor during a torrential rain process is analogous to the valley that supplies water for the river, and the so-called torrential rain is analogous to the flood in the river at a cross-section. The corresponding rainfall of the torrential rain is also determined by the altitude of the inflow space of water vapor. No doubt, the higher the inflow space, the greater the amount of the torrential rainfall, where the altitude can also contain superpositions of strati and convective clouds. With the financial support and human resources of the Department of Hydroelectricity of the central government of China, we conducted over 100 real-life case tests and comparisons. At the end, we summarized the following general rules that can be and have been applied in practical weather forecasts.

8.3.3.1 Predicting the Amounts of General Rainfalls

No doubt, in meeting the needs of agricultural operations and constructions of irrigation works and hydroelectric projects, and in terms of the objects directly involved in the service of weather forecasting, each meaningful precipitation involves a large area or certain regional rainfall of warm strati. The amounts of the rainfall are limited to those magnitudes of light to moderate rains. The digitization of the relevant information or the general characteristics of the V-3θ graphs for the purpose of predicting the amounts of rainfall are as follows:

In general there is a layer of relatively weak ultralow temperature.
The 3 θ-curves of the V-3θ graphs would be quasi-parallel to the T axis at near 850 to 700 hPa that could at most reach the height of 600 hPa or lean to the right slightly so that they form less than 30° angles with the T axis. The quasi-parallel or almost coinciding state of the θ_{sed}, θ^* curves implies that there is developing or developed severe warm stratus in the mid and low atmosphere. The altitude of the quasi-parallel or almost coinciding state of the θ_{sed}, θ^* curves is the height of the warm stratus. If a water vapor source exists at 10 longitudinal/latitudinal distance, then the weather condition belongs to the category of general rainfalls.

On the V-3θ graph of a general precipitation, one can see the 3 θ-curves bend to the left underneath warm stratus. That indicates that the structure of this layer of the atmosphere is instable, where the corresponding θ_{sed}, θ^* curves are close to each other and quasi-parallel to the T axis or form obtuse angles with the T axis.

An overall tropospheric clockwise rolling current centers almost at the stratus.
What needs to be noted is that for practically meaningful precipitations, in which pedestrians on the streets need to use umbrellas, the atmospheric clockwise rolling currents have to be such overall rolling

currents that involve the entire troposphere. On the other hand, to constitute practically meaningful clear weathers that are at least light cloudy weathers, then the entire troposphere is taken by overall counterclockwise rolling currents.

By an "overall rolling current," we mean that in the troposphere there is only one such a rolling current that can be either clockwise or counterclockwise. If in the troposphere there are two or more than two layers of rolling currents, it will be referred to as nonoverall rolling currents or multiple layers of rolling currents. Generally speaking, when there are multiple layers of rolling currents, the corresponding weather will be not too bad and not too good, either. Depending on whether the water vapor supply is sufficient or not, the weather condition will be cloudy or somewhere between cloudy and alternatively clear and cloudy; the corresponding wind speed is also somewhere between breezes and strong winds. If it is the summer season, at most it will be thundershower weather.

The forecasting of the corresponding amounts of rainfalls can be made by following the references below:

1. If in the V-3θ graph, $\theta^* - \theta_{sed} \leq 10 \sim 8$ K and these curves are quasi-parallel to each other and perpendicular to the T axis (this characteristic can appear in the structures of torrential rains, heavy rain gushes, and extraordinarily heavy rain gushes with differences in the altitudes of the bottom layer of warm and humid air, the strengths of the eastern counterclockwise rolling currents of the observation stations, and how long the structural stability can last), it implies that of a general precipitation. In other words, the severity of the convection involved is much weaker than that of large-area, disastrous, torrential rains. That is to say, for a large-area flood causing torrential rains or the regional torrential rains of severe convection, the digitalized V-3θ graphs can clearly indicate the structural difference. That is why this method is very beneficial for the forecasts of torrential rain floods in areas similar to that of the major river systems in China.

2. Attention should be paid to the counterclockwise rolling current existing underneath the 850 hPa (about 1,500 meters) or the flow field of the anticyclone within the 5~10 longitudinal/latitudinal distance to the northeast and east of the predicted area. In this case, the prediction of the amount of rainfall within the area of 200 square kilometers can be made within that of light to moderate rains (about 1~10 mm), with emphasis placed on less than 10 mm. As for the prediction of the special location of the precipitation, one should employ the method of digitization of automatically recorded information. For the prediction of a station without the automatically recorded information, one has to use the rough position of the ultralow temperature to estimate the approximate location of the rainfall.

3. If the thickness of the counterclockwise rolling current on the east of the observation station can reach 700 hPa (about 3,000 meters), the amount of rainfall within 200 square kilometers will be about the magnitude from moderate to heavy rains (that is 11-25 mm) with emphasis placed on less than 25 mm. For the prediction of special rainfall locations, use the method described in (2) for stations with devices for automatically collecting information.

4. If the thickness of the counterclockwise rolling current on the east of the observation station can reach 500 hPa (about 5,500 meters), then the amount of rainfall within 200 square kilometers will be about the magnitude of heavy to torrential rains (about 26-50 mm) with emphasis placed on the magnitude of heavy rains or less than 50 mm.

The above shows that the magnitude of rainfall is closely related to the altitude of the converging layer of water vapor and the previously mentioned spatial flow valleys.

Evidently, the rules listed above in principle are given for meeting the need of ordinary operations of agricultural and irrigation works and involve the prediction of nondisastrous precipitations. For waterfront areas, since the atmosphere contains abundant humidity, if the predicted amounts of rainfalls are upgraded accordingly by one magnitude level, the outcomes basically agree with the actual situations of large area precipitations except for the cases of local torrential rains and high strength rainfalls.

8.3.3.2 Predicting the Precipitations of Torrential Rains

Currently, torrential rains are measured and defined by days (24 hours). The problem of torrential rains involves the frequency of occurrence, the available protection facilities of the area of occurrence, and the properties of the rainfalls. For example, although a heavy rainfall can easily go beyond 50 mm within an hour or several hours and a sustained rainfall can also reach beyond 50 mm in 24 hours, there is no doubt that these two rainfalls have completely different properties. Because each sustained torrential rain in general covers a large area and can easily cause floods leading to loss of lives and properties, it is inappropriate to define torrential rain disasters by just measuring the amount of rainfall at the place of rainfall. Considering the fact that the method of digitized V-3θ graphs can distinguish the properties of these two kinds of torrential rains, in this book, by *torrential rains* we mean those that sustain, while placing the torrential rains with severe rainfalls in the analysis of severe convective weathers. For example, in some countries there are not many incidences of torrential rains that last for over 24 hours. However, for China, torrential rains that last for more than three days are quite frequent.

The prediction of the amount of rainfall from torrential rains and other heavier precipitations and the structural characteristics of their V-3θ graphs can be illustrated as follows:

1. For the great amounts of rainfall for torrential rains and heavy rain gushes, or referred to as disastrous torrential rains, the relevant convergent layer of water vapor is relatively thick; the troposphere above the area of about 5–10 longitudinal/latitudinal distance east of the predicted location is controlled by a counterclockwise rolling current; and the wind direction along the supply line of water vapor agrees with that of the supply line as observed at the adjacent observation stations (in particular, for the station installations of 300 kilometers apart, one should see the right directional winds at three to four stations) in order to guarantee the transportation of water vapor. Here, in general, $\theta^* - \theta_{sed} \leq 8 \sim 10K$; for sustaining torrential rains, $\theta^* - \theta_{sed} \leq 3 \sim 5K$ and the θ^*, θ_{sed} curves are quasi-parallel to each other and perpendicular to the T axis. The thickness of the southwest, south, or southeast wind layer (in the northern hemisphere) leads to rainfalls of different amounts according to 500, 400, or 300 hPa, respectively.
2. An overall clockwise rolling current.
3. The ultralow temperature in the sky above a torrential rain, heavy rain gush, or other heavier rainfall is only a thin layer with sharp turns. In other words, the ultralow temperatures corresponding to torrential rains are relatively thinner when compared to those of severe convections (Figures 8.16 and 8.17).
4. The θ curve looks like a bow facing left, accompanied with irregular turns. One needs to note that the irregular turns along the θ curve means that convective clouds are developing.

The prediction of the amounts of rainfalls can be made using the following guidelines:

1. If the converging layer of water vapor reaches 500–400 hPa and the east of the observation station is controlled by an overall counterclockwise rolling current that covers at least 5–10 longitudinal/latitudinal distance, then one can forecast a forthcoming torrential rain or a heavy rain gush of 500–100 mm. Correspondingly, for inland areas, the precipitation can be less than 100 mm; and for coastal areas, one should mainly consider forecasting 100 mm.
2. If the converging layer of water vapor reaches 400 hPa and above and the east of the observation station has an overall counterclockwise rolling current covering at least 5–10 longitudinal/latitudinal distance, then one can predict a forthcoming heavy rain gush or an extraordinarily heavy rain gush of 101–200 mm. For inland areas, focus on 200 mm; for coastal or low-latitude areas, or areas covered by a typhoon, above which the ultralow temperature is below –70°C, then the predicted amount of rainfall should be doubled.
3. If the thickness of the converging layer of water vapor can reach 300 hPa and the east of the observation station is controlled by an overall counterclockwise rolling current of the size of at least 10 longitudinal/latitudinal

distance, the relevant amount of rainfall can reach an unimaginable level. In this case, the amount of rainfall can easily go beyond 500 mm within 24 hours.

What needs to be pointed out is that our survey of real-life weather forecasts indicates that other than the cases with experienced forecasters, no methods in modern science can computationally predict the rainfalls that are beyond the magnitude of heavy rain gushes. This end in fact is the consequence for the inability of modern science to employ irregular information. Our investigation of over 10,000 real-life cases reveals the fact that realistic disastrous weathers without a single exception occur in the changes of irregular information. Although the method of digitization of irregular information might not be the best way to deal with the irregular information, it has at least illustrated that by introducing irregular information, we are in the door of potentially resolving the problem of disastrous weathers.

As a theoretical recognition, the problem of natural disasters in essence has forced us to correctly understand irregular information and to introduce the relevant method for practical applications. At the very least, this problem has forced us to realize why the system of modern science cannot handle irregular information and why it cannot seriously treat irregular information.

8.3.3.3 Strong Winds and Sand–Dust Storms

Evidently, flying sand and sand–dust storms cannot exist without strong winds. However, strong winds do not automatically imply sand–dust storms. However, these two weather phenomena are closely related; and the V-3θ graphs can reveal the corresponding characteristics that are beneficial for making relevant predictions. For this reason, we will illustrate these two weather phenomena together. No doubt, the primary condition for a sand–dust storm to occur is strong wind. Second, a source of sand must exist. Without such a source, even with strong winds, there will not be any sand–dust storm. However, these two conditions are not entirely separated from each other. This is because the temperature of a bottom cushion without vegetation coverage, under solar radiation, can increase faster than that of a well-covered bottom cushion with vegetation. For the former, it is easier for a great temperature difference to occur along the vertical direction so that the speed of a strong wind will inevitably increase. When the speed of a wind reaches a magnitude of 6 (10.8–13.8 m/s), the wind is referred to as a strong wind. By *sand–dust* weather, we mean that an existing strong wind lifts up the dust from the ground, making the air dusty, so that the visibility at the ground level is poor. Such weathers can be classified into three categories, with the most severe known as black windstorms:

■ *Floating dust* — means the weather phenomenon with a visibility of less than 10 kilometers.

- *Flying dust* — means the weather phenomenon of visibility of about 1–10 kilometers.
- *Sand dust* — means the weather phenomenon of visibility of less than 1 kilometer, accompanied by high-speed winds that might seem to have a spinning spiral structure. For severe sand–dust storms, the instantaneous wind speed can reach 25 m/s, and the wind strength can be over that of magnitude 10.

Traditionally, the forecasts of strong winds and sand–dust storms are carried out by mainly using the density of the isobaric lines on the weather maps of the ground level. That is, increasing gradient forces of pressure are employed to predict the occurrence of strong winds and sand–dust storms. However, the moments when the isobaric lines start to gather densely generally appear at or after the occurrence of the sand–dust weather. And the relevant commercial forecasts are done using the technique of monitoring the already existing sand–dust weather. In essence, this is not a true forecast. This end explains why oftentimes, even without the isobaric lines gathering, sand–dust storms occur.

Evidently, to resolve the problem of forecasting sand–dust storms with lead time, we have employed the V-3θ graph method that can include irregular information. The relevant characteristics can be summarized as follows:

1. Before a strong wind or sand–dust storm occurs, there is very little water vapor in the atmosphere, and the structure of the V-3θ graph is opposite of that of rainfalls. That is, the θ, θ_{sed} curves are very close to each other, or at least the θ_{sed} curve is leaning toward the left between the θ, θ^* curves, which is more obvious for sand–dust storm weathers. And, before the occurrence of the sand–dust storm, the θ, θ_{sed} curves could almost entirely coincide with each other. For the purely strong-wind weathers, there is also very little water vapor in the air; however, the θ_{sed} curve near the ground level leans toward the right between the θ, θ^* curves, or it is close to the θ^* curve. This end implies that the bottom cushion underneath has vegetation coverage instead of a bare, dry desert. That is the difference between a pure strong weather and that of sand–dust storms. What is interesting is that with these V-3θ graphic characteristics, even if the forecaster is from far away, he can still make predictions for the relevant strong wind or sand–dust storm weather.

2. The day before the occurrence of strong winds and a sand–dust storm, one can observe the left leaning of the θ curve, forming an angle with the T axis greater than 45°, approaching 70°–80° or nearly perpendicular to the T axis at individual layers. What makes strong winds differ from sand–dust storms is that before purely strong-wind weather, the numerical value of the θ_{sed} curve near the ground level is greater than that before the occurrence of sand–dust storm weather. In general, the value for purely strong-wind weather is 100–15K greater than that of sand–dust storm weather and the near-ground level θ_{sed} reveals some inversion. In terms of sand–dust storm

weathers, the temperature difference $\theta_{sed} - \theta < 5K$ with θ_{sed} and θ almost coincide; or the θ^* curve forms an obtuse angle with the T axis and $\theta^* - \theta_{sed} \geq 20K$.

3. Before the occurrence of either strong winds or sand–dust storms, there is always an ultralow temperature; and the degree of drop in temperature and the degree of left leaning of the 3 θ-curves in the entire troposphere are the most instable structures among all convective weathers.

4. For the Northern Hemisphere, the clockwise rolling currents of strong winds and sand–dust storms, except the cases of typhoons and hurricanes, are made up of northwest winds in the upper altitudes and southwest winds in the lower altitudes.

5. During the later stage of the process of strong winds and sand–dust storms, as soon as θ_{sed} increases near the ground level or a layer in the mid or low altitude, and leans toward the right in between θ, θ^*, then a local thundershower will appear, indicating the end of the strong wind or sand–dust weather. What one needs to be extremely careful of during this process is that the phenomenon of ultralow temperature combined with an increase in water vapor in a certain altitude layer can trigger lightning, causing fire disasters.

As mentioned earlier, the quality of vegetation coverage of an area can be described by the closeness between the θ and θ_{sed} curves of the V-3θ graphs. If for a certain area the θ and θ_{sed} curves are very close to each other ($\leq 5K$) or completely coincide, this implies that the vegetation coverage of the region has been completely destroyed.

8.3.3.4 High-Temperature Weathers

Although temperature is one of the weather factors mentioned in daily weather reports, due to the lack of sudden drastic changes, the forecasts of temperature do not attract as much attention as those of precipitations. As a matter of fact, forecasting temperature correctly is not an easy task. Besides, temperature is one of the energies over which humans have no control. Of course, how to truly effectively make use of this energy is still an open problem awaiting further investigation. In essence, the forecast of temperature is directly related to our ability to predict the weather. No doubt, if we make a mistake in our forecast of the weather, then we naturally have made a mistake in our prediction of the temperature. As is well known, transitional changes in weather are currently a difficult problem in weather forecasting, which surely involves the forecast of the temperature.

In particular, with the human pursuit of a higher quality of life, meteorological science and technology have moved to the front to provide services to economic development. The purpose of forecasting temperatures is no longer limited to the need of light, heat, and water of the agricultural sector. To this end, temperatures, especially high temperatures, have been involved in the consumption of energies, redistribution, and effective use of energies.

Although high temperatures involve the problem of high consumption of energy, currently, high temperatures have been listed as inconvenient weather conditions that affect people's normal lives. In particular, sustained high-temperature weathers have already been seen as disastrous weather conditions. In general, if the temperature of a day or several consecutive days is 2°~3°C higher than the average high of the month, or the appearance of a day's temperature that is 3°~5°C higher than the previous day's high of over 30°C, then the phenomenon of this sudden increase in temperature is referred to as a high-temperature weather. In other words, the traditional forecasts based on extrapolations of the previous temperatures can neither correctly predict high-temperature weathers with sudden temperature increases nor provide the relevant predictions on how long a high-temperature weather would last.

Accordingly, by *forecasting high-temperature weathers*, we mean how to predict with lead-time the occurrence of sudden temperature increases and how long sustained high-temperature weather could last. To this end, the V-3θ graphs can provide a method different from those of modern science.

1. *Analysis of the counterclockwise rolling currents in the troposphere.* Let us use the Northern Hemisphere as our example. In the Northern Hemisphere, each counterclockwise rolling current mainly takes on one of the two forms: North wind counterclockwise rolling currents or east wind counterclockwise rolling currents. In the spring season, the counterclockwise rolling currents of the Northern Hemisphere are mainly made up of north winds; but during the summer season, the counterclockwise rolling currents in areas of mid and low latitudes are mainly made up of east winds. North wind counterclockwise rolling currents most likely help to alter the properties of cold high pressures of western air streams, creating high-temperature weathers. And, east wind counterclockwise rolling currents most likely make subtropical high pressures stable and stagnant, leading to high-temperature weathers. That is also the basic condition for subtropical high pressures to form, develop, and stabilize.

2. *Analysis of structural states.* For the high-temperature weathers caused by altered properties of cold high pressures, the patterns and forms of distribution of the 3 θ-curves are somehow similar to those of strong winds and sand–dust storms, which directly reveal how dry the atmosphere is, with greater obtuse angles between θ* and the T axis. This is so in particular for areas of dry deserts; and in the entire process, no clockwise rolling current will ever appear (as soon as a clockwise rolling current appears, the high temperature weather will soon end).

For front-line forecasters, it should be noted that the main characteristic of high-temperature weathers is the counterclockwise rolling currents. Do not mistakenly treat a clockwise rolling current of a local layer as being overall. It is important to recognize the counterclockwise rolling currents consisting of north winds and east winds with their speed increasing with altitude; west winds and southwest winds with their speed decreasing with altitude; and the slightly decreasing speed of partially west winds in the mid and high altitudes; or the phenomena and tendency for the speed of north winds and east winds slightly increasing which appear 2–3 days before the occurrence of high-temperature weathers. However, generally speaking, the difference between strong wind and sand–dust storm weathers is that the wind speeds of high-temperature weathers are at least one magnitude level smaller with the key being the difference in the direction of rolling currents, where along with high temperatures it is not possible for clockwise rolling currents to appear.

8.3.3.5 Dense Fog Weathers

Among the various weather conditions, fogs are such main weather phenomena, along with the processes of rainfalls and snowfalls, in which the visibility of the atmosphere decreases. For a country like China, where earlier economic and political prosperities came mainly from agriculture, fogs should not be seen as disasters. In fact, for certain crops, trees, and forests, fogs can provide water and help to reduce ultraviolet light effects. For example, the specific tropical forests in Sipsongpanna, Yunnan, China, have exactly benefited from the nourishment of fogs. However, with the development of industry, the release of polluted industrial materials has greatly increased the density of the fogs and pollution of the atmosphere. In particular, low visibility creates difficulties for transportation and air traffic, which to a degree produces a kind of disaster. It can be said that the danger created by the modern low visibility is not any less than those of wars and pestilences.

In traditional meteorological science, fogs are generally classified into advective fogs (those fogs transported from other places) and radiation fogs. However, our empirical works indicate that those dense fogs that truly affect transportation, flights, and so forth, are actually mixtures of the so-called advective and radiation fogs. In terms of effective forecasting of fogs, the rolling current directions under sufficient water vapor are the key. Here, fogs can be divided into two groups: low-altitude fogs and high-altitude fogs.

By *low-altitude fogs*, we mean the dense fogs that appear on the areas over plains, along the coast, or near the ocean. By *high-altitude fogs*, we mean the fogs that appear in the mountains of 1,500–3,000 meters above sea level. These fogs are in fact low layers of clouds. We must note that low-altitude fogs represent a problem of counterclockwise rolling currents, while high-altitude fogs the clouds of clockwise rolling currents. So, low visibilities, including those appearing during falling rains or snows, and even those of condensated water vapor near the ground level, cannot be generally referred to as fogs. Low-visibility phenomena also include the convective floating dusts before the occurrence of rainfall processes, or flying sand and sand–dust storms. However, by *fog* it is meant in principle to be the fog consisting of the tiny water drops formed by condensated water vapor at the ground level. It is a product of the process of atmospheric counterclockwise rolling currents.

Frankly speaking, it might be difficult for the layman to clearly distinguish the low-altitude clouds in mountainous areas from the fogs in areas of plains. But, when compared with actual case studies, the process of learning can be quite easy. One of the current problems is that for some meteorological stations located in mountainous areas, low-altitude clouds that cover the stations are seen as fogs. To this end, the forecaster only needs to double-check whether or not the low-altitude atmosphere is a counterclockwise rolling current in order to separate clouds from fogs. We emphasize the separation between clouds and fogs because fogs are an indicator of cloudy and rainy weathers to turn into clear weathers, while clouds are an omen for the weather to become cloudy and rainy.

In the past 10-some years, there has been a sign that dense fog weathers are moving north and spreading to the east. In other words, there is a shortage of water resource on the land, while the water vapor in the air is moving to the north. Particularly in Yunnan Province of China, the phenomenon of decreasing fogs in Sipsongpanna has already threatened the well-being of the tropical forests. So, the large-scale north movement of fogs and the phenomenon of fogs spreading eastward surely represent an important problem to investigate.

In this book, we provide a method of forecasting using V-3θ graphs in the sense of meteorology by treating fogs as a problem of change.

The basic characteristics of the atmosphere before the occurrence of a dense fog include the following:

1. The vertical distribution of the 3 θ-curves is basically similar to those of an ending precipitation. That is, the θ-curve leans to the right and forms an acute angle with the T axis, while the θ_{sed}, θ^* curves not only lean to the right but also form large, leaning angles. In particular, if the layer of inversion, as indicated by the θ_{sed}, θ^* curves based on the information of the 20th hour, is more than 25~30K, and the atmosphere beneath 850~700 hPa constitutes a counterclockwise rolling current of north winds, then in general, there will be a dense fog weather at the deep night or in the morning of the next day. In particular, the atmospheric structure after rains is similar to what is just described. It can be said that we have not failed even once in our prediction of dense fog weathers. For front-line forecasters, they should make themselves familiar with their surroundings; in particular, if a river, lake, or rice fields are in the neighborhood, it will be more advantageous for fogs to form.

2. If a concentrated layer of water vapor appears at the altitude of 850~700 hPa (that is, θ_{sed}, θ^* get relatively close to each other and lean to the right together with the θ curve), however, the 3 curves are still apart, then combined with a low-altitude clockwise rolling current one has the phenomenon of low-altitude clouds, also referred to as dense "fog" weather in mountainous areas of the sea level of about 1,500 meters.

It should also be noted that other than the difference between the directions of low-altitude rolling currents, for true fog weathers the wind speeds are generally below 4 m/s.

The difference between the counterclockwise rolling currents of dense fog weathers and those of high-temperature weathers is that the former one pays attention to the low altitudes, and the latter, the upper altitudes. In terms of forecasting dense fog weathers, one should mainly pay attention to the conditions of abundant water vapor near the ground level, counterclockwise rolling current, small wind speed, and a layer of inverted temperature.

In the previous paragraphs, we respectively introduced the analyses and methods of forecasting for the six most often seen disastrous weather conditions. No doubt, forecasting disastrous weathers, as discussed in this book, is not entirely about the problem of weather. From the point of view of materials' evolution, the problem of weather is exactly a typical case about how to analyze changes. Through various tests, we found that the method of digitization of information is indeed a valid and effective way to resolve the problem of transitional changes. At the same time, the involved problem of epistemology touches on how to understand the problem that quantities cannot handle irregular information; and in turn, because quantities cannot handle irregular information, we further considered the problem of how to understand quantities and the analysis power of quantities. Even if the reader only wishes to learn the system of methodology introduced here, he or she should still understand why the digitization of information is different from the quantification of the information. Otherwise, when one faces situations not mentioned in this book, he or she will have difficulty replying with comprehension of what is discussed in this book. Therefore, to fully and truly master the methods studied in this book, one should at least know at the height of epistemology, why events, especially peculiar events, do not comply with the calculus of formal logic of quantities, and why changes in events follow their own rules of transformation. To this end, we suggest not skipping over the previous chapters, because changing an established system of concepts is a difficult matter. In particular, the information we deliver here is from events where we extract the useful information of change, which is different from modern science, where information of invariance is extracted from events.

8.4 The Problem of Mid- and Long-Term Forecasts

Although our basic idea of forecasting is fundamentally different from those of the traditional methods, due to its practical successes, our work has been financially supported by various foundations and government agencies. And to satisfy their needs of producing timely, correct forecasts, several professional stations across China have adopted our methods in their daily weather forecasting operations.

Because of the validity and effectiveness of our methods in the area of short-term weather forecasts, over the years various government officials and front-line weather forecasters inquired about our opinions on mid- and long-term weather forecasts, hoping that we could produce relevant forecasts about major droughts and floods without relying on the methods of statistics. The reason for this last requirement was due to experiences like this one: One time, a forecast for the situation of the coming flood season from a very reputable organization indicated a normal and smooth season without much complication. However, the actual happening was that the month of July experienced major episodes of disastrous floods and the month of August suffered from severe droughts. That is, the authoritative forecasts using statistics did not provide any useful information and the needed lead-time for preparing for the necessary protective measures against the flood and drought disasters. In terms of the formality, the tendency forecast of statistics was correct, since the average of floods and droughts is indeed a normal season without any disaster. However, due to such kinds of forecasts, people have suffered losses both times. Because of numerous similar experiences, relevant government officials and front-line forecasters really wanted us to look into the possibility of using our methods to address the problem of mid- and long-term weather forecasts with the specific requirement that hopefully we could predict forthcoming disasters that would occur within half a month or one month; they no longer like to see any statistical-tendency prediction of the forthcoming season of floods and droughts.

Based on what has been discussed in this book, it can be seen that the fundamental reason for atmospheric movements is the unevenness of solar radiation and changes in the radiation. So, major drought and flood disasters should not only be dynamics problems of the bottom cushion. To address the problem of forecasting these disasters, we have to start to look for the possible solutions from the distribution of heat and relevant changes in the distribution. And, the only factor that can directly affect the heat and its distribution has to be the rotation of the earth in its movement so that the position of direct solar radiation changes. However, traditional meteorological science studies the mid- and long-term weather forecasts by focusing only on the atmosphere itself without involving such other factors as the earth's nutations, movements of the poles, the speed of rotation, and so forth. In the following two sections, we will briefly illustrate our relevant ideas by looking at two specific events: the high-temperature drought of the summer of 2006, and the snow and ice disaster of January 2008.

8.4.1 Analysis and Forecast of the High-Temperature Drought of Summer 2006

Currently, the study of atmospheric phenomena, especially the appearance and development of major disastrous weathers, belongs to the area of weather systems. For example, if we asked why last year the torrential rains along the reaches of

Yangtze River were heavy, then the commonly seen explanation accompanying the relevant weather forecasts would be, a subtropical high pressure stayed on the south side of the Yangtze River reaches. As for the extreme hot temperature during the summer of 2006 and the hot drought that occurred in Sichuan and Chongqing in the same time frame, the given meteorological explanations would be, the subtropical high pressures moved further north (or west); the activities of the cold air are relatively weak; heat on the plateau is relatively too strong; since the winter and spring seasons of 2005–2006 the amount of snow accumulation on the Qinghai-Tibet plateau has been reduced, causing increased plateau thermal effect that creates strong summer winds; and so forth. If we further questioned why subtropical high pressures have stayed along the Yangtze River reaches, or why subtropical high pressures have gone further to the north or west, we would hardly see any analysis on the mechanism for the formation and development of subtropical high pressures. In 20th-century meteorological science, it has been a common practice for scholars to investigate the effect of the bottom cushion on the atmosphere, leading to the studies on El Niño and La Niña phenomena. No doubt, uneven heat distribution and changes in the distribution on the bottom cushion can influence the activities of the atmosphere. However, we believe that the main cause of atmospheric movements is created by the varying positions of the direct solar radiation on the earth. Therefore, to investigate the problem of major disastrous weathers, we should start with the factors that affect the positions of direct solar radiation.

Since the 1990s, we have employed changes in the earth's nutation, polar movement, and rotational speed to make mid- and long-term weather forecasts. We have provided each month's drought and flood predictions 3 months ahead of time with good accuracy. In terms of the basic rainfall situation along the reaches of Yangtze River for the 2006 flood season, we provided the following forecast in the start of June 2006: There will not be much plum rains. The actual weathers of the season proved that our forecast was correct. This work employed the most recent data published on the Internet by the authoritative international agency — International Earth Rotation and Reference Systems (IERS). We analyzed the earth's nutation and changes in the earth's rotation and their impacts on atmospheric activities. When we analyze abnormally high-temperature weathers using the V-3θ graphs, we employ the appearance, strengthening/weakening, and movement of east winds at the altitude level of 300 to 100 hPa to predict the occurrence, strengthening/weakening, and movement of subtropical high pressures. What is practically meaningful is that the appearance, movement, and development of the layer of east winds are ahead of the occurrence, development, and movement of the relevant subtropical high pressures. So, this layer of east winds can be directly applied to forecast high-temperature weathers, reaching such a rate of accuracy that as long as the provided information is correct, our forecasts will be correct.

For reasons like different belief systems, very few observation stations over an ocean, and so forth, forecasting subtropical high pressures by using the traditional methods has been a relatively difficult task. For China, all major weather-related disasters are basically related to subtropical high pressures. That explains why for the Chinese community of meteorologists, subtropical high pressures are often the "ultimate reason" for analyzing disastrous weathers. This end most likely originated from the practice of limiting the research of meteorological science in the area of the earthly atmosphere. It also explains why atmospheric science is currently seen as a part of geoscience. In this section, we intend to enlarge our scope of activities to the relevant effects of the earth's nutation, polar movement, and rotational speed from the angles of mutually interacting factors and practical applications.

8.4.1.1 Layers of East Winds and Subtropical High Pressures

8.4.1.1.1 Forecasts of Subtropical High Pressures

Earlier in this section we mentioned that by using the V-3θ graphs, we can reveal the high-altitude structure of east wind layers of subtropical high pressures. The strong east-wind layers of subtropical high pressures can spread from 100 hPa to below 400 hPa. The mid-altitude atmosphere below 500 hPa is a layer of low humidity. The corresponding weather will have light clouds and soft breezes; however, at the near-ground level or the ground surface, it is steaming hot. If high-temperature weathers are also seen as disastrous, then forecasting such weathers can be based on whether there is a high-altitude layer of east winds or not; and the thickness of this layer and the strength of the east winds are also directly related to how hot the weather will be.

Figure 8.20 through Figure 8.25 provide the V-3θ graphic characteristics of the east-wind layers for Fuzhou station of July 2003; Nanjing (Jiangpu) station of August 2003; and Chongqing, Lhasa, Hongyuan, Xi'an stations for August 18, 2006; and the situation of spread for the year of 2006. A close analysis indicates that, first, Figures 8.20 and 8.21 depict that for a high-temperature weather of a subtropical high-pressure system to appear, there must be a high-altitude layer of east winds; and second, from Figure 8.22 through Figure 8.25, an expansion of the east-wind layer for the year of 2006 can be seen, where the north end (Figures 8.22 and 8.23) could have reached Hongyuan and Xi'an, and the west side not only passed Chongqing but also reached Lhasa; the thickness was far greater than that of 2003, spreading from 100 hPa to below 400 hPa (Figure 8.22 through Figure 8.25).

What is practically meaningful is that this east-wind layer appears prior to subtropical high pressures. So, now we can explain why traditionally, meteorologists could analyze weather phenomena until a subtropical high pressure appeared, and improve the situation to that of forecasting weathers by using changes in the east-wind layer. This improves the current live weather reports of the form of monitoring pressure systems. Figure 8.26 is the V-3θ graph of Chongqing on August 1, 2006. It indicates that the high-altitude east winds at 100 hPa are ahead of those at 500 hPa, further confirming the significance and practicality of our discovery for the purpose of weather forecasts.

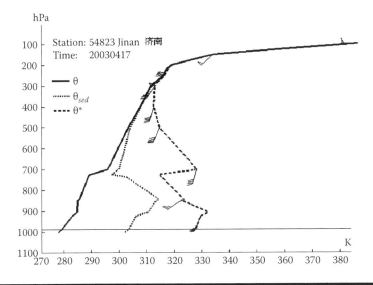

Figure 8.19   The 8th hour on April 17, 2003, Jinan.

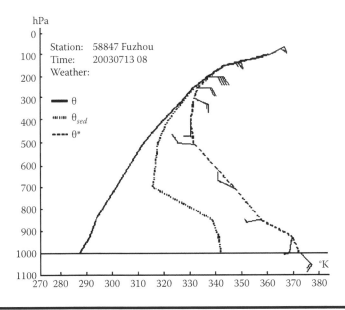

Figure 8.20   The V-3θ graph of Fuzhou station at the 8th hour, July 13, 2003.

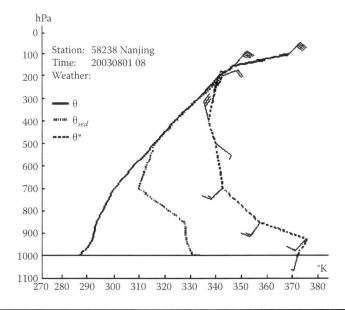

Figure 8.21   The V-3θ graph of Nanjing station at the 8th hour, August 1, 2003.

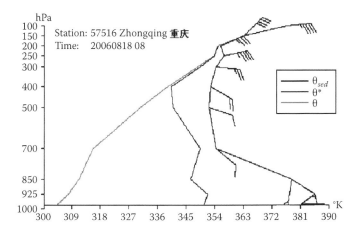

Figure 8.22   The V-3θ graph of Zhongqing station at the 8th hour, August 18, 2006.

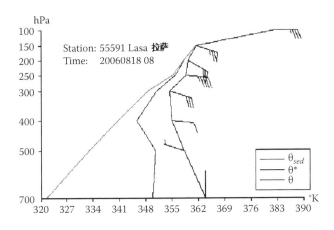

Figure 8.23 The V-3θ graph of Lasa station at the 8th hour, August 18, 2006.

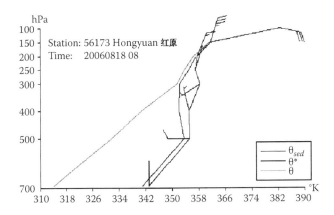

Figure 8.24 The V-3θ graph of Hongyuan station at the 8th hour, August 18, 2006.

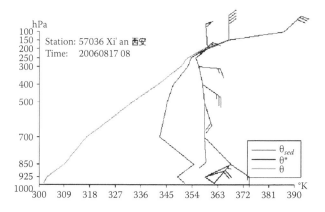

Figure 8.25   The V-3θ graph of Xi'an station at the 8th hour, August 18, 2006.

By using the V-3θ graphs to make weather forecasts, one can skip over the subtropical high-pressure systems and make use of the characteristics of and changes in the east-wind layer to directly predict the weather factors. This of course poses the following problems: Where is the origin of the high-altitude east-wind layer? What causes it to change?

According to the supplementary principle of airstreams and the returning west-wind principle of the deviation forces, one can use the V-3θ graphs to trace the changes in location and strength of high-latitudinal west-wind zones and analyze the changes in the mid-latitudinal east-wind zones. Figure 8.27a,b already show that the west-wind zone withdrew to the north of 50°N; the strength of its wind speed is basically the same as that the of east-wind zone at near 30°N (Figures 8.22 and 8.23). The 36096 station is located at about 600 kilometers west of Lake Baikal, which is located at 52° to 55°N; and the 30554 station about 300 kilometers east of the lake. According to our experience of many years, if the winds at 3–5 stations on the south and north sides of Lake Baikal turn into similar strength northwest winds, then in the forthcoming 2–3 days, the subtropical high-pressure system at the area of 30°N will weaken (Figures 8.28 and 8.29). The temperature of Chongqing station in this case study dropped from above 39°C on the 18th day to 33°C on the 20th day. However, the ultralow temperature still existed, and Hongyuan station showed an extraordinarily stable structure (Figure 8.24). So, within the coming half month, the high temperature would be sustained. The basic idea of our forecasts is not whether a cold airstream presses south; instead, it is how the northwest winds weakened the support of the west-wind zone so that its returning flow also causes the east-wind layer at the mid-latitudinal areas to also weaken correspondingly. These are forecasts of informational digitization, and the reason why they possess the lead-time needed for any weather forecast and are more accurate than the forecasting method using extrapolation. It emphasizes the key points on why subtropical high pressures may withdraw eastward.

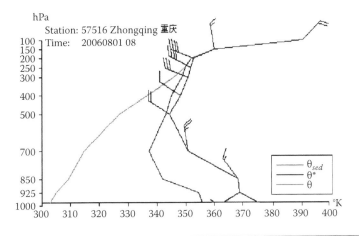

Figure 8.26 **The V-3θ graph of Zhongqing station at the 8th hour, August 1, 2006.**

Considering that the sweltering hot weathers and droughts of the year of 2006 were shared by many regions across the world and possessed different characteristics when compared with the historical records, in this section, we comparatively analyzed the high-temperature disastrous weather of 2006 based on our earlier works of using the rotational speed of the earth and publicly available information published on the Internet to make long-term weather forecasts.

8.4.1.1.2 Analysis of the Information on the Earth's Rotation

In this work, we employ the authoritative information on the Internet published and regularly updated by the IERS.

8.4.1.1.2.1 The Information of Nutation of the Earth's Rotational Axis —
Nutation is a term from astronomy; it means the not fixed and constantly varying direction in space of the rotational axis of the earth. Since the rotational axis of the earth wobbles along with the earth's movement, it makes the poles of the earth deviate from their normal locations, which causes the earth to wobble along its orbit of motion. It consequently affects the speed of rotation of the earth. Here, when the wobbling of the rotational axis is stabilized at a certain angle, the speed of rotation of the earth will be increased; frequent changes in the direction of the axis can make the rotation of the earth slow down, leading to irregular movements in the atmosphere, oceans, and even the crust. If the axis continues to move in a certain way for a prolonged period of time, some abnormal natural phenomena can also occur over a sustained period of time. Although nutation itself is very small, due to the large radius of the earth, the nutation moments can be quite great. The corresponding speed of rotation of the earth also goes through changes. Combined with the scale

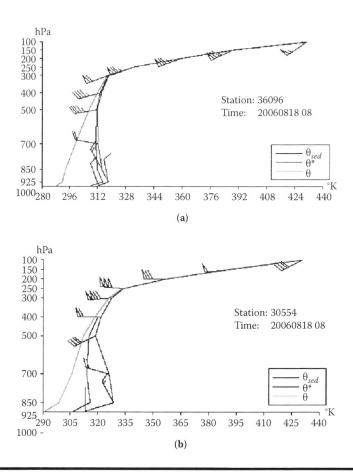

Figure 8.27 (a) The V-3θ graph of station 36096 at the 8th hour, August 18, 2006. (b) The V-3θ graph of station 30554 at the 8th hour, August 18, 2006.

of measurement of the radius of the earth, the variable moments of nutation can no longer be ignored. That is how we have held a different opinion from treating the rotational speed of the earth as a constant as done in traditional meteorological science. In fact, it has been our long-held belief that the investigations of the problems of weathers and earthquakes should not be stopped at mechanics but instead should at least involve the studies of moments of forces. Because moments of forces touch on the concept of structures, it led us to consider using changes in structures to make forecasts, where the fundamental idea in the design of the V-3θ graphs is to first establish the atmospheric structures of the affected disastrous regions by using irregular-sounding information. The difference between the method of structures and the traditional methods is that we carried the traditional system of dynamics to the depth of structures of moments of forces, including structural changes in the moments of forces. So, when considering applying the rotational speed of the earth,

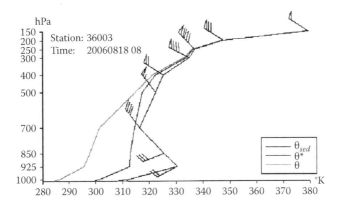

Figure 8.28   The V-3θ graph of station 36003 at the 8th hour, August 18, 2006.

we had to modify the original constant of traditional mechanics into a variable of moment of force. The irregular time series information of the earth's nutation and speed of rotation need to be transformed into digitized structural information. Due to the current situation that many colleagues in the profession of weather forecasts still have trouble comprehending that time does not occupy any material dimension and the digitization of irregular information of phase spaces, in the following, we will only apply the form of time series information hoping that at the very least, people could realize that behind changes in the atmosphere there are other factors that constantly cause these changes.

Figure 8.30 is a time series graph of the earthly nutation from January 1, 2005, to November 9, 2006. From this graph, it can be seen that the changes in the magnitude and frequency of the nutation during 2005–2006 are quite clear. Since the start of 2006, the nutation magnitude not only weakens but also approaches stability. In particular, starting in April and May, the changes in the magnitude and frequency experienced a sudden drop. A decrease in the nutation magnitude implies that the amount of wobbling in the earth's axis is reduced so that the airstreams in the atmosphere along the latitudinal direction are strengthened; here the latitudinal direction stands for the upper-layer atmosphere of the troposphere. From the supplementary principle of airstreams combined with the deviation force from the earth's rotation, the high-altitude east-wind layer has to be formed due to returning flows of west winds. The east-wind layer of returning flows, because of the existing counterclockwise rolling current, triggers the formation of a subtropical high pressure in the mid and low atmosphere. No doubt, this new understanding has pushed the common belief of meteorology of treating pressure systems as the cause of weather phenomena to a higher level of using changes in the earth's nutation and rotational speed as indicators for forecasting pressure systems. Because of this improvement, the relevant forecasts for high-temperature weathers can be made at least 3–4 months ahead of the actual happenings. And by using the location of an east-wind layer, we have

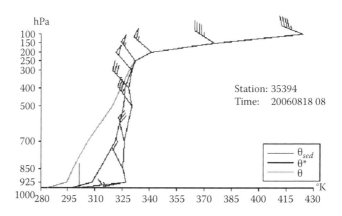

Figure 8.29   The V-3θ graph of station 35394 at the 8th hour, August 18, 2006.

established an effective method for practically predicting the advance or retreat and strengthening or weakening of, and other changes in, a subtropical high-pressure system. We can further trace the cause of changes in the east-wind layer to the nutation of the earth and the effects of changes in the earth's rotational speed (see below for more details). Currently, many scholars still have not sufficiently recognized the significance and effects of irregular information so that this information has been seen as random or indeterminate and ignored in relevant studies.

8.4.1.1.2.2 The Length of Day (LOD) Data — Figure 8.31 shows the changes in the LOD from January 2, 1973, to August 23, 2006. *LOD* means the difference of the actually observed length of day and the average daily length that is equal to 86,400 seconds. The unit of LOD is a millisecond (ms). From this figure, it can be seen that during the 33 years of the coverage, the changes in the LOD have not only an overall periodicity but also the nonperiodicity over any chosen period of time. From 2000 to 2005, the LOD seems to be relatively shorter. And during 2006, the LOD shows a tendency of prolonging. The LOD is inversely proportional to the rotational speed of the earth; its computational formula is

$$\Omega = 72\ 921\ 151.467064 - 0.843994809\ LOD \qquad (8.1)$$

where Ω stands for the rotational speed of the earth; its unit is 10^{-12} arc length/second). The unit of the LOD is millisecond (ms).

   From a comparison between Figures 8.30 and 8.31, it follows that in 2006, the nutation magnitude changes only slightly or does not change at all. So, the corresponding LOD increases slightly and the earth's rotation slows a little bit. Constrained by the quasi-conservation of the angular momentum of the system of the earth and its atmosphere, changes in the atmosphere tend to be slow so that airstreams along the latitudinal direction would relatively increase. Considering the

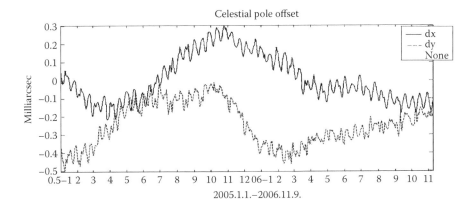

Figure 8.30 **The earth's nutation from January 1, 2005 to November 9, 2006, where *dx* and *dy* stand for the deviation amounts of the rotational axis along the *x*- and *y*-directions in the celestial polar coordinate system; the unit of the vertical axis is one-millionth arc second.**

temporary stagnation of the atmosphere caused by the rotational speed, in about 10-15 days, the position of direct solar radiation will change. This end has been often applied to our mid- and long-term weather forecasts, producing success time and again. Based on the information of August 18, 2006, the most recent data available before this writing, and previous information, we test-forecasted that from the end of August 2006 to the early part of September, the existing east-wind layer would not completely withdraw from Chongqing area so that there would still be the high-temperature weather of over 35°C in the region. The actual happening afterward showed that at the 20th hour on September 4, 2006, the subtropical high-pressure system withdrew from Chongqing area. This example along many others shows that the main factors that affect changes in the atmosphere should be changes in the position of the direct solar radiation and that major disastrous weathers are also caused by changes in heat distribution.

What is left open in our work is that we are still not sure about the causal relationship between the earth's nutation and the rotational speed, or whether or not they are more or less determined by the relative positions of other celestial bodies.

Although the earth's nutation, pole movement, and rotational speed are factors affecting the activities of the atmosphere, in terms of the overall changes of the atmosphere, they should be important contributing factors. They trigger the occurrence of disastrous weather conditions through various combinations with changes in the atmospheric structure. And, varying atmospheric structures determine specific weather characteristics through rolling currents. So, in order to effectively apply our method for weather forecasting, at the same time when using irregular information, one should master what information corresponds to what kind of atmospheric structure and what tendency of change could be expected. That is how we applied the earth's nutation, pole movement, and rotational speed, combined with the structure of the temperature field of the ultralow temperature at the high altitude, to make the forecast that in early June 2006, there would be almost no plum rains along the reaches of Yangtze River. (This forecast was completely opposite of that made by one of the most prestigious agencies of meteorology in China). During our entire funded research period from 2001 to 2004, our research group twice correctly made the long-term prediction with 2 months lead-time that there would not be large tidewater in Qiantang River; in 2001, we predicted 3 months ahead of time that in September Sichuan would suffer from an autumn flood; in 2002, we correctly predicted that there would not be any major disaster during the flood season around China; and in 2003, we correctly predicted the regional torrential rain in Chengdu (the actual happening that four heavy torrential rains occurred in the month of August); and in 2004, we correctly predicted with 3 months of lead-time the occurrence of a high-temperature hot autumn; among others.

As is well known, compared with the current short-term weather forecasts, long-term forecasts are real predictions, because any forecast of over half a month cannot be derived by using extrapolations of the initial values and by blaming the atmospheric systems of pressures and altitude fields. Currently, the community of meteorologists classifies long-term weather forecasts as the problem of climates, and the statistical concept of averages once populated the study of climates. That is why since long ago, climatology has been established on the basis of statistics. With the birth and development of systems science, although studies of climatology have been seen as part of systems science, its basic methodology is still limited by the methods of statistics. As a matter of fact, long-term weather forecasts also stand for the problem of transformation of irregular information except that the irregular information is no longer limited to the atmosphere itself; instead, it involves information of change in celestial bodies and the structural transformations of the information of change.

8.4.1.1.2.3 Comparison between the Summer LODs of 2005 and 2006 — Under the constraint of the quasi-conservation of angular momentum of the system consisting of the earth and its atmosphere, slowing rotational speed, that is, the decreasing angular momentum of the earth, can be dynamically balanced by increasing angular momentum of the atmosphere. So, the decreased tangent speed of the solid earth can cause the momentum along the latitudinal direction of the fluid atmosphere to increase so that the movement of the atmosphere at the midlatitude area will speed up leading to rampant west winds. That is why the information of the earth's nutation, pole movement, and rotational speed can be used for long-term weather forecasting due to their lead-time of appearance ahead of the relevant weather phenomena.

Figure 8.32 pictures the changes in the LOD from January 1, 2005, to August 23, 2006. A careful comparative analysis reveals that in general the distribution of the LOD for the summer of 2006 is increased when compared with that of the

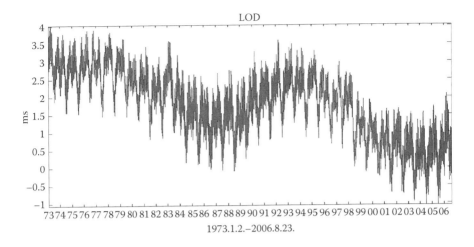

Figure 8.31 **Changes in the earthly days from January 2, 1973, to August 23, 2006, where the unit of the vertical axis is millisecond (ms).**

summer of 2005. Correspondingly, the rotational speed of the earth in 2006 is slower than that of 2005. So, the wind speed in 2006 along the latitudinal directions in the atmosphere has to be greater and the relevant east-wind returns stronger so that the corresponding subtropical high pressures are also more powerful, more stable, and stretch further to the west. This end leads to sustained drought in Sichuan, Chongqing, and surrounding areas. What can be verified is that the east-wind layers in Figures 8.22 and 8.23 not only are thick but also contain faster-speed winds. Such structures have been rarely seen since 1981 when we started using the V-3θ graphs. Figures 8.22 and 8.23 not only reveal the thickness of the east-wind layers, but also show the strengths of the east winds.

Because the published information on the Internet does not contain those of the earth's nutation, pole movement, and rotational speed of over 50 years ago, we are unable to further analyze the practical effects of these factors. Although according to the concept of irregular information that history cannot entirely repeat itself, as an exploration on the underlying mechanism it is still necessary for us to explore the relevant mechanism in the hope of modifying the well-accepted knowledge in order to achieve much better forecasting results.

From the irregularities in the nutation, pole movement, and rotational speed of the earth, it is not hard to see the irregularities in the changes of weather and the atmosphere. This end at least indicates that for the investigation of problems of change, one should focus on the study of information of changes. In other words, problems of constant change cannot be successfully understood by using the theories and methods established for invariant systems. Science is the process of exploration; it cannot be completely identified with the system of empirical knowledge acquired through practical activities. It is because each process of exploration is biased due to the interference of well-accepted beliefs. So, only such pieces of experience that have stood the test of practically resolving problems can be collected into the system of knowledge. In this work, some of our opinions and realizations might be different from those commonly accepted. Our purpose of presenting them is not for that of pursuing theoretical arguments. Instead, we would like to see whether these different understandings can actually resolve the problems of truthfully making weather forecasts. When seeing through history, it can be said that the existence of open problems stands for challenges to traditional theories and accepted concepts.

8.4.1.1.2.4 Some Explanations

1. Using the east-wind layers to forecast subtropical high pressures has been shown to possess the desired causal relationship. This end has deepened the traditional treatment of subtropical high-pressure systems as the cause of weather phenomena to a different level of changes in the east-wind layer and in the west-wind zones. By doing so, we can at least be able to see further out in time than the monitoring, extrapolating kind of "forecasts." In terms of epistemology and methodology, we have truly materialized the purpose of foretelling what is forthcoming before it actually happens.

2. Considering that adjustments in the atmosphere are posterior to the nutation, pole movement, and rotational speed of the earth, the information on the nutation and rotational speed constitutes such information that is more advanced than changes in the atmosphere. Our analysis of data indicates that the year of 2006 has shown a clear decrease in the magnitudes of the earth's nutation and rotational speed. From this discovery, we successfully predicted that the west winds at the midlatitudes would be more rampant than the past years so that the relevant subtropical high pressures would be more powerful and stable. This end has been confirmed by the actual happenings. If this discovery can be repeatedly verified by future weather forecasting practice, then the investigation of the problem of weather forecasts will have a better vision into the future, which in fact has entered the realm of evolution science.

3. Limited by the supply of information, the work presented here is only explorative in nature, although our earlier long-term weather forecasts have achieved better-than-expected results. As of this writing, the publication of information is still based on the practice of regularization or quasi-linearization. The way we can correctly understand irregular information still needs to be improved. Our claim that irregular information stands for information of change goes directly against the concept of randomness of the probabilistic universe and that the invariant problems investigated by using initial-value extrapolation established since the time of Newton cannot really be listed as the determinacy of the fatalism. Each concept of determinacy has to be established on the basis of changing materials. Studies of the problems of invariant materials

or events do not have much significance in terms of making predictions. So, evolutions of materials belong to the problem of prediction of evolution science, and directly challenge the concepts of determinacy and indeterminacy of the system of modern science.

8.4.2 The January 2008 Snow–Ice Disaster Experienced in Southern China

8.4.2.1 Some Explanations

This snow calamity and ice disaster should be seen as well-known worldwide, or at least a snow and ice process of over 43°C or below 41°C that we have never experienced in our long careers. Evidently, for this process, it was not because of the low temperature involved that made it rarely seen. Instead, it was because it mixed with both snow and ice and happened over the half of China that is south of the reaches of Yangtze River and lasted a long period of time. We were dragged into the prediction analysis of this disastrous snow and ice process because our former students who are currently working at Chengdu Bureau of Meteorology sent us the meteorological sounding data of January 11, 2008, and asked us to provide relevant results of analysis. The initial request was only about whether or not there would be precipitation during the 11th night (the opinion of the forecasters of the meteorological station was that yes, there would be) without involving the entire snow and ice process. It can be said that until January 15, 2008, no one in the bureau noticed or expected the appearance of such a severe process of snow and ice. Speaking frankly, even we only expected a major snow process that could last over 15 days without any anticipation of such a serious ice disaster. No doubt, this expectation has something to do with our earlier working experience in the northern parts of the country, where snow disasters that were mostly heavy snows at below –10° or –20°, or the snowfalls with which melting snows occurred at around freezing point are only some brief phenomena. So, this long-lasting snow calamity and ice disaster was only the first such weather process we had seen in our professional careers.

8.4.2.2 Key Points for Forecasting the 2008 Snow–Ice Disaster

What needs to be emphasized is that we noted earlier in our careers that before a snowy weather, not only will clouds (warm layers of clouds) fall lower, but also the phenomenon of lowered jet stream layers of ultralow temperatures appears, which coordinates with the temperature at the layer of 850~700 hPa reaching below 0°C. So, when we noticed at the 8th hour on the 11th day that the altitude of the jet stream layer of ultralow temperatures was lowered and reaching the south over 30°N, we stayed on high alert. Our experience and analysis indicate that the main factors that contribute to the formation and sustained maintenance of icy snow weathers include those in the following three sections.

8.4.2.2.1 Lowering Jet Streams of Ultralow Temperature

The jet stream layer of ultralow temperature at Dingri station located on the south hill of the Tibetan plateau reached 300 hPa. Corresponding to this time, the jet stream layers at several stations — Ganzi station, which is located on the northeast hill of the same plateau; Chengdu station, which is located in a basin; and other stations, such as Xichang — were also lowered to the level of 300 hPa (Figure 8.33 through Figure 8.36). This phenomenon was even stretched over to Hongzhou station, which is located on the lower reaches of Yangtze River (Figure 8.37). And, the west winds were rampant and powerful and showed a tendency of lowering (in the actual process, the jet stream layer of ultralow temperature was even lowered to 400 hPa), reaching a level about 2 to 3 kilometers below the normal altitude. Due to the lowered thickness of the troposphere, the speed of the northwest or partially west winds was increased at the mid-altitude levels of the atmosphere. When these strengthened northwest or partially west winds met the northeast winds that carried water vapor and traveled south along the coastal line, a low-altitude warm layer of clouds was formed, creating a snowfall covering a large area.

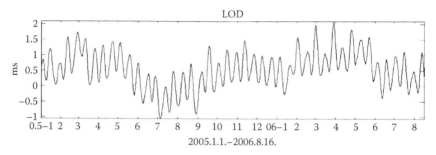

Figure 8.32   Changes in LOD from January 1, 2005, to August 23, 2006, where the unit of the vertical axis is a millisecond (ms).

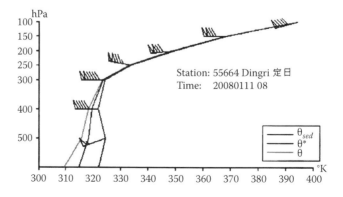

Figure 8.33   Dingri at the 8th hour on January 11, 2008.

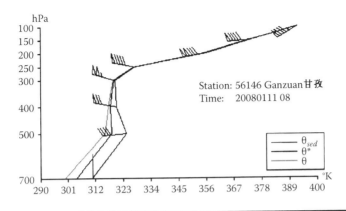

Figure 8.34   Ganzi at the 8th hour on January 11, 2008.

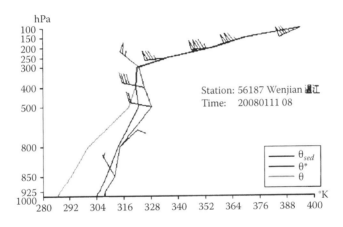

Figure 8.35   Wenjian (Chengdu) on January 11, 2008.

8.4.2.2.2 Southward Movement and Spread of Jet Streams of Ultralow Temperature

The jet stream layer of ultralow temperature was stretched to the south of 30°C, making the cold air of the north hill of Qinghai and the Tibetan plateau (or the north of the jet stream layer) able to directly reach the middle and lower reaches of Yangtze River. The jet stream layer of ultralow temperature moved south in large scales for a sustained, long period of time. Based on the graphic structures of the Lijiang and Tengchong stations (Figures 8.38 and 8.39), even based on the stability principle of heat distributions, one could have also predicted that the stability of this disastrous process could last 15~20 days. In order to provide our reader a deeper impression, please pay attention to this kind of extraordinarily stable structure of the atmosphere, where the angle formed by the distribution of the potential

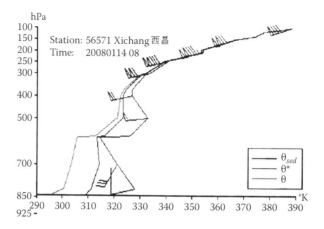

Figure 8.36   Xichang on January 14, 2008.

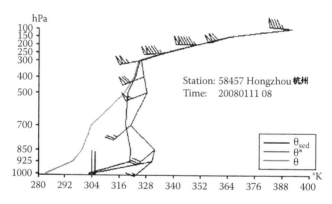

Figure 8.37   Hongzhou at the 8th hour on January 11, 2008.

temperature of the θ curve and the T axis approaches 45° or slightly less than 45°, which represents an extraordinarily stable structure of the atmosphere. However, the overall extraordinarily stable structure is not the same as the stabilities of specific atmospheric layers. Figure 8.37 through Figure 8.41 all show an overall extraordinarily stable structure; however, some of the individual layers in the troposphere were instable. No doubt, Figure 8.40 (a subtropical counterclockwise rolling current) represents an extraordinarily stable structure, which is an overall stability and also the stability of each specific layer. The overall extraordinarily stable structure possesses the characteristics of sustainability. Even without any hint from the information of astronomy, if the radiosonde stations that are 300 kilometers apart all indicate extraordinarily stable structures, one can predict that the structure would sustain for about 15 days. Considering the contrast of the atmosphere against the solid earth, the predicted maintenance could be behind 3–5 days.

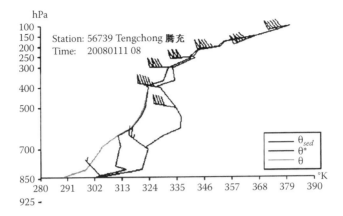

Figure 8.38   Tengchong at the 8th hour on January 11, 2008.

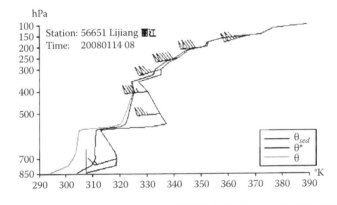

Figure 8.39   Lijiang at the 8th hour on January 14, 2008.

No doubt, the problem of sustained stability absolutely cannot be controlled by the atmosphere itself or the "evil siblings" of El Niño and La Niña, which exist in the bottom cushion of the atmosphere. Otherwise, there would not be the alternating changes between winters and summers. In other words, neither El Niño nor La Niña is powerful enough to resist the changes in the direct solar radiation. So, this process of large-scale snow calamity and ice disaster in southern China must involve the effects of the earth's nutation, pole movement, and rotational speed that affect the direct solar radiation. According to the relevant works done by Dr. Wei Ming, it was shown that this process was mainly affected by the pole movement of the earth's axis of rotation. At this junction we would like to mention that in the latter half of the 20th century, there was once a well-circulated speculation in the community of meteorologists that the atmosphere has a memory of about 15 days. In fact, the direct solar radiation at one location lasts about 15 days. So, based on the structure and state of the atmosphere, one can produce weather forecasts for about 15 days. This end is one of the basic methods used in our many years of mid- and long-term weather forecasts. On February 18, when we (and our students) were preparing for a speech to be delivered at a conference to summarize the experiences learned from this process of snow calamity and ice disaster, an open question was when the temperature would start to rebound. Based on the available information on that day, we predicted that only after another 15 days, the temperature would start to rise substantially. Later the actual happening confirmed our prediction.

8.4.2.2.3 The Practical Benefits of Forecasts

According to a lowered jet stream layer of ultralow temperature and its southward expansion, the east winds near ground level that were returning to the inland of southern China along the southeast coastline that carried abundant water vapor, and the accumulated air temperature in the low altitude of the south was higher than that of the north, we predicted that snowfalls and snow melting would occur simultaneously, leading to the snow calamity and ice disaster. This forecast was telephoned to the relevant departments of meteorology on the morning of the 11th day. Specific to the Chengdu area, our forecast was as follows: In the night of January 11, there would be strong winds; a precipitation would occur after noon of the 12th day; after that the weather would officially enter an icy snow weather that would last about 15–20 days. The actual happening was that from the evening of the 11th day to the noon hour of the 12th day, a sustained strong wind appeared; and a precipitation began at the 13th hour on the 12th day. After all that, the weather condition indeed entered the icy snow process. However, the Bureau of Meteorology only directly accepted our short-term forecasts, ignored the forecasted precipitation, and lacked the courage to announce the mid- and long-term forecast.

It should be admitted that this snow calamity and ice disaster has not been seen by the professionals of weather forecasting in the past 100 years. What is amazing is that even after the event not a single factor that could be plausibly used to explain the event was located. In principle, since the 12th day, the forecasts made by using the method of extrapolation had done their best, although they were useless. In comparison, if one masters the epistemological concept of digitizing the information of structures and states, at the very least on the 14th day, he would be able to make the prediction of a large-area, sustained snow calamity and ice disaster. No doubt, such an outcome would be able to help reduce major societal and economic losses.

8.4.2.2.4 Brief Explanations of the Relevant Methods

1. This method of forecasting mid- and long-term disastrous weathers comes from our earlier analysis of major weather disasters many years ago and how the nutation, pole movement, and rotational speed of the earth in its celestial

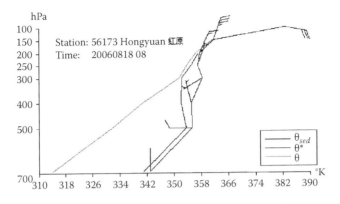

Figure 8.40   Hongyuan at the 8th hour on August 18, 2006.

movement can affect the direct solar radiation. Based on these studies, we suggested the causes for large-area changes in atmospheric heat. Later our work was financially supported by the Bureau of Science and Technology of Sichuan Province for 3 years in order for us to practically test our ideas. The basic reason underlying our work was mentioned earlier — that the activities of the atmosphere are mainly caused by the changes and uneven distributions of the direction of solar radiation. Other than one time when our forecast of the rainfall location of a large-scale torrential rain was off by over 200 kilometers, all the rest have been within the range of 200 kilometers of what we had forecasted.

2. In Figure 8.33 through Figure 8.46, notice that Dingri station is located on the south side of the Tibetan Plateau; Ganzi, a station on the northeast side; and Chengdu (Wenjiang), Hongzhou, and other stations are either located along 30°N or south of 30°N. From these figures, it can be seen that the jet stream layers of ultralow temperature had been lowered by as much as 1–1.5 kilometers (their normal altitude should be at 200–250 hPa). The most seriously affected location by this disastrous process was Chenzhou City, Hunan Province. We can see (Figures 8.41 and 8.42) how much the jet stream layers of ultralow temperature had dropped (to 400 hPa), the thick layer of warm clouds formed by returning northeast winds at the low altitudes, and the substantial significance of the overall extraordinarily stable structure of the right-leaning heat distribution. All these patterns can be utilized together to improve the quality of future weather forecasting services.

It can be said that based on our comprehensive analysis by using the characteristics of the atmospheric structures to the nutation, pole movement, and rotational speed of the earth, and other relevant factors, this episode of snow calamity and ice disaster indeed has been rarely seen in the past 100 some years.

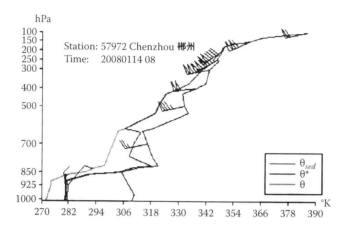

Figure 8.41 **Chenzhou at the 8th hour on January 14, 2008.**

3. As for Henan Province, what should be noted is that the layer of ultralow temperature at Nanyang station at the 8th hour on January 11 (Figure 8.43) had already started to drop; and at the 20th hour of the same day (Figure 8.44) the wind speed at 300 hPa increased and the process of lowering ultralow temperature layer could be vividly seen; combined with the lowering layer of ultralow temperature and increased wind speed at the neighboring Wuhan station (Figure 8.46), it could be forecasted that there would be a snow disaster in Henan. For the front-line forecaster, for forthcoming snows, he should look for lowered skies, where *lowered sky* means the lowered layers of ultralow temperature and thick, warm clouds and increasing wind speed. In other words, the extraordinarily low layers of ultralow temperature before

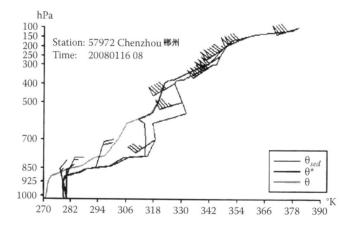

Figure 8.42 **Chenzhou at the 8th hour on January 16, 2008.**

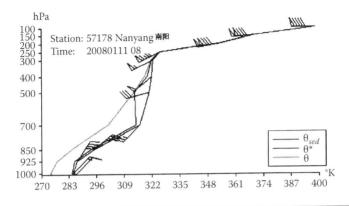

Figure 8.43 **Nanyang at the 8th hour on January 11, 2008.**

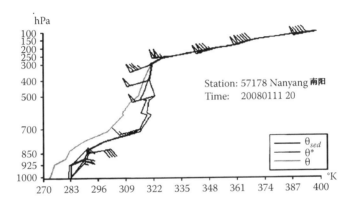

Figure 8.44 **Nanyang at the 20th hour on January 11, 2008.**

major snowfalls are lower than those of summer rainfalls for an average of 50–100 hPa. In order to effectively deal with severe natural disasters, it is worth emphasizing the correct forecasting lead-time of about 15 days. With such a period of lead-time, reasonable amounts of preventive measures could be applied. This end has been a weak link existing in the current forecasting practice of disastrous weathers.

Once again, investigations on evolutions and problems of change reveal the incompleteness of the system of quantities and the fact that our method of digitizing information can help resolve some of the problems, which have been seen as some of the extremely difficult problems of modern science, that affect our daily lives. Since science is about challenging human wisdom, and open problems are the sources of the wisdom, we should value the appearance of unsettled problems and be good at analyzing these problems. From our investigations of unsettled problems, we

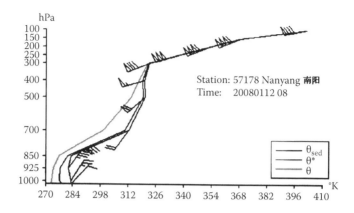

Figure 8.45 Nanyang at the 8th hour on January 12, 2008.

recognize additional pressing problems; in our attempt to resolve problems, we develop the science further.

8.5 Examples of Case Studies

8.5.1 *The Windstorm in the Bay of Bengal on May 3, 2008*

Based on what is available to us, we found that this windstorm had its characteristics: There was a well-developed powerful west wind layer of ultralow temperature, which is similar to that of black windstorms of west jet streams. Its structural characteristics do not belong in the category of traditional tropical windstorms or hurricanes. As for the forecast of this disastrous process, the clear signals could have been seen on May 1, 2008, at the latest. So, we would not say that it happened without any prior warning.

Because this windstorm at the Bay of Bengal, Burma, appeared suddenly and was "unexpected," the loss was extremely high and large scale. Many colleagues requested us to apply our methods to see what could be the problem and whether or not we could actually predict this disastrous event ahead of time. What we have to make clear is that we have very limited access to the relevant information, in particular, we could not obtain any information from within Burma other than those news reports on the severity of the disaster. Also, among the obtained information, we also missed station details and relevant times of collection. All the information available to us has been employed in the construction of Figure 8.47 through Figure 8.54. So, our discussion here is only limited to these V-3θ graphs.

Although we did not have any relevant sounding information from within Burma, based on the information of the observation stations on the west bank of Bay of Bengal and that within the Andaman Sea of the Bay, we can see the structural characteristics of this windstorm as follows:

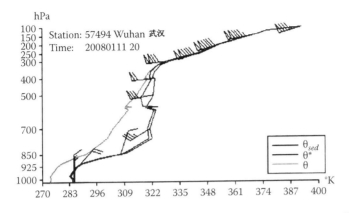

Figure 8.46   Wuhan at the 20th hour on January 11, 2008.

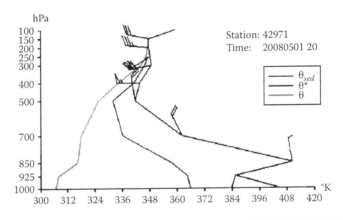

Figure 8.47   An observation station on the northwest bank of the Bay of Bengal (the 20th hour on May 1).

1. The jet stream of west winds was rampant and powerful; it was indeed a rarely seen phenomenon in all the available historical records for such a powerful jet stream to sustain in the month of May.

2. The entire atmospheric structure is analogous to the characteristics of severe sand–dust storms at mid- and high-latitudinal areas of the Northern Hemisphere, or those of black windstorms. The vertical atmospheric structure of the water vapor condition is the instable angle of greater than $70°–80°$ with the T axis.

3. The layer of ultralow temperature is thick and rampant. On May 1, areas near the windstorm had already seen a layer of ultralow temperature that was similar to or even more powerful than those of disastrous hailstone weathers or tornado conditions (Figures 8.47 and 8.48 or Figures 8.49 and 8.50). Here,

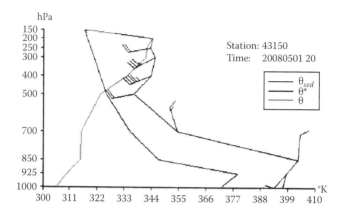

Figure 8.48 **An observation station on the west bank of the Bay of Bengal (the 20th hour on May 1).**

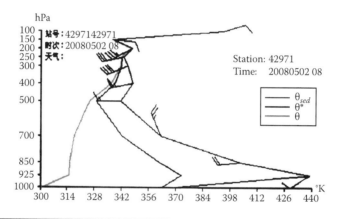

Figure 8.49 **An observation station on the northwest bank of the Bay of Bengal (the 8th hour on May 2).**

the missing data, as shown in Figure 8.50, exactly indicates the extraordinary strength of the layer of ultralow temperature.

4. What needs to be noted is that no matter whether it is a tropical windstorm or typhoon or hurricane, its atmospheric structure belongs to the tropical system, where not only is there no layer of ultralow temperature of the west wind kind, but also are the θ_{sed} and θ^* curves most likely quasi-parallel to each other and perpendicular to the T axis or forming obtuse angles with the T axis of less than 120°. However, for this specific process, the angles formed by the θ_{sed} and θ^* curves and the T axis almost reached 150° and the θ curve was always perpendicular to the T axis. In particular, in Figures 8.52 and 8.53, the degree of how steep the θ curve stood is a characteristic that appears

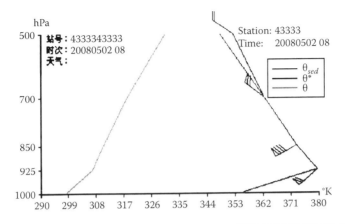

Figure 8.50   An observation station in the Andaman Sea in the Bay of Bengal (the 8th hour on May 2).

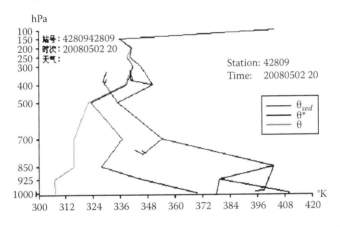

Figure 8.51   An observation station on the northwest bank of the Bay of Bengal (the 20th hour on May 2).

in high-latitudinal droughts, deserts, or plateau areas. So, this windstorm at Bay of Bengal, Burma, should be mainly about strong winds. Although Figure 8.50 also indicated strong rainfalls, the key was the disaster caused by strong winds.

8.5.1.2 The Problem of Prediction

No doubt, this windstorm had its causality and process. So, it could not be seen as suddenly appearing or being unexpected. The key questions are how we can comprehend it and why the system of modern science sees it as coming unexpectedly.

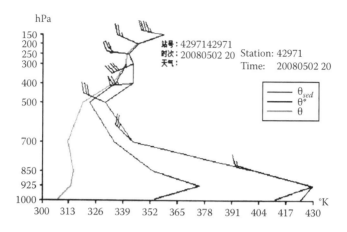

Figure 8.52   An observation station on the northwest bank of the Bay of Bengal (the 20th hour on May 2).

The fact that modern science cannot make forecasts for "suddenly appearing" disastrous weathers is no longer a problem of once or several times. The key here is that scholars and professionals have not seriously thought about what the problem is with the sudden appearance. So, in order to be able to predict the unexpected sudden appearances, it is time for us to check over what we know in modern science and how we have to think differently.

The central problem here is that this windstorm at Bay of Bengal, Burma, possessed the characteristics of black windstorms of the tornado form of west winds with powerful damaging effects. So, it could not be seen as a traditional tropical windstorm. Next, studies of weather problems cannot be limited to the atmosphere

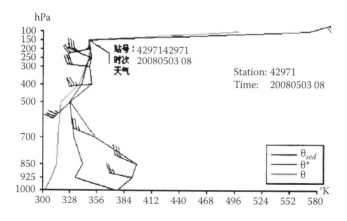

Figure 8.53   An observation station on the northwest bank of the Bay of Bengal (the 8th hour on May 3).

itself. For such a May weather, because there was even a zone of west winds sustained in the low-latitudinal area, there must be some other more essential reasons than simple explanations about the atmosphere only.

So, even if we did not acquire the least amount of necessary information, from Figure 8.47 through Figure 8.54 alone, this windstorm process could still be accurately forecasted ahead of the occurrence. This case study once again shows that the central task of the prediction science should be about the disasters caused by transitional changes. And epistemologically, we should recognize that irregular events represent information of change. Otherwise, "unexpected sudden appearance" would forever be an unsolvable scientific problem. Besides, quantities are after all only postevent formal measurements. Continuing to play the fun game of quantities can only make us unprepared and make us wait for more natural disasters to visit and suffer from the severe aftermaths.

8.5.2 Changes in Atmospheric Structures for the Sichuan Earthquake on May 12, 2008

In this section, we make our analysis of the atmospheric structures based on the meteorological information collected both before and after the major and shocking earthquake of Wenchuan, Sichuan, on May 12, 2008. No doubt, the atmospheric structure went through some clear changes from before the earthquake to after the earthquake. Once again, this analysis is different from the traditional monitoring and extrapolation kind of weather forecast. Similar analysis of earthquake information should not be done using the traditional methods of monitoring, either, where the so-called calmness before the occurrence of an earthquake is a piece of important information for the prediction of major earthquakes.

Considering the severity in damage of earthquakes and the difficulty of obtaining earthquake information, we had to make use of the knowledge about how to separate nonatmosphere factors in order to correctly forecast weather conditions using meteorological data. In other words, to correctly forecast weathers, one has to know how to successfully distinguish the misguided cases caused by nonmeteorological factors. Here, among these nonmeteorological factors are the effects of earthquakes, wars, nuclear explosions, and so forth. As for the exclusion of misguided cases in weather analysis, it involves such changes in the atmospheric structural processes as that of before, during, and after the major earthquake at Wenchuan, Sichuan, on May 12, 2008. Although our initial purpose of this work was not about the earthquake (it was about how to separate earthquake information that affects the atmospheric structure from the meteorological observations), it did involve the problem of predicting the underlying earthquake. That was how this work caught the attention of the relevant students and professionals. They enthusiastically encouraged us to record the details in order to explain the matter more clearly. All the data employed in this section were provided by Zhou Lirong, Zhang

Kui, and Hao Liping of Chengdu Bureau of Meteorology. At the same time, they also took part in the related works.

8.5.2.1 Changes in Atmospheric Structures before the Earthquake

Within the area of the earthquake, there was no radiosonde station. So, the station involved in this work is the only station located near the epicenter. As for the analysis of the information, the listed atmospheric structures all indicated that from northwest Sichuan to the south of Gansu the atmosphere was extraordinarily instable. For Ganzi, Hongyuan, and Wudu stations (Figure 8.55), the angles between the θ curve and the T axis were about 90° or beyond; the quasi-vertical portions of the θ curve reached above 500 hPa for the altitudes of about 5,500 to 7,000 meters. That structure is exactly the opposite of that of large-area, heavy snow weathers. As a piece of meteorological information, if there is not strong wind of at least a magnitude of 6 appearing within the 3-hour interval before and after the 20th hour, this kind of extraordinarily instable structure is an omen for a forthcoming major earthquake. Even nonprofessional readers can see from the serial comparisons of the graphs in Figure 8.55 the changes in the atmospheric structures. That is, from the transformation of the convective airs from the midaltitude to the lower altitude, what is meaningful for weather forecasts is that such a severe structural instability of the atmosphere did not lead to the appearance of strong winds. Speaking differently, in terms of the problem of weather, during the transformation of energy in the previous atmospheric structure, a strong wind phenomenon has to be created. If within 6 hours after the graphic structures in Figure 8.55a appeared, the phenomenon of strong wind of at least a magnitude of 4 did not appear, then it implies that the extraordinarily instable structure in the atmosphere is caused by released energies of

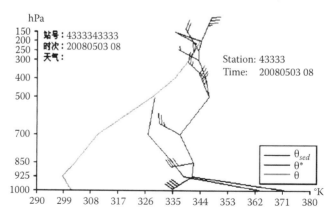

Figure 8.54 An observation station in the Andaman Sea in the Bay of Bengal (the 8th hour on May 3).

a nonmolecular form from the earth's crust so that the energy in the air is not only from the accumulation of the atmosphere itself. This end is referred to as a nonmeteorological factor.

As for the difference between earthquakes and weather phenomena, in terms of the earthquake in Wenchuan, one needs to note the following main characteristics:

1. The condition of vegetation of the Gansu and Sichuan areas south of Qing Mountain is better than that of the northwestern dry regions, which explains why in the figures the θ_{sed} curve that contains water vapor cannot lean to the left between the θ and θ^* curves (the situation of leaning to the left can only appear in the desert areas of the northwest; also note that near the ground level at Hongyuan station, due to sufficient water vapor, the left-leaning distribution appeared, where only after entering the free atmosphere, the air suddenly became dry).

2. It is exactly because the area south of Qing Mountain is not a desert that the θ curve should not show a severe instability by being almost perpendicular to the T axis.

3. Out of such a powerful, instable distribution of heat there were no strong winds appearing, which means that the severe unevenness of this process of heat distribution was not a problem of the atmosphere. That was how Zhou Lirong recognized the strange characteristic of this atmospheric structure.

4. The key method used to make the final distinction is that the heat radiation from the earth's crust is carried out in the infrared form, which explains why in the information of the 8th hour, such heat radiation could not be seen. If it appeared at the 20th hour of the evening instead of the daylight at the 8th hour, it would then be observed and judged as a piece of information of earthquake.

So, as a factor that interferes with the weather forecast, the heat distribution of the atmosphere itself can be clearly separated from the heat distribution with external influence. In particular, the θ curves of the Ganzi, Hongyuan, and Wudu stations already showed the heat distributions of quasi-90° or beyond, which were impossible to be heat distributions of the atmosphere itself.

Although for this forecast, we were first focusing on the problem of weather, because Shoucheng OuYang, one of the authors of this book, had first-hand experience of analyzing major earthquakes, he immediately realized the problem of a forthcoming major earthquake.

As for the key points for prediction, one should note the following:

1. At the time when an extraordinarily instable structure appears, pay attention to the area with better vegetation coverage to see whether or not the θ_{sed} curve leans to the left between the θ and θ^* curves and close to the θ curve;

2. At the time when such an extraordinarily instable atmospheric structure appears, there is no strong wind of at least magnitude 4 appearing within the 3 hours before and after the appearance of the instable structure;

3. Near 250 hPa there is ultralow temperature of the folding kind, or the cloud chart of the corresponding altitude shows a twist-up or crossover of cloud strips;

4. The distribution of the water vapor of the entire atmosphere is dry on the bottom and wet on the top; this end sufficiently shows the essential difference of earthquake structures from that of rainfalls, where the bottom is wet and the top is dry. In other words, for problems of meteorology, bottom dry and top wet should be a stable structure. When this meteorological stability is combined with an instable atmospheric structure, one has a problem of nonmeteorological factor(s).

8.5.2.2 Severely Instable Atmospheric Structures and Characteristics of the Atmospheric Structures along the Edge of the Epicenter

Dachuan station is located in the northeast direction of the epicenter. No doubt, its atmospheric structure also became instable and was clearly different from that shown in Figure 8.55. That is, the degree of instability was obviously smaller than those in the serial graphs in Figure 8.55.

1. The θ curve that reflects the heat distribution already leaned to the right, and whose instable angle was only about 80°. That indicates that the severity of instability is far smaller than those shown in the serial graphs in Figure 8.55.

2. The altitude of instability was relatively low, which was at least 100 hPa lower than those in the serial graphs in Figure 8.55. For Wenjiang (Chengdu) station, there appeared to be a clear right turn at 600 hPa (Figure 8.57a), indicating that the severity of the possible earthquake in Chengdu would not be less than that of the epicenter. And, after the earthquake, signs of rainfall also appeared (Figure 8.57b).

3. Although the structure of the atmosphere was also that of dry bottom and wet top, its degree was far less than those shown in the serial graphs in Figure 8.55.

4. Although the severity of instability in Figures 8.56 and 8.57 was less than those shown in Figure 8.55, they still belonged to the category of instable atmospheric structures. So, even if they were seen as pure meteorological phenomena, the corresponding strong winds still would have to appear. However, in Chengdu no wind reaching even a magnitude of 4 appeared.

5. In analysis, one should also note the problem of twist-ups at the upper altitudes. It can be analyzed by combined use of cloud charts, where layers

of ultralow temperatures correspond to the twisting clouds or crossovers of cloud strips on the cloud charts. In particular, the twisting clouds or crossovers of cloud strips of the atmosphere of bottom dry and top wet are signs of the so-called earthquake clouds. And the dry bottom and wet top of the atmosphere also signals a major nonmeteorological disaster. That explains why Chengdu downtown was located at the edge of the epicenter without experiencing major up and down vibrations except some left and right shakes.

So, our presentation here shows that recognizing pure meteorological problems and the atmospheric structural characteristics of nonmeteorological interferences touches on not only how to correctly forecast weathers but also the problem of how to predict earthquakes. This is why we believe that the current separation of meteorology and seismology suffers from at least some epistemological problems, which at a degree has delayed the healthy development of prediction science.

8.5.2.3 Aftershock Rainfalls and Weakening of Instable Energies

8.5.2.3.1 Aftershock Rainfalls

Figure 8.58 is the graph constructed using the sounding data at the 8th hour on May 13, 2008. It showed that the instable atmospheric structure was changing to a more stable structure. That signaled the start of a relatively heavy rainfall process in Chengdu City, and by our method, Chengdu Bureau of Meteorology successfully forecasted the heavy rain gush process of May 13–14.

8.5.2.3.2 Reduction in Instable Energies

During the writing of this section, Zhou Lirong sent over the sounding information of the 20th hour of May 16 for Dachuan and Wenjiang stations, which is shown in Figure 8.59. By comparing Figure 8.56 and Figure 8.59, it can be seen even just from the θ curve alone that the altitude of instability had dropped to below 700 hPa (about 3,000 meters); in particular, for Dachuan station it went beneath 850 hPa (about 150 meters). In comparison with Figure 8.56, the degree of drop of Dachuan station was more than that of Wenjiang station.

As for Hongyuan and Wudu stations, which are very close to the epicenter, we can also compare Figure 8.60 with those graphs in Figure 8.55. It can be seen that their extraordinarily instable energies had obviously lowered; and the degree of right leaning of the curves at above 500 hPa was completely analogous to those after the earthquake in the graphs in Figure 8.55. This end indicated that there would be aftershocks coming; however, the severity of the aftershocks would not be

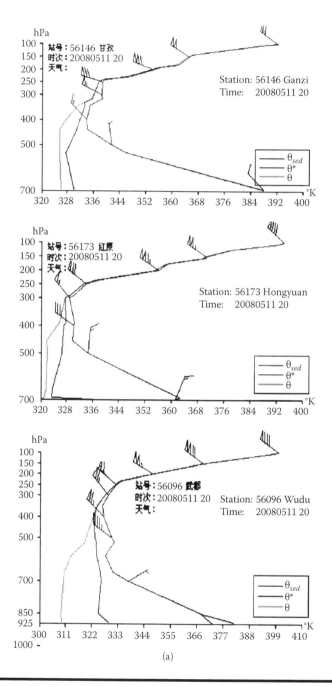

Figure 8.55 The extreme instable atomosphere structure over the northwest Sichuan Province. (a) The graphs above are for the 20th hour on May 11 before the earthquake.

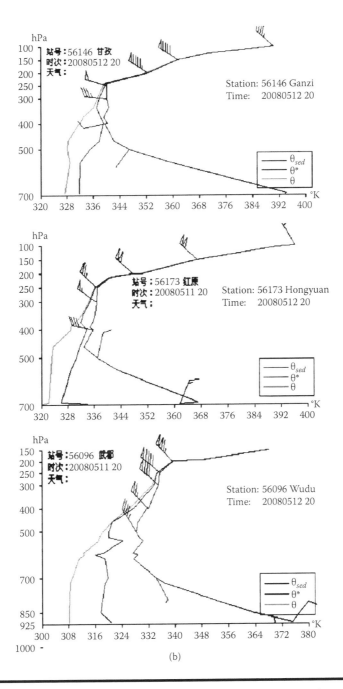

Figure 8.55 **The extreme instable atomosphere structure over the northwest Sichuan Province. (b) The graphs above are for the 20th hour on May 12 after the earthquake.**

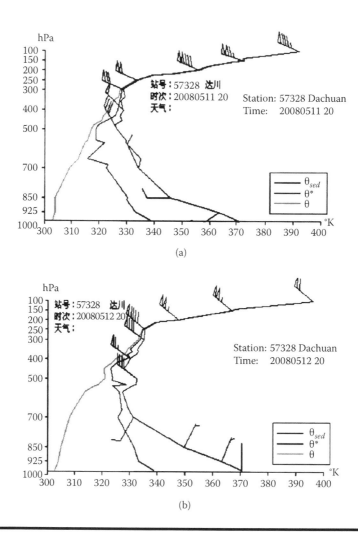

Figure 8.56 (a) The atmospheric structure of Dachuan station at the 20th hour, May 11, 2008, before the earthquake. (b) The atmospheric structure of Dachuan station at the 20th hour, May 12, 2008, after the earthquake.

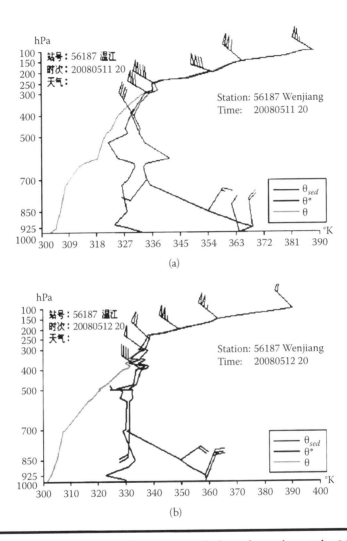

Figure 8.57 (a) The atmospheric structure of Chengdu station at the 20th hour, May 11, 2008, before the earthquake. (b) The atmospheric structure of Chengdu station at the 20th hour, May 12, 2008, after the earthquake.

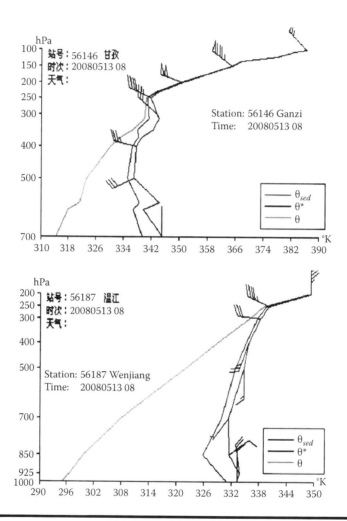

Figure 8.58 **Changes in the atmospheric structures after the earthquake.**

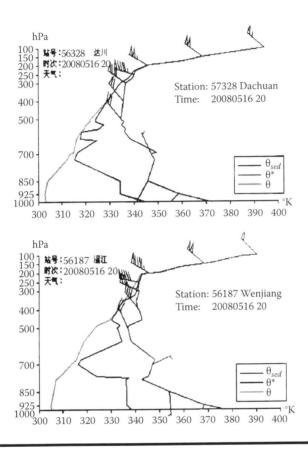

Figure 8.59 **The atmospheric distributions of heat at Dachuan and Wenjiang (Chengdu) stations.**

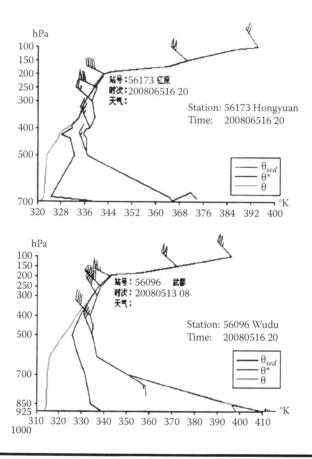

Figure 8.60 **The atmospheric distributions of heat at the Hongyuan and Wudu stations.**

more than or equivalent to that of the first wave of major quakes in the afternoon of May 12. This end was well confirmed by the actual happenings.

8.5.2.4 Some Simplified Explanations

It can be said that prediction science has substantially challenged the theoretical system of modern science; or at least after experiencing this major earthquake in Sichuan, China, we should clearly recognize the need to modify how we think and solve problems. We believe that modern science itself belongs to the design and manufacturing of durable goods and cannot be employed in the studies of the problem of prediction.

Due to the severity of this earthquake in Sichuan, we are no longer constrained by the unspoken rules of our profession. It revealed how we analyze misguided cases in weather forecasts, and we hope that our presentation here can play the role of a flying pebble to attract many beautiful and practically useful gems.

For the professions of earthquake prediction and weather forecasting, it seems to be necessary to make the most recent and updated observational information accessible to the public so that different methods can be employed to analyze these data.

For the research and practical implementation of major natural disasters, there is a need to reorganize the relevant departments and bureaus. When the division is too fine and no different opinion is allowed, the reality is that these professions are monopolized by a privileged few. It is not good for the advancement of science and technology.

Besides, we disagree with the claim that for this earthquake no prior sign appeared. Such major earthquakes may not follow the common belief that when minor shakes are felt, the major earthquake is forthcoming. The so-called calmness before each major earthquake is exactly a piece of important, advance, ominous information of the forthcoming disaster. The key is what method those involved use to analyze the relevant information.

Acknowledgments

The presentation of this chapter is based on Bergson (1963), Capitaine et al. (2003), Chen et al. (2003), Einstein (1976), Guo (1995), IERS conventions (2000), Lin (1998), OuYang (1998c, 2000, 2006a,b), OuYang and Chen (2006), OuYang and Lin (in press A), OuYang et al. (2001b, 2002a, 2005b), OuYang and Wei (2006), Prigogine (1980), Sabelli (2000), Sabelli and Kauffman (1999), Seidelmann (1982), Shannon and Weaver (1949), Weng (1984), Zeng and Lin (2006), and Zhang (2005).

Chapter 9

Digital Transformation of Automatically Recorded Information

In terms of modern science, the answer to the question of whether or not numbers can be employed to substitute for materials or events or changes in materials and events is definitely a yes, without any doubt; and in the past 300-plus years modern science has been in operation in the form of quantities. After reconsidering this question from the angle of predicting the unknown future, it is no longer that natural for us to replace materials or events by numbers. No doubt, if quantities cannot substitute for events, there must be the question of whether or not quantities can be employed to substitute for information. And, for information, people can clearly see the difference between before and after in the order of time. This difference reveals that time is embedded in the changes of materials. Evidently, without any change in materials, the concept of time would become meaningless. It has been an indisputable fact that modern science has not resolved the problem of what time is. At the very basic level, modern science has avoided the fundamental problem of change in materials and events. What is unfortunate is that the existence of this problem of modern science was not realized until after the 1960s.

Because change stands for the fundamental problem of the material world, one naturally sees the fact that physics should be a science of investigating changes in materials and the relevant processes. The development of science is encouraged by the need of solving existing unsettled problems. In other words, it can be said that the development of science comes from the calls of problems.

Due to the need for survival, the ancestors of Chinese people long ago realized the processes and periodicities in material changes (for more details, see the *Book of Changes*; Wilhalm and Baynes, 1967). However, it is exactly at this problem that modern science loses its glory without being able to provide the corresponding theories and methods. It can be said that the ancient Chinese comprehended objective things using figures and structures, known as *Yi*, or change, leading to the indistinguishable mixture of science and culture, and constituting the scientific system that characterizes the human knowledge. Since the ancient time of China, there has been a sequence of six elementary skills, known as *Li, Le, She, Yu, Shu* (first tone), and *Shu* (fourth tone). If we illustrate these six skills using the modern language, then *Li* stands for the mastery of social ethics and moralities; *Le*, the mastery of the ability of knowing; *She*, the spirit of exploration and innovation; *Yu*, the leadership ability, *Shu* (first tone), the creation of abstract thoughts and the ability to describe; and *Shu* (fourth tone), the training of formal logical reasoning. The reason formal logical reasoning is ranked the last among the six elementary skills is that numbers are not the fundamental essence of materials and events. That is why throughout the history of China, quantitative comparisons have never become a standard for scientificality.

Modern science stays at the formality of quantities without noting that quantities can betray the underlying reality and are in essence not the same as events themselves. As for the dominance of the current quantitative culture, it can be traced back to the Pythagorean philosophy that numbers are the origin of all things, which is completely different from the dominance of materials and events found in ancient Chinese culture.

9.1 Some Briefings

In our previous book, *Entering the Era of Irregularity* (OuYang et al. 2002a), we pointed out the nonevolutionality of the first push system, from which one has to naturally think about evolution science and its relevant system of methods. Evolution is about changes in materials and events so that it involves how to comprehend variable events and information of change and how information of change can be handled. No doubt, finding ways to comprehend and deal with information of change is about addressing problems of change, which directly touches on the human desire of foretelling the unknown future. So, the true meaning of prediction is about variable events and predetermining what to expect ahead of time, including specific time and location of occurrence. As for problems of constant change, one has to investigate changing events.

In Chapter 7 of our book *Entering the Era of Irregularity*, entitled "Physical Quantities and the Problem of Time," we especially emphasized Leibniz's events being the essence (of physical reality). What is implied in Leibniz's statement is that events are not quantities.

From the systemic yoyo model (Lin, 2008b), it can be naturally seen that materials exist first, before any postevent quantity, so that from quantities one will not be able to locate the causes in the studies of evolutions.

In terms of the practical benefits of scientific research, first, one should focus on how to clearly understand the essence of the events or problems instead of determining what quasi-problems are, because they cannot be addressed within the Newtonian system. Second, one should clearly comprehend the basic characteristics of the theories involved and what kinds of problems these theories can help to resolve. What is specifically important is that we need to know what problems modern science cannot resolve and why this is so. After that, we will be able to establish the relevant theories and methods. To this end, we first correctly understood why weathers have been seen as a problem of constant change, because modern science stands for theories that maintain the invariance of materials and events. Third, we questioned what causes modern science's inability to deal with problems of constant changes. To this end, we discovered that the core of the problem is the tool of the quantitative analysis that cannot handle irregular information. It is exactly because quantities cannot deal with irregular information that we realized irregular information cannot be simply ignored or disregarded, as what has been done throughout the entire spectrum of modern science. This understanding motivated us to think about how to comprehend irregular information and led to our recognition that irregular information is about changes. Because the system of quantitative analysis cannot deal with information of change, we had to invent different ways to keep and apply the information of change.

9.1.1 Quantification of Events and the Problem of Digitization

There is no denying that all the glorious achievements of science in the past 300-plus years have come from the quantification of events. However, in the area of predictions, quantifications of events have met with the challenge of irregular information. Since C. Shannon and Weaver (1949) introduced the concept of information and employed the probability to define the quantity of information and the relevant operations of information, the concept of information, due to its quantification, has been associated with uncertainties. When asked why this has been so, we found that no matter whether it is the quantitative analysis of the dynamics or statistics, irregular information cannot be dealt with. This end in essence has implied that events are not quantities, and irregular events are information of change, from which the concept of evolution science is derived, because of the nonquantifiability of events and information.

Newton's second law of motion stands for a system of quantitative average numbers. Even if we treat averages as the generality, they still cannot substitute for the specifics of events. Theories of statistics are also being limited to stable series so that peculiarities are eliminated. So, from logical reasoning, it follows that if the generality studied by using average numbers represents invariance, then the peculiarities have to represent variability. Correspondingly, irregular information is that of change. That is why in the investigation of problems of evolution, there should be not any randomness, and events and information cannot be processed or modified in the fashion of numbers. From this understanding, we introduced the method of digitization of irregular information. Because modern science did not realize that the digitization of events is itself a method of analysis of the events, it implies that to fully analyze events, one has to walk out of the realm of quantification. The reason why in China quantitative formal logic has been ranked after the figurative structures of events is that these figurative structures are originated directly from the fundamental characteristics of the events, while quantities are merely the attribute of postevent formality. However, for a long time, people have been confused about digitization and quantification. One of the direct reasons for the confusion is that events can be denoted using the symbols of quantities. In applications, one should be clear that the quantitative symbols used for representing events are only employed to distinguish the events from each other; they are not the same as the underlying quantities. It is important to note that even when events are labeled using quantitative symbols, these quantities do not in general comply with the quantitative calculus of the formal logic. This end is the essential difference between digitization and quantification. Next, what is more important is that in the past 300-plus years, people seem to have been accustomed to using quantities to substitute for events, mistakenly believing that these quantitative symbols or quantified events follow the rules of the quantitative calculus of formal logic, just the same as the regular quantities. Fortunately, after over 300 years of facing the difficulty of dealing with irregular events, we finally uncovered the problem that exists along with quantitative formal logic. In terms of quantities, we have to always be careful because quantities can be employed to compute nonexistent matters more accurately than those that actually exist. This end reveals the nonrealisticity characteristic of quantities and explains why by using quantities, we have to lose the specifics of events. No doubt, without specifics there would not be any practically meaningful accuracy.

Digitization is not the same as numericalization or quantification. Digits, numbers, and quantities are not the same concept. Here, digits should be understood as a set of terminology of the information science useful for describing specifics of events; and digitization is about a refinement of events with regard to their differences. Unfortunately, in the current world of learning, results, produced out of quantitative interpolations, averages, and so forth, are commonly seen as having the desired accuracy. For example, by using quantitative interpolation, average, or other operations, the actual sunshine in the east and the torrential rain in the west could be quantitatively refined to be nonsunshine and nonrain weather.

No doubt, the significance of our era of digitization is about walking out of the realm of quantification. So, the digitization engineering of events is not about quantitative or numerical "engineering." That explains why digital forecast is not the same as numerical forecast. Digits do not comply with the rules of operations of quantities. Similar to the symbols used in the *Book of Changes*, digits cannot be directly added to, subtracted from, multiplied by, or divided by each other. They are merely labels used to truthfully pass on the information of concern.

Digital information is a development from symbolic information to refined information; the latter is not the same as the earlier stage of symbolization of information. Although the current digitization is mainly about visual, figurative structuralization, including our method of digitization, it still did not walk out of the category of visible figures of Euclidean spaces and can only be seen as elementary digitization. Based on the meaning of this concept of elementary digitization, our method of digital (or digitized) forecasts is exactly a method of prediction of digitized events. In terms of the future development of digitization, it should also include the structuralization in non-Euclidean spaces of the dynamic senses of seeing, hearing, smelling, and touching. The general concept of digitization should reflect the structural characteristics of recognizing information from multiple angles. Its purpose is to delicately describe the process of change of events so that specifics of events can be easily recognized without pursuing the formal unit in the name of quantification. And, the essence of digitization is generalized structuralization. Although mathematical symbols for quantities can be employed to label the digitized information, these symbolic digits cannot be mistakenly seen as numerals or numerical values. In short, digitization is about accurately describing the differences of concrete events, and complies with the transformation principle of events themselves (for more details, see Section 7.3). That is why digitization is more in line with objective reality.

In particular, information of change is about the digitization of processes. Its practical and theoretical significance has gone way beyond our expectations. Its applications and development will definitely alter the system of modern science, leading to resolutions of many currently unsolvable problems. It has to be pointed out that the reason modern science did not recognize peculiar events as being events of change is that, since long ago, the world of learning has been misguided by the physics of physical quantities or by the practice of substituting events with quantities.

9.1.2 Variable Events and Time

Maybe time is a problem that could make fools out of people. Other than in our book *Entering the Era of Irregularity*, we have also studied this problem in other publications. There is no doubt that modern science has not successfully answered what time is. In our study of the *Book of Changes*, besides our specific understanding of the concept of digitization and our introduction of the relevant methods, what we also learned is that this book has clearly answered what time is, which has motivated the fundamental changes in our thoughts and systems of understanding.

The core of the question of what time is, is at the evolutionality of materials, and it stands for a signature concept about changes in events. Modern science avoids changes in materials, so that time in modern science becomes a parameter that has nothing to do with change. The *Book of Changes* points out that time is given an expression in materials and in the changes of events; time is neither a static parameter nor a static physical quantity. In other words, if we continue to use the accustomed language of modern science, even if "time" is seen as a physical quantity, it is still different from other static physical quantities, such as force, mass, and so forth, because it is embedded in the changes of events. In other words, no matter whether the underlying materials change or not, force, mass, and other physical quantities exist on top of the materials. However, *time* is different; it only exists along with the changes in the underlying events. Without change, time will not exist. So, time exists in the differences between the before and after of events. That is why *time* can no longer be treated quantitatively as a physical quantity or parametric dimension, as it has been in modern science.

Corresponding to this recognition are the time–order differences of events. These timely differences can contain the irregularities of information of change. So, the method of digitization can be employed to reveal the changeability of materials and events. No doubt, this end will provide a brand-new method of analyzing digitized, automatically recorded irregular information.

Also, modern science not only treats time as a quantity but also identifies parametric dimensions with material dimensions. Therefore, time does not occupy any material dimension and cannot be identified as a physical quantity. Instead, time can be employed to reveal characteristics of change in events, where the method of digitization can be applied to uncover the changeability of events. Although the concept of time and the applications of digitization have challenged modern science at the fundamental levels in terms of concepts and methods, the essence is about solving practical problems. Because time is given an expression in the changes of events, evolution science has to answer what time is. Change is the true characteristic of the objective world and is the common foundation for the development of natural and social sciences. One of the important reasons why social sciences have departed from natural sciences is that modern science did not resolve the problem of time and changes in events. It is our belief, as well shown in Lin (2008), that if natural sciences successfully resolve the problem of time and changes in events, then the investigations of natural and social sciences can be unified under the same roof of methodology. To this end, in this chapter we will use practical case studies to illustrate how the digitization of automatically recorded information can be beneficially applied.

9.2 Digitization of Automatically Recorded Information of Disastrous Weathers

Currently, various professional fields have their own automatically recorded "oscillating" information. Various electronic gauges and equipment used in different

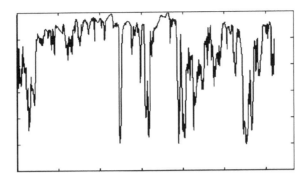

Figure 9.1   Irregular information of the oscillation kind.

fields have their respective ways of collecting data. For instance, from electrocardiograms of medicine to automatic weather stations of meteorology, a great many different kinds of equipment designed for the desired observations and purposes are in existence. The collected data can generally be plotted as the commonly known irregular waves of the oscillation kind (Figure 9.1). Due to habit, they are also referred to as mixed waves, such as the concepts of mixed waves of terrains and objects, mixed ocean waves, and so forth.

Most commonly applied methods of traditional quantitative analysis are about selecting the effective pieces of waves by using filtering, averaging, or other methods so that the irregularities contained in mixed waves are eliminated. However, when seen from the angle of evolution science, the irregular data shown in Figure 9.1 stand for those data of some irregular events. From the epistemological point of view of evolutions of events, it follows that data of irregular events cannot be seen as meaningless. The irregularity of the corresponding information cannot be artificially eliminated, or "cooked" using such methods as the quantitative averaging, or investigated for its overall characteristics. As the analysis of evolutionary events, one has to uncover the significance and effects of the specifics of the events.

Irregular information of the oscillating kind, such as that shown in Figure 9.1, can be seen in all human endeavors. Conventionally, the horizontal axis in Figure 9.1 stands for the traditional concept of time, and the vertical axis, the physical quantity or the data of a specific event of interest.

The central problem for now is how we can understand and make use of this kind of information. If the observation errors are within the allowed range, then the data, as shown in Figure 9.1, will stand for some kind of realization of changes in the underlying objective event. So, when people use smoothing, filtering, or other methods to eliminate the irregular information as in Figure 9.1, it is equivalent to artificially deleting the involved changing event. We need to note the difference between the designing and manufacturing of durable goods and analysis of changes in materials and events. For the latter, we have to analyze the irregular information of change. So, we have to directly face irregular information that cannot be dealt with by using the quantitative analysis system of modern science so that we will be able to establish methods on how to process irregular information and reveal the physical significance of it. To this end, the recognition that time is given an expression in the changes of events, we conclude that time cannot be treated as a parametric dimension like other physical quantities. In other words, changes in events have to be reflected in changes of information. When the events are invariant, the corresponding information has to be static. Second, when information of events is shown by numerical quantities, these data in fact are not the pure quantities of the traditional sense.

Using the quantitative calculus of formal logic cannot truthfully reveal changes in these data; and what is shown in the frequent changes existing in the data embodies the phenomenon of frequent transitionality of the events. Considering the distributional unevenness of changes in the events, distributional gradient has to exist. So, we can make use of the azimuth of the gradient as a variable and employ the mathematical tool of reciprocating functions to represent the changes in the information of events without "cooking" the information of change using the quantitative calculus of formal logic. By doing so, we can transform any oscillating time series data into digitized figures, reflecting the changes in the underlying events. At the same time, the realistic information is not injured, while the changeability of events is revealed, where the invariance of static events is also kept. In other words, the purpose of this method is to make use of irregular information as much as possible so that the underlying changeability can be realistically reflected without damaging the invariance of static events. So, methodologically, we depart from modern science where only Plato's static figures can be utilized and enter the field of change in events and materials of evolution science.

In order to confirm the practical validity of this method of digital transformation of irregular information, we have collaborated with many colleagues from a wide range of scientific disciplines and organized our students in a huge amount of experimental works. What is interesting is that, as of this writing, in all kinds of experiments, as long as the observational information is realistic, our results have shown the general laws of change.

9.2.1 Digital Transformation of Automatically Recorded Information on Locations of Torrential Rains

In terms of the sounding data, the automatically recorded meteorological observations can be considered a piece of original information without human interference other than some potential machine errors. However, for a long, long time, these automatic records have only been treated as observations without being practically applied. According to the nonparticlized principle of evolution of materials that under heaven there is nothing but the Tao of images (OuYang et al., 2002a), irregularities are the determinism of nature instead of informational uncertainties, as introduced by Shannon and others (Shannon and Weaver, 1949). Since information is about an event that has already occurred, the information has to be deterministic (OuYang and Lin, 2006b). Even based on the currently available knowledge, all materials' movements in the universe are originated from the uneven distribution of "heat," causing Brownian (irregular) motions of the stirring type. Therefore, Lao Tzu's "extraordinary Tao" is the law of the universe, which must touch on how to deal with irregular, small-probability information. Since thermodynamics was established after particle dynamics, while inheriting the form of particle dynamics, this new theory was named thermodynamics. Later, this new theory became the "objective" foundation for the concept of randomness and the method of statistics to form.

The automatic records reveal precisely the naturally existing irregularities. In order to show the practical significance of these irregularities in the prediction of evolution, in this chapter, we employ the V-3θ graphs of the blown-up system to pinpoint the time and amount of precipitations. Then, by using structural transformations of these automatic records, we predict the location of precipitation. By doing so, we not only utilize the resource of meteorological information but also materialize the goal of understanding irregular information. Since the automatic records are quantitative time series information, they inherit the tradition of treating time t as the parameter of a physical quantity. The timely orderliness of the corresponding barometric pressures, temperatures, humidity, and wind directions and speeds is shown as irregular, complex "wave motions." (It should be clear that irregular disturbances are not the same as wave motions.) See Figure 9.2, which is the automatically recorded humidity information. Obviously, similar time series information can be observed in other areas of learning, such as earthquakes, aging of products, medicines, and so forth. Since the irregular disturbance shown in Figure 9.2 is in general extremely difficult to recognize to the ordinary eyes, it is rarely directly applied for any practical purposes. To make use of such information on the basis of randomness, a method of regularization is employed, causing injuries to the data or quantitatively fabricating the information, leading to erroneous analysis of the underlying evolution. Also, studying irregularity of an event is exactly the weakness of the current method of quantitative analysis. To this end, the present, fashionable quantitative analysis damages or fabricates events by using mostly "filtering" or "smoothing."

For problems of evolution, we do not treat time as a physical quantity because it does not occupy any material dimension and it reflects changes in materials'

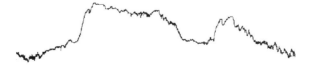

Figure 9.2 Automatically recorded time series information.

structures. By transforming quantitative time series information into structural information in the positional phase space, we can beautifully convert the difficulty of recognizing patterns in time series quantitative information into easy recognizability of positional structures. This method has been successfully employed in the study of aging and pathological analysis in medicine and the nonlinearity of dynamics, leading to some clear outcomes (Sabelli, 2000; OuYang et al., 2002). In this section, we will only use problems of meteorology as examples to illustrate the easy recognizability of structural transformations, proving that problems of meteorology are those of evolution science. Our work shows how to correctly understand changes in important meteorological factors and related procedures of operation.

Before making an official weather forecast, one could first predict where the disastrous weather phenomena would occur by using the digitized V-3θ graphs of the blown-up system; for more details, see Chapter 8. With this method, we have achieved such small-scale predictions as those of the area sizes between 200~300 km^2.

Forecasting local torrential rains for such a small area as 200 to 300 km^2 has been extremely difficult, refined work. For the longest time, by employing the conventional methods, evaluations for local weather stations to do such small-area forecasting have mostly been about 40 points out of a total of 100 points possible. Very few forecasters can or dare to make predictions for such major torrential rains of above 100 mm. However, by utilizing the structural method of our V-3θ graphs, where V stands for a vector, to forecast torrential rains, the frontline meteorologists have achieved over 70~80% accuracy rate, where cases with erroneous information are included. To this end, we will summarize the V-3θ graphic characteristics of torrential rain weathers.

The V-3θ graph analysis mainly includes the atmospheric wind direction, the vertical structure of the wind speeds, water-vapor distribution, and the intensity of the ultralow temperature. It is a structural analysis of the relative intensities or strengths of this special, key information. The V-3θ graphic characteristics for the process of torrential rain weather are the following:

1. For such disastrous weathers of major precipitations as at least torrential rains or heavy rain gush, the layer of water vapor is relatively thick, and an anticyclone exists within 5 to 10 or more longitudinal and latitudinal distances to the east of the observation station, where the anticyclone contains a horizontal circulation field helping to transport water vapor. The θ_{sed} and θ^* curves are quasi-parallel, located near each other within 8–10K, and perpendicular to the T-axis, consisting of southwestern, southern, or southeastern winds of such a thick layer as from the ground level to 500, 400, even 300 hPa. The θ curve is in the shape of a bow facing left, accompanied with irregular zigzag turns, which represent the development of convective clouds (see Chapter 8 for more details).
2. An overall clockwise rolling current.

3. For the precipitation of at least a heavy rain gush, a thin layer of ultralow temperature exists near the top of the troposphere.

Here, what needs to be noted is that the basic characteristics of torrential rains are that the quasi-parallel bottoms of the θ_{sed}, θ^* curves are perpendicular to the T axis.

9.2.1.1 Case Studies of Informational Digitization

During July 30–31, 2002, Chengdu experienced a historically rarely seen regional torrential rain process. By employing the sounding data of Chengdu station and those nearby within 5 to 10 longitudinal and latitudinal distances of Chengdu station collected on July 29, 2002, we analyzed the structure of the forthcoming torrential rain and heavy rain gush for Chengdu area and made the forecast for the specific time and location of the disastrous precipitation. In Figure 9.3a,b the V-3θ graph at the 8th hour and the 20th hour, respectively, on July 29, 2002. Based on the structural irregularity and the direction of the rolling current, we successfully made the forecast for the forthcoming regional torrential rain and heavy rain gush.

9.2.1.2 Digital Transformation of the Information of Rainfall Locations

For the current practice of predicting rainfall locations, the outcome is checked by using the rain gauge at one of the observation stations within the region of 200 km², where the diameter of the rain gauge is 20 cm and no rainfalls outside the gauge are considered. So, the difficulty faced by the front-line forecasters can be felt vividly. Based on our concept of digitization, the accuracy of predictions is determined by the frequency of observation, the distribution density of the observation stations, and the accuracy of the irregular information. No doubt, there is a need to constantly improve the technology of observation and the theory on how to employ irregular information.

According to the principle of evolution, irregular information stands for a problem of transitionality. So, let us employ reciprocating functions as our method to deal with the phase space structure of irregular information, where the positional distributions of the variables are used as our phase space. Specifically, we take the physical quantity of the transitionally significant factors as the positional angle variable and then use their sin$A$ and cos$A$ as the positional phase space, where A can be any physical quantity that occupies a material dimension. Here, time does not occupy any material dimension and cannot be used as a physical quantity. For any physical quantity that occupies a material dimension, it must have a spatially uneven distribution so that the quantity possesses a gradient. With its gradient, the variable must have its directionality and positionality. So, we can make use of the

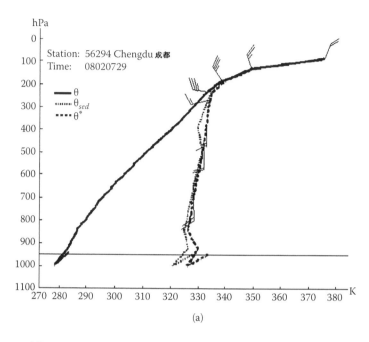

(a)

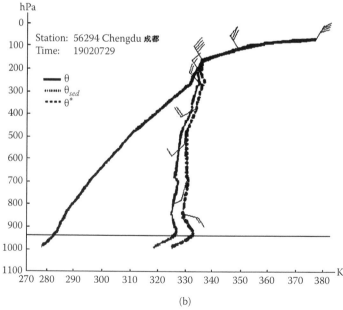

(b)

Figure 9.3 (a) Chengdu at the 8th hour on July 29. (b) Chengdu at the 20th hour on July 29.

structural characteristics of the frequency of changes of the physical quantity to distinguish between process and rainfall location. Also, the constructed structural graphs contain both the specifics of information and visibility, constituting an easy-to-use tool where the messier the graph of the irregular information the better.

The conventional method uses time as the horizontal axis to index the changes of the physical quantity over time. However, the outcome is mostly an unrecognizable complexity. Let us use barometric pressure as an example. Because of the 1,000 hPa base value, changes in barometric pressure when shown as a time series graph can be difficult to recognize. If we transform the conventional time series into a distribution in a phase space, we can apply $\sin A$ as the vertical axis and $\cos A$ the horizontal axis, where $A = V$ (the wind direction), $|V|$ (the wind speed), Td (the humidity), T (the temperature), or P (the barometric pressure). By doing so, at the same time when we show that time does not occupy any material dimension, we can apply the density of reciprocating line segments in our graph to express the irregularity of our information due to the reciprocating nature of these trigonometric functions. By taking the departures in the barometric pressure readings collected before, during, and after a precipitation, we can easily spot structural differences on the graphs. Since the blown-up principle requires the information of inversed order structure, we construct the corresponding spatial graphs for the significant weather factors respectively in the order of wind direction, wind speed, humidity, temperature, and barometric pressure. Based on the automatic records collected at each 15-minute time interval, each of the significant factors shows definite recognizability, with the best results in wind speed and humidity. Considering the length of this presentation, we will only list the processes' structural distribution graphs for wind speed and humidity. Here, what deserves our attention is that irregular information is not about complexity, and the cause of complexity cannot be addressed scientifically by using the concept of complexity. The key in complex information is its peculiarity, such as in our case, the specific structures of the ultralow temperature, wind, and distribution changes of water vapor. In the following, we will respectively provide the digitization of wind speed and humidity.

9.2.1.3 Digitalized Comparison between the Humidity Evolutional Processes of the Chengdu and Longquan Stations

The digitization of the humidity factor reveals the convenient human visual ability to distinguish objects and patterns, involving the chaoticity existing before the occurrence of a special event, the quasi-invariance existing during the occurrence of the event, and the regularity of the mixed waves of terrains and objects. So, this method alters how we learn and reminds us why irregular information could not be disregarded, leading to the nonrandomness of information of change. Due to the differences in the functions and properties of various factors, the sampling periods for these factors are varied. For automatic weather stations, the time interval of sampling is set at 10 seconds per collection. According to our experiments, in order to make the patterns recognizable by the ordinary Joe, the sampling time interval for the humidity factor should not be smaller than 15 minutes; for wind direction and speed, it should not be smaller than 1 minute.

The overall structural characteristics for the humidity evolution of a precipitation process are the following:

■ Before the precipitation, the phase space is closed with frequent reciprocation.
■ During the precipitation, the smaller the humidity, the fewer reciprocations and the heavier the rainfall.

From Figure 9.4c,d,e, it can be seen that as the precipitation ends, the number of reciprocating line segments increases gradually, and the line segments eventually close up again.

From a comparison between the phase space graphs of Chengdu and Longquan, it can be seen that when Figure 9.4b1,b2 are compared with Figure 9.4a1,a2, the number of reciprocating segments is decreasing; where on July 29, Longquan suffered more of a decrease (Figure 9.4b2) than Chengdu (Figure 9.4b1). That is, Chengdu is "messier" than Longquan. That implies that the amount of rainfall in Chengdu would be more than that of Longquan. The specific location of the heavier rainfall would be in Chengdu instead of Longquan. And this prediction was later confirmed by the actual happenings. That is, this method can show the process of change as well as differences in rainfall locations.

What needs to be clear is that the acute changes in wind speed before rainfalls agree well with the experience of weather forecasters. However, as for the major changes in humidity at the ground level before precipitation and the phenomenon of stabilizing changes in humidity during the precipitation process, the transitional theory of meteorology does not have any exploration. Or, it can be said that the traditional theory does not at all know about the process of change in humidity. This end sufficiently shows that if it were not for our serious investigation of irregular information, this phenomenon would not have been discovered. That is also the reason why we time and again mentioned the concept of physics of processes.

Since the observation station at Longquan did not have any record on the relevant wind direction and speed, we cannot produce the related process graphs.

Our previous digitization analysis of basic factors indicates that the digitization of information possesses not only the differences easily distinguishable for processes, but also the separability between spatial regions. What is important is that these graphs reveal the following:

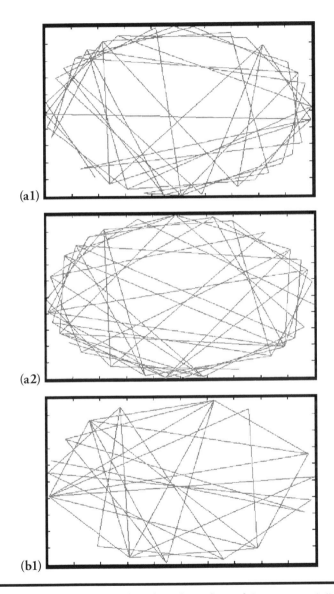

Figure 9.4 The digitized humidity for Chengdu and Longquan. (a1) Chengdu humidity, July 28. Precipitation: none. (a2) Longquan humidity, July 28. Precipitation: none. (b1) Chengdu humidity, July 29. Precipitation: 12.1 mm.

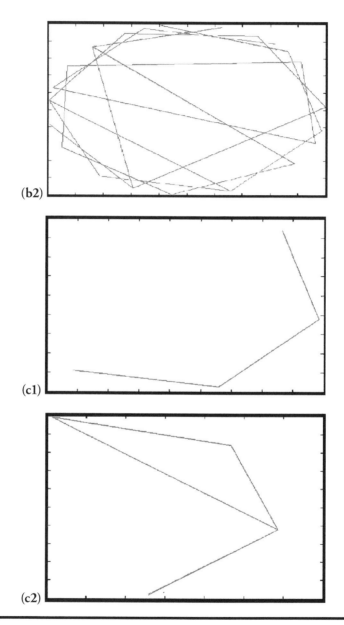

Figure 9.4 The digitized humidity for Chengdu and Longquan. (b2) Longquan humidity, July 29. Precipitation: 21.5 mm. **(c1)** Chengdu humidity, July 30. Precipitation: 127.7 mm. **(c2)** Longquan humidity, July 30. Precipitation: 14.2 mm.

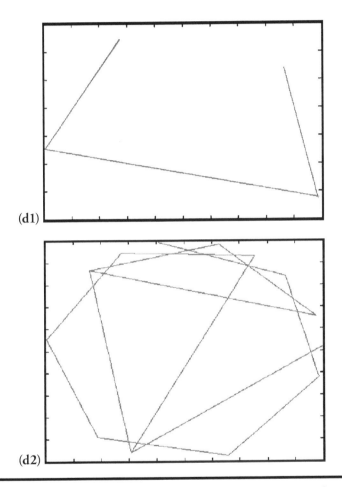

Figure 9.4 The digitized humidity for Chengdu and Longquan. (d1) Chengdu humidity, July 31. Precipitation: 63.6 mm. (d2) Longquan humidity, July 31. Precipitation: 25.7 mm.

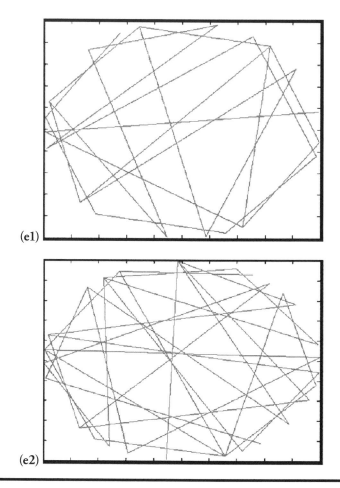

Figure 9.4   The digitized humidity for Chengdu and Longquan. (e1) Chengdu humidity, August 1. Precipitation: none. (e2) Longquan humidity, August 1. Precipitation: 0.3 mm.

1. Meteorological science is a part of the process physics of evolution. So, we cannot continue to treat it as a theory of modern science, or a branch of inertial system physics.

2. The method of digitization provides a system of methodology for evolution science. It shows that irregular, small-probability information is not a problem of uncertainty without a cause and that this information cannot be ignored or eliminated in the name of stability and smoothness of such methods as filtering, smoothing, and so forth. In terms of evolution science, both smoothing and filtering limit the underlying evolution by fabricating false information. In other words, studies of changing events using the quantitative system of modern science are an epistemological mistake.

3. "When chaos appears, change is nearby" or "the extraordinary Tao" is not only a philosophical principle, but also an evolutionary theory of process physics. And, each irregular "chaos" is about the key peculiarity instead of the conventional large probability. That is, every chaos has its law or regularity of chaos. And, the regularity is about the physical mechanism of the peculiarity instead of the stochastic quantities.

In short, the method of digitization also reveals epistemologically that the quantitative analysis of the inertial system established in the past 300-plus years essentially is quantitatively invariant without touching on problems of evolution. Its realizable applications can only be in the area of estimation useful for the design of durable products. It cannot play the role of basic science of evolution. In terms of methodology, the application of structural transformations of time series information is not limited to the refined forecasting of specific locations of precipitation. It can be applied to any evolutionary time series information collected in such areas as engineering and aging of products.

9.2.2 Digitalized Information and Applications in the Forecasting of Thunderstorms

The certainty and uncertainty of information have been a focal point of debate in the world of learning for nearly 100 years. During the latter half of the 20th century, there once appeared a so-called chaos theory, leading to the debatable concept of probabilistic universe. Our work presented in this book indicates that instead of endlessly arguing about determinacy or indeterminacy, it would be more practically beneficial for us to rigorously study and analyze what irregular information is and how to make use of it. To this end, we employ the previously described method to specifically analyze the forecast of thunderstorm weathers, a world-class difficult problem as seen by modern science. Our work can be practically applied in commercial or military flight operations to produce tangible economic results.

Convective weathers are one of the greatest hidden dangers for the safety of air traffic. The disconnected but nonstop thunderstorm weather that occurred during July 19–21, 2004, in a certain undisclosed region of Sichuan province, forced the airport of the area to cancel all its scheduled flights for over 20 hours. In particular, for all commercial or military airports, where either no radiosonde station or no radar system exists, forecasting such dangerous weathers is very difficult and needs to be resolved urgently.

Before digitizing the automatically recorded information, one should employ the V-3θ graphs to predict regional (for about 200 to 300 km^2) disastrous weathers. Since radiosonde stations are constructed using the large-scale space distribution of meteorology, the time and space distances are 12 hours and 200 to 300 km, respectively. So, the current radiosonde station network has missed out most of the convective weathers of less than 100 km and shorter than 6 hours. However, the ultralow temperature phenomenon, which causes strong convective weathers, can reach such a range as the mesoscale (500 to 1000 km²). So, we can make use of V-3θ graphs to recognize where and when the phenomenon of ultralow temperature exists before a thunderstorm actually occurs. Without an ultralow temperature, convective weather would not appear (OuYang et al., 2000). Even as a prediction index, ultralow temperature is a piece of very important advanced information that becomes available ahead of the convective weather event.

The real-life case, presented here, was the actual thunderstorm convective weather that occurred over an undisclosed training airport in Sichuan province. The airport is about 20 km from the radiosonde station in Chengdu City and belongs to the region within 150 km that the sounding data cover. So, the atmospheric structure over this region can or has to be analyzed by using the V-3θ graphs of the Chengdu radiosonde station.

9.2.2.1 The V-3θ Graphic Characteristics of Strong Convective Thunderstorm Weathers

According to the principle of blown-ups (OuYang et al., 2000) and real-life case studies, the V-3θ graph's structural characteristics for before the appearance of a convection can be summarized as follows:

1. The distribution of the θ curve unstably tilts to the left with ultralow temperature exits (Figures 9.5 and 9.6).
2. Both θ_{sed} and θ^* curves are strongly instable while leaning to the left, and they are close to each other, indicating the existence of water vapor layers; or wet at lower levels and dry in upper levels; or wet in the middle with upper and lower levels dry. That is, they most likely form the instable structure of a bee's waist.
3. The troposphere is an overall clockwise rolling current.

For our comparison purpose, Figure 9.5 shows that at the 8th hour on July 18, 2004, no ultralow temperature existed in the V-3θ graph at Chengdu station, and the troposphere was taken up by a counterclockwise rolling current made up of northwest winds. Until the 8th hour of July 19, no thunderstorm appeared. This end indicates the importance of ultralow temperatures. However, in the V-3θ graph of the 20th hour on the 18th (Figure 9.6), an ultralow temperature and instable structures had already appeared in the V-3θ graph. At 700 hPa, southwest winds appeared; and the amount of water vapor also started to increase at 500 hPa. That is, signs of changes in the atmospheric structure had already appeared except that near ground level, the winds still blew from the northwest direction without forming an overall clockwise rolling current.

Figure 9.7 shows the typical structure for convective weathers, where the upper space contains an ultralow temperature; the lower level atmosphere is instable; and at the middle levels, the θ_{sed} and θ^* curves are close to each other (between 700 to

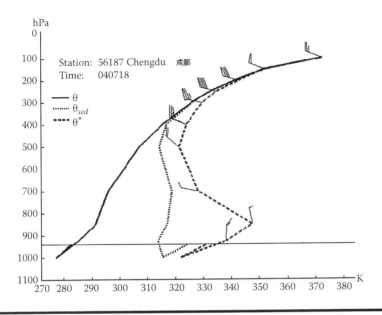

Figure 9.5   Chengdu at the 8th hour on July 18, 2004.

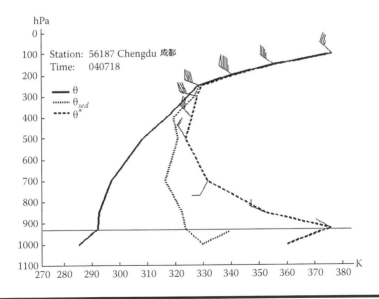

Figure 9.6   Chengdu at the 20th hour on July 18, 2004.

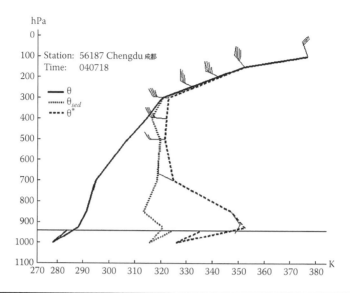

Figure 9.7   The V-3θ graph of Chengdu at the 8th hour on July 19, 2004.

500 hPa), forming the structure of a bee's waist. This is exactly what is necessary for a strong convective weather to form — instable atmosphere and existence of a water vapor layer. What is different about the structure of torrential rains is that the condition of abundant water vapor is not needed here. As the key for the purpose of forecasting, we need to have southbound winds in the bottom atmospheric layer. For our case in point here, there were southeast winds at 850 hPa. When making forecast, we need to pay attention to the wind source of the existing southbound winds so that we have to look at the wind directions at the upper-reach stations so that a clockwise rolling current is formed.

Since Figure 9.7 had contained all of what we need, we could forecast that in the greater Chengdu city area, thunderstorm weather would appear. The actual happening was that at the said airport, a thunderstorm occurred and lasted for nearly 6 hours. As for predicting the actual location of the thunderstorm, using radiosonde stations only cannot help much. That is why we introduced the method of digital transformation of automatically recorded irregular information. However, for the overall evolution analysis of the atmospheric structure, we cannot walk away from the V-3θ graphs of Chengdu's radiosonde station.

Figure 9.8 describes the V-3θ graphic characteristics of Chengdu station at the 20th hour on July 19, 2004. In general, for a strong convective thunderstorm weather to finish, the atmospheric clockwise rolling current structure has to be changed into a counterclockwise rolling current structure. However, after the nearly-3-hour thunderstorm on July 19 temporarily stopped, the instable structure in the atmosphere, as shown in Figure 9.8, still existed and became more instable than that at the 8th hour earlier. The ultralow temperature did not disappear, and near the

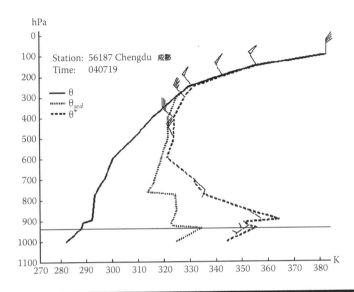

Figure 9.8   The V-3θ graph of Chengdu at the 20th hour on July 19, 2004.

ground level, southwest winds still existed. So, the true and more intensive thunderstorm was still in the making. To begin with, those who learn how to analyze convective weathers, the situation of missing data should be noticed in Figure 9.10. In particular, in Figure 9.10, the data above 500 hPa appeared to be incorrect. It can be said that such situation occurs quite often with convective weathers. The reason such misguided situations occur is that the ultralow temperature (ionized states) located at the top layer of the troposphere interferes with the telecommunications signals. One method of remedy is to analyze whether the atmospheric structure at the mid and lower levels is becoming stable. The bottom layer of the atmosphere in Figure 9.10 is instable, implying that a truly severe thunderstorm is in its development. From Figure 9.9, the V-3θ graph at the 8th hour on July 20, 2004, it can be seen that the layer of the ultralow temperature became thicker and stronger. The amount of water vapor near the ground level was increased and θ_{sed} and θ^* curves zigzagged many times, indicating the development of convective clouds. And the upper levels of the atmosphere became dryer. So, we predicted that on July 20, a sustained thunderstorm would occur. The actual happening was that the undisclosed area suffered from a 6-hour sustained thunderstorm on the specific day.

Although two thunderstorms appeared on July 19 and 20, 2004, the 8th hour V-3θ graph of July 21 (Figure 9.11) still showed the existence of an ultralow temperature and a clockwise rolling current. Combining the atmospheric structures of nearby stations — if these structures also contain ultralow temperatures or instable clockwise rolling currents, then we can be sure that the thunderstorm process at the station in question has not yet finished. In terms of our case study, Figure 9.11 was such a situation. So, the ultralow temperature over Chengdu at the 8th hour on

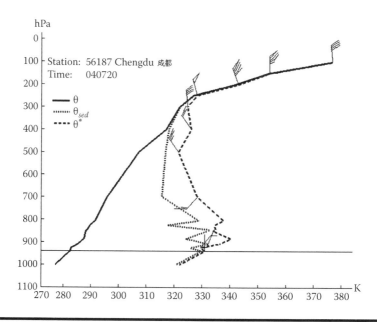

Figure 9.9   The V-3θ graph of Chengdu at the 8th hour on July 20, 2004.

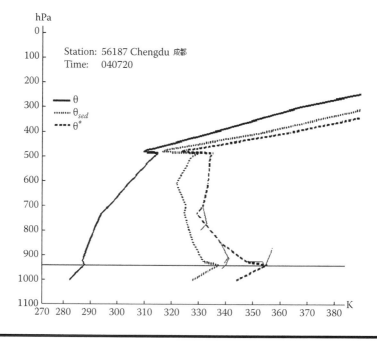

Figure 9.10   The V-3θ graph of Chengdu at the 20th hour on July 20, 2004.

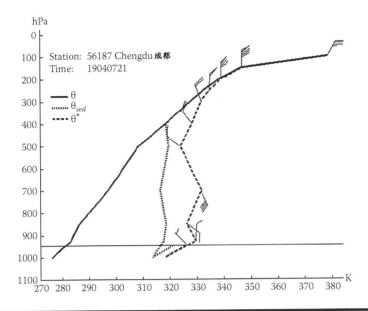

Figure 9.11   The V-3θ graph of Chengdu at the 8th hour on July 21, 2004.

the 21st day of July was still a sign of convective weather. Hence, July 21 would still experience another process of thunderstorm.

However, in the analysis, we should notice that the atmospheric structure at this hour had already gone through fundamental changes. That is, near the ground level, north winds had appeared, and the temperature dropped more than 20°. The θ curve and the $θ_{sed}$ and $θ^*$ curves near the ground level had tilted to the right. Although 850 to 700 hPa were still taken by southbound winds and above 500 hPa northbound winds, at levels beneath 850 hPa there had been a counterclockwise rolling current. That is, the overall clockwise rolling current had been broken. So, even if there appeared a thunderstorm, it would not be severe. The actual happening was a half-hour of thunders. Although each radiosonde station can only present the overall situation of the region it covers, it can also reveal the background of local thunderstorms. However, as for specific locations of the appearance of thunderstorms, it will be difficult to predict by only relying on the radiosonde station, because after all, it covers a huge area in terms of time and space.

9.2.2.2 Digital Transformation of Automatically Recorded Time Series Information of Humidity for Predicting Rainfall Locations

In this section, let us look at the digital transformation of the automatically recorded information collected at the training airport. The underlying method is the same as discussed earlier (for more details see the analysis of torrential rains).

Figure 9.12 provides the phase space graphs of humidity for the thunderstorm process on July 21. What needs to be explained is that this thunderstorm was the last part of the entire weather process. It lasted for only half an hour, from the 16th hour to half past the hour. However, on the phase space graphs, we can still see evident reflections. To this end, the traditional methods are truly helpless for such a short-lived thunderstorm, which has been seen as a world-class difficult problem. From Figure 9.12, it can be seen very clearly that before the thunderstorm, the phase space graphs of humidity were characterized by frequent reciprocations. During the thunderstorm, it became stable jumps. And after the thunderstorm, the phase space structure recovered back to a reciprocating pattern but different from that before the thunderstorm. It can be said that the method of digitization can differentiate the changes appearing throughout the entire process of evolution and that without any dispute, people without any background in meteorology can recognize these differences in structures.

In terms of the specific thunderstorm of our case study at the training airport, we can foretell, based on the phase space graphs in Figure 9.12, that in the afternoon of July 21, the entire process of the thunderstorm would finish.

What we need to point out is that no matter whether it is from the V-3θ graphs or the phase space graphs of automatic records, it can be seen that weather changes are a problem of evolution, that events are not quantities, and that the phenomenon of thunderstorms cannot be predicted by using the method of extrapolation. The reason why in this section we provided the phase space graphs for 6 hours and 12 hours ahead of the appearance of thunderstorms is to show the time-effectiveness of advanced predictions. Our results illustrate that for the so-called intensively strong convective weathers, it is possible for us to make 12-hour advance forecasts, which is way beyond the currently believed 6-hour limit of predictability. This discovery is very significant for commercial and military flight operations. It not only involves quite a simple method but also taps deeper into the readily available information resource.

9.2.3 Digital Forecasts for the Locations of Fogs and Hazes

In Chapter 8 we discussed the problem of forecasting regional low visibilities. In principle, the area covered by a low-visibility weather condition is much greater than that of convective hailstones, heavy rainfalls, and other phenomena. Generally, the ordinary method of forecasting regional weather phenomena can be employed. However, along with the development of economies and the increasing density of the transportation networks, the phenomenon of low visibility itself contains unevenness. And, fogs and hazes are different. Although they are all weather phenomena of low visibility, there are differences in their properties and ways of affecting human lives.

As a phenomenon of changing weather, thick or heavy fogs mainly appear along with the transitional changes in the weather after rainfalls or snowfalls. Of course,

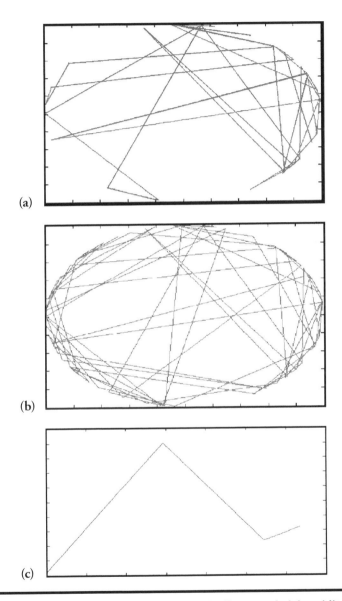

Figure 9.12 The digitization of the automatically recorded humidity on July 21 for the undisclosed training airport. (a) Before the process (info of 6 hours). (b) Before the process (info of 12 hours). (c) During the process on the 21st day.

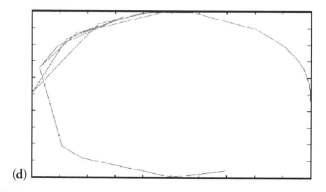

Figure 9.12 The digitization of the automatically recorded humidity on July 21 for the undisclosed training airport. (d) After the process of the 21st day.

there are also fogs before rainfalls or during rainfalls. In particular, for mountainous areas, one needs to pay attention to the fogs of clouds (in fact, they are clouds instead of fogs). Even for the forecasting method of weather factors, the currently widely employed method of weather maps of the pressure systems (or the potential altitude systems) (as a matter of fact, the current numerical forecasting method is also developed on pressure systems) lags behind the weather phenomena (or factors) to be predicted. In the current daily practice of these methods, this problem of lagging behind has been experienced time and again. However, it has not caught the attention of the theoretical and epistemological researchers. Since the 1950s and 1960s when we first learned how front-line forecasters were frustrated with this problem, we once systematically investigated this problem and found that the reason for the holdup is that uneven distribution of heat causes changes in the atmospheric density, then, and only after then, the barometric pressures start to change. Second, in the currently available observation data of the meteorological science, the actual observations are "modified" by using static approximations, where the atmospheric density is assumed to be constant, so that the variable pressure is limited to be invariant. That is why the "pressure systems" always lag behind the actual weather phenomena. In particular, for sudden weather conditions, we have often seen the appearance of the "pressure systems" after the precipitation. That empirical fact motivated us to employ the inverse information order of the key factors in our forecast method of digitization without even involving the pressure factor. No doubt, the rising movement of a clockwise rolling current has to cause a low-value pressure system; and the sinking of a counterclockwise rolling current is accompanied with a high-value pressure system. That explains why the effects of rolling currents can be ahead of the relevant pressure systems and the weather phenomena. This end illustrates why the method of digitization can produce true weather forecasts without extrapolating the "pressure systems." Specific to the forecasts of fogs, this method provides the needed features of unevenness in punctuality, changeability, and regionality.

No doubt, for the peculiar information of change, the traditional system of quantitative analysis has faced difficulty. This system of analysis is developed on the basis of modern science. In principle, it is a system using averages and is good at dealing with generalities. However, after we deepened our investigation of information, we discovered that peculiar events represent information of change, which touches on a needed change in our accustomed epistemology. In fact, even for the purpose of sufficiently making use of the available information resources, we should also excavate the automatically recorded information that has already been collected but not used. In this section, we use the fog weather that occurred during the latter half of December 2006 as our example to show how to employ our method of digitization to deal with fogs and hazes in order to produce the timely forecasts. Our work indicates that this method indeed possesses the clarity of analysis and practicality of application; and the rate of accuracy for forecasting the locations and magnitudes of fog weathers is also greatly improved.

9.2.3.1 About the Weather Conditions of the Case Study

On December 21, 24, 26, 27, 2006, the city area of Chengdu experienced multiple heavy fogs, where the fog condition on the 24th day was very thick and lasted a long time, and the condition on the 26th day covered the greatest land area.

The fog distribution for the 24th day was as follows: The visibility in Wenjiang, Xinjing, Jintang, and Pujiang was between 0.1 to 0.6 kilometers; the visibility in Xindu, Shuangliu, and Pi County was less than 100 meters; and the daytime visibility at the Wenjiang station was once less than 70 meters and lasted nearly 6 hours.

The fog distribution for the 26th day was as follows: The entire city of Chengdu other than Dujiangyan experienced a heavy fog weather. The visibility for Wenjiang, Chongzhou, Pi County, Longquanyi, Xindu, Xinjing, Qionglai, Jintang, was between 0.1 to 0.6 kilometers. The visibility for the stations at Pujiang, Shuangliu, Guzhou, and Dayi was less than 100 meters.

Because in Chapter 8 we have already provided the in-depth explanation on visibility weathers, at this junction we will only detail the process analysis for the regional low visibility that occurred on December 24 and 26 in order to illustrate the differences and the basic points for prediction. In principle, one should note the following:

1. The low-altitude counterclockwise rolling current and the low wind speed near the ground level, which is fundamentally different from the high-altitude counterclockwise rolling current of high temperature weathers.
2. The sufficient water vapor at low altitudes or the bottom cushion. So, the forecaster should be familiarize himself with the regional terrains and other specific conditions of change.

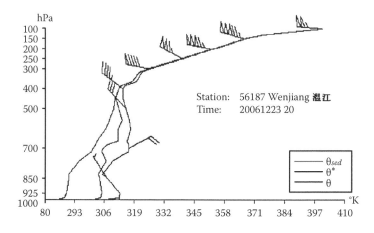

Figure 9.13   The V-3θ graph at the 20th hour on December 23, 2006, Wenjiang.

3. The existence of a layer of radiant inversion near the ground level, which is one of the important criteria for forecasting low visibility. One should pay particular attention to such a phenomenon for plains and areas along the coast. For mountainous basins, the daily temperature differences are relatively small; the forecaster should more carefully analyze the problem of inversion. For example, regarding for the low-visibility weather on December 24, 2006, (Wenjiang City) that occurred in Chengdu, which is located in Sichuan Basin, if not careful enough, it would be very easy for the forecaster to miss the layer of inversion of south winds near the ground level in the V-3θ graph of the 20th hour on the 23rd day, which was the day before the low-visibility weather appeared. In general, after experiencing an entire day's solar radiation, the layer of inversion would disappear. However, a layer of inversion in the sounding data at the 20th hour was still maintained (note that in Figure 9.13, the 3 θ curves all tilted to the right near the ground level, which implied the existence of a layer of inversion even though their angles of tilt were very small). The reason for us to cite this example here is we like to get the reader's attention and caution him that the low visibilities ahead of rains are relatively difficult to predict. This difficulty in predicting the low visibilities before rains is due to the atmospheric instable structure under a counterclockwise rolling current, which could be easily confused with the instability that appears before the precipitation, because counterclockwise rolling currents should in general correspond to stable structures. The mismatch in this specific case indicated that the atmosphere was about to undergo changes. For such conditions, one should not just focus his attention on the area within 300 kilometers around Chengdu (Wenjiang) station; instead he should also pay attention to the atmospheric structures of the surrounding stations. With a concrete analysis, it was found that the stations at

Xichang, Chongqing, Dachuan, and Hongyuan had all developed counter-clockwise rolling currents and ultralow temperatures while providing water vapor for the Chengdu plain (all the graphs are omitted). This discovery indicated that the weather condition in the Chengdu plain would change to that of precipitation with a low visibility first appearing ahead of the rainfall. Although the relevant forecasters had successfully predicted this low visibility, heavy fog condition using the V-3θ graphs, according to the analysis of the automatically recorded information using the method of digital transformation, it was found that this low-visibility weather was mainly caused by haze instead of fog. The low visibility weather on December 26, 2006, was truly due to heavy fogs.

9.2.3.2 Digital Transformation of Automatically Recorded Information and Forecast of Locations

No doubt, the automatically recorded data belong to those of the familiar oscillating information (Figure 9.14). Due to their oscillations along with time, they are conventionally known as wave motions. Evidently, from Figure 9.14, even with many years of training it is still very difficult for anyone to clearly obtain much of a meaningful message.

Because of this problem of practical difficulty, the method of digitization of information helps to make the hidden message readily recognizable. Of course, for different weather phenomena, the relevant predicting factors would show different patterns. In general, precipitations are closely related to wind speed and humidity; but for fogs, they would not appear without the effect of water vapor. In this specific example, we first applied the digitization of the humidity information, and then the digitization analyses of the wind direction and wind speed. From the analyses it was found that for the fogs before and after the precipitation, there are clear differences in terms of the digitized wind direction and wind speed. For these fog conditions, although the digitization of the humidity data showed similar formalities, an obvious difference in the levels of water vapor concentration for the fog condition ahead of the precipitation (December 24) and that after the precipitation (December 26) was seen.

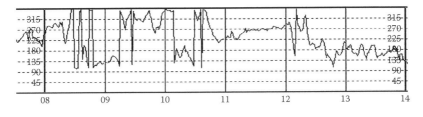

Figure 9.14 The automatically recorded wind-direction time series for December 26, 2006.

9.2.3.2.1 Digitization Analysis for Location Using Humidity (Wenjiang, December 24)

In the following figures, the notes of *per minute* or *per 15 minutes* in parentheses stand for the sampling time intervals. Per minute means a sampling point is collected each minute; and per 15 minutes, each 15 minutes. The purpose for us to use two different sampling time intervals is to show the reader how the degree of irregularity in the sampling can reflect the degree of recognizability for what is going to happen. The underlying principle is that unusual information should be a reflection of an irregular event. One should not alter the sampling scheme to just distort or alter the essential characteristics of the event. So, for each predicting factor employed, one will need to test different sampling time intervals. In other words, testing for the feasible sampling time intervals is also part of the work of digitization. Its purpose is to find the time interval that can reflect the objective reality of the information, where one can also make use of comparative analysis. No doubt, that represents a piece of the basic work associated with each application of the method of information digitization. We need to note that for different problems, the forms of expression and outcomes are different.

From the serial figures of December 24, Figure 9.15 through Figure 9.17, it can be seen that in terms of the fog conditions that appear before precipitations, the choice of the sampling time intervals does not make much difference. And, although there were differences between the digitizations for before and after the event, the differences were not clear.

9.2.3.2.2 Digitization for Location Using Predictor Humidity (Wenjiang, December 26)

For the digitizations of the automatically recorded information on December 26, the serial graphs, Figure 9.18 through Figure 9.20, the predictor humidity showed a clear difference for before, during, and after the process, where the minute sampling interval was the best. To satisfy the need of forecasting, we purposely applied 6 hours of information.

Although the choice of the sampling interval for forecasting the low visibility prior to the appearance of precipitation does not make much difference, from our other case studies, it is found that the minute sampling interval seems to be the best for practical purposes. In particular, in terms of the physical meanings, the phenomenon of low visibility prior to precipitations appears not entirely due to the concentration of water vapor; instead, it is mainly caused by convective air streams, which also carry other non-water-drop objects, such as dust. So, for the low visibility condition of December 24, it would be appropriate to call it a haze, as shown by the serial figures.

The message delivered by the serial graphs of the low visibility on December 26, after the precipitation, Figure 9.18 through Figure 9.20, is that the condition was a

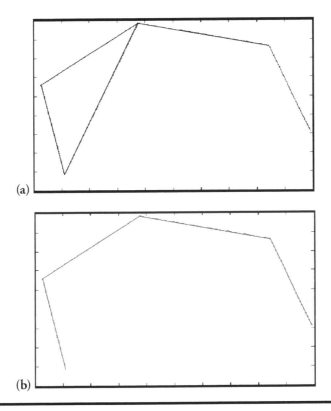

Figure 9.15 (a) The 6 hours before the fog of the 24th day (per minute). (b) The 6 hours before the fog of the 24th day (per 15 minutes).

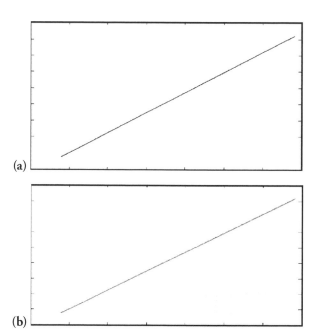

Figure 9.16 (a) During the fog of the 24th day (per minute). (b) During the fog of the 24th day (per 15 minutes).

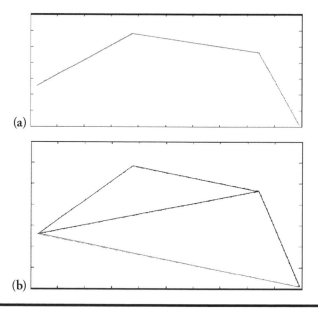

Figure 9.17 (a) The 6 hours after the fog of the 24th day (per minute). (b) The 6 hours after the fog of the 24th day (per 15 minutes).

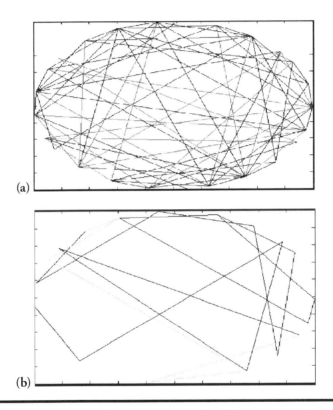

Figure 9.18 (a) The 6 hours before the fog of the 26th day (per minute). (b) The 6 hours before the fog of the 26th day (per 15 minutes).

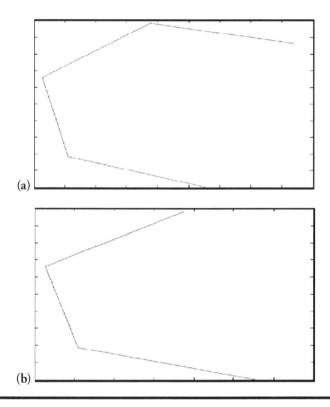

Figure 9.19 (a) During the fog of the 26th day (per minute). (b) During the fog of the 26th day (per 15 minutes).

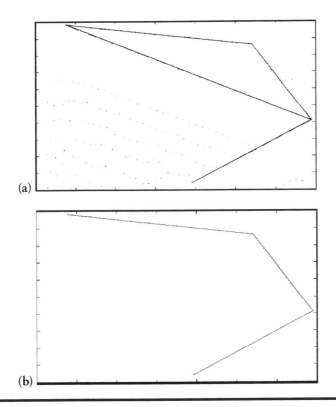

Figure 9.20 (a) The 6 hours after the fog of the 26th day (per minute). (b) The 6 hours after the fog of the 26th day (per 15 minutes).

fog formed by concentrated water vapor. At this junction, we would like to mention that the digitization scheme of automatically recorded data is designed to pinpoint the effect of irregular information. So, it is sensitive to irregular disturbances or eddy flow information. So, the serial figures' insensitivity to the humidity on December 24 implies that the low-visibility weather that day was mainly a haze. This difference between fogs and hazes can also be given by using wind direction and wind speed.

9.2.3.2.3 The Digitization of the Wind Direction on December 24

Refer to Figure 9.21 through Figure 9.23.

9.2.3.2.4 The Digitization of the Wind Speed on December 24

From the serial figures of December 24 (Figure 9.24 through Figure 9.26), it is readily seen that no matter whether it is before, during, or after the low visibility, changes in the wind direction and speed were quite frequent with much difference. The reason is that although the wind speed before the precipitation was

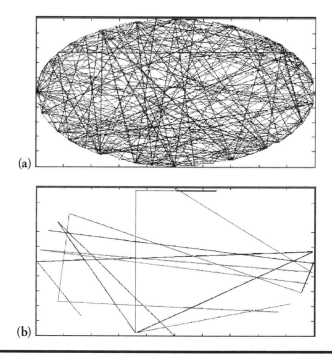

Figure 9.21 (a) The wind direction of the 6 hours before the process of the 24th day (per minute). (b) The wind direction of the 6 hours before the process of the 24th day (per 15 minutes).

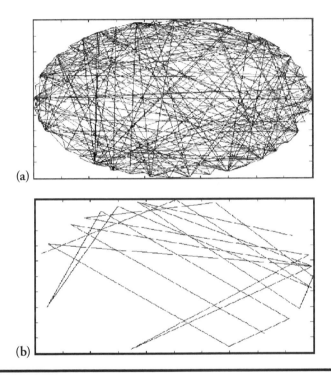

Figure 9.22 (a) The wind direction during the process of the 24th day (per minute). (b) The wind direction during the process of the 24th day (per 15 minutes).

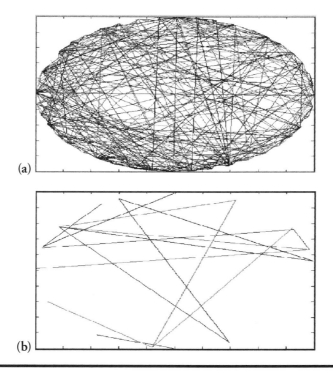

Figure 9.23 (a) The wind direction 6 hours after the process of the 24th day (per minute). (b) The wind direction 6 hours after the process of the 24th day (per 15 minutes).

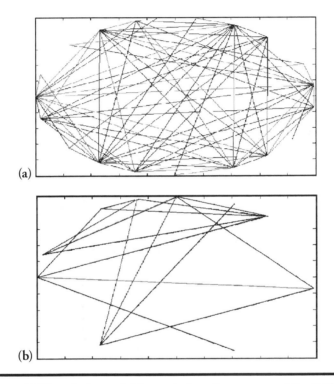

Figure 9.24 (a) The wind speed 6 hours before the process of the 24th day (per minute). (b) The wind speed before the process of the 24th day (per 15 minutes).

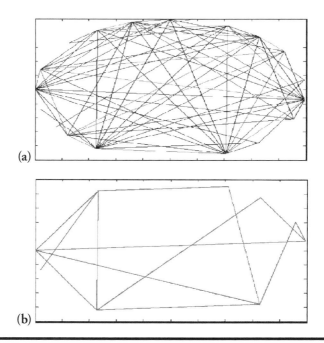

(a)

(b)

Figure 9.25 (a) The wind speed during the process of the 24th day (per minute). (b) The wind speed during the process of the 24th day (per 15 minutes).

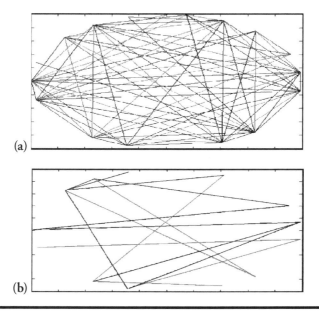

(a)

(b)

Figure 9.26 (a) The wind speed after the process of the 24th day (per minute). (b) The wind speed after the process of the 24th day (per 15 minutes).

relatively small, due to the convective activities in the atmosphere, both wind direction and speed went through frequent changes. This end very well reveals the sensitive capability of describing the irregular or chaotic flows of the method of digitization. That is how it shows that the low visibility before the precipitation was haze instead of fog.

9.2.3.2.5 Digitization of the Wind Direction and Speed for Heavy Fog Process on December 26 (Wenjiang)

9.2.3.2.5.1 The Digitization of the Wind Direction of December 26 — Refer to Figure 9.27 through Figure 9.29.

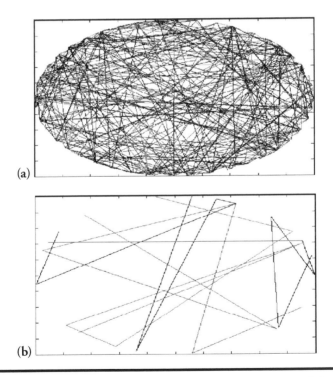

Figure 9.27 (a) The wind direction 6 hours before the process of the 26th day (per minute). (b) The wind direction 6 hours before the process of the 26th day (per 15 minutes).

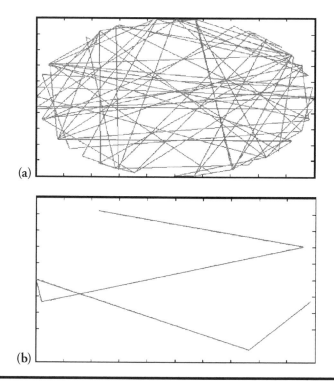

Figure 9.28 (a) The wind direction during the process of the 26th day (per minute). (b) The wind direction during the process of the 26th day (per 15 minutes).

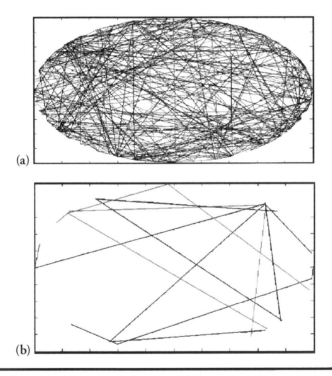

Figure 9.29 (a) The wind direction after the process of the 26th day (per minute). (b) The wind direction after the process of the 26th day (per 15 minutes).

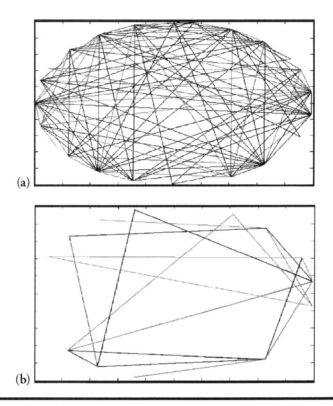

Figure 9.30 (a) The wind speed 6 hours before the process of the 26th day (per minute). (b) The wind speed 6 hours before the process of the 26th day (per 15 minutes).

9.2.3.2.5.2 The Digitization of the Wind Speed of December 26 — No doubt, the serial figures (see Figure 9.30 through Figure 9.32) of December 26 completely revealed the process of the fog condition from frequent changes (before the condition) to changes of decreasing frequency (during the condition) to changes of increasing frequency (after the condition). Based on these figures, there is no longer any need to further deliberate on the difference between the low visibility of haze before precipitations and that of the fog before precipitations.

9.2.3.2.6 Discussions

Our analyses using the V-3θ and the digitization of automatically recorded information indicate that these methods can maintain the objectivity of the data and make the underlying message of the information easily recognizable. They can be employed at places the traditional methods could not be relied upon. Our examples have shown that what is different in the method of digitization from

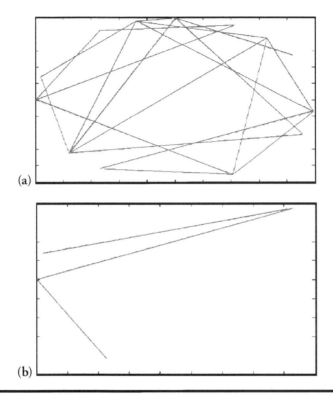

Figure 9.31 (a) The wind speed during the process of the 26th day (per minute). (b) The wind speed during the process of the 26th day (per 15 minutes).

modern science is that the more complicated the available information, the better for practical purposes. The current world of learning is still entangled in randomness and complexities without fundamentally comprehending what events are and what the physical significance of irregular or complex events is. It has been a widespread belief that the automatically recorded meteorological data are too complicated to be of much use. However, our work shows that chaos exactly reveals transitional changes in the underlying evolution, and the method of structural transformation can resolve practical problems cleanly, intuitively, and conveniently.

1. The meteorological science cannot be merely seen as an invariant extrapolation of the conventional inertial system. As for disastrous weathers, they should at least be treated as part of the physics of processes of evolution problems. That is why we said earlier that physicists should be foretellers of what is forthcoming in terms of changes in materials and events.
2. Information is deterministic instead of random. To maintain the objectivity of events and information, the method of digitizing events and information

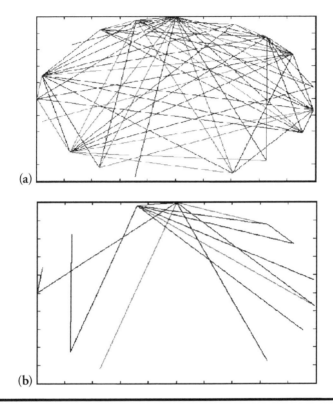

Figure 9.32 (a) The wind speed after the process of the 26th day (per minute). (b) The wind speed after the process of the 26th day (per 15 minutes).

is very meaningful; and information is not the same as quantities, which involves a reform in the system of knowledge.

3. The V-3θ graphs and the method of digitization of automatically recorded information provide a system of methods for investigating evolutions. Their applicability is not limited to the area of meteorological science. Also, there is a need to bring new understandings to the concept of information, which should be a problem for the future development of information science.

Besides, the presentation here illustrates that other than making the automatically recorded data available, we should also excavate the irregular information hidden in the massive amounts of data. In fact, information is a resource that cannot be processed only by using quantities. Our studies suggest additional "assets" other than the conventional regularized information. That is the reason we repeatedly said that if over 300 years ago identifying events with quantities was the human wisdom, then continuing to equate quantities with events 300-plus years later would be an intelligent deficiency of man.

9.3 Examples of Digitizing Seismological Oscillating Information and Prediction of Earthquakes

When Zeng Xiaoping and Lin Yunfang, both from the Institute of Geophysics of the Chinese Bureau of Seismology, heard about our method of digitization of information, they immediately applied our method using the available information of geomagnetism to investigate earthquakes and other related natural disasters. For a more detailed presentation of their work, please consult with their original works. In this section, we will only highlight their main results.

9.3.1 Digitization Characteristics of Geomagnetic Information

Geomagnetic information is naturally created in the evolution of the earth when it spins and resolves around the sun and possesses the signals about changes in the underlying structural attributes. These signals change along with changes in time and space between the sun and earth. Practical analysis indicates that changes in geomagnetic information often occur before major disasters or events on the earth or in its environment, indicating their value of predicting disasters and events. What needs to be clear is that the reason materials' evolution is different from the central problems of the modern scientific system is that materials change themselves in the form of rotational movements, creating materials with new attributes and states different from the original initial value system. In this process of creation, major disasters and events are also caused. As a problem of epistemology, the informational irregularities on the traditional time series plots cannot be understood as superpositions of different waves. Instead, they are signals about injuries and damage to the underlying materials caused by subeddies. Hence, information is deterministic; and the injuries and damage in materials and changes in events do not follow the rules of the formal logical calculus of quantities, which constitutes a major challenge to the system of modern science and the contemporary beliefs. As a reform of the methodological system, not only is there the need to include irregular information, but also a need to establish the corresponding method of practical analysis.

There is no doubt that earthquakes are a geophysical phenomenon under multiple interactions existing in the movement of the earth and are developed in the form of rotation and occur in some specific process of evolution. According to the blown-up principles of evolution science, the essence of the phenomenon of irregularity is the information of change, and the essence of change is about transitional changes in events. So, in terms of the method of analysis, we cannot continue to employ the system of quantitative analysis. Instead, we will make use of the method of digitization of irregular events. To this end, Zeng Xiaoping and Lin Yunfang test-employed our method of digital transformation of automatically recorded information. Their work indicates that this method can pinpoint the structural characteristics of frequency of change in geomagnetism before earthquakes so that the epicenters of forthcoming earthquakes can be located ahead of time. Other major natural disasters can be handled similarly.

From the traditional time series information of the day-to-day changes in geomagnetism, it is generally impossible to discover any abnormality in the constantly varying data. On the other hand, by employing the method of digitizing information, abnormal behaviors can be easily spotted. These structural abnormalities are closely related to those of disastrous events, such as earthquakes, torrential rains, and extreme weather conditions. Let us look at, for example, the earthquake of magnitude 8.1 that occurred on November 14, 2001, along the border area between Qinghai and Xinjiang provinces. The hourly geomagnetism Z's values, collected at Geermu station in Qinghai about 380 km from the epicenter, are plotted in Figure 9.33 on the left-hand side. From these time series curves, no major abnormality prior to the earthquake can be sorted out. However, from the corresponding phase space graphs (on the right-hand side of Figure 9.33), we can easily see that something major is in the making.

In Figure 9.33, the time series disturbance in the magnetic field is analyzed with one digit per 3 hours. So, each day's disturbance is shown with a k number consisting of 8 digits, where k = 0, 1, 2, ..., 8, 9, representing 10 different scales. The interval 0–2 is classified as calm; 3–4 as general disturbance; 5–6 as disturbance; 7–8 as strong disturbance, and 9 as extremely strong disturbance.

From the time series plots on the left-hand side in Figure 9.33, it is clearly seen that the changes in the curves are not obvious. Even for the case of extremely strong disturbance, the time series plot (2nd on the left-hand side) contains only one individual point that jumped away from the rest. And, the day before the major earthquake and on the day of aftershocks (the bottom two plots in Figure 9.31 on the left-hand side), no deviation appears on the time series plots. On the other hand, corresponding to the case of extremely strong disturbance, the phase space structure graph (second on the right-hand side) had already shown a change in structure. If a warning for the imminent earthquake were given on November 6, there would be about a week's time of foresight. If an imminent earthquake forecast were given on November 13, there would still be about 24 hours of advance warning. The prediction of imminent earthquakes has been an international problem of great difficulty, and as of this writing, there still does not seem to be any adequate starting point for analysis so that prediction is beyond reach. The idea of digital transformation of irregular information has shown, as in our example, a substantial improvement in the prediction techniques with practical significance. It also implies a major change in the underlying epistemology. That was why our idea attracted Zeng Xiaoping and Lin Yunfang's great interest. They believe that there is a need for furthering the understanding of this method. In such a study, there will involve such problems as how to understand materials' evolution and differences in materials' existence, how to comprehend irregular information, and why the traditional method of regularization cannot resolve the problem of predicting devastating natural disasters. In terms of analyzing the characteristics of earthquakes, one should pay particular

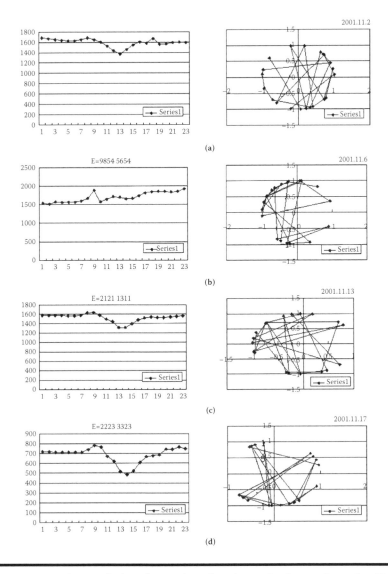

Figure 9.33 Before and after the earthquake of magnitude 8.1 along the Qinghai and Xinjiang border on November 14, 2004. The graphs on the left-hand side are the daily time series plots of the hourly geomagnetism Z values; the graphs on the right-hand side the digitized figures for the time series on the left. The observation station is of the distance L = 380 km away from the epicenter. From top to bottom: (a) Stands for the magnetic field's hourly k = 3332 4212 disturbance on November 2: general. (b) The magnetic field's hourly k = 9854 5654 disturbance on November 6: extremely strong. (c) The magnetic field's hourly k = 2121 1311 disturbance on November 13: calm. (d) The magnetic field's hourly k = 2121 1311 disturbance on November 17: calm.

attention to the calmness after the extremely strong disturbance of November 6, which is similar to the placidity right before a torrential rain and the ebb-tide phenomenon right before a tsunami. Correspondingly, the expanding abnormality on the digitized phase space structures of the irregular information indeed has clearly shown the arrival of the disaster.

On both November 13 (the day before the earthquake, third graph on the left-hand side) and November 17 (the third day after the earthquake, the bottom graph on the left-hand side), the time series curves of the geomagnetic field are very calm, while the corresponding digitized structures show that there were obvious abnormalities in the geomagnetic field, where the circular structure was seriously deformed. These deformations indicate that during November 13–17 the geomagnetic field at the said location had gone through irregular changes. These changes cannot be easily recognized from the time series plots. That is, the method of information digitization can conveniently reveal changes in the geomagnetic information and shows its applicability in the area of imminent earthquake predictions.

9.3.2 Digitization Characteristics of the Information of Normal Conditions

Let us focus on some of the stations located at, for example, Huhhot (HH), Geermu (GM), Wuhan (WH), and Xinyang (XY), and analyze the phase space graphs for the absolute values of the overall strength F and its horizontal and vertical components H and Z of the geomagnetism measured since 1991. It can be seen that under normal conditions, the monthly data plots are basically distributed evenly over a circle in the phase space (Figure 9.34).

9.3.3 Digitization Characteristics of Abnormal Conditions

Let us take the daily nuclear torsion data, which are the absolute values of the overall strength F and its horizontal and vertical components H and Z of the geomagnetism, observed at the 21st hour, Beijing time, of the following four stations: Huhhot (from January 1994 to December 1998, 4 years total), Xinyang (from October 1990 to December 2002 for a total of 12 years and 3 months), Wuhan (from October 1990 to December 2002 for a total of 12 years and 3 months), Geermu (from January 2000 to December 2001 for a total of 2 years), and other stations around China (from January 2004 to May 2005 for a total of 1 year and 5 months).

Classes of disasters considered include earthquakes, floods caused by torrential rains, high-temperature droughts, sandstorms, low-temperature freezes, and other abnormal weather events or incidents near an observation station. From the digitized figures of the absolute values of geomagnetic data, three types of abnormal digitization structures are found to relate to natural disasters.

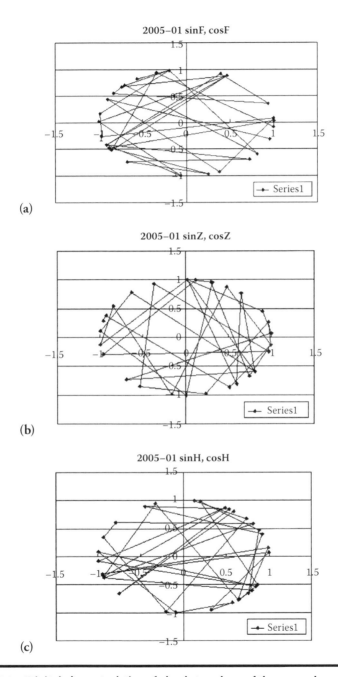

Figure 9.34   Digital characteristics of absolute values of the normal geomagnetic field. (a) Wenan station F, January 2005. (b) Jiayuguan station Z, January 2005. (c) Luoyang station H, January, 2005.

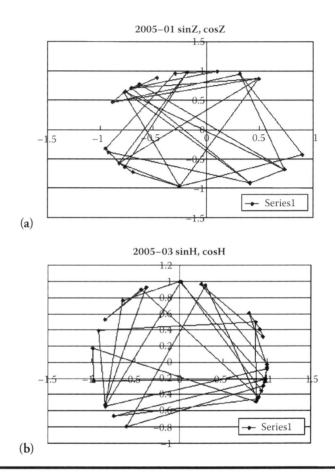

(a)

(b)

Figure 9.35 Type A abnormalities in the geomagnetism. (a) A break-off at the 3 o'clock location (Wuhan Z, January 2005). (b) A break-off at the 6 o'clock location (Tanggu H, March 2005).

Type A abnormalities: The distribution on the phase space circle breaks off on the locations at roughly 3, 6, 9, and 12 o'clock positions for as long as about $\pi/2$ arch-length. And the circle breaks off and deforms in the east–west direction or quasi-south–north direction (Figure 9.35).

Type B abnormalities: The circle of the phase structure breaks off in the first quadrant for $\geq \pi/2$ arch-length and is obviously deformed and damaged (Figure 9.36).

Type C abnormalities: The phase graphs contain only a few reciprocating line segments. The shape of the circle is basically maintained with slight deformation (Figure 9.37).

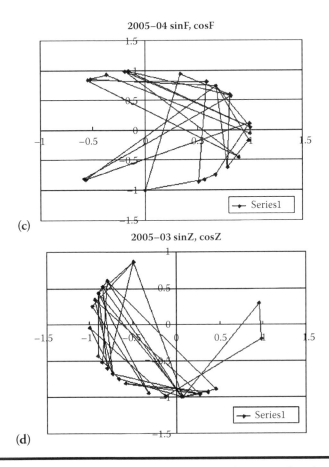

Figure 9.35 Type A abnormalities in the geomagnetism. (c) A break-off at the 9 o'clock location (Wulanhua F, April 2005). (d) A break-off at the 12 o'clock location (Hangzhou Z, March 2005).

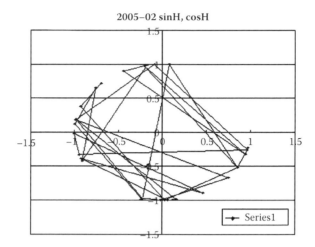

2005–02 sinH, cosH

Figure 9.36 Type B abnormalities on the digitization of the geomagnetism (Ninghe H, February 2005).

Types A and B abnormalities are mostly related to earthquake disasters, revealing the specifics of these disasters. For weather-related disasters, the digitization graphs are mostly symmetric and reciprocatingly instable, which are similar to what we observed in analyzing weather disasters. This end indicates that for different types of disasters, their digitization graph representations are different. This fact needs to be investigated further.

Type C abnormalities correspond respectively to earthquake and/or extreme weather disasters at the nearby area. If Types A and B abnormalities, the structure deforms severely, they mostly signal such disasters as earthquakes and other various kinds of disasters, including severe weather conditions. For weather-related events, Type C abnormalities mostly correspond to high-temperature droughts or flood disasters caused by torrential rains.

9.3.4 Analysis of Disastrous Events

9.3.4.1 Relationship between Abnormality of Information Digitization and Earthquakes

We will use the earthquake with a magnitude of 8.1 that occurred along the border of Xinjiang and Qinghai provinces on November 14, 2001, as our example. We take the exact-hour values of the vertical component Z collected at Geermu station, located about 380 km from the epicenter, and the absolute values of the overall geomagnetic strength F and its vertical and horizontal components Z and H to check the prediction capability of the digitization of irregular information.

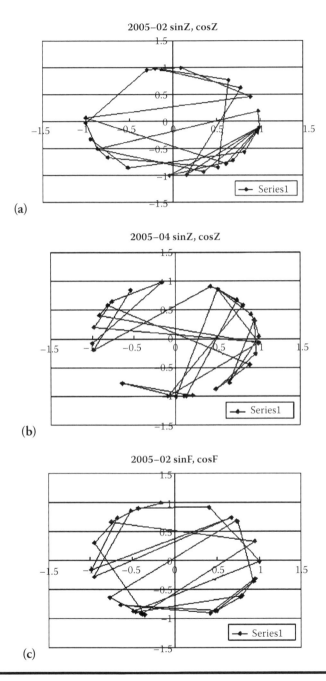

Figure 9.37 **Type C abnormalities on the phase graphs of the geomagnetism (a few crossing lines). (a) Malinshan Z, February 2005. (b) Lianyungang Z, April 2005. (c) Nantong F, February 2005.**

Table 9.1 The Months and Geomagnetic Readings Showing an Abnormality

Year/Month	2000/01	2001/01	2001/02	2001/05	2001/09	2001/10
Magnetic Intensity	Z and H	Z	Z	H	F and H	F

1. According to the abnormality criteria given above, the phase graphs of each day's exact-hour geomagnetic Z values from October 2001 to November 2001 show that abnormalities appeared on the following days: On the 4th, 9th, 14th, 25th, and 30th in October; and the 6th, 8th, 10th, 11th, 12th and 13th in November (these were the 41st, 36th, 31st, 20th, 15th, 8th, 6th, 4th, 3rd, and 2nd, and last days, respectively, before the earthquake); and the 17th (three days after the earthquake). What is shown is that as the disastrous event approached, the frequency of abnormalities in the geomagnetic components Z increased accordingly.

2. From the monthly phase structure graphs of the absolute values of the geomagnetic readings F, Z, and H of the Geermu station, we can see the months and the geomagnetic readings when an abnormality occurred during 2000–2001 (see Table 9.1).

9.3.4.2 Structural Abnormality in Geomagnetic Readings and Several Kinds of Disasters

1. Abnormal geomagnetism corresponds to 80% of the disasters occurring in the coming 6 months and 94% of the disasters within the coming 8 months.

2. Types A and B abnormalities are mostly related to earthquakes and other kinds of disasters (including disastrous weathers). Type C abnormalities are mainly connected with high-temperature droughts or floods of torrential rains.

Besides, in order to verify the prediction capability of this method, relevant predictions were made using available information.

9.3.5 Test with the Earthquakes of 2005 (as of November 30, 2005)

9.3.5.1 The Data Collection

Once in early November 2004 and then another time in early June 2005, Zeng and Lin (2006; in press) respectively obtained the daily F/Z/H readings collected at the 21st hour, Beijing time, from 37 geomagnetic stations from across China for

January–October 2004 and from 67 geomagnetic stations for January–May 2005. Then, they calculated and produced the digitization graphs for each month. On two separate occasions, they forecasted potential earthquakes and other natural disasters for regions across China for the time periods November 2004 to June 2005 and June–December 2005. For details, see Table 9.2.

Table 9.2   Predicted Disasters of Various Kinds

Area	Station	Type A	Type B	Type C	Forecasted Disaster	Actual Happening
Lower reaches of Yellow River	Dashan	0	1	6	Weather	High temperature
	Guangping	1	1	5	Weather	High temperature
	Jinan	3	2	7	Weather	High temperature
	Liaochen	3	2	6	Weather	High temperature
Tri-province area of Jin, Yu, and Shaan	Linfen	0	0	3	M4-5 Weather	Mine disaster
	Luoyang	2	0	3	M4-5 Weather	Torrential rain Geology
	Lushi	2	1	4	M4-5 Weather	Torrential rain Geology
	Jingyang	0	0	4	M4-5 Weather	Torrential rain Geology
Three-gorges, Chongqing, midor north of Gizhou	Chongqing	3	0	3	M5 Weather Geology	M4.1, torrential rain Geology
	Gizhou	0	3	3	Weather Geology	Torrential rain Geology

9.3.5.2 Tests of Actual Forecasts (as of the End of November 2005)

9.3.5.2.1 Earthquakes

Earthquakes missed in their forecasts: Based on the data collection done right before the completion of their work, there were three earthquakes of about magnitude 6 that occurred in the remote areas of Xinjiang and Tibet, and two of magnitude < 5 that occurred at the northern tip of the northwest part of the region. The cause for the missed forecasts was that in these remote areas, there were no installations for geomagnetic observation, and the severity of these events was relatively minor.

Earthquakes predicted in their forecasts: In four predicted areas, five earthquakes occurred. The error in their prediction of magnitudes ranges $\Delta M = 0.7$–1.4. The earthquakes that occurred within or near the forecasted regions include the following:

- Shizhu of Chongqing, M4.1, on November 21, 2004
- Chuxiong — Shuangbai, Yunnan, M5.0, on December 26, 2004
- Simao, Yunnan, M5.0, on January 26, 2005
- Huidong — Huize on border of Chuan-Dian, M5.3, on August 5, 2005
- Wenshan, Yunnan, M5.3, August 3, 2005

Within or near the four regions forecasting the occurrence of other disasters, five minor earthquakes appeared:

- Wulatezhongqi, Inner Mongolia, M4.1, on January 27, 2005
- Yantai — northern sea area of Weihai Sea, Shandong, M4.0, on May 9, 2005
- Lindian, Heilongjiang, M5.1, on July 25, 2005
- Pingguo, Guangxi, M4.4, on October 27, 2005
- Ruichang — Jiujiang, Jiangxi, M5.7, on November 29, 2005

9.3.5.2.2 Other Disasters

According to the statistics reported on the Earthquake Information Network of China, for most of the disastrous areas as forecasted by using our method, the corresponding disasters appeared. In some regions, even multiple, various kinds of disasters appeared. Here, their forecasts were accurate for 12 areas, there were false warning for 4 regions, and 1 forecast was missed. Also, there appeared 12 large-scale disasters that they could not predict due to a lack of information. So, the success rate of their forecasts was 70.6%; the rate of false warnings 23.5%; and the rate of missing forecasts 5.9%.

9.3.5.2.2.1 Correctly Predicted Major Disasters — Dedu area, Heilongjiang, suffered from multiple disasters. In Lindian, an M5.1 earthquake occurred (a

historically rarely seen event of such magnitude) on July 25, 2005, together with multiple forest fires. Zalandun experienced a torrential rain of hailstones on July 15, 2005. Zalaiteqi, Inner Mongolia, saw such a flood disaster of torrential rains on July 7–8, 2005, that the loss it caused was near the level of the extraordinary flood in 1998.

Changchun and neighboring areas, Jilin, experienced multiple disastrous events, among which was a flood disaster of a heavy rain gush at Ningan and Salan of Heilongjiang on June 10, 2005. The scale of that event was seen only once in the past 200 years.

Chongqing and its neighboring areas suffered from multiple disasters. For example, on June 21, 2005, the torrential rain at Wuxi, Chongqing, caused a landslide and created a barrier lake. On June 23, the barrier dam collapsed. On July 6–9, a nonstop torrential rain hit Dazhou City, Sichuan, in which both the severity and amount of the rainfall surpassed the levels of the flood disaster in early September 2004, which occurred on such a scale only recorded once in the past 100 years.

The four provinces Shanxi, Hebei, Shandong, and Henan, and the areas from the middle to lower reaches of Yellow River suffered high temperature conditions.

There appeared nonstop torrential rains in South Shaanxi, east of Gansu, and northwest of Hubei. The torrential rain in south Shaanxi during July 5–8 and September 19–October 10 caused the lower reaches of Wei River and branches of Han River to experience the greatest flood peaks since 1981 and 1983, respectively, flooding 44 rivers and causing over 80 landslides of different mountains. The ignited disaster of mud–rock flows was greater than that in 2003. During August 14–29, northwest Hubei was attacked by a historically rare, extremely heavy torrential rain, causing the Han River to rise and a huge area of landslides, which involved 202 large mountains.

Frequent mine calamities in Shanxi, Shaanxi, and Henan were correctly predicted.

In the Xi River valley of Guangxi, an extremely heavy torrential rain, seen only once in the past 20 years, caused floods, mud–rock flows, and landslides.

There appeared torrential rains in the mid-reaches of Yangtze River. During September 1–4, Xinyang, Wuhan, Hefai, Jiujiang, and other areas suffered from the aftermath of the 13th typhoon, named Qingli. Jinzhai of Anhui had a violent windstorm and torrential rain, leading to such a severe mountain torrent that was recorded only once in over 100 years. On October 5, 2005, the speed of the Han River current at Longwangmiao station was over 5 m/s, the fastest in the past 20 years. On October 6, the greatest flood peaks for the past 22 years occurred in Hankou, Hanchuan, and Xiantao of Hubei.

The 9th typhoon Maisha and the 15th typhoon Kanu attacked north Zhejiang, Tai Lake, and the end area of Yangtze River.

Rarely seen tornadoes and torrential rains of hailstones appeared in Jiangsu province on April 20 and 25 and June 14–15.

Four earthquakes of magnitude M5.0–5.3 occurred in Yunnan. The errors in our predictions were $\Delta M = 0.2–0.5$.

Torrential rains, mud–rock flows, and landslides appeared in Chengdu, Sichuan, and neighboring areas.

9.3.5.2.2.2 Areas of False Warning — *Northwest area of Gansu*. A predicted earthquake of M5–6 did not occur.

Lanzhou area. The predicted M6± earthquake did not occur.

At the tri-province area of Shanxi, Hebei, and Inner Mongolia. The predicted M5± earthquake did not occur.

East Liaoning peninsula. The predicted M5± earthquake did not occur.

None of these predicted earthquakes actually occurred, indicating that the once-a-day geomagnetic information used in the study was still not certain enough to provide the needed message on minor earthquakes. Also, it is possible that the information itself might not have provided the truthful information of irregularity. Also, some misguided cases were discovered in the available information.

9.3.5.2.2.3 Missed Areas in the Forecasts — The disastrous drought and typhoon in Hainandao Island were missed.

9.3.5.2.2.4 Disastrous Events in Areas with No Information — Since the relevant information on the geomagnetic field was not available, any prediction on the following severe or devastating disasters was not made. As for weather-related problems, these can be resolved by using automatically recorded information at meteorological stations.

During August 14–29, 2005, northwest Hubei suffered from the attack of a historically rarely seen, extremely heavy torrential rain, causing the water level in Han River to rise and a large area of over 22 landslides.

Nanyang, Henan, had a tornado of scale 10 on August 2, 2005. No such high-scale tornado was recorded in the past 100 years.

A typhoon landed in Fujian, Zhejiang, and Guangdong.

Seven tornadoes appeared in Liaoning, as well as a flood disaster at the upper reaches of Hun River.

There were two earthquakes and a major forest fire in Northern Heilongjiang.

The Talimu River valley, Xinjiang, experienced a severe drought.

A drought covered Yunnan, Gizhou, and Guangxi in April–May 2005.

The delta of the Pearl River experienced a high-temperature drought and torrential rains.

There were 4 M5.2–6.2 earthquakes in Tibet.

Several snow disasters occurred in Tibet: On March 2, 2005, Ali area was hit by a strong snow disaster. On March 2, Cuona had the greatest snowfall of the last century. On April 19, in Linzhi-Bomi a huge-scale avalanche happened.

The purpose for Zeng Xiaoping and Lin Yunfang to analyze and calculate various disasters of different magnitudes was to practically confirm the prediction capability of our method. From the listed results, the reader can see the validity of our method and where there are still problems. As we have said before, the development of science is in answering the call of problems.

After the previous was concluded, both Zeng Xiaoping and Lin Yunfang once again did their prediction tests for another 3 years from 2005 to 2007. Comparing all the major disasters from across China during the year of 2007, their work verifies that:

1. The Type A, B, and C abnormalities in geomagnetism are respectively the omens for earthquakes, torrential rains and related floods, and high-temperature droughts and their secondary disasters. Type BJ abnormalities for heavy rain gushes and Type J abnormalities for extraordinarily heavy rain gushes are discovered (Figure 9.38).
2. The index GW for high-temperature droughts can be applied in reference to the prediction of drought and flood disasters.
3. For the complicated abnormal spatial distributions of the geomagnetism of such areas as Beijing, the mouth of Yangtze River, and Bohai Bay, a plausible explanation — the thermal island effect — is developed.

Besides, these authors also investigated the reasons for failed predictions. Their work includes the following two main reasons: First, further comprehensions of various types of abnormalities are needed; and second, they experienced shortage of observation points and data and difficulties of obtaining timely information.

Using abnormalities existing in the digitization of geomagnetism to predict disasters across the entire country of China has been the method employed for forecasting disasters in the past 3 years. Through the computations of the available data of geomagnetism over most parts of China from October 1990 to December 2003, it has been discovered that some structural abnormalities in geomagnetic information correspond to the occurrences at or near the geomagnetic stations of earthquakes of over magnitude 5, extreme weather conditions, and geological calamities, such as mine disasters, in the following 6–8 months with the respective rates of accuracy of 80 to 94%. For the test predictions made during the 3 years from 2005 to 2007, the rate of success is lower than 80%. For the actual happenings of natural disasters (as of October 15, 2007) from across the entire country of China, the success rate of predictions using our method for the year of 2007 is 70–78%.

In terms of methodology, applications of the digitization of time series information are not limited to the problems of change in meteorology. As long as the available time series information is about evolution or changes of events, this method can be applied.

No doubt, the method of digitization is originated from our new understandings of irregular information. On the basis of these new understandings, we utilize the method of digitization to extract the information of change in events, which is different from the idea of modern science that invariant information is retrieved

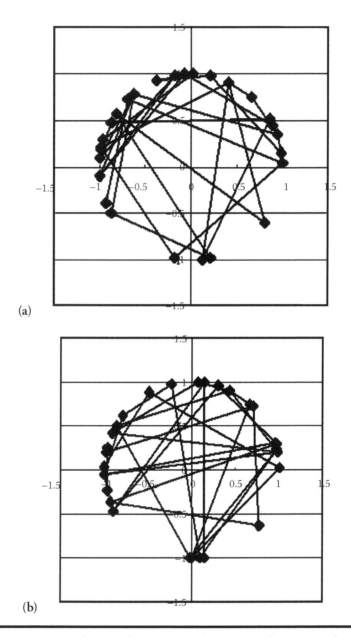

(a)

(b)

Figure 9.38 Type BJ abnormality at Chongqing (CQ) and type J+C abnormality at Jinan (JN) before the appearance of the extraordinarily heavy rain gush in July 2007. (a) CQ0703-Z (b) CQ0704-F.

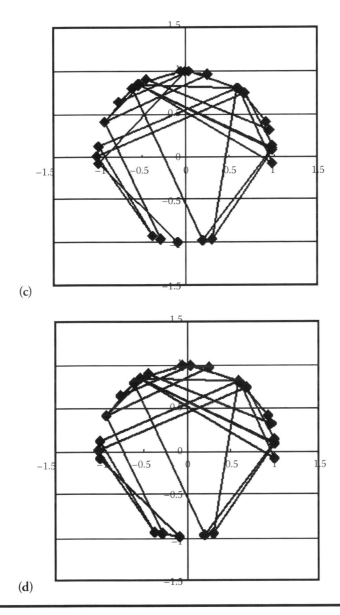

(c)

(d)

Figure 9.38 Type BJ abnormality at Chongqing (CQ) and type J+C abnormality at Jinan (JN) before the appearance of the extraordinarily heavy rain gush in July 2007. (c) JN0612-F (d) JN0612-Z.

from events. Our examples presented in this and the previous chapter very well show the fact that successfully saving ourselves (mankind) from natural or man-made dangers might just be a slight twist away from our accustomed way of thinking. However, at the very moment of the needed twist, most of us hesitate, wander, and eventually decide to return to the conventional wisdom, leading to more and unnecessary heavy losses.

Acknowledgments

The presentation in this chapter is based on Bergson (1963), Chen et al. (2005b), Chen et al. (2005c), Chen et al. (2003), Einstein (1976), Lin (2008b), OuYang (2006b), OuYang and Chen (2006), OuYang et al. (2005), OuYang and Lin (in press A), OuYang et al. (2001b), OuYang and Peng (2005), OuYang, McNeil et al. (2002a), OuYang, Sabelli et al. (2002b), Prigogine (1980), Sabelli (2000), Sabelli and Kauffman (1999). Shannon and Weaver (1949), Wang (1981), Weng (1984), Zeng, Hayakawa et al. (2002), Zeng, Lin et al. (2001), Zeng and Lin (2006, in press).

Afterword

Because quantities are postevent formal measurements, quantitative or numerical analysis cannot predict what is forthcoming. Because modern science experiences difficulties or cannot deal with irregular information, the system of modern science, established since the time of Newton, meets with a major challenge. In particular, in this book, we employed real-life cases and scenarios to present our success stories in the area of forecasting disastrous weathers, which have been considered world-class difficult problems in modern science, using the prediction method of digitization of information. These well-documented case studies show perfectly the validity of our theory and method. Considering that the true value of a scientific theory is its impact on practice and what brand-new knowledge about nature it can bring forward, the validity of applications in real-life situations presented in this book has presented an urgent need to reform the accepted concepts of epistemology, which naturally forces us to rethink the artificiality of the quantitatively conscientious behaviors of man since the time of Pythagoras of ancient Greece. That is to say, Pythagoras's numbers, being the origin of all things, and Newton's *Mathematical Principles of Natural Philosophy* have revealed such a fact: The essence of the philosophical law that changes in quantities lead to changes in qualities is the consequence of the quantitative culture of human conscientious behaviors instead of an event of science. Establishing and developing the science of events is the next important goal of contemporary and future generations of scholars.

Even if we do not talk about whether beliefs are idealism or materialism, just in terms of physics, it would not investigate nonexistent objects or matters; otherwise, it would not be referred to as physics. So, from elementary formal logic, one can conclude that without objects there will not be quantities; when the objects do not change, the corresponding quantities will not vary. However, such a statement as "changes in quantities lead to changes in qualities" has been well professed as a law of philosophy in university classrooms and academic lecture halls throughout the world, and followed by many great scientific minds of history as their criterion of thinking and reasoning. Looking back in history, what seems to be very odd is that since at least the 16th century, studies in science and philosophy have not suffered much from constraints of national policies and regulations; however, even among those scholars who lived in liberty and freedom, the law of philosophy that "changes in quantities lead to changes in qualities" has been followed strictly. From this historical fact, the incorrigibility of conscientious behaviors and the controlling power and authority of consciousness can be clearly seen. So, the philosophical law that changes in quantities lead to changes in qualities is a piece of the quantitative culture of human conscientious behaviors instead of the science of events. And, the conclusions derived from our works presented in this book that events are not quantities, events do not comply with the calculus of formal logic, events do not contain randomness, and so forth, are all very important results of the epistemology.

Speaking frankly, there must be an epistemological source of origin for modern science to become the conscientious behavior of an era. The essence of this source of origin is to substitute the quantitative culture with the science of events. The reason we use the phrase *quantitative culture* instead of *quantitative science* is that quantities are formal concepts that appear after events and reflect the one-sided wish of human conscientious behaviors. However, quantities are not the same as the operational and transformational behaviors of realistic or objective events. For example, the familiar quantity "0" is a human invention instead of a discovery. In the corresponding materials or events, there does not exist any 0. Mathematically, when 0 is multiplied with an arbitrarily chosen number, the answer is always 0; and dividing a nonzero number by 0 always leads to the quantitative infinity (in modern science, the quantitative infinity is referred to as ill-posed; however, in ancient China, it meant that quantities cannot deal with events or cannot describe events). In these conventions of number operations, there are logic contradictions. For example, the quantity 1 is not the same as 2. However, when these are multiplied by 0, the answer is 0 in each case, leading to $1 = 2$, which contradicts with $1 \neq 2$, a contradiction of logic (this end is the key of the Berkeley paradox of 300 years ago and its recent reappearance (Lin, 2008)). At the same time, we suffer from the following fact: The formal logical rules of operation of quantities are not the same as the principles of transformation and operation of materials and realistic events. Presently, people should note the fact that quantities can be employed to calculate nonexistent things more accurately than those that actually exist. So, in practical applications, one cannot avoid losing valuable information and the behaviors of fabricating events when solely relying on the formal logical manipulations of quantities. This end reveals the possible misleading effects of conscientious behaviors of man (OuYang and Chen, 2006). For example, there is artificiality in the basic assumptions of calculus, which was established over 300 years ago; at the same time, the conclusion which is not relevant to the specific path taken does not agree with objective reality. Although mentioned here are just some simple explanations, they still well reveal the existences of the artificiality contained in quantitative analysis and the problems of disagreement with the reality created by substituting quantities for events. So, right now is the time for us to develop the science of events and use it to replace the quantitative culture.

Almost the entire 20th century was seen as an era of wave motions. In these 100 years, almost all scientific theories came from wave motions. For example, the entire field of physics of the 20th century was filled with many different kinds of waves, such as probabilistic waves in quantum mechanics, gravitational waves in general relativity theory, the solitary waves of the KdV equation and various consequent solitons, the shock waves for Burgers' equation, trapped waves, and others. However, what needs to be clear is that since the 1960s, the research of wave motions entered the so-called nonlinear systems. Combined with the development of computers, computational mathematics was also involved in the study of nonlinearity, leading to once well-sounded chaos theory. Corresponding to this period of time was the era of nonlinearity, with chaos theory and solitary waves being the representatives.

The essence of the fashionable era of nonlinearity was to push for the indeterminate concept of a probabilistic universe. During that period of time, works along the indeterminacy of nonlinearity were seen as the newest development and major breakthroughs of science, while criticizing the determinacy of classical mechanics and others. However, people did not truly understand that modern science in its entirety is a system of invariance so that it cannot be referred to as deterministic. The true deterministicity should be judged against the variability of materials and events. Making invariance as deterministicity itself reveals a confusion of the fundamental concepts. In essence, if materials and events are treated as invariant eternity, then there is no longer any need to mention deterministicity. Evidently, the reason the indeterminate concept of a probabilistic universe could become a hot topic of study was that modern science cannot deal with realistically irregular information and people did not truly understand that irregular information stands for information of change. Some scholars wanted to challenge modern science. However, when faced with practical problems, they backed off due to their lack of ability (that is, they did not recognize that irregular information means information of change and did not have an operational way to deal with irregular information). That is, the scientific community experienced a time period of chaos.

Evidently, for the scientific community to walk out of this chaotic situation, the primary task is to reveal the essence of nonlinear problems with numerical computations being the secondary. In this book, we clearly proved that nonlinearity is the rotationality of solenoidal fields by employing vector analysis and the field theory based on the formula of scalar product of gradients. This work shows that nonlinear mathematical models of dynamics cannot be linearized and that there is a need to revisit all the results and theories, including Einstein's general relativity theory, established by using linearization. Based on all these works, we completely revealed the important fact that nonlinearity stands for rotationality and affected many accomplishments established in the 20th century by linearizing or quasi-linearizing nonlinear problems in various disciplines. One of the theories, which are affected greatly, is of Rossby's long waves, which has been seen as a fundamental theory in meteorological science for over half a century, leading to other concepts, such as probabilistic waves and gravitational waves of physics. In this respect, our work is only the beginning. We surely hope that our colleagues will produce more magnificent results in the years to come.

After modern science formed its first push system in the form of quantities, it has ignored the problem of mutual interactions. Speaking more specifically, the quantitative analysis system does not provide a method useful to analyze and deal with mutual reactions. The kinetic energy in the form of speed of the first push is introduced as follows:

$$e = \frac{1}{2} m V^2$$

Einstein's mass–energy formula

$$E = m C^2$$

is also a formula of energy in the form of speed. What needs to be pointed out is that the familiar conservation of kinetic energy in the form of speed cannot limit transfers of energy (OuYang et al., 2002a) and Einstein's assumption that the speed of light in a vacuum is constant can only make energy transferred to the void (Einstein, 1976). So, the law of conservation of energy of modern science is an incomplete law of conservation of energy. So, even the durable goods designed and manufactured in the system of the first push do not possess the needed space for the transfer and transformation of energy, leading to easy damage to these supposedly durable goods. That is the reason why, in the current design and manufacture of the so-called durable goods, the risk values obtained empirically or statistically are used to substitute for the relevant theoretical values, and each formal application of modern science has to be based on the data obtained from repeated experiments or the statistical analysis of the historical information. The essence of the problem is about how to comprehend the materialisticity of energy. To this end, if the square of speed can constitute kinetic energy, then the square of angular speed can also be kinetic energy. Because each realistic physical space is curved, we introduced the concept of stirring motion and the conservation of stirring energy (OuYang et al., 2000; Lin, 2008). What is practically meaningful is that the conservation of stirring energy can be transferred and stored through the secondary circulations and transporting materials so that the conservation of energy is perfected. This work provides a new way of thinking for the design and manufacture of commercial goods and for the planning and construction of major engineering projects. It also makes the problems existing in modern science stand out more clearly. In particular, when we investigate problems of material evolutions, the concept and the conservation of stirring energy lay down the theoretical foundation for evolution science to become a scientific system.

The extinction of the ancient kingdom of Angkor in Asia and the disappearance of Mayan culture in America in essence have warned us about how to comprehend nature and how to "reshape" nature. It has been well described in Lao Tzu (time unknown) and Liang (1996), "If crooked, then it will be straightened. Only when it curves, it will be complete." What is unfortunate is that contemporary people are still competing with each other using so-called high tech to make the highest elevated artificial lakes using concrete dams and construct the tallest buildings in the world. Do all of these really stand for the advancement of science and human wisdom or not? Evidently, if high tech pushed for ignorance, then the degree of loss and damage would be far beyond those experienced by the disappearance of the ancient kingdom of Angkor and the Mayan culture. So, let us truly pray for the benevolent to see benevolence and the wise to see wisdom!

What is discussed in this book we are now familiar with; quantities have the attributes of formality of materials and measurements of events. Under the general conditions, quantities cannot illustrate the differences in the properties and states of materials or events. So, at the same time when we are fond of the strengths of quantities, we should also well understand the weaknesses of quantities. As the next stage in the development of science, we should be well versed in both the strengths and weaknesses of the manipulations of quantities. Only by doing so can we appropriately apply the theories and methods of quantitative analysis to resolve practical problems.

As for the problem of direct manipulations, quantitative comparability is widely seen as the only standard for scientificality in modern science. However, in applications this standard meets with the challenge of inaccurate computations of quasi-equal quantities. The inaccurate computations of quasi-equal quantities are also the mathematically infinitesimal differences of large quantities, which exactly represent some incomputable quantitative problems. For example, if $X \approx Y$, that is X is approximately equal to Y, then the result $X - Y \approx 0$ is naturally unreliable, because the computational results in general fall in the category of error values, becoming meaningless. It is our new understanding of this elementary knowledge that we successfully resolved the mystery of chaos and nonlinearity. In scientific history, there used to be a fondness for mathematical equations, creating the belief that equations are eternal (Kline, 1972). However, the inaccurate computations of quasi-equal quantities illustrate that equations are not eternal. We truly hope that this discovery can provide the relevant theoretical guidance for the currently popular works on stable numerical computations in both theory and applications (see the relevant chapters and sections in this book). At the same time, this discovery also leads to the following results: Nonisomorphic quantitative computations of equal quantities are indistinguishable (although the computations are reasonable in terms of the quantitative formal logic, they are works that can be misleading in specific practical problems); the widely applied method in numerical analysis of smoothing and filtering designed for the purpose of eliminating quantitative instability leads to modification or fabrication of the relevant information and events at the same time that quantitative stability is reached; as a new understanding of the quantitative infinity, in particular the nonlinearity of nonsmall quantities can approach infinity much quicker than linear quantitative computations, which make us realize that the method of smoothing used in numerical computations greatly limits the increase of quantities so that they do not approach infinity. The essence is not allowing the quantities to vary so that the corresponding changes in materials become zero.

Modern science treats the quantitative infinity as ill-posed without specifically explaining its physical significance. From our discussions in this book, it has been seen that the quantitative infinity stands for the infinity of Euclidean spaces and corresponds with transitionalities in curvature spaces. It can be said that in applications, the quantitative infinity is not the same as any physical reality. That is why ancient Chinese during the time period of Warring States knew that infinity is a concept of quantities, and in objective reality it represents a problem of events that quantities cannot handle (*wu, liang wu qiong*, from the chapter of "Autumn Water" of Zhuang Zi (Watson, 1964), materials or events cannot be dealt with using quantities). Starting from quantitative instability, our work went on to studying irregular information, illustrating the fundamental differences between Euclidean spaces and curvature spaces, revealing the problems met by the quantitative analysis of modern science, and addressing the epistemological issues of ill-posed concepts that have been bothering quantitative analysis for a long time. Besides, in the current mathematics, there are also the problems of not agreeing with reality in terms of the rules of manipulations and procedures of operation. For example, in the current system of calculus, the fundamental assumption of continuity cannot be carried out in practice. That is why in the history of mathematics, a problem in the foundation of calculus had to be listed as the second crisis of mathematics (Lin, 2008). In practical applications, operations of calculus, in particular the operation of differences, has never complied with the assumption of continuity. Next, the rules of operations of calculus have nothing to do with the specific path taken. However, having nothing to do with the specific path taken itself has lost the practical significance of applications for integrals. Its essence is nothing but maintaining the invariant initial value systems with allowance for only very minor vibrations.

Modern science is established on various physical quantities, quantitative phase spaces, and parametric dimensions. So, a very natural question arises: What are these physical quantities, parametric dimensions, and phase spaces from? The functional relationships consisting of physical quantities are about describing the connections of the relevant variables. In physics, the most fundamental physical quantities are those of mass and force. All other physical quantities are constructed on top of these two concepts using conceptual cycles and quantitative formal logic. For example, for Newton's second law, $f = ma$, both f and a exist within the inside of m, and f comes from m. These kinds of cyclic concepts are often unavoidably employed in the system of quantitative analysis. In particular, time is seen as a quantitative parametric dimension and limits changes in quantities in the name of well-posedness. The consequence is that the changes under investigation become irrelevant to time. So, time becomes a parameter that has nothing to do with change.

This end in essence reveals that modern science studies movements without time. In particular, in applications of physical quantities and parametric dimensions, in order to make use of the calculus of quantitative formal logic indiscriminately, the outcomes are derived in such a way that they have nothing to do with the original problem or mathematical model. For example, one of the original mathematical models studied in chaos theory came from the convective movements of the atmosphere in the x- and z- spaces. After introducing physical quantities, it became a model of the stratified parametric dimension X, Y, and Z space of the strength of motion, linearity, and nonlinearity that it no longer had any connection with the original x- and z- spaces. We surely hope that our discussion in this book can provide some reference value for the future investigations on how to employ physical quantities and parametric dimensions.

This book tries to distinguish the word *digit* from *quantity*. In China, *digit* stands for the concept of yao (the yin and yang lines) of the *Book of Changes*, which is different from the concept of quantities or numerical values. So, we cannot limit "digits" to only quantities or numerical values. According to our discussion in this book, digits contain the multidimensional information of materials and events, including states, structures, attributes, colors, smells, tastes, locations, functions, quantities, and so forth. The practical significance of digitization contains at least that of figures and should not be constrained to that of the currently popular numerical computations. To fully understand and deal with information, we must recognize that in collected data there is irregular information — the information of change. It does not stand for randomness and cannot be modified. In other words, in the era of digitization, irregular information can be dealt with and should not be treated as random or uncertain. So, the core of our discussion in this book is about going beyond quantities and quantification, and at the same time, giving digitization back its original characteristics. After several decades of practice and thinking, we established the specific method of handling information using digitization. The most noticeable advancement of this method is that it can be employed to process both regularized information and irregular information (the information of change) so that it makes it possible for us to analyze changes in events. Considering that current computer technology can still be developed further, our method of digitization can only be seen as at its infant stage, waiting to be truly advanced into curvature spaces so that it will possess the function to represent the overall multidimensionality of materials and events with colors, tastes, smell, and the like. It can be reasonably predicted that in the foreseeable future, computers, which were initially developed on the basis of quantitative computations, can help to avoid many weaknesses of the system of quantitative analysis and provide us a colorful and vibrant world.

In short, this book is written to satisfy the requests of thousands of our readers and colleagues. It contains our many years of practice and our understanding of the scientificality of ancient civilizations. And, it opens a new page of epistemology and creates the methodology system of digitization of events. In terms of the practical use value, this book presents a brand-new realm of knowing the natural world. Our work shows that the digitization of events and the relevant theories have a widely open field for development. We truly hope that all the interested readers can test this method by using their own problems so that this method can be further improved and developed. The message we would like to pass on in this book is that the deterministicity of modern science is about extracting invariant information out of events; not only is the system of statistics unable to collect any random information, but also does it not help the deterministicity to eliminate randomness. At the same time, what is declared in this book is that when the method of digitization extracts the information of change out of events, it does not lose the inherent invariant information. This end challenges the system of modern science both epistemologically and methodologically. We truly hope that all the interested readers can join hands with us to construct the next higher level of modern science.

References

A. Akimov and G. Shypov, Torsion field and experimental manifestations, *J. New Energy* 1, 1997, 67.

H. G. Apostle (translator), *Aristotle's Metaphysics*, Bloomington: Indiana U. Press, 1966.

H. Bergson, L'evolution creatrice, in: *Oeuvres*, Editions du Centenaire, Paris: PUF, 1963.

M. Born, Zurquantenmechanik der stoβvorgange, *Zeitschrift fur Physik*, 37, 1926, 37, 863–867.

N. Capitaine, J. Chapront, S. Lambert, and P. Wallace, Expressions for the celestial intermediate pole and celestial ephemeris origin consistent with the IAU 2000A precession-nutation model, *Astron. Astrophys.* 400, 2003, 1145–1154.

G. Y. Chen and S. C. OuYang, Lorenz model and "chaos" problem, *Scientific Research Monthly* (Hong Kong), 2005, no. 7–8 (serial 11), 51–53.

G. Y. Chen, B. OuYang, and T. Y. Peng, System stability and instability: An extended discussion on significance and function of stirring energy conservation law, *Engineering Science of China* 3, 2005a, 44–51.

G. Y. Chen, B. L. OuYang, D. S. Yuan, L. P. Hao, and L. R. Zhou, Systems stability and instability—an expanded discussion on the significance and effects of the conservation law of stirring energy, *Engingeering Science of China* 7, 2005b, 41–65.

H. Z. Chen, N. Xie, and Q. Wang, Application of the blown-ups principle to thunderstorm forecast, *Applied Geophysics* 2, 2005c, 188–193.

Z. L. Chen, L. P. Hao, and L. R. Zhou, The blown-up analysis of the regional torrential rain and heavy rain gush on July 20 in Chengdu, *Sichuan Meteorology* 83, 2003, 7–9.

L. de Broglie, *Ann. D. Phy. Sér.* 10, 1925, 2.

A. Einstein, *Complete Collection of Papers by Albert Einstein*, Beijing: Commercial Press, 1976.

A. Einstein, *Collected Papers of Albert Einstein*, Princeton: Princeton University Press, 1997.

J. English and G. F. Feng, *Tao De Ching*, New York: Vintage Books, 1972.

H. Eves, *An Introduction to the History of Mathematics*, Shanxi: Shanxi People's Press, 1986.

P. K. Feyerabend, *Against Method: Outline of an Anarchistic Theory of Knowledge*, Shanghai: Shanghai Press of Translation, 1992.

R. Fjörtoft, On the changes in the spectral distribution of kinetic energy for two dimensional non-divergent flows, *Tellus* 5, 1953, 225–230.

C. S. Gardner, at el., Method for solving the Korteweg de Vries equation, *Phy. Review Lett.* 19, 1967, 1095–1097.

K. Gödel, Über formal unentscheidbare Sätze der Principia mathematica und verwandter Systeme I, *Monatshefte für Mathematik und Physik* 38, 1931, 173–198.

C. H. Gu, Some whole solutions and their asymptotic properties of quasi-linear symmetric hyperbolic equations, *Mathematics Annuls*, 1978, no. 21, 130–134.

F. H. Guo, Seismological mechanics, *Sichuan Science and Technology* 20, 1995, 354–359.

S. W. Hawking, *A Brief History of Time: From the Big Bang to Black Holes*, Bantam Dell Publishing Group, 1988.

W. Heisenberg, Uberden anschulichen Inhalt der quantentheoretischen kinematic und mechanic, 43, *Zeitschrift fur Physik*, 1927, 127–198.

S. L. Hess, *Introduction to Theoretical Meteorology*, New York: Holt, Rinehart and Winston, 1959.

R. Hide, Some experiments on thermal convection in a rotating liquid, *Quart. J. Roy. Meteorol. Soc.*, 79, 1953, 161.

IERS conventions, 2000, accessed June 2000, http://www.iers.org/products/9/1379/orig/finals2000A.all

M. Kline, *Mathematical Thought from Ancient to Modern Times*, Oxford: Oxford University Press, 1972.

A. Koyré, *Études Newtoniennes*, Paris: Gallimard, 1968.

Lao Tzu (time unknown), *Tao Te Ching: a new translation by Gia-fu Feng and Jane English*, New York: Vintage Books, 1972.

G. P. Li and Y. Z. Guo, *Lecture on Mathematical Physics 1A: Problems in Mathematical Physics*, Wuhan: Wuhan University Press, 1985.

H. M. Liang, *Lao Tzu* (in Chinese), Liaoning: Liaoning People's Press, 1996.

J. Lin and S. C. OuYang, Exploration of the mystery of nonlinearity, *Research of Natural Dialectics* 12, 1996, 34–37.

Y. Lin, Multi-level systems, *International J. Systems Sciences* 20, 1989, 1875–1889.

Y. Lin, A few systems-colored views of the world, in: *Mathematics and Science*, edited by R. E. Mickens, Singapore and New Jersey: World Scientific Press, 1990.

Y. Lin, Mystery of nonlinearity and Lorenz's chaos, *Kybernetes: The International Journal of Cybernetics and Systems* 27, 1998, 605–854.

Y. Lin, Systemic yoyo model and applications in Newton's, Kepler's laws, and others, *Kybernetes: The International Journal of Systems, Cybernetics, and Management Sciences* 36, 2007, 484–516.

Y. Lin, Systemic Studies: The infinity problem in modern mathematics, *Kybernetes: The International Journal of Cybernetics, Systems, and Management Science* 37, 2008a, 387–578.

Y. Lin, *Systemic Yoyos: Some Impacts of the Second Dimension*. New York: Auerbach Publications, an imprint of Taylor & Francis, 2008b.

Y. Lin, S. C. OuYang, M. J. Li, H. W. Zhang, and L. J. Jiang, On fundamental problems of the "chaos" doctrine, *International Journal of Applied Mathematics* 5, 2001, 37–64.

Y. Lin and Y. P. Qiu, Systems analysis of the morphology of polymers, *Mathematical Modeling: An International J.* 9, 1987, 493–498.

Y. Lin, W. J. Zhu, N. S. Gong, and G. P. Du, Systemic yoyo structure in human thoughts and the fourth crisis in mathematics, *Kybernetes: The International Journal of Systems, Cybernetics, and Management Science* 37, 2008, 387–423.

S. F. Mason, *A History of the Sciences*, Shanghai: Shanghai People's Press, 1989.

Mathematics Manual Group, *Mathematics Manual*, Beijing: Press of People's Education, 1979.

Newton, *Philosophiae Naturalis Principia Mathematica*, Cambridge, MA: Harvard University Press, 1972.

Q. Z. Mei, *Mechanics of Water Wave Motions*, Beijing: Scientific Press, 1985.

S. C. OuYang, *Methods of Applied Mathematics*, Beijing: Meteorological Press, 1984.

S. C. OuYang, Solitary waves in atmospheric disturbances and interactions, *Journal of Chengdu College of Meteorology*, 1986, no. 1, 34–47.

S. C. OuYang, Numerical experiments on the stability of quasi-linear advection equations, *Computer and Applications*, 1992, no. 1, 1–10.

S. C. OuYang, *Numerical Situation Forecasts and Applications*, Beijing: Press of Meteorological Science, 1993.

S. C. OuYang, *"Break-offs" of Moving Fluids and Several Problems about Weather Forecast*, Chengdu: Chengdu University of Science and Technology Press, 1994.

S. C. OuYang, Pansystems view of prediction and blown-ups of fluids, *Applied Mathematics and Mechanics* 16, 1995, 255–262.

S. C. OuYang, Explosive growth of general nonlinear evolution equation and related problems, *Applied Mathematics and Mechanics* 19, 1998a, 165–173.

S. C. OuYang, The blown-up of fluid and two-phaseness of currents and waves, *Kybernetes: The International Journal of Cybernetics, Systems and Management Science* 27, 1998b, 636–646.

S. C. OuYang, *Weather Forecast and Structural Prediction*, Beijing: Meteorological Press, 1998c.

S. C OuYang, Linearization and the lost world of the formal analysis, *Meteorological Science* 20, 2000, 354–359.

S. C. OuYang, Media waves and broken media waves, *Scientific Research Monthly* 18, 2006a, 1–4.

S. C. OuYang, "Twists-up," transform of spin flows and physics of process, *Engineering Sciences* 4, 2006b, 81–88.

S. C. OuYang, Mass-energy-gravitational waves and solenoidal field-universal gravitation, *Scientific Inquiry*, in press.

S. C. OuYang, Second stir and conservation of stirring energy, *Scientific Inquiry*, in press.

S. C. OuYang et al., Physics properties of Schrödinger equation and excessive expansion of the concept of wave motions, *Advances in Systems Science and Applications* 1, 2001, 112–116.

S. C. OuYang et al., "Probabilistic waves" and torsions of quantum effects, *Engineering Science of China* 7, 2005a, 1–6.

S. C. OuYang and G. Y. Chen, End of stochastics and quantitative comparability, *Scientific Research Monthly* 14, 2006, 141–143.

S. C. OuYang and G. Y. Chen, Evolution science and problem of digitization, *Journal of Contemporary Scholarly Research*, vol. 7, 2007, 23 – 28.

S. C. OuYang and Y. Lin, On spinning of matters and some important studies absent in contemporary science, *Scientific Inquiry* 7, 2006a, 55–62.

S. C. OuYang and Y. Lin, Disillusion of randomness and end quantitative comparabilities, *Scientific Inquiry* 7, 2006b, 171–180.

S. C. OuYang and Y. Lin, V. Bjerknes circulation theorem, universal gravitation and Rossby's waves, *Scientific Inquiry*, in press.

S. C. OuYang and Y. Lin, Physical quantities, time and parametric dimensions, *Scientific Inquiry*, in press.

S. C. OuYang, Y. Lin, Z. Wang, T. Y. Peng, Blown-up theory of evolution science and fundamental problems of the first push, *Kybernetes: The International Journal of Cybernetics, Systems and Management Science* 30, 2001a, 448–462.

S. C. OuYang, Y. Lin, Z. Wang, and T. Y. Peng, Evolution science and infrastructural analysis of the second stir, *Kybernetes: The International Journal of Cybernetics, Systems and Management Science* 30, 2001b, 463–479.

S. C. OuYang, Y. Lin, and Y. Wu, Fundamental properties of nonlinear equations and eight related theorems, *Scientific Inquiry*, in press.

S. C. OuYang, D. H. McNeil, and Y. Lin, *Entering the Era of Irregularity* (in Chinese), Beijing: Meteorological Press, 2002a.

S. C. OuYang, J. H. Miao, Y. Wu, Y. Lin, T. Y. Peng, and T. G. Xiao, The second stir and incompleteness of quantitative analysis, *Kybernetes: The International Journal of Systems and Cybernetics* 29, 2000, 53–70.

S. C. OuYang and T. Y. Peng, Non-modifiability of information and some problems in contemporary science, *Applied Geophysics* 2, 2005, 56–63.

S. C. OuYang, H. Sabelli, Z. Wang, Y. Lu, Y. Lin and D. McNeil, Evolutionary "aging" and "death" and quantitative instability, *International J. of Computational and Numerical Analysis and Application* 1, 2002b, 413–437.

S. C. OuYang and M. Wei, Principle question of harmonic quantitative analysis of spectrum expansion, *Engineering Sciences* 3, 2005, 40–43.

S. C. OuYang and M. Wei, Discussion on super high-temperature weather and city construction, *Engineering Science* 4, 2006, 56–64.

S. C. OuYang and S. Y. Wei, *Situational Numerical Weather Forecast and Applications* (Appendix A: Basic Concepts and Operations of Tensors), Beijing: Meteorological Press, 1993.

S. C. OuYang, Y. Wu, Y. Lin, and C. Li, The discontinuity problem and "chaos" of Lorenz's model, *Kybernetes: The International Journal of Systems and Cybernetics* 27, 1998, 621–635.

S. C. OuYang, Y. Wu and Y. Lin, Probabilisticity of Schrodinger equation and transmutations of high speed flows, *Scientific Inquiry*, in press.

S. C. OuYang, K. Zhang, L. P. Hao, and L. R. Zhou, Structural transformation of irregular time series information and refined analysis of evolution, *Engineering Science of China* 7, 2005b, 36–41.

N. A. Phillips, A coordinate system having some special advantages for numerical forecasting, *J. Meteorology*, 14, 1957, 184–185.

I. Prigogine, *From Being to Becoming: Time and Complexity in the Physical Science*, W. H. Freeman and Company, 1980.

H. Sabelli, Complement plots: Analyzing opposites reveals mandala-like patterns in human heartbeats, *International Journal of General Systems* 29, 2000, 799–830.

H. Sabelli and L. Kauffman, The process equation: Formulating and testing the process theory of systems, *Cybernetics and Systems* 33, 1999, 261–294.

E. Schrödinger, Quantisierung als eigenwertproblem, *Annalen der Physik*, 79, 1926, 501.

P. K. Seidelmann, 1980 IAU Nutation: The final report of the IAU Working Group on nutation, *Celest. Mech.* 27, 1982, 79–106.

C. E. Shannon and W. Weaver, *The Mathematical Theory of Communication*, Champaign, IL: Illinois University Press, 1949.

B. Sun, *Mo Zi*. Beijing: Hua Xia Press, 2000.

C. D. Wang, Development and significance of information theory, *Communications in Natural Dialectics*, 1981, no. 2, 35.

M. L. Wang, Initial value stability and solitary wave solution of Burgers type equations, *Research and Comments of Mathematics*, 1982, no. 2, 67–97.

R. H. Wang and Z. Q. Wu, Several problems regarding two-dimensional system of quasilinear hyperbolic equations—existence and uniqueness, *Jilin University Journal*, 1963, no. 2, 459–502.

B. Watson (translator), *Chuang Tzu: Basic Writings*, New York: Columbia University Press, 1964.

W. B. Weng, *The Foundations of Prediction Theory*, Beijing: Press of Petroleum Industry, 1984.

W. B. Weng, *Foundations of Prediction Theory*, Beijing: Press of Petroleum Industry, 1996.

R. Wilhalm and C. Baynes, *The I Ching or Book of Changes* (3rd edition), Princeton, NJ: Princeton University Press, 1967.

G. B. Whitham, *Linear and Nonlinear Waves*, Beijing: Scientific Press, 1986.

G. B. Whitham, *Linear and Nonlinear Waves*, New York: Wiley-Interscience, 1999.

G. W. Whittan, Nonlinear dispersive waves, *Proc. Roy. Soc. London*, 283A, 1965, 238–261.

A. Wolf, *A History of Science, Technology and Philosophy in the 16th & 17th Centuries* (in Chinese), Beijing: Commercial Press, 1985.

Y. Wu and Y. Lin, *Beyond Nonstructural Quantitative Analysis: Blown-ups, Spinning Currents, and Modern Science*, River Edge, NJ: World Scientific, 2002.

Y. Wu and S. C. OuYang, About the nonlinear stability of two-dimensional fluids with horizontal and vertical shears, *Journal of Meteorology* 47, 1989, 324–331.

N. J. Zabusky and M. D. Kruskal, Interaction of solitons in a collisionless plasma and recurrence of initial states, *Phys. Rev. Letter*, 15, 1965, 240–243.

X. P. Zeng, M. Hayakawa, Y. F. Lin, and C. Xu, Infrastructural analysis of geomagnetic field and earthquake prediction, in: *Seismo Electromagnetics: Lithosphere-Atmosphere Coupling*, edited by M. Hayakawa and O. A. Molchanov, 463–468, Tokyo: TERRAPUB, 2002.

X. P. Zeng and Y. F. Lin, Structural transformation of irregular geomagnetic information and prediction of disasters and accidents, *Scientific Research Monthly* 14, 2006, 126–129.

X. P. Zeng and Y. F. Lin, Testing the forecasts of 2007 disasters—new understandings of abnormal phase space geomagnetic structures, in press.

X. P. Zeng, Y. F. Lin, C. Xu and S. C. OuYang, Turning changes in evolution of geomagnetic field and infrastructural analysis of earthquake prediction, *Kybernetes: The International Journal of Cybernetics, Systems and Management Science* 30, 2001, 365–377.

H. Z. Zhang, Structural analysis of irregular information and the forecast of regional fogs, *Journal of Natural Disasters* 14, 2005, 43–48.

Index

"f" indicates material in figures. "t" indicates material in tables.